T0331750

Ore Deposit Geology

Mapping closely to how ore deposit geology is now taught, this textbook systematically describes and illustrates the major ore deposit types, and links them to their settings in the crust and the geological factors behind their formation.

- Written for advanced undergraduate and graduate students with a basic background in the geosciences, it provides a balance of practical information and coverage of the relevant geological sciences, including petrological, geochemical, hydrological and tectonic processes.
- Important theory is summarised without unnecessary detail and integrated with students' learning in other topics, including magmatic processes and sedimentary geology, enabling students to make links across the geosciences.
- Students are supported by pointers to further reading, a comprehensive glossary, and problems and review questions that test the application of theoretical approaches and encourage students to use what they have learnt.
- A website includes visual resources and combines with the book to provide students and instructors with a complete learning package.

John Ridley is the Malcolm McCallum Chair of Economic Geology at Colorado State University where he has taught ore deposit geology and field geology for the past seven years. His earlier academic positions spanned three other continents, with positions at universities in Australia, Switzerland and Zimbabwe, and over his career he has taught courses in geochemistry, petrology and structural geology. Professor Ridley has published over seventy articles in refereed journals and books, in addition to geological maps and reports for companies and groups within the mining and minerals industries.

"This is ore deposit geology the way most professional economic geologists think, using deposit descriptions as a basis for understanding genetic processes. It is clearly structured, simply illustrated and lucidly explained. This book will be appreciated by students, teachers and professional geologists for its clarity of expression and scholarship of content."
Dr Noel C. White – *Consulting Economic Geologist and Honorary Research Professor, CODES, University of Tasmania*

"If you've been searching for a modern textbook on metallic ore deposits, this is a good choice. Incorporating recent observational and theoretical advances, excellent graphics, an accessible treatment of chemical processes, and end of chapter questions, this book appears ideal for undergraduate geology majors."
Professor Donald M. Burt – *School of Earth and Space Exploration, Arizona State University*

"In this textbook, the author combines up-to-date scientific literature with well-structured discussions on ore-forming processes. This results in stimulating and insightful chapters, which provide students and teachers with an extremely useful tool. The end-of-chapter boxes and questions provide great pedagogic support for courses."
Dr. Paolo S. Garofalo – *Geological & Environmental Sciences Department, Università di Bologna*

Ore Deposit Geology

JOHN RIDLEY

Colorado State University

Shaftesbury Road, Cambridge CB2 8EA, United Kingdom

One Liberty Plaza, 20th Floor, New York, NY 10006, USA

477 Williamstown Road, Port Melbourne, VIC 3207, Australia

314–321, 3rd Floor, Plot 3, Splendor Forum, Jasola District Centre, New Delhi – 110025, India

103 Penang Road, #05–06/07, Visioncrest Commercial, Singapore 238467

Cambridge University Press is part of Cambridge University Press & Assessment, a department of the University of Cambridge.

We share the University's mission to contribute to society through the pursuit of education, learning and research at the highest international levels of excellence.

www.cambridge.org
Information on this title: www.cambridge.org/9781107022225

First published 2013
9th printing 2022

A catalogue record for this publication is available from the British Library

Library of Congress Cataloging-in-Publication data
Ridley, John, 1957–
Ore deposit geology / John Ridley, Colorado State University.
pages cm
ISBN 978-1-107-02222-5 (Hardback)
1. Ore deposits–Textbooks. I. Title.
QE390.R53 2013
553–dc23

2012046334

ISBN 978-1-107-02222-5 Hardback

Additional resources for this publication at www.cambridge.org/oredeposit

CONTENTS

BOXES

PREFACE

Ore deposit geology may be taught late in a student's undergraduate programme, or even at the beginning of a graduate programme. At this stage the student is able to link many of the concepts involved in ore genesis into his or her knowledge of the building blocks of the geosciences. Or it may be taught early, to show students potential applications of their emerging knowledge. This book is designed for the former case and also to be a reference if a student chooses to work professionally in the field. Knowledge typical of standard undergraduate courses in mineralogy, igneous, metamorphic and sedimentary geology, structural geology and tectonics, and geomorphology are assumed, together with some basic chemistry. It is not assumed that students have had courses in isotope geochemistry or in geochemical thermodynamics.

The subject is scientifically expansive. There are few areas of the Earth sciences which are not touched upon in a thorough study of the subject. A balance needs however to be reached in teaching between detail and breadth. Some geologists work on one deposit type all their lives; others need to have enough knowledge of all types to make suitable judgements when they are called to evaluate them. A balance also needs to be reached between the local and the global. A practising ore deposit geologist may remain in one area or work over the world. The subject thus needs to be taught with case studies from globally, but also tailored to local geology.

It was a truism while I wrote this book, over the years before, and hopefully for at least a few years after, that the professional community of ore deposit geologists needs to become scientifically more sophisticated to be successful in our search for resources to replace those that are being mined. The easy to find, outcropping ores are almost all found. We need to be able to predict the likely positions of ores, and find them and evaluated them with a reasonable success rate through drilling to depths of a few hundred metres. Accurate prediction will require good knowledge of the nature of ore, of empirical markers of ore systems, and accurate conceptual models of how ores form.

It is with the aim of training with all the ends and needs listed above this book was written. Description, genetic analysis and discussions of advances in understanding of other sub-disciplines of the geosciences are combined. About 30 deposit types are described and discussed in varying depths.

I made two specific choices in format that warrant some explanation: First, for the benefit of simplicity, I felt the need to take stands on some of the conflicting interpretations and debates in the community of ore deposit geologists and to write some of the deposit descriptions and analyses from the viewpoint of one 'side'. Debate is an integral part of science. Descriptions run the risk of being supplanted as new discoveries are made. Genetic analysis runs an even greater risk. We need to keep aware that at least some of our interpretations, especially of the unknowable of ore genesis in the geological past, are at all times 'castles in the air'. We need further to be aware that our interpretations can be intensely personal.

My intention throughout is to present deposit descriptions and data as we know them, and leave a teacher enough room to discuss with students where and why other interpretations may be held. The book would have been much longer and the text less transparent had multiple interpretations been weighed up against each other in every instance at which they could be weighed. I have probably offended people by taking sides. However, at least one of two conflicting models must be wrong.

A second choice was to minimise citation in the text. This choice was also made for conciseness and flow. I have in general cited articles which I judge to be the best recent descriptions of classic deposits of each type, rather than work that supports genetic models. Ore deposit geology is a warren of detailed observations and data. It is easy, however, to search (in for instance Web of Science or Google) for original papers that report specific data and their interpretation at deposits. It is much more difficult to find syntheses of ore forming processes and discussions of commonalities of deposits.

Some notes on the components of the text:

A word or phrase in bold in the text is a signal that the word is defined in the glossary.

Boxes serve various purposes:

- Samplers of research techniques which are commonly referred to in ore deposit geology literature. The intention is to tell students that these research techniques exist and to give sufficient background that research results can be understood in broad terms.
- Discussions of results of recent research in the Earth sciences which is relevant to ore deposit geology.
- Discussions of debates on the genesis of ore deposit types, specifically those debates which inform on how debates develop and are resolved.

Questions and exercises are in a range of forms:

- Sample arithmetic calculations given with the intention to put sizes, rates of processes, etc., into context. My experience in teaching is that some students learn through such quantification, whereas for others the questions would leave no lasting understanding.
- Review questions. Many of these are designed to link themes from different sections of a chapter or from different chapters. Students cannot answer these questions by simply reading through a single subsection of the book.
- Discussion question. These include both questions that have an answer and others that specialists could debate and not resolve.

Further readings are an introduction into selected themes from the literature. They are chosen for accessibility. Some give an introduction to lines of research on background topics.

ACKNOWLEDGEMENTS

Many people have contributed directly or indirectly to the writing and production of this book. Tim Nutt, David Groves, and Christoph Heinrich were my mentors in the science of Ore Deposit Geology. The concept and content of this book grew out of course notes that I prepared for students at Macquarie University and were subsequently expanded, developed and updated at the University of Bern and at Colorado State University. The input and support of colleagues at these universities in the development of these notes, especially John Lusk, Simon Jackson and Larryn Diamond is acknowledged. Students were ever critical readers and pushed for greater clarity. Current and past students at Colorado State University helped additionally with proof-reading, figure drafting, and paperwork, in particular, Kyle Basler-Reeder, Elaine Jacobs, Anne Ji and Crystal Rauch. Importantly and lastly I would like to thank all the geologists at mine and exploration sites and their companies who have donated precious time and resources, provided guided tours, facilitated access for me, and have supported research projects of my students. It would have been impossible to put together a textbook such as this without having field notes, images and memories collected at many deposits of many types around the world.

1 What is an ore deposit?

1.1 Definition and scope of ore deposit geology

We extract many types of commodities from the Earth; minerals and rocks from mines, hydrocarbon liquids and gases and groundwater through pumping or where they rise to the surface under their own pressure, the heat of rocks as geothermal energy. An ore deposit is 'what is mined'. Precise definitions of the term are based on economics rather than geology, for instance:

> ore is rock that may be, is hoped to be, will be, is or has been mined; and from which something of value may be (or has been) extracted.
>
> (Taylor, 1989, Ore reserves – a general overview. *Mining Industry International*, vol. 990, pp. 5–12.)

Which commodities are included by the definition of ore deposits?

The economic definition of an ore deposit given above would include:

- ores of metals
- ores of gemstones
- ores of minerals used as feedstock for production of industrial chemicals
- ores of minerals used in industrial products
- rock used as aggregate, for building stone
- coal and oil shale

This book does not cover all mined commodities. Traditionally, the study of ore deposit geology has been concerned with mineral resources in which the product of economic interest is one or more mineral, specifically either those minerals from which a metal is extracted or gemstones. A practising ore deposit geologist may be expected to have knowledge of sources of these commodities. The study does not include either the mining of rocks or of coal. Nor does it generally include the mining of minerals that are used exclusively as feedstock in industrial processing or products (**industrial minerals**).

It is, however, not in all cases possible to make a clear-cut distinction between the categories of commodities in the list above: there are some ores from which both a metal and one or more industrial minerals are extracted, for instance products from bauxite are both aluminium metal and feedstocks for Al-bearing abrasives and refractory products, and diamond is both a gemstone and an industrial mineral. Further, not all mineral resources are mined. Some uranium is obtained by *in situ* leach extraction involving pumping solvent through uranium-bearing rocks, specifically sandstones, and some metals are extracted from saline brines pumped from sediments that underlie salars or salt lakes. Despite these caveats, the tradition of ore deposit geology as being concerned with the sources of metals and of gemstones is closely followed in this book, although some industrial minerals are discussed.

What are the important mined commodities?

By tonnage, the major mined commodities are not ores but are coal and construction materials (Table 1.1). The production by tonnage of a number of industrial minerals is also much greater than all metal ores except iron ore. The extraction of a number of metals (e.g. Fe, Al, Cu, Zn, Ni, Au), however, are within the top few commodities in terms of value and each contributes to the order of 0.1% of the dollar value in the world's economy. Metals are low-tonnage–high-value products relative to many larger-tonnage mined commodities (Figure 1.1). Because of this characteristic they are transported and traded worldwide. Other than diamonds, for which industrial uses make up about half the market, gemstones are very minor commodities by both tonnage and value.

What constitutes an ore deposit?

An ore deposit is made up of one or more **ore bodies**. These are the masses of rock that contain ore and from which the commodity of value will be extracted. Not all ore within an ore body will be extracted. Ore bodies are divided into **reserves** and **resources**. These terms have precise legal meanings in many countries, but as a generality the difference is that reserves are ore that is economically feasible to mine and for which there are no legal or engineering impediments to mining, while resources are ores that may potentially be extracted at some time in the future. Engineering constraints on excavation form one set of factors that will influence what ore is economic to mine (Figure 1.2). The mining of an ore body may be from an open pit, an underground mine, or a combination of the two.

1.2 Ore deposit geology and related sciences

The extraction of an economic commodity from ore

Ore contains **ore minerals** intermixed with other minerals (**gangue**) from which they must be separated through milling, to break down ore into constituent minerals, followed, typically, by flotation to separate the minerals of economic interest from the gangue. Most ore minerals of metals are not 'native' metals, but are compounds in which the metals are bonded and from which they must be extracted. Likewise, many industrial minerals need to be refined or processed (**beneficiated**) before sale. Methods of ore processing through flotation, refining and extraction are chosen based on ore mineralogy and on the physical characteristics of the ore. The science of extraction involves aspects of chemistry and engineering and forms a field study called extractive **metallurgy** (Figure 1.3).

The mineralogy of ores

Many ore minerals are relatively rare minerals. Some are widespread as accessory minerals in many types of igneous, metamorphic and sedimentary rock. Others are present

Table 1.1

Commodity	Energy mineral, Gem, Metal ore, Industrial mineral	Worldwide production, Mt a^{-1}	Per annum increase % (20-year average)	Approx. price per ton, US$	World value of industry US$ $\times 10^6$/a
Aggregates	**I**	15 000	3	5	**45 000**
Coal	**E**	7000	1	35	**245 000**
Limestone (cement, lime)	**I**	2300	6	6	**13 800**
Pig iron (iron ore)	**M**	2200	7	60	**132 000**
Common clays	I	500	1.5	6	3000
Salt (halite)	**I**	260	3	60	**15 600**
Phosphates	**I**	160	1	127	**20 320**
Gypsum	I	150	5	9	1350
Industrial sand	**I**	112		250	**28 000**
Sulfur	**I**	70	2	260	**18 200**
Aluminium	**M (I)**	37	4	1800	**66 600**
Kaolin	I	30	3	135	4050
Potash	**I**	25	1.5	300	**7500**
Manganese (Fe–Mn)	**M**	23	0	1000	**23 000**
Magnesite	**I**	19	4	400	**7600**
Feldspar	I	18		70	1260
Copper	**M**	15.5	3	7000	**108 500**
Bentonite/ Fullers Earth	I	13.5	2	70	945
Trona/soda ash	I	12	6	140	1680
Zinc	**M**	11	2	1600	**17 600**
Baryte	I	8	4	50	400
Talc	I	7.2	0	110	800
Chromium (Fe–Cr)	**M**	7	4	4000	**28 000**
Borates	**I**	6		1100	**6600**
Titania	I (M)	5.8	3	400	2320
Fluorite	I	5	0	180	900
Lead	M	3.8	0.5	1100	4180
Diatomite	I	2.2	0	250	550
Asbestos	I	2	−3	1100	2200
Nepheline syenite	I	2	−2	100	200

Table 1.1 (*cont.*)

Commodity	Energy mineral, Gem, Metal ore, Industrial mineral	Worldwide production, Mt a^{-1}	Per annum increase % (20-year average)	Approx. price per ton, US$	World value of industry US$ $\times 10^6$/a
Perlite	I	1.7	−1.5	300	510
Nickel	**M**	1.4	3	27 000	**37 800**
Zirconium minerals	I (M)	1.2	2	300	480
Graphite	I	1.1	3	3000	330
Magnesium	M	0.7	5		
Tin	M	0.3	3	13 000	3900
Molybdenum	**M**	0.2	4	44 000	**8800**
Antimony	M	0.19	6	5300	1000
Tungsten	M	0.058	3	30 000	1740
Vanadium (V_2O_5)	M	0.054	3	5500	300
Uranium (U_3O_8)	E	0.051	2	60 000	3060
Silver	**M**	0.021	2	450 000	**9450**
Lithium	M/I	0.015	5	75 000	1100
Gold	**M**	0.0023	−0.5	36 000 000	**82 800**
Mercury	M	0.0013	−8	18 000	25
Platinum-group elements	**M**	0.0005	2	42 000 000	**8400**
Diamonds	**G (I)**	14 t	3	9×10^9	**13 000**

Approximate worldwide annual production of major products of the mining industry, rate of increase of production, unit price, and value. Commodities of most value are highlighted in bold. For production less than 1 Mt a^{-1}, only selected important metals and diamond are listed. From Kogel *et al.*, 2006, and United States Geological Survey Minerals Yearbooks 2008 and 2009. Prices of many industrial feedstocks and minerals vary greatly with product quality. In these case weighted average prices are given where these are possible to estimate.

only in ores. Important ore minerals include native metals, sulfides, **sulfosalts**, oxides and hydroxides, and specific silicates, carbonates and minerals of other classes (Table 1.2).

Goldschmidt's four-fold geochemical classification of the elements is based on predominant bonding affiliation and was originally formulated to explain the chemical differentiation of the Earth into core, silicate mantle and atmosphere:

- Lithophile – elements that combine with oxygen, dominantly in oxide and silicate minerals;
- Chalcophile – elements that combine with sulfur, dominantly in sulfide minerals;
- Siderophile – elements that occur as native metals or as alloys;
- Atmophile – elements forming elemental gases.

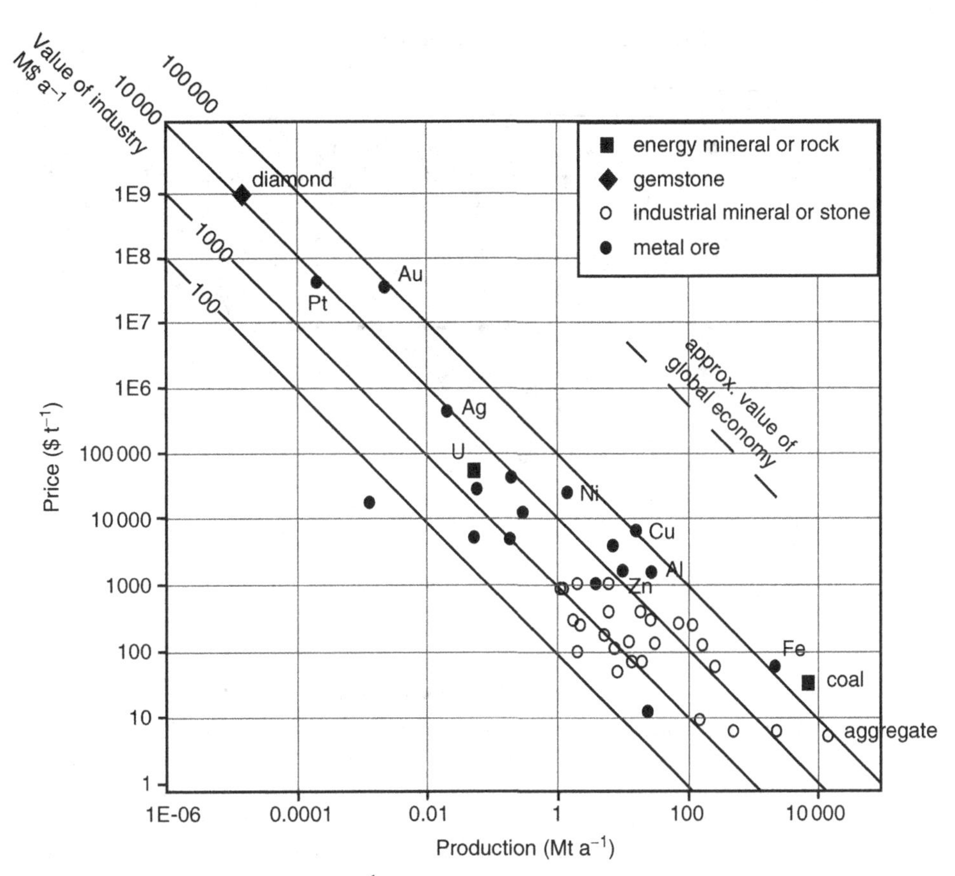

Figure 1.1 World production (Mt a^{-1}) versus price per tonne of mineral and rock commodities (2009 data from USGS Mineral Resources Program publications and various additional sources). Most metals are low-tonnage–high-value commodities. The global value of production of a metal is shown by the diagonal lines and for a number of metals is in the top rank of commodities.

The classification can be modified to reflect chemical behaviour at the Earth's surface, in the crust and in the upper mantle (Figure 1.4), and as such is a guide to the mineral class that each element most commonly occurs in. Each of these groups of elements is systematically positioned within the periodic table (Figure 1.4). Some elements can have two or three bonding affiliations, depending on the geochemical environment, and as indicated by secondary affiliations shown in Figure 1.4. Of particular significance in this respect are high valence states of semi-metals in surface and near-surface geological environments.

Comparison of Figure 1.4 with Table 1.1 shows that many of the economically important metals are chalcophile or siderophile elements (e.g. Ag, Au, Cu, Pb, Pt). Historically, mining of native metals and sulfide minerals dominated the industry because of the relative ease and low costs of extraction. Chemical bonding in sulfide minerals is largely covalent, and is such that the amount of energy required to extract the metal is relatively small. In contrast, chemical bonding in many oxides and in silicate minerals

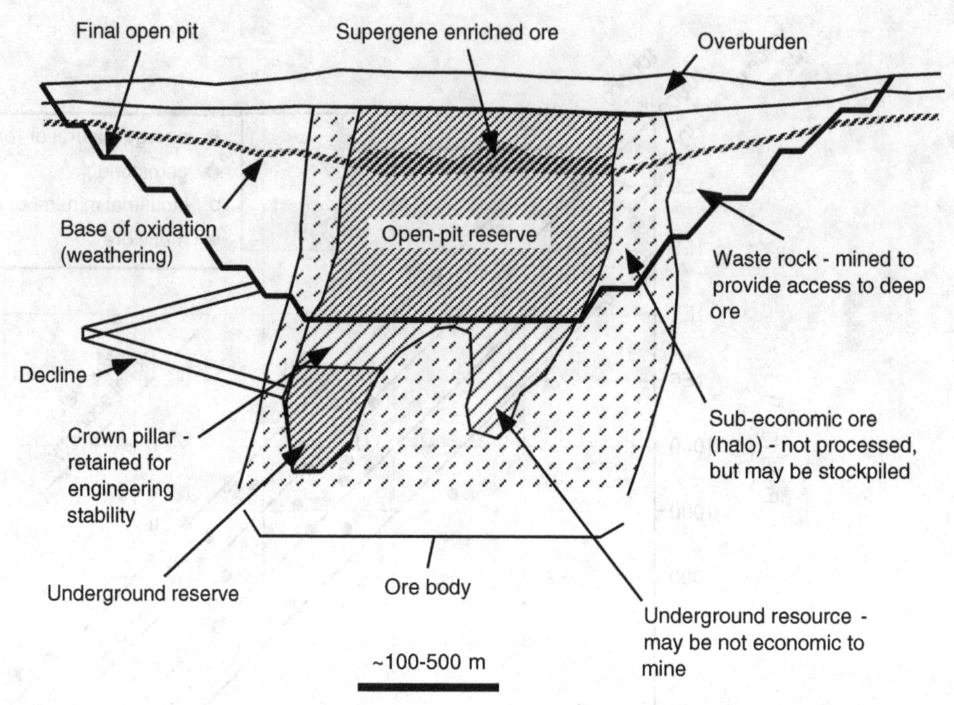

Figure 1.2 Schematic cross section through an open-pit mine, illustrating the geological, economic and engineering definitions of ore, modelled on the case of a porphyry copper deposit. The ore body and the reserves in many deposits have much more irregular shapes than shown here. The scale bar indicates only the order of magnitude scale: a pit could be from hundreds of metres to a couple of kilometres across and would most likely be sub-circular or elliptical in plan view.

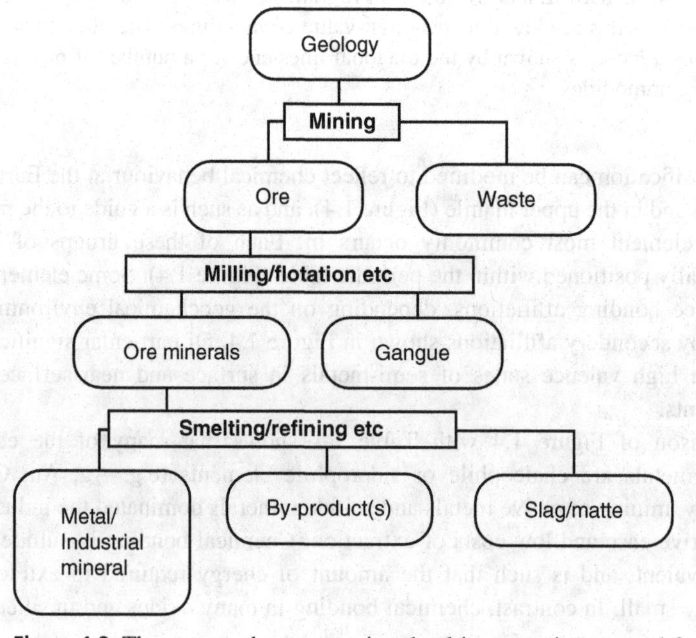

Figure 1.3 The steps and processes involved in extracting a metal from an ore deposit.

Table 1.2

Element	Native metals, alloys	Sulfides, sulfosalts, arsenides, etc.	Oxides, hydroxides	Silicates, tungstates, carbonates, etc.
Fe		*Pyrite* *FeS_2*	Haematite Fe_2O_3	Siderite $FeCO_3$
		Pyrrhotite *FeS*	*Magnetite* *Fe_3O_4*	
			Goethite $FeO(OH)$	
Mn			Pyrolusite MnO_2	Rhodochrosite $MnCO_3$
Al			Gibbsite $Al(OH)_3$	
			Boehmite $AlO(OH)$	
Cr			Chromite $FeCr_2O_4$	
Cu		Chalcopyrite $CuFeS_2$		
		Bornite Cu_5FeS_4		
		Chalcocite Cu_2S		
Zn		Sphalerite ZnS		
Ti			Ilmenite $FeTiO_3$	
			Rutile TiO_2	
Pb		Galena PbS		
Ni		Pentlandite $(Ni,Fe)_9S_8$		
Mg				Magnesite $MgCO_3$
Sn		Stannite Cu_2FeSn_4	Cassiterite SnO_2	
Mo		Molybdenite MoS_2		
U			Uraninite UO_2	Carnotite $K(UO_2)(VO_4){\cdot}1.5H_2O$
Ag	Silver Ag	Argentite Ag_2S		
Au	Gold Au			
PGEs	Platinum Pt	Sperrylite $PtAs_2$		
		Laurite RuS_2		

Common ore minerals of economically important metals, classified by mineral class. (Minerals in italics are not currently significant ore minerals.) Some metals are extracted as trace components of ore minerals of more abundant metals, e.g. an important source of silver is as a trace component (solid solution) in galena (100–1000 ppm). PGEs are the platinum-group elements Ru, Rh, Pd, Os, Ir and Pt.

is dominantly ionic, and the technological barriers to the extraction of metals from these minerals were only overcome in the twentieth century.

The geochemistry of ores

In consideration of what is available as Earth resources we are interested in the Earth's crust, and most particularly the continental crust. Any rock in the Earth's crust contains all the 92 naturally occurring elements. The chemical composition of the crust as a whole (known as average crust) has been estimated by averaging chemical analyses of a large number of samples of crustal rock. The concentrations of the elements in the crust vary over many orders of magnitude from major elements present at per cent levels, to trace elements at ppm (parts per million by weight) to ppb (parts per billion) range (Table 1.3),

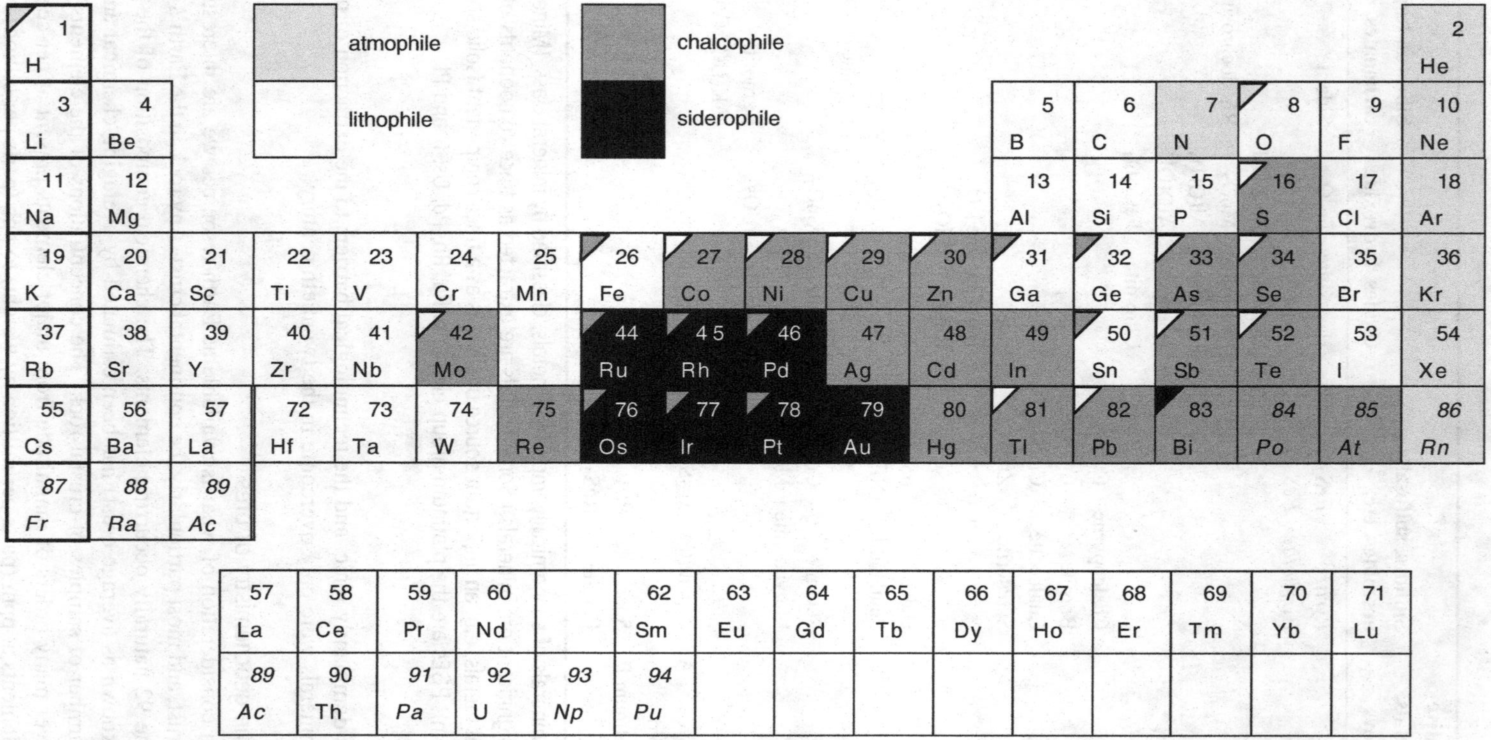

atmophile

lithophile

chalcophile

siderophile

1	2	3	4	5	6	7	8	9	10	11	12	13	14	15	16	17	18
1 H																	2 He
3 Li	4 Be											5 B	6 C	7 N	8 O	9 F	10 Ne
11 Na	12 Mg											13 Al	14 Si	15 P	16 S	17 Cl	18 Ar
19 K	20 Ca	21 Sc	22 Ti	23 V	24 Cr	25 Mn	26 Fe	27 Co	28 Ni	29 Cu	30 Zn	31 Ga	32 Ge	33 As	34 Se	35 Br	36 Kr
37 Rb	38 Sr	39 Y	40 Zr	41 Nb	42 Mo		44 Ru	45 Rh	46 Pd	47 Ag	48 Cd	49 In	50 Sn	51 Sb	52 Te	53 I	54 Xe
55 Cs	56 Ba	57 La	72 Hf	73 Ta	74 W	75 Re	76 Os	77 Ir	78 Pt	79 Au	80 Hg	81 Tl	82 Pb	83 Bi	*84 Po*	*85 At*	*86 Rn*
87 Fr	*88 Ra*	*89 Ac*															

57 La	58 Ce	59 Pr	60 Nd		62 Sm	63 Eu	64 Gd	65 Tb	66 Dy	67 Ho	68 Er	69 Tm	70 Yb	71 Lu
89 Ac	90 Th	*91 Pa*	92 U	*93 Np*	*94 Pu*									

Figure 1.4 Periodic table showing classification of the elements following Goldschmidt, but modified to match the bonding behaviour of the elements in geological environments of the crust and upper mantle. Triangles in the upper left of a cell indicate a secondary affiliation of an element. Short-lived elements of radiogenic decay chains are shown in italics.

Table 1.3 Chemical composition of average crust (approximately equivalent to diorite) in ppm (parts per million = g t^{-1} weight). Economically significant elements are highlighted in bold, see Table 1.1.

Element	ppm	Element	ppm	Element	ppm
O	464 000	La	30	Ge	2
Si	281 500	Nd	28	Ho	1
Al	82 300	Co	25	Eu	1
Fe	56 300	Sc	22	Tb	0.9
Ca	41 500	N	20	Tl	0.5
Na	23 600	Li	20	Lu	0.5
Mg	23 300	Nb	20	I	0.5
K	20 900	Ga	15	Tm	0.5
Ti	5700	**Pb**	13	Sb	0.2
H	1400	B	10	Bi	0.2
P	1050	Th	10	Cd	0.2
Mn	950	Pr	8	In	0.1
F	625	Sm	6	Hg	0.08
Ba	425	Gd	5	**Ag**	0.07
Sr	375	Dy	3	Se	0.05
S	260	Hf	3	Ar	0.04
C	200	Cs	3	Pd	0.01
Zr	165	Yb	3	**Pt**	0.005
V	135	Er	3	**Au**	0.004
Cl	130	Br	3	He	0.003
Cr	100	**U**	3	Te	0.002
Rb	90	Be	3	Rh	0.001
Ni	75	As	2	Re	0.001
Zn	70	Sn	2	Ir	0.001
Ce	60	Ta	2	Os	0.001
Cu	55	W	2	Ru	0.001
Y	33	**Mo**	2		

to short-lived radiogenic decay products, for instance radium, which are present at much less than parts per trillion (ppt).

These patterns of elemental concentration in average crust reflect abundances in the Solar System (cosmic abundances) and chemical partitioning during condensation of the Earth from the cloud of matter from which the Sun formed. In general, element abundances in the Solar System decrease with increasing atomic number Z, most strongly so for atomic numbers greater than Fe ($Z = 26$). There is also a pattern that elements with even atomic number are more abundant than elements of similar mass but odd atomic number (Fe, $Z = 26$, is more abundant than Mn, $Z = 25$, or Co, $Z = 27$). Since condensation of the Earth as a rocky planet, the elements have been internally partitioned on a global scale as a result of two processes. The separation of the metallic core from the silicate component that made up the primitive mantle concentrated siderophile elements

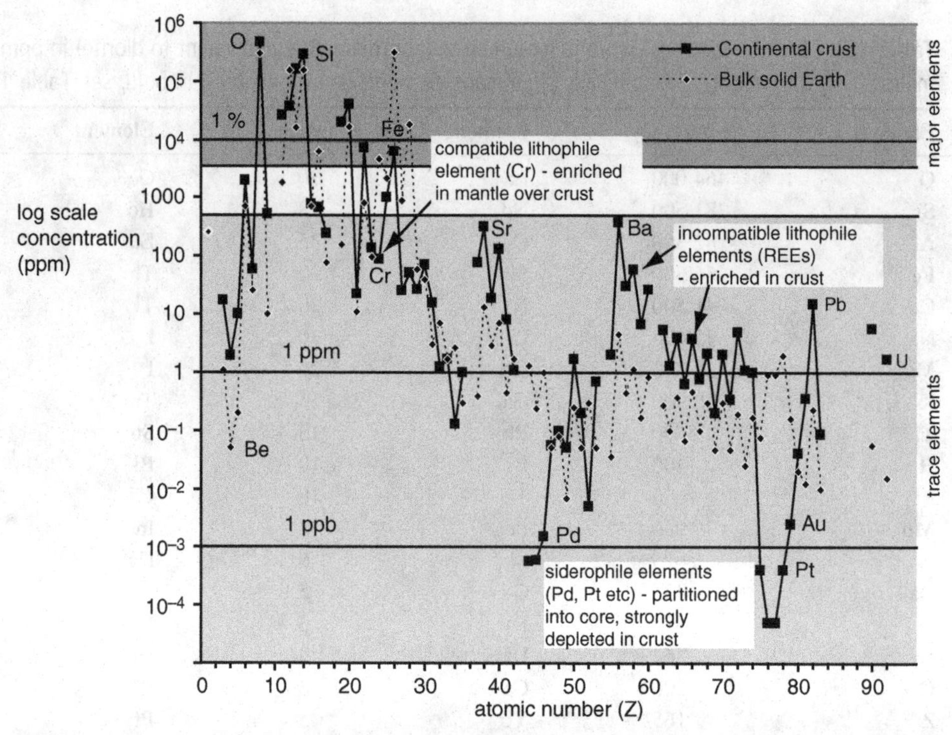

Figure 1.5 Element concentrations by atomic number in bulk solid Earth, as a proxy for its composition at the time of condensation, and in continental crust showing the range of element abundances over about nine orders of magnitude, and examples of relative enrichment and depletion into the crust that result from separation of the core and from formation of the crust through partial melting of mantle. Concentration data from GERM database at http://earthref.org/GERM.

in the core. Melting in the upper mantle progressively through Earth's history has produced a chemically distinct buoyant long-lived crust. Some elements are more abundant in the crust than in the Earth as a whole, while others are less abundant (Figure 1.5).

Except for Fe, Al and Ti, a typical crustal rock contains economically valuable metals in trace concentrations (Table 1.3). These trace elements are present either as minor components in solid solution in silicate and oxide minerals (e.g. Cu in mafic silicates) or as major components in accessory minerals (e.g. Zr in zircon).

1.3 The geology of ore deposits

A geological definition of an ore deposit

As ore is rock from which it is economic to extract a commodity, we need to consider how costs of extraction vary with different rocks and different rock mineralogy. The major costs of the operation of a large open-pit mine are rock haulage (e.g. transport of rock from the mine to the mill, and of waste rock to a dump) and rock crushing. Labour costs normally dominate operational costs of an underground mine, but the underlying cause is

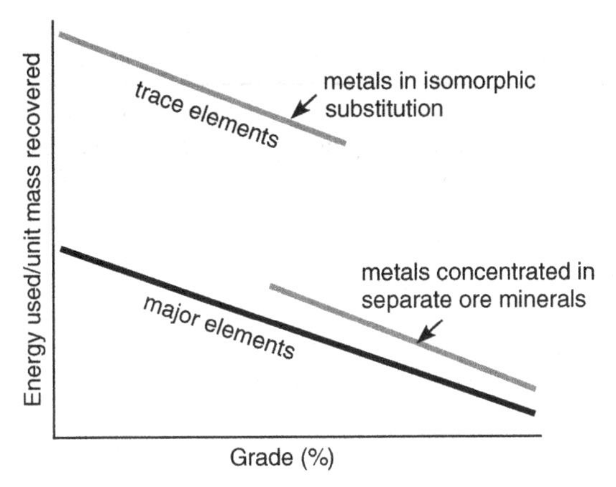

Figure 1.6 Schematic of the energy required to extract metals from ores and rocks of different grade and mineralogy (after Skinner, 1976).

similar as for open-pit mines – the costs of mining are proportional to the tons of rock moved and processed. The reward is the mass of commodity sold. If all other factors (e.g. mineralogy) are constant, the costs of extracting a commodity per mass are to a first approximation inversely proportional to the concentration (**grade**) of the element in the rock (Figure 1.6). The high energy costs of extracting trace elements from isomorphous solution in common minerals, especially from silicate minerals, mean that there are few ores from which such extraction is made; a point that is also apparent in the listing of ore minerals in Table 1.2. In general, the most cost effective extraction is from minerals that have a high concentration of the metal mined from rocks of high grade.

Approximate present-day economic grades of ore bodies for a number of metals are listed in Table 1.4. For most metals, the economic grades for mining have decreased through history, largely as a result of advances in the technology of mining and extraction, and decreasing costs of haulage. The **enrichment** factor over average concentrations (**Clarke of concentration**) required for an ore to be economic ranges from a few times for some major elements up to many orders of magnitude for trace metals such as Au, Pt and Hg. The enrichment factor generally increases with decreasing abundance of the element in the crust.

In general, ore deposits of a metal are rocks in which the concentration of the metal is significantly higher than in average crust. They are natural enrichments of the metal in the Earth's crust, they are geochemical **anomalies**. A similar definition applies for deposits of industrial minerals – these are natural concentrations of the mineral of interest.

A mixed economic–geological definition of an ore deposit in the case of metal ores would thus be:

> an ore deposit is a rock body in which there is a naturally enriched concentration of one or more metals and from which it is economic to extract these metals.

The significance of ore deposit size

The smallest commercial mine operations, for instance on veins, are typically of ore bodies of about 1 Mt, which is equivalent to a cube of rock about 75 m across, the exact volume depending on rock density. The largest ore deposits are a few gigatonnes, equivalent to an open pit a few kilometres long and several hundred metres deep. For

Table 1.4

Metal	Clarke = average concentration in upper crust	Grade in typical ore	Clarke of concentration = enrichment factor average crust → ore
Al	8%	30%	4
Fe	5%	60%	12
Ti	5700	5%	10
Mn	950	5%	50
Cr	100	5%	500
Li	20	1%	500
U	3	0.1%	300
Sn	2	1%	5000
W	1.5	0.3%	2000
Ni	75	1%	100
Zn	70	10%	1000
Cu	55	1%	200
Pb	12	10%	10 000
Mo	1.5	0.3%	2000
Ag	0.1	100	1000
Hg	0.1	1%	100 000
Au	0.004	5	1200
Pt	0.002	5	2500

Concentrations of some economically important metals in average upper crust, and typical **grades** and enrichment factors of ores. Compositions are in ppm, except where indicated. Note that the list gives average grades of ore bodies. In any ore body there will be a range of ore grades, ore will be mined at lower grades than average, and a mine will have internal **cut-off grades** below which rock is considered sub-economic ore or waste (compare Figure 1.2).

any metal and ore deposit type, mined ore bodies vary in size, typically by about two to three orders of magnitude. A common graphical representation of ore deposit grade and size is of log grade versus log tonnage of deposits (Figure 1.7), with diagonal lines showing the contained mass of the commodity metal.

Ore deposit size plays two roles in the economics of mining. First, economies of scale mean it is generally cheaper to mine a ton of ore in a large deposit than in a small one. Large, open-pit mines can generally profitably mine lower-grade rocks than indicated in Table 1.4. These economies of scale are implied from the inclined lower cut-off of the distribution of deposits by grade and tonnage that is apparent in Figure 1.7.

The lower limit of ore tonnage in plots such as Figure 1.7 is also largely controlled by economic factors. There are accumulations of ore minerals that are smaller, but these are not plotted. Such accumulations are too small to be economical to extract and are known as **occurrences**. They are of scientific interest, but are not listed in inventories or deposits. **Prospects** are known accumulations of ore minerals which have the potential to be identified as ore deposits with thorough exploration, for instance drilling. The upper limits of the ranges of grade and tonnage are geological. Large, high-grade deposits are rare.

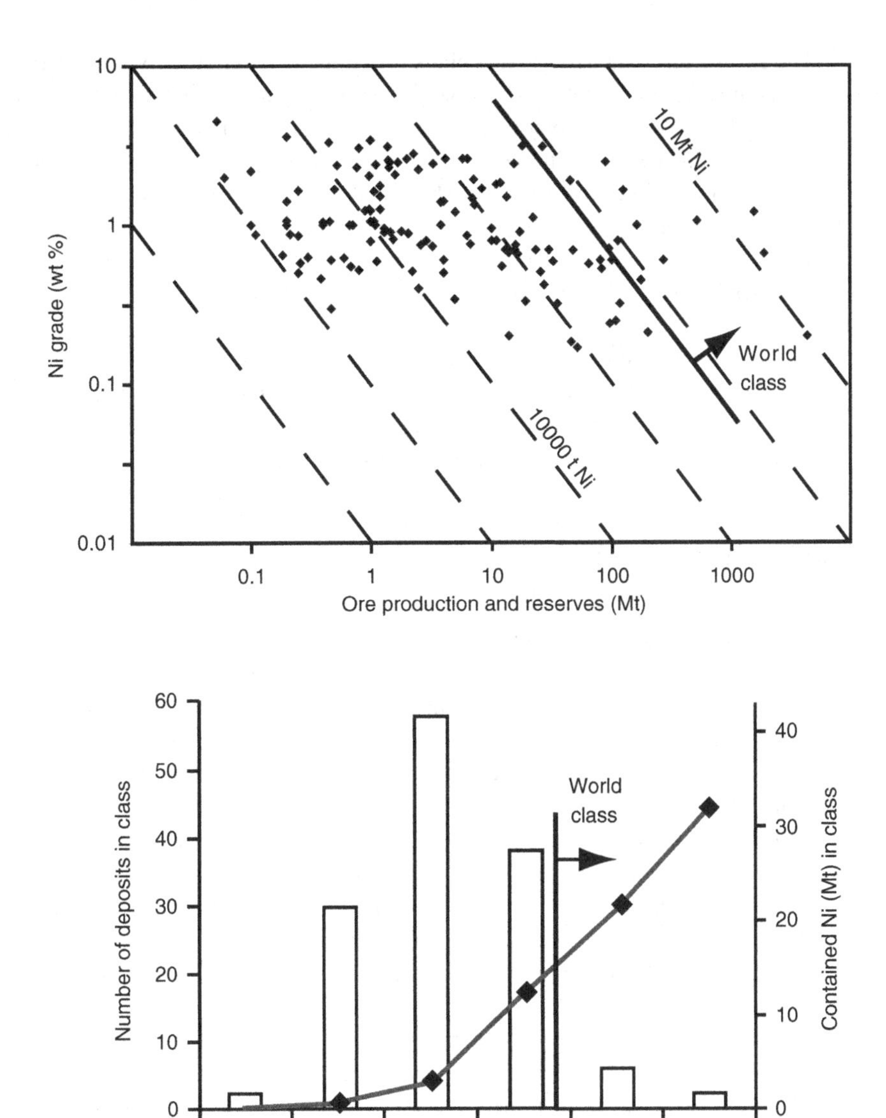

Figure 1.7 Above: Grade–tonnage plot of 136 worldwide known economic and potentially economic sulfide Ni deposits (see Section 2.3). The oblique dashed lines delineate total contained Ni. Data from the Geological Survey of Canada reported in Eckstrand and Hulbert (2007). Note that the quoted tonnage and grade of any deposit is based on cut-off grades, and may change somewhat with exploration and changing economics. Below: Number of deposits in different size categories as measured by contained Ni, and total contained Ni in each size class. The cut-off for world-class deposits is shown, i.e. the top 10% of deposits by contained metal.

The second role that deposit size plays is that the rare abnormally large deposits can be overwhelmingly important for the world's supply of a metal. For the 136 deposits plotted in Figure 1.7, 45% of the total nickel is in the two largest deposits. **World-class** deposits have been defined as those in the top 10% of any category with respect to metal contained. For many commodities this small number of world-class deposits contain between 60 and 90% of global resources, in the case of Ni about 85% (Figure 1.7). World-class deposits are thus critical both for global supply and for the financial health of major mining companies, as ownership provides a reliable long-term revenue stream.

Geological factors affecting economics of ore extraction

In addition to ore deposit grade and size there are other factors, both geological and societal, that affect whether an ore deposit is mined, and the cost of extracting a commodity will vary somewhat from deposit to deposit. The role of an ore deposit geologist is as much to describe and evaluate the geological factors that control the costs of extraction of a commodity as it is to find ore. One consideration in the economics of mining is the amount of waste rock that is required to be mined in order to access the ore bodies. Geological factors that affect economics include:

- Shape and depth of the deposit. Per tonne of rock mined, the cheapest ore body to mine is a flat-lying body at the Earth's surface. It is generally cheaper to mine an ore body that has an overall sub-spherical shape than a thin vein. Similarly, per tonne of rock mined, open-pit mining is cheaper than underground mining. However, in order to access ore, a higher ratio of waste rock to ore (**stripping ratio**) will generally be mined from an open-pit than from an underground operation, and the relative costs of the two methods of mining reverse for the deeper parts of many ore bodies (Figure 1.2) and for deeply buried ore bodies.
- Mineralogy and texture of the ore. These factors affect the cost of mining, the costs of crushing and milling of ore, and the cost of metallurgical extraction of a metal from ore minerals. In some cases we need to consider also the presence of deleterious elements and the costs of removing these from the ore before processing, for instance, phosphorus in iron ores.
- The presence of multiple extractable products. We distinguish **co-products**, as those additional products which control the economic feasibility of a mine, and **by-products**, which are extracted from mined and milled ore or waste if the costs of metallurgical extraction are favourable, but which do not significantly affect the economics of the whole mining operation. The distinction between these two categories is often arbitrary, but many mines produce multiple commodities, and some metals which are only extracted for specialist small-volume markets are extracted entirely as by-products (e.g. Sc, Te).

1.4 Ore deposit geology as a science: classifications and deposit models

Ore deposit geology or economic geology is the science behind the discovery and evaluation of mineral resources. One of the long-term research aims of ore deposit geologists is to understand the genesis and formation of ores in the geological history of the Earth sufficiently well to predict the locations and nature of deposits.

Economics does not care what geological processes were involved in the formation of the natural geochemical anomalies that are ore. The processes of ore formation can be magmatic, metamorphic, sedimentary or hydrogeological. The processes can occur in the mantle, in the crust, at the Earth's surface; they can be affected by tectonic setting, climate, etc.... Ore deposit geology is thus concerned with, and builds on, all parts of geology.

Classifications of ore deposits

There are many thousands of known ore deposits. In view of this large number, we need a scheme of categorisation into ore deposit types. Categorisation both simplifies the process of developing understanding and allows estimation of statistical measures of ore deposit spatial distribution, ore grade and ore tonnages, which can be input into estimates of total resources in a given region and used in strategic planning. The description of the common and typical characteristics of ore deposit types is an integral part of this book.

The classification of ore deposits into types or classes is based on combinations of their characteristics, including which economically extractable commodity or commodities they contain; the nature of their ore, and whether the ore occurs in **veins, lodes**, or is **disseminated** through large masses of rock; their mineralogy; their **host-rocks;** and their geological settings. The number of ore deposit types is not fixed. Different classification schemes divide and combine in different ways. The current scheme of the Canadian Geological Survey, for instance, recognises 85 types of metal ores. Different classification schemes of ore deposits are based on different features as the primary attribute for categorisation. None is perfect. Commonly used attributes are:

- element or mineral extracted, e.g. Cu, Au, Fe etc;
- host-rock type, e.g. large ultramafic–mafic intrusive bodies, carbonate-bearing sedimentary rocks;
- tectonic setting and/or geological age, e.g. back-arc basin, intracratonic sedimentary basin, Proterozoic intracratonic basins;
- major genetic process of enrichment, e.g. magmatic processes, sedimentary processes.

This book is structured around the last of these methods. The chapters are divided so as to describe ore deposits that form in the following situations:

- magmatic – concentration as a result of chemical and mineralogical processes in magmas;
- hydrothermal – concentration as a result of precipitation from heated aqueous fluids migrating through crustal rocks;
- sedimentary – concentration by mechanical or chemical processes at the time of sedimentation;
- **regolith** – enrichment as a result of weathering processes.

Ore deposit types and models

Over time, ore deposit types defined by empirical observations have become ore genetic types as a result of research, geological observations and interpretation of the geological processes of their formation. Ore deposit types are thus ore genetic 'models'.

> An ore deposit model is a conceptual and/or empirical standard, ideally a population of natural phenomena, embodying both the descriptive features of the deposit type, the larger ore-bearing environment, and an explanation of these features in terms of geological, and hence of chemical and physical, processes.
>
> Hodgson, 1987

Genetic models have been developed in order to explain how deposits form. Because they are the result of rationalisation of knowledge, they are a powerful means of organising data in a form that enhances understanding, prediction and communication.

What type of deposit have we found and how did it form? Although the mechanisms of deposit genesis rarely affect the exploitation of an ore body once its position, grade, tonnage and mineralogy is known, these are questions that are asked in the exploration industry and the minerals-extraction industries during exploration and deposit evaluation. Ore deposit models are an important component of communication in the industry and serve to aid exploration and deposit evaluation. For instance:

- A full genetic model explains why a deposit forms in a specific geological and tectonic setting, and the geochemical and structural processes involved in its formation. Models guide where to search for a deposit type within the Earth as a whole and, when combined with knowledge of local geology and geological history, on a much smaller scale within an exploration lease.
- A model describes and explains the shapes and forms of ore bodies. These are parameters that guide efficient evaluation of a prospect, such as optimal positioning of drill holes.
- A model considers the mineralogy of ore and guides the observations needed to evaluate, for instance, what co-products may be present, what ore grades can be expected, and what method of metal extraction may be best.

However, the answer to the question 'what type of deposit have we found?' may be ambivalent and reflect a degree of doubt. Ore deposit genetic models are geological interpretations and are thus uncertain in specifics or even as a whole. They are our best interpretation at the current state of knowledge. This is true both of assignments of deposits to a model and to our interpretation of the processes of genesis of an ore deposit type. Our understanding of the genesis of ore deposits is subject to debate and revision and improves with new data, new observations, new exposures, new geochemical analytical capabilities, and theoretical analysis. With better knowledge and the discovery of more deposits, models become revised and better formulated.

Disputes about the genesis both of specific ore deposits and of ore deposit types are common. In fact there are few ore deposit types for which there has not been some degree of disagreement over some aspects of ore genesis. Ongoing development, discussion and refinement of ore genetic models is an integral pursuit of ore deposit geology, and hence also of discussion in this book.

We can observe and take direct measurements on ore formation in only a few cases where the processes are in action at the Earth's surface (for instance, on the sea floor, and in mineral precipitates in volcanic fumaroles), but even in these cases we cannot observe all the processes and events that lead to the formation of ore, and must make interpretations of unseen processes that are taking place at depth. The complexities of the geological record make interpretation of ancient ore deposits even more uncertain. Parts of a deposit may become disrupted as a result of deformation, mineral textures may be overgrown with no preservation of earlier textures. For these reasons, the most persistent debates are generally over the genesis of deposits in deformed and metamorphosed rocks.

1.5 The future of ore deposit geology

The mining and exploration industry is renowned for boom-and-bust cycles. Historic records back to about the twelfth century suggest that cycles of boom and bust have always occurred. There is, however, a long-term ongoing need for discovery and for deposit evaluation, and hence for ore deposit geologists. Mining depletes reserves. The depleted reserves must be replaced through discovery and development in order to continue mining. This point is true both for the world as a whole and for individual mining companies. The wealth of a mining company is largely its 'yet-to-be-mined' ore reserves. To maintain its wealth it needs to replace what it mines through discovery or through purchase of resources discovered by other groups.

Although the use of commodities is affected by economic cycles, there are long-term trends of increasing rates of use. The global use of most metals has increased by two to three orders of magnitude in the last century. Average annual percentage increases in production of many important mineral commodities over the past two decades are listed in Table 1.1. These increases are driven by increased consumption as a result of a combination of global population increase and the overall increased standards of living (per capita wealth). So long as there is an increase in global wealth and population, we expect demand for most commodities to increase.

Increases in mine production have occurred despite growing efforts towards recycling. Recycling can never be to 100%. There will always be losses from abrasion during product use and at each stage during processing for recycling. A recycling rate of up to 70% may be reasonably achievable for many metals. Recycling shortfalls and also any increase in demand over time must be covered by mined metals.

Box 1.1 How much ore is there in the world?

Although the global rate of extraction of most metals has increased by two or more orders of magnitude over the last century, the tonnages of reserves have also increased such that we are apparently no nearer to having extracted all known resources of any metal. The question of how much ore is available is important for assessing if and when resources may become scarce, and ultimately whether there are limits to extraction in the world. Strategic planning at national or regional levels should also be based on estimates of total available resources and lifetimes of mining operations.

There can be no fixed answer to the question of 'how much ore is there?'; what rock is ore depends on economics. Lower-grade ore in low-grade ore bodies, or in the low-grade haloes of economic ore bodies, is economic to mine when prices are higher or when the costs of extraction are lower. As rocks with grades higher than a certain value are increasingly rare with increasing grade the tonnage of available ore is increased if the grade at which we can economically extract a metal is reduced.

Mathematical expressions of the relations between economic grade and tonnage of ore have been proposed. One approach is to consider the statistical distribution of tonnage of rock of different grades and to assume that what has been sampled and analysed is representative of the whole crust or a part of the crust.

Lasky (1950) derived a relationship for ore grade within explored ore bodies and clusters of ore bodies for which a large amount of geochemical data was available. The relationship indicates a near exponential increase in available ore with decreasing 'cut-off' grades. As typical cut-off grades have decreased over the long term, for instance from about 3% to 0.5% over the past century for copper, this is one reason why reserves have not diminished over the past century. (For a survey of long-term trends of grades and tonnages mined for a number of commodities, see for instance, Mudd, 2009.) In an independent study, Ahrens (1954) demonstrated that concentrations of trace elements in suites of magmatic rocks have log-normal distributions, and although the data sets he used were not extensive enough to include samples of ore-grade rock his results are of interest to understanding of the abundance of ores and, like the relations of Lasky, indicate increasing ore tonnages with decreasing grades of economic mining.

Can the historical trend to mining lower and lower-grade ores be continued indefinitely? Skinner (1976) discussed one factor that should be considered in answering this question. Where they occur at near-average crustal concentrations, many metals are present at trace concentrations substituting in the lattice sites of major elements in common rock-forming minerals; at ore concentrations they are present predominantly as an essential component of sulfide or oxide minerals. This mineralogical contrast between ores and average rock raises the costs of metal extraction from lower-grade rock (see Figure 1.6). Much more energy is required to extract Cu from a Cu-bearing silicate mineral than from a Cu-rich sulfide mineral, even at the same grade. Skinner therefore proposed that there is a mineralogical barrier that will prevent mining grades from continually decreasing. Further, because of element partitioning between phases, there may be two overlapping log-normal distributions of grade of the type proposed by Ahrens. In the case of Cu for instance, there will be a log-normal distribution of grades of rocks in which Cu is held in silicates and a distribution for ores in which Cu is held in sulfide minerals (Figure 1.8). If this is the case, the rule that ore tonnages increase with decreasing extractable grades may not universally apply.

We do not know whether grades are bimodally distributed by mineralogy as suggested in Figure 1.8. We do know, however, that some sulfide minerals are accessory minerals in magmatic rocks which are not ore, but have metal concentrations that are close to average crustal concentrations, e.g. iron–copper sulfide minerals (Stimac and Hickmott, 1996) and molybdenite (MoS_2) (Audétat *et al.*, 2011).

Alternative lines of logic and independent data can be used to estimate tonnages of ore in the world and have been investigated. Kesler and Wilkinson (2008, 2009) start their analysis of the question of the abundance of ore deposits with the premise that to date we have only exploited those deposits which outcrop at the surface. A deposit

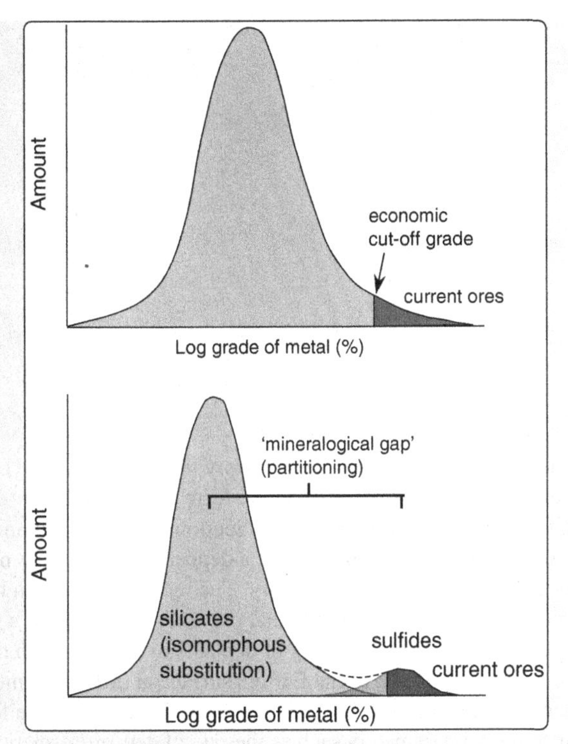

Figure 1.8 Schematic representations of the amount of a trace element in the Earth's crust by grade of element in rock, differentiating ores as rocks of highest grade. The upper graph shows a unimodal, approximately log-normal distribution. In this case, an exponentially increasing amount of metal is available if the economic cut-off grade of ore is reduced. The lower graph shows a bimodal distribution of sulfide ores and the metal as a trace component in solution in silicate minerals. In this case the tonnage of available ore is only marginally increased if the economic cut-off grade is reduced. (Modified from Skinner, 1976.)

formed in the past may be at the present surface, be buried, or have been eroded away. They argue that if deposits of any type have been forming at a uniform rate through time, then the numbers of deposits of different ages now exposed at the surface reflects the rate of production through geological time, the depth of formation, and the rate of burial and erosion of rocks. The last factor is what they call 'tectonic diffusion'. If we know the age distribution of deposits at the present surface, their depth of formation, and make estimates of the likelihood and rates of burial or erosion, we can estimate the rate of deposit production. Although the modelling does not take account of many geological processes, including tectonic controls on when and where erosion or burial takes place, a conclusion of the analysis is that we have to date exploited only a small fraction of ore in the crust. The majority of ore is as yet undiscovered, and much of it is at potentially mineable depths of less than about 3 km (Figure 1.9).

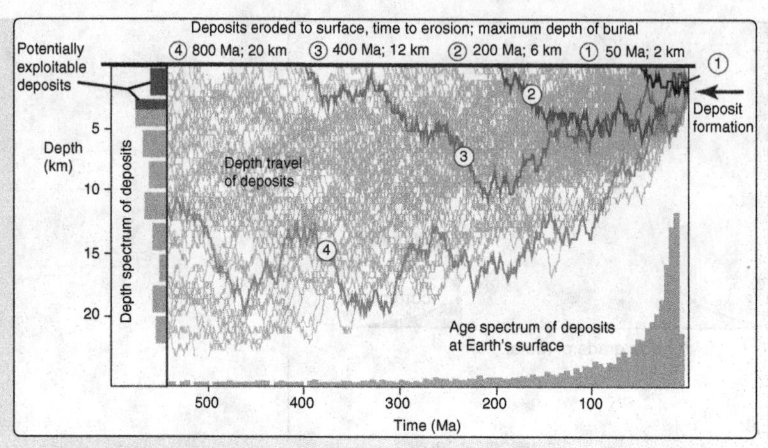

Figure 1.9 Numerical model simulation of the depth history through burial and erosion in the crust of a large number of ore deposits assumed to form at 1.9 km depth – as an approximation of the case of porphyry deposits (see Section 3.1.1) (Kesler and Wilkinson, 2008). Over each model time step of 1 Ma, a deposit moves up 468 m, down 468 m or stays at the same depth with equal probability. The movement at each time step is independent of the preceding time step. Deposits that move above the Earth's surface are assumed to be eroded away. Note the scattering of deposits over depth with time. The number of deposits that are predicted to be at the Earth's surface at different times is given at the bottom of the diagram. To estimate the number of deposits that are formed and preserved in the Earth, model parameters such as the rate of deposit formation and the rate of burial and erosion are tuned so as to provide the best match between the predicted spectrum of ages of deposits at the surface and the age spectrum of known deposits in the world. Note the large number of deposits at a few kilometres depth at all stages in the model: any deposit which is at a depth of less than 3 km may be economically extractable.

Questions and exercises

Exercises

1.1 Quantification of the relationship between ore grade and energy required to extract a metal illustrated schematically in Figure 1.6.

The extraction of Cu from sulfide ore requires about 70 kWh (kilowatt hours) of energy per tonne of ore. Extraction from ore in which Cu is in silicate minerals requires about 500 kWh per tonne of ore. Research local current approximate costs of energy (per kWh) for industry and the market price of Cu. Calculate the costs of the energy that would be required to produce a tonne of Cu from sulfide ore of 1% and 0.1% grade and from silicate ore of the same grades. How do these costs compare with the market price of the metal?

1.2 Using data from Tables 1.1 and 1.4, and assuming an average global ratio of mined waste rock to ore of 5 (= stripping ratio), make an estimate of the total mass of rock that is mined per year to extract Cu.

If consumption continues to increase at the annual rate given in Table 1.1, and ore grades on average decrease by 2% per year, what mass of rock will be moved per annum in 100 years time to extract Cu?

Estimate the area of land that would be taken up for mining every year, assuming that mining is on average to 100 m depth?

Discussion questions

1.3 The different schemes of classifying ore deposits have different purposes and value. In addition to classification based on genetic processes of ore deposit formation, classifications are available based on commodity, host-rock type and tectonic environment. Under what practical situations would each of these classification schemes be of use?

1.4 Using the scientific literature, mining industry and government reports, and material on the internet, write a report on one of the minor metal and gem commodities in the list below. None of these commodities are specifically covered or discussed in this book.

Antimony	Indium	Thorium
Arsenic	Mercury	Vanadium
Bismuth	Scandium	Emerald
Boron	Selenium	Opal
Gallium	Silicon (Silica metal)	Ruby and sapphire
Germanium	Strontium	
Helium	Tellurium	

The report should include:

(i) A summary of the geochemical behaviour of the element or mineral, whether the commodity is recovered as a by-product or co-product during mining of another commodity, and the grade required for economic extraction or mining.

(ii) A summary of the uses of the commodity, the history of its use, and its current importance in the world economy.

Further readings

The following are recommended for additional information and discussion of background economic and societal questions about ore deposits and their extraction:

Kesler, S.E. (1994). *Mineral Resource Economics and the Environment*, New York, MacMillan College Publishers.

Kogel, J.E., Trivedi, N.C., Barker, J.M., and Krukowski, S.T. (eds.) (2006). *Industrial Minerals and Rocks, Commodities, Markets, and Uses, 7th Edition*, Littleton, Society for Mining, Metallurgy and Exploration Inc.

Mudd, G.M. (2009). *The Sustainability of Mining in Australia: Key Production Trends and their Environmental Impacts for the Future*. Research Report no. RR5, Department of Civil Engineering, Monash University and Mineral Policy Institute.

Skinner, B.J. (1976). A second iron age ahead? *American Scientist* **64**, 258–269.

2 Magmatic ore deposits

Magmatic ore deposits, also known as **orthomagmatic** ore deposits, are deposits within igneous rocks or along their contacts in which ore minerals crystallised from a melt or were transported in a melt. Ore deposits that form in and around igneous rock units as a result of mineral precipitation from aqueous solutions or **hydrothermal** fluids are hydrothermal ore deposits. **Magmatic-hydrothermal** ore deposits are hydrothermal ore deposits in which the aqueous solutions were derived from magma. These ore deposits are described further in Chapter 3.

2.1 Petrological and geochemical background to magmatic ore formation

2.1.1 Processes of magma development

Magmas evolve in magmatic systems. A magmatic system is the site of melting, the pathway of magma migration and the site of crystallisation within a specific geological and tectonic environment. A magmatic system will extend from the mantle to the upper crust in most cases (Figure 2.1). The composition of an igneous rock is the result of the complex set of petrological processes in the development of the magmatic system from the site of melt formation in the mantle, during melt transport through the upper mantle and crust, to the site of final crystallisation. The processes that control magma compositions and compositions of igneous rock that crystallise from the magmas include: partial melting at the source and interaction between the melt and unmelted, **restite** minerals; interaction with **wall-rocks** the magma comes in contact with; assimilation of partial melts of wall-rocks; mixing of magmas from different sources; separation of a magma into two or more immiscible melts; fractional crystallisation of minerals; and mineral accumulation.

Most geological melts are partial melts of a source rock. Rocks melt progressively as temperature is increased above the respective solidus. When melt first forms, it is dispersed through rock and rates of upward transport are low. The processes of melt collection and more rapid upward migration to form magma that reaches the upper crust are complex, and for melts that are progressively formed with increasing temperature, probably involve variable degrees of separation of melt from restite as well as variable rates of transport. Typical magmas are likely mixtures of multiple aliquots of melt that may have variably interacted with rocks after their formation as they rise through the mantle and crust. Despite this complexity it is possible to distinguish magmas by their relative degrees of partial melting. These relative degrees of melting are expressed

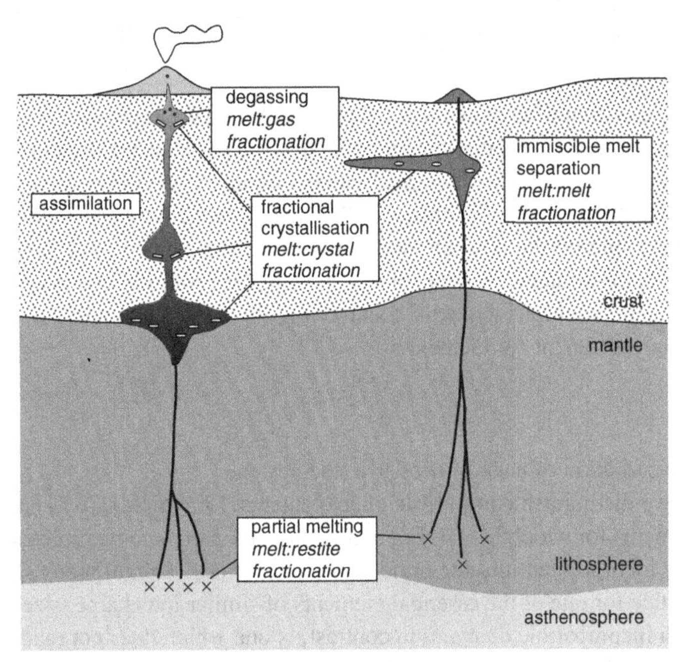

Figure 2.1 Sites and mechanisms of compositional evolution of magmas and the development of magmatic chemical variability within a schematic of magmatic systems extending from mantle lithosphere or asthenosphere into the upper crust.

through a simplified 'batch melting and extraction' model whereby a magma is assumed to be a result of melting of a source rock to a specific percentage and, once this percentage is reached, the magma is extracted and does not further interact with the source rock or minerals as it rises. Because this is a simplified model, relative degrees of partial melting are termed 'apparent percentages of melting' in the following sections.

2.1.2 Partitioning of elements between phases – the common process of chemical evolution in magmatic systems

In geochemistry, a **phase** is formally defined as matter in a specific form. Magma typically includes more than one phase and is thus heterogeneous. Multiple phases in a magma can include one or more melt phase, multiple mineral phases, and in some cases gas phases. The minerals interacting with a melt may be unmelted restite minerals in the source rock, minerals lining conduits and magma chambers, or minerals that have crystallised from the melt. The gases may be **exsolved** (released) from the melt or be incorporated from the surrounding wall-rocks.

At each stage in the history of a magmatic system, there are chemical interactions between phases. Minor and trace elements are partitioned, or distributed, between the coexisting phases based on the thermodynamic principle of minimisation of chemical energy. Nernst **partition coefficients** (K values), give the equilibrium ratios of concentration of the element between any two coexisting phases (two minerals, a mineral and melt etc.) and are defined by:

$$K_i^{a/b} = \frac{c_i^a}{c_i^b} \quad (2.1)$$

where c_i^x is the concentration of element i in phase x. When we are considering distribution between a mineral and a melt, it is conventional to write partition coefficients with a as the mineral and b as the melt. The numerical values of partition coefficients for any pair of phases depend on temperature (T), pressure (P) and the chemical composition of the phases, and also in principle on the absolute concentration of the element, but generally can be assumed to be approximately constant at any stage in the evolution of a magmatic system.

Where there are multiple minerals coexisting with a melt we define a bulk solid to liquid partition coefficient for element i:

$$D_i = \sum_{1 \to x} f^x K_i^x \quad (2.2)$$

where f^x is the fraction of each mineral x in the system.

In general we distinguish **compatible** trace elements, for which $K_i > 1$, and **incompatible** trace elements for which $K_i < 1$. For a mineral–melt pair, a compatible element is one that is readily incorporated into the mineral because of easy isomorphous substitution in the crystal lattice for one of the essential elements of similar ion charge, size and bonding behaviour. An incompatible element, in contrast, is one which does not readily substitute into the mineral and 'prefers' residing in the melt. The concentration of the compatible elements is greater in the mineral than in the melt, while the concentration of incompatible elements is lower in the mineral. Ionic radii and charges of many metals are shown in Figure 2.2. Many of the elements with similar charge and radius to the major mineral-forming elements such as Al, Mg and Fe will have compatible behaviour in the mantle and in mafic magmas in the crust, whereas incompatible elements mostly either have a higher charge (the high-field-strength elements – HFSEs) or larger radii (the large-ion lithophile elements – LILEs) compared to the major mineral-forming elements.

Compatible elements can become concentrated in restite or early formed cumulate minerals, and in general have highest concentrations in ultramafic and mafic rocks (e.g. peridotite, gabbro). Incompatible elements in general have higher concentrations in more felsic rocks (upper continental crust), and become progressively more enriched in rocks with higher silica content. Note, however, that different minerals are present in an evolved felsic melt than in a mafic melt, and the crystal lattice sites available for substitution are thus different. In consequence, some elements switch behaviour from incompatible to compatible or vice versa during magma evolution. Elements that switch from incompatible to compatible behaviour have the highest concentration in magmas of intermediate composition while those that switch from compatible to incompatible behaviour have the lowest concentration in magmas of intermediate composition.

2.2 Types of magmatic ore deposits

Ores are rocks with unusually high degrees of element concentration. Despite the complexity of petrological evolution of magmatic systems, common igneous rocks have a limited range of chemical compositions. With respect to copper content, for instance,

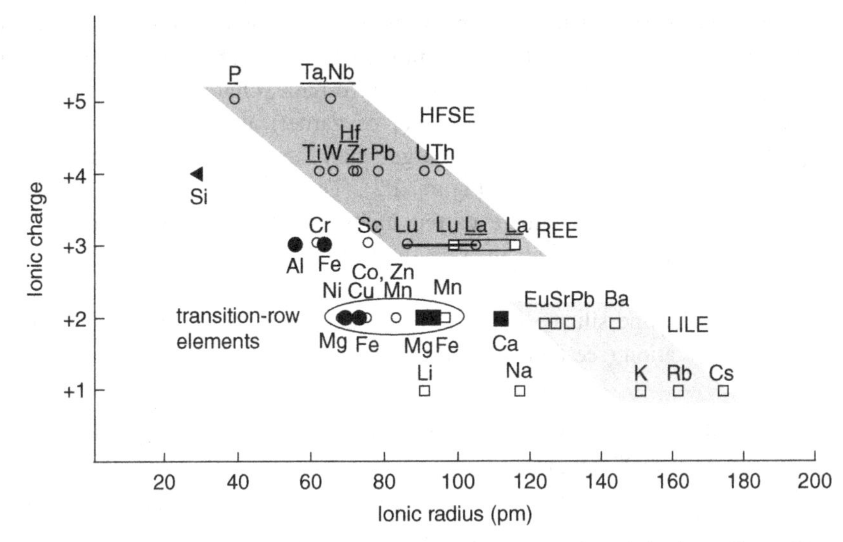

Figure 2.2 Classifications of cations based on ionic charge and ionic radii. HFSE = high-field-strength elements; LILE = large-ion lithophile elements; REE = rare-earth elements. The REEs form a homologous series from light (La) to heavy (Lu) which are not plotted separately but fall along the horizontal single and double lines on the diagram. The major elements in mantle silicate minerals are plotted with large solid symbols, trace elements with open symbols. The underlined elements are those lithophile elements which are extracted or may in the future be extracted from carbonatites (see Section 2.2.1). These are all HFSEs with ionic radii significantly larger than those of major elements of the same charge. Radii are Shannon ionic radii (Shannon, 1976, *Acta Crystallographica* **A32**, 751–767). Squares are radii of ions in eight-fold or higher coordination; circles are radii of ions in six-fold coordination and triangles are radii of ions in four-fold coordination. Radii of both six-fold and eight-fold coordination are shown for the REEs.

concentrations in igneous rocks range from about 50 ppm in average ultramafic rocks, through around 100 ppm in mafic rocks to 25 ppm in felsic granites and rhyolites. Specific magmatic processes in specific geological environments concentrate elements so as to form ore.

Five processes of ore formation are distinguished in magmatic systems: (1) concentration of ore elements as a result of very low degrees of partial melting, (2) accumulation and concentration of ore minerals in magma chambers during progressive crystallisation of magmas, (3) separation of two immiscible melts in a magma, (4) extreme fractionation during progressive crystallisation of a magma, and (5) incorporation of a mineral that occurs at a specific depth in the Earth. The first four processes can be related to the sequence of steps of magma evolution from melting in the mantle to final crystallisation in the crust as shown in Figure 2.1. At each of these steps, chemical partitioning of elements between phases is central to ore formation. Each process is introduced below and is discussed in more detail in the following sections.

(1) Concentration of ore elements as a result of very low degrees of partial melting

Partial melting can be a process of ore formation because the concentrations of incompatible elements will be higher in the melt relative to the source rock. The smaller the percentage of partial melting, the more enriched the melt is in incompatible elements. In the mantle, incompatible elements are those that do not substitute into the crystal structure of any of the major minerals of the mantle (olivine, clinopyroxene, orthopyroxene and, depending on depth, plagioclase, spinel or garnet) and are those elements that have significantly different ion radii or charges than the essential **components** of these minerals (Mg, Fe, Si and Al, see Figure 2.2). The strong concentration of incompatible elements into small-percentage partial melts is central to the formation of ores of a number of trace elements in rare and unusual magmatic rocks, particularly in carbonatites, as described in Section 2.2.1. It is also important in the formation of ores in strongly alkaline silicate igneous rocks such as in the Kola alkaline province of the Russian Federation (see Petrov, 2004).

(2) Accumulation and concentration of ore minerals in magma chambers during progressive crystallisation of magmas

There are a small number of ore minerals which accumulate directly from silicate magmas during fractional crystallisation in a magma chamber, including chromite as a source of chromium, and magnetite as a source of vanadium. Both of these minerals typically occur as minor or accessory phases disseminated through commonly occurring igneous rocks, including mafic and ultramafic cumulates. However, given specific conditions, concentrated accumulations of these ore minerals do occur, for instance in almost monomineralic thin layers in cumulate piles of thick layered intrusions. These unusual accumulations are discussed in Section 2.2.2 and can be explained by cumulate processes in combination with specific additional physical and chemical processes in magma chambers.

(3) Separation of two immiscible melts in magma

This process is most important for the development of concentrations of sulfide minerals in igneous rocks (Sections 2.2.3 and 2.2.4). Sulfide minerals are a minor component of most rocks, including those of the mantle. On partial melting, sulfide minerals dissolve into the silicate melt up to the limits of sulfide solubility, which in silicate magma is typically a few hundred ppm. As the silicate magma migrates and evolves the limits of saturation change, and some of the dissolved sulfide may become **immiscible** and separate out as droplets. These droplets may segregate to form a separate sulfide melt phase that will crystallise on cooling. Chalcophile and siderophile elements, particularly Cu, Ni, Pt and related elements, are markedly more compatible in solution in sulfide melt than in silicate melt, and will thus be strongly concentrated in an immiscible sulfide melt.

Ore may be also formed more rarely in other immiscible melt–melt pairs. Immiscible separation of iron oxide-rich melt from aluminous felsic and anorthositic silicate magmas has been suggested as the mechanism for the formation of some magnetite–ilmenite Fe–Ti ores in intrusions in some cases known as nelsonites which are mined at Tellnes in Norway as a source of titanium minerals (e.g. Charlier *et al.*, 2006), and immiscible separation of phosphorous-rich melts from carbonatite magmas to form an apatite-rich igneous rock called phoscorite at zoned alkaline intrusions such as in the Kola alkaline province (Petrov, 2004).

(4) Extreme fractionation during progressive crystallisation of magma

As a melt progressively crystallises, the concentration of incompatible elements which are not partitioned into crystallising minerals increases. Enrichment of incompatible trace elements by orders of magnitude is possible in the last remaining few per cent or less of melt. These elements may be major components of rare minerals that crystallise from the last remaining melt

This is the process proposed for the genesis of many granitic pegmatites (Section 2.2.5). Pegmatites are ores for many so-called rare metals, for instance, Li, Be, Nb, Ta, Sn and U, all of which are incompatible in common rock-forming minerals but are essential components of one or more rare minerals that can crystallise from a melt, and also for gem minerals which include an incompatible element in their structure, such as emerald (with Be) and topaz (with F).

(5) Incorporation of a mineral that occurs at a specific depth in the Earth

This is the case of diamond deposits in kimberlite and lamproite which form from magmas that are generated below the depth of the graphite–diamond transition in the mantle.

Diamond deposits are a special case of ore deposit because the critical factor in their formation is not concentration of diamonds but is rather carriage of relatively low concentrations from depth of greater than about 140 km in the mantle, where diamond is a stable mineral, to the surface (see Section 2.2.6).

2.2.1 Deposits formed from small-fraction partial melts: light rare-earth element (LREE) ores in carbonatites

Carbonatites are rare igneous rocks which are composed of more than 50% primary carbonate minerals, generally mixtures of calcite, dolomite and siderite. About 500 carbonatite intrusions and a small number of localities of extrusive carbonatites are known worldwide.

Many carbonatite intrusions host one or more separate ore bodies, containing different combinations of incompatible trace elements, and in some cases also of some major elements. Important examples of carbonatite-hosted deposits include: ores of LREEs, Nb and Fe at Bayan Obo, Inner Mongolia, China (Yang *et al.*, 2011); resources of Ti and Th at Powderhorn (Iron Hill), Colorado, USA (Van Gosen, 2009); ores of LREEs at Mountain Pass, California, USA (Haxel, 2005; Long *et al.*, 2010); ores of Cu, Co, Zr, Hf, Fe, apatite, and vermiculite, and also by-products Au, Ag, Ni and Pt at Palabora, South Africa (Palabora Mining Company staff, 1976); and of LREEs and Nb at Mt Weld, Western Australia (Lottermoser, 1995). Tantalum is present at ore grade in some carbonatites. Carbonatites and genetically related rocks are the primary global source of the LREEs.

Geochemical nature of LREEs

Many of the trace elements which are extracted from carbonatites and from strongly alkaline silicate igneous rocks are HFSEs. These are trace metals which have a large ionic charge (+3 to +5) and relatively small ionic radius, and are consequently incompatible in common silicate minerals in the mantle (Figure 2.2). The REEs are the row of elements of atomic number $Z = 57$ (La) to $Z = 71$ (Lu), the LREEs are the lower-atomic-number elements in this row (La, Ce, Pr, Nd, Sm and Eu) and the heavy rare-earth elements (HREEs) are those from Gd to Lu. Because they have the same outer-shell electron configuration, the REEs exhibit very similar chemical behaviour in natural environments and readily substitute for each other in minerals. All form ions with +3 charge, although two, Eu and Ce, can also be present in natural environments with +2 and +4 charge, respectively. The increasing atomic number is taken up by addition of electrons to the inner 4f electron shell; this causes progressive reduction in ion size across the row, and hence differences in partitioning constants which allow partitioning of the LREEs from the HREEs. The HREEs are less abundant than LREEs in the crust, and are most strongly concentrated in different types of ores than the LREEs (see page 337 in Chapter 6).

The nature of typical carbonatites and hosted LREE ores

Intrusive carbonatites occur most typically as roughly cylindrical, composite, approximately pipe-shaped, steeply plunging stocks up to about 3 km in diameter, although they can also occur as irregular lenticular bodies (Figure 2.3). Swarms of carbonatite dykes are often present within a few kilometres of larger intrusions. Some carbonatites are isolated intrusions, but most are adjacent to or within complex zoned intrusions of mafic and ultramafic alkaline intrusive rocks. The rocks located within a few hundred metres of the contacts of carbonatite intrusions and of zoned intrusions of carbonatite and alkaline silicate igneous rocks are in almost all cases converted to fenite. This is a **metasomatic** rock type characterised by high potassium content, such that one or more of K-feldspar, riebeckite, and biotite are important minerals.

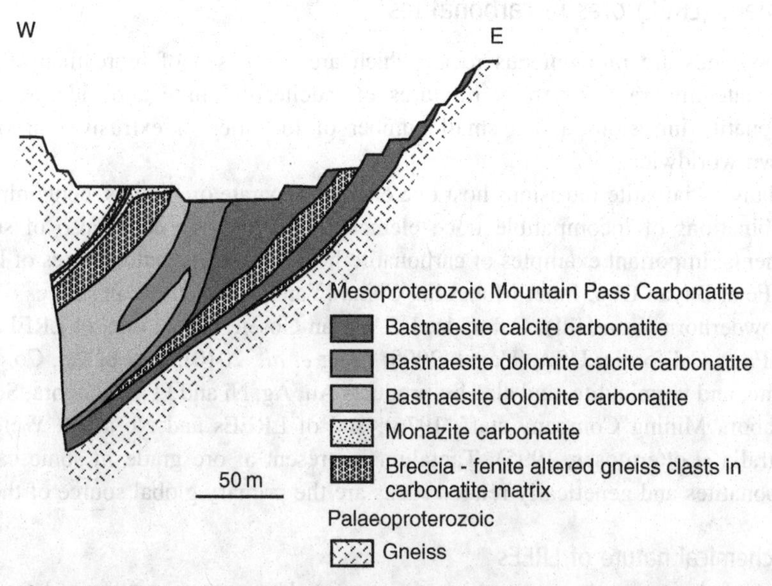

Figure 2.3 Cross section through the open pit of the Mesoproterozoic Sulfide Queen carbonatite intrusion at Mountain Pass, California, USA (after Castor, 2008) showing characteristic heterogeneity of these intrusions. The ore zones are essentially the three bastnaesite-bearing units within the intrusion.

On average, carbonatites have the highest concentrations of rare-earth element oxides (REOs) of all crustal rock types. Summed REO concentration is about 0.5%, which is around 30 times that of average continental crust. In unweathered REE-rich igneous carbonatite, the REEs are hosted in different minerals in which they are essential components, especially the REE fluorocarbonates bastnaesite [$(La,Ce)CO_3(F,OH)$] or more rarely parisite [$Ca(Ce,La)_2(CO_3)_2F_2$], or the phosphate monazite [$(Ce,La,Th)PO_4$]. The ores have igneous textures with the ore minerals either disseminated relatively uniformly or as concentrations in clusters with other minerals through the carbonatite.

Economic ores of LREEs are hosted in the small number of carbonatites that have significantly higher concentrations in part or all of the intrusion, with grades of up to a few weight per cent (wt %) combined LREEs. At Mt Weld, ore is in weathered carbonatite and the enrichment to ore grade is a result of lateritic weathering (see Section 6.1). In contrast, the high concentrations of LREEs at Bayan Obo and Mountain Pass occur in rocks with essentially fresh igneous minerals and textures and the ores are thus interpreted as 'primary' magmatic ores. These two LREE ore-bearing carbonatites are unusual compared to other carbonatites. At Mountain Pass (Figure 2.3), ore is in calcitic and dolomitic carbonatite with barite as an important gangue mineral. The carbonatite is unusually associated with an ultrapotassic alkaline intrusion. At Bayan Obo, ore is present in calcite carbonatite, but the highest grades are in composite lenses of unique iron oxide–fluorite–aegerine-augite rock which is hosted within a large (10 by 2 km outcrop area) carbonatite intrusion. Other carbonatites with high primary REE concentrations are known but many have not been fully evaluated as possible deposits of REEs, for instance, along the East African Rift and at Kanneshin (Afghanistan).

Genesis of LREE ores in carbonatites

The environments of melting to form carbonatite magmas are inferred using geochemical reasoning and through comparison of rock compositions with results of petrological experiments of melting at high pressures. At pressures greater than about 2.5 GPa, hence depths greater than about 90 km, dolomite can be a stable mineral in mantle peridotite. Up to a few per cent of carbonate is present in some mantle-derived xenoliths, hence suggesting that it is a minor component of some mantle peridotite. Carbonatite melts are produced in experiments by low percentages of partial melting of carbonate-bearing peridotite at between about 2.5 GPa and 6 GPa. The carbonate minerals are fully melted after a few per cent partial melting, and the maximum percentage of melting that will produce a carbonatite melt from dolomite-bearing mantle peridotite is thus estimated to be less than about 3%. With slightly higher temperatures and degrees of melting, the same source rock will produce carbonate-bearing alkaline magmas (Figure 2.4). As the solidus is at a higher temperature at lower pressures (Figure 2.4) carbonatite melts may in most cases freeze on rising through the upper mantle and may only reach the surface in unusual circumstances of rapid rise from the depths of melting.

The geochemical behaviour and mineralogical setting of REEs in mantle environments is not fully known; however, their likely behaviour on melting can be interpreted. All REEs have ionic radii too large for easy substitution in crystal lattice sites in major mantle minerals, and unless there is a minor mineral in which they are compatible they will be incompatible and will be concentrated in a melt. Partition coefficients are lower for

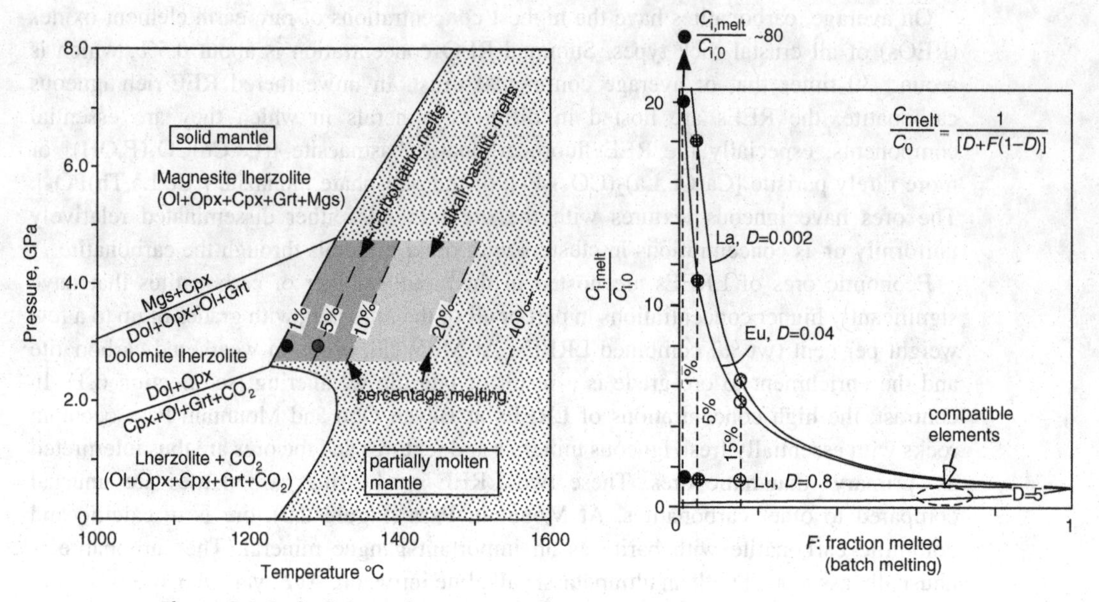

Figure 2.4 Left: Schematic pressure–temperature diagram of the environment of formation of carbonatitic and alkali–basaltic melts from carbonate-bearing mantle peridotite (dolomite and magnesite lherzolite), after Dalton and Presnall (1998). Numerical percentages of partial melts at any pressure and temperature, and the boundary between carbonatitic and basaltic melts are based on experiments with 2.5 wt % CO_2 of Dasgupta *et al.* (2007). The exact positions of these isotherms will depend on the modal per cent carbonate in the mantle and are shown to illustrate trends only. Right: Element enrichment in melt compared to source rock as functions of fraction melted (F) and bulk partition coefficient (D_i) for the rare-earth elements La, Eu and Lu, based on the equation for batch (one-step) partial melting, as shown. The shaded circles correspond to melt extraction at the conditions shown on the P–T diagram. Partition coefficients are taken from Dasgupta *et al.* (2009).

LREEs than for HREEs, hence the former will be more strongly concentrated in melts. The concentrations of LREEs in typical carbonatites and the relative concentration of LREEs over HREEs are of the order that is expected in low-percentage batch partial melts of carbonate-bearing mantle peridotite (Figure 2.5).

Small degrees of partial melting of mantle with average REE concentrations can produce melts with approximately 0.5% REOs. Additional processes are, however, required to produce enrichment from about 0.5% to per cent levels of LREEs in ores. Most carbonatites in the upper crust are not primitive magmas whose composition is unchanged between formation of melt in the mantle and final crystallisation in the upper crust, but have evolved through fractional crystallisation and other petrogenetic processes, probably at various depths in the upper mantle and in the crust. Processes that could cause enrichment to per cent levels of LREEs include fractionation during crystallisation, immiscible separation of alkaline silicate melts and carbonatite melts, and hydrothermal processes. At both Bayan Obo and Mountain Pass there are multiple phases of carbonatite dykes; later dykes have higher concentrations of REEs, a fact that suggests fractional crystallisation was a contributing process in the formation of these ores (see discussion of this process in Section 2.2.5).

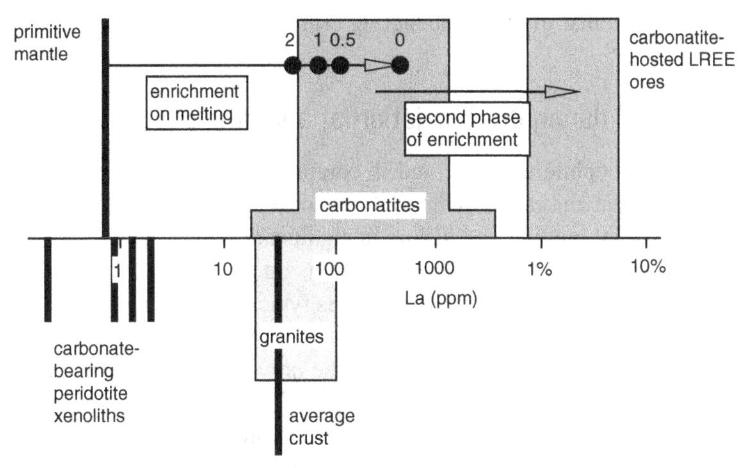

Figure 2.5 Lanthanum concentrations (ppm) in rocks and reservoirs in the mantle and in carbonatites and ores in carbonatites: La is shown here as proxy for LREEs. Partial batch melts of primitive mantle of percentages appropriate to form carbonatites will have La concentrations shown by the circles, which correspond well with concentrations in carbonatites. A second phase of enrichment is required to produce concentrations of per cent level of ores. Bulk peridotite–carbonatite melt partition coefficients are those of Dasgupta *et al.* (2009). Lanthanum concentrations in carbonate-bearing peridotite xenoliths were measured by Ionov *et al.* (1993) and indicate that carbonate-bearing peridotite is not unusually rich in REEs compared to normal mantle.

Mantle will melt to a low degree where temperatures are not significantly elevated above normal temperatures at that depth in the mantle. Were temperatures more elevated, more voluminous silicate melts would form, in the first instance alkaline mafic to ultramafic rocks. Some of the geologically most recent carbonatites are along the East African Rift, which is an incipient continental rift cutting relatively cold lithosphere of a part of a continent that has otherwise been tectonically stable over time periods of hundreds of millions of years, as is typical of many continental shields. This is thus one possible environment of subtle, minor heating of the mantle to produce low-temperature, low-volume melts. Other carbonatites, for instance those in the Kola magmatic province of the Russian Federation, intruded at the same time as large-volume extensive intraplate mafic magmatism in Eastern Europe, but at the edges of the main area of magmatism and many hundreds of kilometres from its centre. This mafic magmatism is an example of a large igneous province (LIP) in which large volumes of mantle-derived magma formed and intruded into and through continental crust in an intraplate setting over a period of at most a few million years. The carbonatites may thus have formed at the cooler edges of a volume of mantle that was undergoing large-scale melting.

A critical factor for the genesis of carbonatites is the presence of a small amount of carbonate in mantle peridotite. We do not know how widespread carbonate is in the mantle. One possible origin of carbonate in the mantle is that it is added metasomatically from fluids that are released in metamorphic mineral reactions in sediments or carbonate-bearing ocean crust that are being subducted and which percolate through the mantle wedge of peridotite that overlies subducting crust.

2.2.2 Deposits formed during differentiation of a silicate melt: chromite deposits

Chromium is a lithophile element, and is compatible in both spinel and clinopyroxene relative to ultramafic and mafic melts in the mantle. It is therefore largely retained in the mantle during partial melting and occurs in significantly lower concentrations in average crust (~ 100 ppm) than in mantle (concentrations up to 1%). Within the crust, ultramafic layers of large mafic–ultramafic igneous bodies typically have the highest concentrations of Cr.

The spinel mineral chromite ($FeCr_2O_4$) is the only ore mineral of Cr. It contains up to ≈ 55% Cr_2O_3, and is an accessory or minor mineral in most ultramafic rocks. Chromites in ultramafic rocks are solid solutions with varying amounts of Al^{3+} and Fe^{3+} replacing Cr^{3+}, and Mg^{2+} replacing Fe^{2+}. Chromite ores are bodies with greater than about 30% chromite, in many cases almost monomineralic chromitite (= chromite rock), in mafic and ultramafic intrusive rocks.

Chromite has different economic value and different uses in the economy depending on its chemical composition. Very little chromium metal is extracted from chromite. The main uses are as feedstock for Cr-bearing chemicals, the manufacture of refractories, and refining to ferro-chrom alloy, which is added to iron in the smelting of Cr-bearing stainless steel. There is a premium on the price of chromite which is of metallurgical grade with a composition such that it can be directly reduced in a smelter to low-melting-point ferro-chrom alloy. The chromite required has a Cr:Fe ratio of about 2.8:1.

Two common types of chromite deposit are recognised on the basis of ore-body form and their geological environment: **stratiform** chromite deposits in large, layered ultramafic–mafic intrusions; and podiform chromite ores in **ophiolites** or 'Alpine peridotites'.

Stratiform chromite deposits

Important examples include the ores in the Great Dyke, Zimbabwe (Prendergast and Wilson, 1989), see map and section, Figure 2.6; the Bushveld Complex, South Africa, Figure 2.25; the Stillwater Complex, Montana, USA (e.g. Spandler *et al.*, 2005); and the Kemi intrusion of Finland (Alapieti *et al.*, 1989). The host intrusions are all abnormally large, sheet-like to upward-flaring intrusions of Archaean or Palaeoproterozoic age that intruded into crystalline rocks of cratons. Intrusion sizes based on outcrop area range from about 30 km^2 at Kemi to 65 000 km^2 at the Bushveld Complex.

The host intrusions are up to a few kilometres thick and are characteristically layered with overall ultramafic lower cumulate units that grade up to mafic units:

- Cumulate sequences onlap onto the base of the intrusion in such a way that the intrusion extends over a greater area at the level of the mafic layers than at the level of the lowest ultramafic layers.
- The cumulate sequences of the intrusions are complex with repeated multiple cyclic units each between about 1 m and 100 m in thickness.

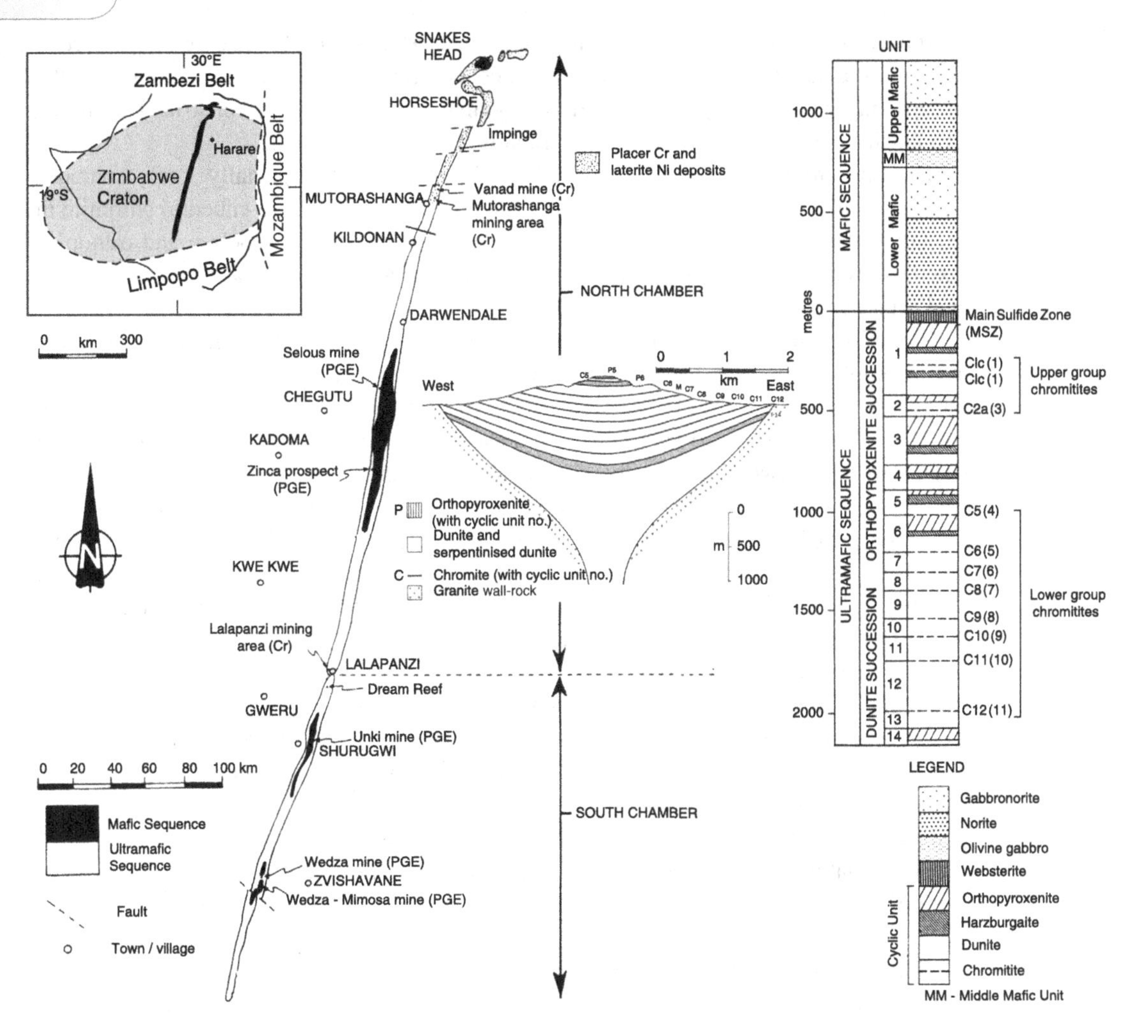

Figure 2.6 Simplified geological map and cumulate stratigraphy of the Great Dyke, Zimbabwe showing the major divisions of the intrusion. The chromitite seams (C numbers) are within the ultramafic successions of the cumulates, and the Main Sulfide Zone as the host to platinum group ore (MSZ) is just below the mafic cumulate sequence (see Section 2.2.4). Locations of principal mineral deposits and prospects (Cr – chromite; PGE – platinum group elements) are shown on the map. Insets show the siting of the Great Dyke in the Archaean Zimbabwe Craton, and the 'champagne-glass' cross-sectional shape of the intrusion that is interpreted from gravity surveys and from the shapes of the layered units. After Prendergast and Wilson (1989).

- Despite the complexity of the cumulate layering in the intrusions, the overall sequence of minerals, as for instance is recorded by the height above the base of first appearance of each mineral, conforms to the sequence of crystallisation predicted from Bowen's reaction series and from experimentally determined liquidus phase relations during progressive crystallisation of mafic magma. The typical sequence of appearance from base upwards is olivine, chromite, orthopyroxene, plagioclase and clinopyroxene. Chromite makes up approximately 1% of modal mineralogy in the lower parts of the succession.

Chromite ores occur within the cumulate sequences as thin, stratiform, chromite-rich rock or chromitite layers (seams) between tens of centimetres to about 1 m thick, parallel to the igneous layering. The layers are generally continuous over kilometres of strike length, but are locally branched, or are offset by structures resembling faults.

The thin chromitite layers typically have sharp contacts, especially lower contacts, against normal cumulate rock types, and have cumulus subhedral to euhedral chromite in an intercumulus matrix composed of one or more of olivine, plagioclase and orthopyroxene, or secondary hydrated minerals, e.g. serpentine and talc, that have replaced olivine and orthopyroxene.

Chromitite layers are best developed in the largest ultramafic–mafic intrusions with the most complex sequence of layers. They are hosted in different rock types in the different intrusions; for instance, in feldspathic orthopyroxenites in the Bushveld Complex, but in peridotite in the Kemi intrusion. Overall, however, they are concentrated at the levels of each intrusion in which orthopyroxene is a cumulus phase, that is, below and up to the boundary between ultramafic-dominated and mafic-dominated levels of the intrusion. They are generally absent from the lowest peridotitic layers and from gabbroic layers that form the upper part of the intrusions. Larger intrusions have multiple parallel chromitite layers: Eleven major and continuous chromitite layers are recognised in the Great Dyke of Zimbabwe, for instance (Figure 2.6).

Many chromitite layers coincide with levels in the cumulate sequence at which there are abrupt changes in mineralogy, often where there are reversals of the sequence of layers that are expected from progressive crystallisation. Where the layers occur within a single rock type, there are reversals in mineral composition trends with height, for instance both the nickel content and the forsterite content of olivine may be higher immediately above a layer than below. These reversals are chemical and mineralogical evidence that there was disturbance to crystallisation of the cumulate pile just before or at the time of chromitite accumulation.

Podiform chromite ores in ophiolites or 'Alpine peridotites'

Ophiolites are fragments of oceanic crust and upper mantle that have been tectonically emplaced onto continental crust. Important examples that host chromite ores include Kempirsai, Kazakhstan (Melcher *et al.*, 1999; Distler *et al.*, 2008), and those in New Caledonia (LeBlanc and Nicolas, 1992), Albania and Turkey.

An ideal complete ophiolite succession is shown in Figure 2.7; the ultramafic tectonites are residual upper mantle from which basaltic magmas (e.g. mid-ocean ridge basalt, MORB) have been extracted. These tectonites have been deformed as a result of solid-state flow of the mantle during continuous sea-floor spreading, during which a foliation and lineation is formed as the residue of melting flows upwards to below the ridge axis and then laterally away from the ridge. The lineation is interpreted as the direction of mantle flow away from the spreading ridge. Cumulates of ocean-ridge crustal magma chambers have a lower ultramafic peridotite sequence and a mafic, gabbroic unit. Basaltic sheeted dykes and overlying pillow lavas formed by repeated outflow of magma from the magma chamber.

Ophiolite sequences and their processes of formation are similar for crust formed in slightly differing tectonic environments in the oceans, including mid-ocean ridges, back-arc basins and primitive oceanic arcs. Mature ocean crust is unlikely to be emplaced onto

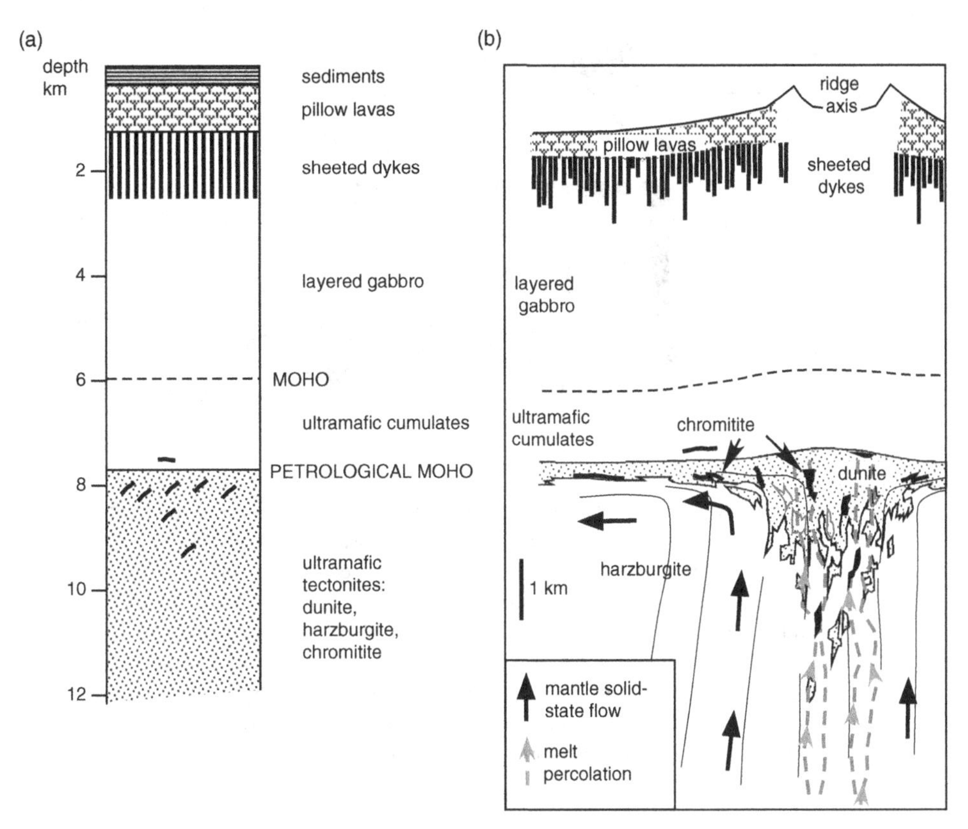

Figure 2.7 (a) Ideal ophiolite succession showing typical unit thicknesses before any tectonic disruption (after Duke, 1983). The Moho (Mohorovičic discontinuity) is the seismically detected boundary between mafic and ultramafic rocks: the petrological Moho is the boundary between ultramafic cumulates that formed in a crustal magma chamber and underlying and older ultramafic rocks of the mantle. (b) Schematic of the crust and upper mantle below a spreading ridge showing melt percolation and solid-state mantle flow and the distribution of dunite and chromitite in the mantle section. Chromitites form below a ridge axis in upwelling mantle and become smeared by solid-state flow of mantle away from the axis.

continental crust, and most ophiolites are fragments of ocean crust and upper mantle that formed in back-arc basins and in oceanic island arcs. However, podiform chromitites have been dredged from mature mid-ocean ridges and we can thus infer similar processes of formation in this environment.

Nature of chromite ores in ophiolites

Chromite ore bodies in ophiolites are typically relatively complex pipe-like- to podiform-shaped (Figure 2.8) bodies of chromitite or chromite-rich dunite with greater than 25% chromite intergrown with olivine or with serpentine that formed as a result of low-temperature hydration of olivine. They are mostly small, from a few kilogrammes to a few million tonnes, often in clusters which are mined in one operation. Areas that are significant producers have a number of relatively large deposits, notably Kempersai in an ophiolite unit

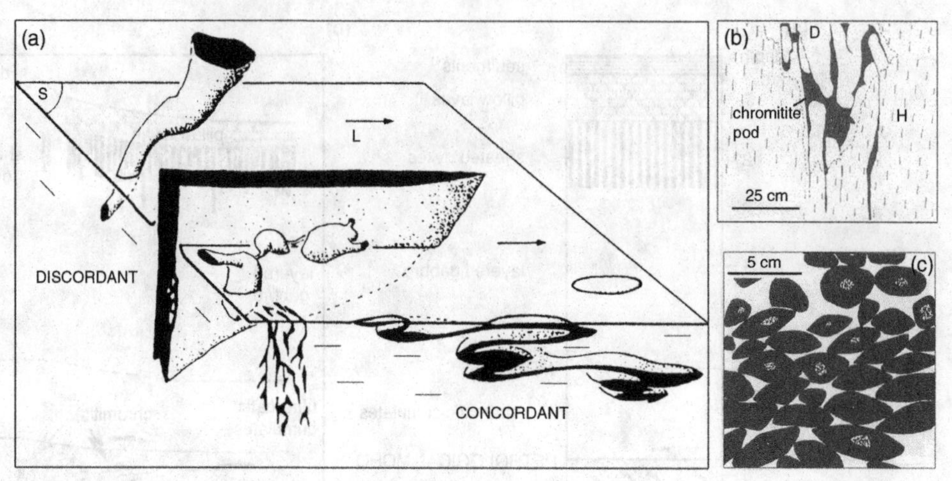

Figure 2.8 Morphology and textures of podiform ophiolite chromite deposits, mainly based on examples from New Caledonia and Oman (Leblanc and Nicolas, 1992). (a) Schematic shapes of chromitite ore bodies in the mantle-tectonised peridotite of the uppermost mantle, divided into discordant and concordant bodies. The horizontal plane (S) is the foliation plane and is parallel to the Moho. The arrow (L) is the stretching lineation. (b) Detail of ore-body shape on an outcrop scale of a small massive irregular concordant chromitite pod hosted in harzburgite (H) but with an envelope of dunite (D) around the chromitite. (c) Nodular texture of a massive chromitite showing flattened and stacked polygranular chromitite nodules in serpentinised dunite. The cores of some chromitite nodules have chromite intergrown with olivine.

of the Southern Ural Mountains. Although small, many podiform deposits have chromite of metallurgical grade, and because of their value were until recently mined in preference to the enormous reserves of chromite in stratiform deposits of the Bushveld Complex.

The pods are not uniformly developed in ophiolites around the world, and are absent or rare in many. Most economic ores are in harzburgite-type ophiolites, in which harzburgite (olivine–orthopyroxene rock) dominates the mantle section. They are rare and poorly developed in lherzolite-type ophiolites (with olivine–orthopyroxene–clinopyroxene ± plagioclase assemblages).

Where they are developed, the majority of ore bodies occur within the ultramafic tectonites between 500 m and 1000 m below the top of the mantle, that is, just below the petrological Moho (Figure 2.7a), although a few are hosted either deeper in the tectonites or in the lowest levels of the overlying ultramafic cumulates.

The chromitites in the mantle tectonites are hosted in irregular bodies of dunite, or where hosted in harzburgite are separated from harzburgite by a centimetre- to metre-thick envelope zone of dunite (Figure 2.8b). Irregular pods and dyke-like bodies of dunite are common in the mantle sections of many ophiolites and form up to 10% of the mantle part of the sequence. These are interpreted as 'replacive dunites', which mark conduits where mafic (MORB) melt percolated through the uppermost mantle and lower crust beneath the ridge (Figure 2.7b).

Many pipes are elongate oblique to the mantle–crust boundary and to the tectonite fabric of the host ophiolite, although most are elongate parallel to the foliation and to the lineation (Figure 2.8a). Ore is characterised by unusual mineral textures, including nodular and 'leopard-skin' textures in which elongate orbicular concentrations of chromite grains occur in clusters up to a few centimetres across in an olivine-rich matrix, and in which the nodules are sometimes cracked, impressed by adjacent nodules and sorted by size (Figure 2.8c). These textures are interpreted to be the result of crystallisation during flow of a composite magma in which bubbles of one magma were suspended in a second magma. The orbicular textures, for instance, were formed where chromite nucleated preferentially at bubble interfaces.

Genesis of chromitites

The processes of formation of both chromitite layers in intrusions and pods in ultramafic rocks of ophiolites were for many years not understood and were debated. Chromite crystallises during progressive cooling of many mafic magmas, most typically as a minor component (~ 1%) in dunite and harzburgite at the base of a cumulate sequence. Processes other than those of normal fractional crystallisation in a magma chamber are required to explain chromitites.

(a) Stratiform chromitites

A number of different chemical and physical processes have been suggested as explanations for the formation of chromitite layers in large, layered intrusions. Some of these proposals have since been discounted, for instance:

(i) Periodic oxidation of the magma, as a result, for instance of addition of water into the magma. This would give rise to short periods of chromite crystallisation. However, gradual and systematic changes in chromite composition across the multiple layers in the Great Dyke, for instance, are difficult to explain if each chromitite layer formed after addition of water to the chamber.

(ii) Differential settling of chromite because of greater density to separate it from other minerals in the magma chamber. However, many chromitite layers are hosted in orthopyroxenite, which includes minor chromite but there is no gradual increase or decrease in either the modal abundance of chromite towards the layers, or a decrease in the grain size of chromite upwards through the layers. Both of these features would be expected if accumulation was a result of settling in a magma chamber.

The interpretations of chromitite formation stem from the multiple lines of evidence which indicate that the large magma chambers that host chromitite layers were not filled by one pulse of magma, but were filled episodically, most likely with multiple episodes of magma input. The evidence for multiple episodes of fill includes the mineralogical and chemical cyclicity of the cumulate layers, the onlap of cumulate layers onto the chamber base, and also evidence from isotope data (see Box 2.3). The explanation for chromitite layers calls on complex physical and chemical processes that are inferred to take place with repeated magma input into partially crystallised large magma chambers, in particular where magmas of different compositions mix and interact chemically. The critical features of liquidus phase petrology that can lead to crystallisation of chromite through magma mixing are illustrated in Figure 2.9.

The cumulus sequence of minerals in the chemical sub-system (Mg,Fe)O–SiO_2–Cr_2O_3 produced by crystallisation of a primitive tholeiitic mafic magma would be: olivine,

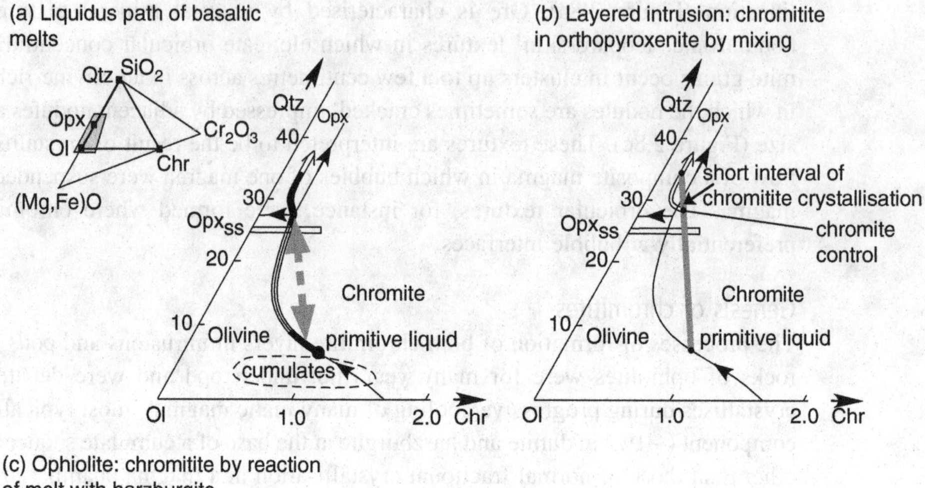

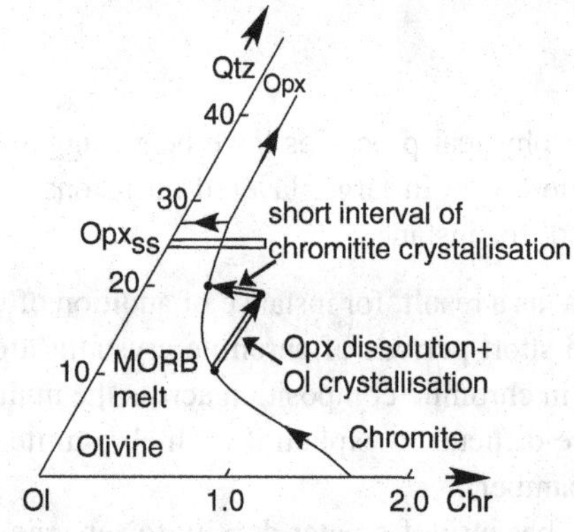

Figure 2.9 (a) Liquidus phase relations and melt evolution of mafic melts in the system SiO_2–(Mg,Fe)O–Cr_2O_3 (after Irvine, 1977) shown for Cr-poor compositions – see insert for compositional field. During undisturbed fractional crystallisation, melt will follow the thickened line along the olivine–chromite eutectic until orthopyroxene begins to crystallise, after which the melt will move into the orthopyroxene field. Interaction between melts that lie along the solid lines can produce mixtures that lie in the chromite field and hence crystallise chromite. (b) Example of environment in a layered ultramafic–mafic intrusion that can produce a chromitite layer hosted in orthopyroxenite (e.g. in the Bushveld Complex). The mixing line is shown by the grey solid line. After mixing there is a short interval of chromite crystallisation before recommencement of orthopyroxene crystallisation. (c) Example of situation in an ophiolite that produces chromitite due to increased silica content of a melt on reaction between the melt and orthopyroxene-bearing wall-rock.

olivine–(minor) chromite, olivine–(minor) chromite–orthopyroxene, orthopyroxene (Figure 2.9). Plagioclase and clinopyroxene are neglected because they incorporate additional oxide components. The evolution of the liquid while olivine and chromite are crystallising cotectically traces a curved compositional path in such a way that a mixture of two melts which have compositions at specific points along the cotectic would produce a melt that would fall in the chromite field and crystallise chromite alone to produce chromitite (Figure 2.9a and b).

Note however, that magma mixing is not the only possible mechanism to form melts in the chromite field. The addition of silica to a primitive magma, for instance as a result of assimilation of felsic **country rock** into the melt, would also move the liquid composition into the chromite field (Figure 2.9c).

The possible history to form stratiform chromitite layers in the orthopyroxenite sequence of the Bushveld Igneous Complex or the Great Dyke is thus (Figure 2.10):

(i) Input of a primitive mafic magma into a large, hence slowly cooling, magma chamber.
(ii) Formation of cumulate dunite and harzburgite and fractionation of the melt by crystallisation along the cotectic line of olivine and chromite to a more silicic composition to a degree at which orthopyroxene is crystallising (Figure 2.9b).
(iii) Input of new primitive mafic magma into the magma chamber above the early cumulates. This may involve lateral expansion of the magma chamber such that higher cumulate layers onlap onto the contact with underlying country rock.
(iv) Mixing along the interface between the new magma and the evolved magma to give compositions along tie lines (Figure 2.9b). Where the mixture lies within the chromite field of the liquidus diagram, a small mass of chromitite will crystallise along the interface between the two magmas, or where they have been effectively mixed.
(v) The accumulation of thin chromitite layers may be due to settling through the magma column or more likely due to deposition during convection in the magma chamber (dynamic settling). The crystallisation of chromite drives the mixed liquid composition back to either the olivine–chromite or the orthopyroxene–chromite cotectic, at which time olivine or orthopyroxene would crystallise, for instance as the intercumulus phase in the chromitite layers.

Multiple layers of stratiform chromitite are interpreted in terms of multiple episodes of input of new magma, in each case of a similar composition.

(b) Podiform chromitites

The explanation for the formation of podiform chromitites in ophiolites is based on the same petrological phase diagram, but invokes different petrological processes. The harzburgite of the upper mantle in typical ophiolites is a residue from melting and extraction of MORB melt at depths of a few tens of kilometres. After melt extraction, this residue is carried upwards by solid-state mantle flow (Figure 2.7). Continually produced MORB melts sourced from deeper in the mantle percolate along channelways through this residuum while it is still hot. This percolating melt forms at depth in equilibrium with orthopyroxene and olivine, but the effect of pressure on mineral–melt equilibrium means that it is no longer in equilibrium with orthopyroxene in the shallow mantle. The upward-percolating deeply sourced primitive MORB melt reacts with adjacent harzburgite, dissolving orthopyroxene and replacing it with olivine to produce the bodies of replacive dunite. The effect of this reaction is to increase the silica content of the melt. There is a range of compositions of MORB melts for which the orthopyroxene to

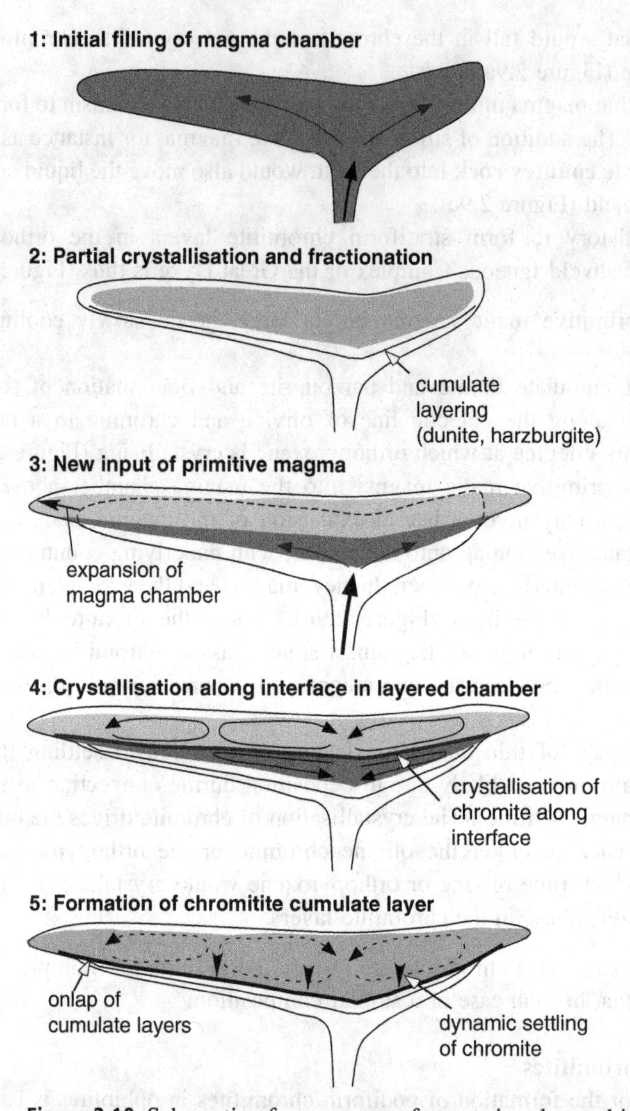

Figure 2.10 Schematic of a sequence of events in a layered intrusion that will give rise to the formation of a thin cumulate chromitite layer. See text for more description.

olivine reaction causes the melt composition to move into the chromite field, and hence causes crystallisation of chromitite (Figure 2.9c). Relicts of melt included in chromite grains indicate that they crystallise from mafic rather than ultramafic melt even though no mafic rocks are otherwise preserved in chromitite or enveloping dunite in mantle tectonites. The concentration of chromitite pods just below the petrological Moho (Figure 2.7) may be a result of ponding of rising magma at this level.

The irregular shapes of the pods are the result of solid-state mantle flow after formation (Figure 2.8). The nodular texture is formed where there was a mixture of two magmas, with globules of one melt suspended in the second melt. Chromite preferentially

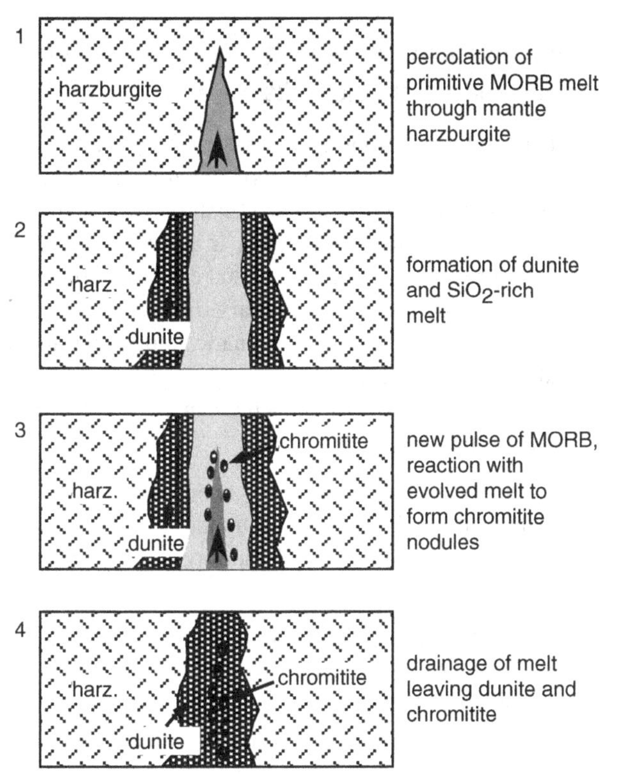

Figure 2.11 Explanation of chromitites, their lithological relations in the mantle and their nodular texture in ophiolitic deposits (after Arai, 1997), based on the phase relations shown in Figure 2.9. Dykes of primitive gabbroic melt react with hot, possibly partially molten, depleted mantle harzburgite while rising along a channelway through the uppermost mantle to form dunite and a more silica-rich melt. A second packet of primitive magma input into the channelway mingles with the silica-rich melt to form an emulsion-like mixture with bubbles of one magma in the second. Chromite preferentially nucleates along the bubble interfaces to form the preserved nodular texture as illustrated in Figure 2.8c.

crystallises at globule surfaces to give the nodules with olivine cores as illustrated in Figure 2.8c. The two melts may be aliquots of the rising magma that have reacted with wall-rock harzburgite to different degrees (Figure 2.11).

Differences in petrological evolution of the mantle at different ocean spreading centres may explain why some ophiolites host podiform chromitites and others do not. The degree of melting of mantle peridotite controls whether the residue is harzburgite (high degree of melting) or lherzolite (low degree of melting). This difference is controlled by the rate of spreading at the ridge or arc. Harzburgites tend to form at fast spreading ridges at which temperatures are hotter at any depth, more basalt melt is formed and extracted and the residue becomes more strongly depleted of less compatible elements than lherzolite. Both the melt that is percolating upwards and the composition of the residue from earlier melting will be different in the two end-member cases and these differences will affect the likelihood of the melt becoming in the chromite field through resorption of wall-rock orthopyroxene into the melt.

2.2.3 Deposits formed from immiscible sulfide melt phases: base-metal Ni–Cu sulfide deposits in mafic and ultramafic rocks

Magmatic sulfide deposits in mafic and ultramafic rocks provide the majority of the global supply of nickel and platinum-group elements (PGEs), and are a major source of copper. The PGEs are the six geochemically similar heavy transition-row elements which have siderophile to chalcophile behaviour (Ru, Rh, Pd, Os, Ir and Pt). All six elements have very low concentrations (< 10 ppb) in average continental crust, in part because they were strongly partitioned into the metallic core in the primary differentiation of the Earth (see Figure 1.5). Cobalt and gold are also enriched in magmatic sulfide ores and are recovered at some mines of magmatic sulfide deposits.

The economically important magmatic sulfide ores can be divided based on major economic commodities and setting into:

(1) **base-metal**, Ni–Cu magmatic sulfide deposits in gabbroic intrusions
(2) base-metal, Ni-dominated sulfide deposits in ultramafic lava flows (**komatiites**)
(3) **precious metal**, PGE magmatic sulfide deposits in large layered ultramafic to mafic intrusions.

Despite this division, Ni–Cu and Ni ores are enriched in PGEs, and PGE ores are enriched in Ni and Cu. Base-metal ores are disseminated to massive concentrations of sulfide minerals with up to 80% sulfide minerals, in and adjacent to igneous rock bodies. The precious-metal ores are disseminated ores in igneous rocks with concentrations of up to at most only a few percentage sulfide minerals.

The geneses of the two types of base-metal sulfide ores are described together because of similarities of process.

Geochemical principles of ore formation through sulfide melt immiscibility

(a) Incorporation of sulfide and chalcophile elements into a silicate melt

Mafic or ultramafic melts derived from the mantle incorporate up to about 1000 ppm, (0.1 wt %) S in solution, dominantly in the reduced form as sulfide with an oxidation number of –2. The sulfide is incorporated through melting of sulfide minerals that are minor disseminated minerals in mantle peridotite. The low solubility of sulfide in silicate magma relative to its abundance in many mantle rocks means that, in contrast to carbonate, which is rapidly depleted in the mantle with low degrees of partial melting (see Section 2.2.1), available sulfide is incorporated progressively into magmas in the mantle with progressively increasing degrees of partial melting. Between 10 to 20% batch partial melting of typical mantle peridotite with about 200 ppm sulfur is required to fully melt sulfide into the mafic melt derived (Figure 2.12).

Chalcophile and siderophile elements will be hosted to a large degree in the trace sulfide minerals in mantle peridotite. The different partition coefficients of Ni, Cu and PGEs between silicate melt and sulfide melt and between silicate minerals and melt mean that these elements reach high concentrations in the melt at different degrees of progressive melting of mantle peridotite (Figure 2.12). Assuming batch melting, PGEs and, to a less marked degree, Cu are expected to have highest concentrations in melts into which the last sulfide minerals have just melted. Concentrations in the melt will drop with

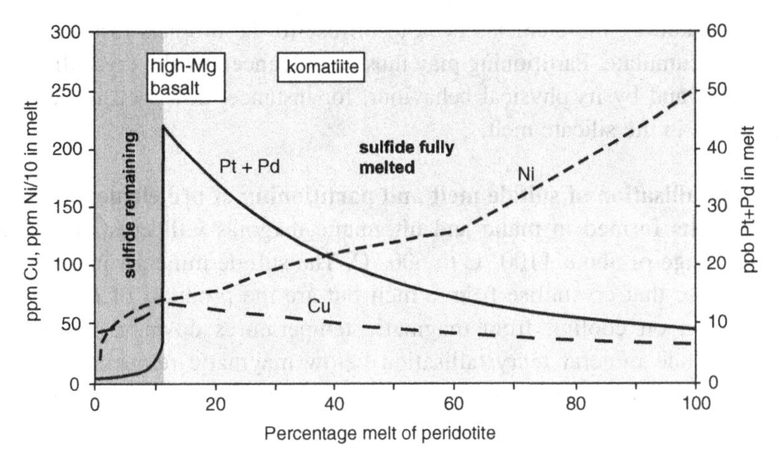

Figure 2.12 Calculated concentrations of Ni, Cu and PGEs in different percentage partial melts of mantle peridotite with 200 ppm S, 2500 ppm Ni, 40 ppm Cu and 5 ppb Pt + Pd (Naldrett, 2010). At up to about 12% partial melting some of the sulfide remains in minerals or in a separate sulfide melt. Sulfide is fully in solution in the silicate melt at higher degrees of melting. The apparent (batch melt model) partial melt percentages for the generation of high-magnesium basaltic melts and komatiitic melts are shown.

further melting because of dilution. In contrast, because Ni is partitioned into olivine, the concentration of Ni in the melt increases progressively with increasing percentages of partial melting.

(b) Sulfur saturation – formation of immiscible sulfide melt

Sulfide solubility in a silicate melt varies with pressure and temperature, and also with the chemical parameters of melt, most importantly with silica and FeO content of the melt. So long as sulfide minerals remain unmelted in the source, the magma is sulfide-saturated at source. A sulfide-bearing magma will, however, become undersaturated or oversaturated with respect to a sulfidic phase during magma migration, cooling, crystallisation and fractionation.

At the temperatures at which ultramafic and mafic magmas crystallise in the crust and at the Earth's surface (1100–1600 °C), the sulfide phase will be molten or largely molten. If a mafic magma becomes sulfide-saturated, drops of immiscible sulfide liquid (melt) form rather than sulfide minerals. Partition coefficients of Ni and Cu between sulfide melt and mafic silicate melt are measured to be about 200 and 1500, respectively. The values for PGEs are difficult to measure in experiments but are estimated to be at least an order of magnitude higher. The chalcophile and siderophile elements (Ni, Cu, Co, PGEs, Au) can thus be strongly partitioned into a sulfide melt phase. These drops or droplets will be dense ($\rho \sim 4.5$ g cm^{-3}, compared to ~ 2.8 g cm^{-3} for a mafic silicate melt) and will tend to settle and hence may accumulate.

The partitioning of the chalcophile and siderophile elements into the sulfide melt will be less efficient if the elements were incorporated into silicate minerals before the immiscible separation of a sulfide liquid. This effect is likely most significant for Ni, which is compatible in olivine; olivine contains up to approximately 1% NiO in ultramafic rocks. In addition, the elements need to diffuse to the droplets faster than the droplets settle or accumulate. Partitioning may thus be influenced by the crystallisation history of the magma and by its physical behaviour, for instance, how well dispersed the sulfide droplets are in the silicate melt.

(c) Crystallisation of sulfide melt and partitioning of ore elements

Sulfide melts formed in mafic and ultramafic magmas will crystallise over the temperature range of about 1100 °C to 600 °C. The sulfide minerals preserved in the ores are not those that crystallise from a melt but are the products of recrystallisation and equilibration on cooling from magmatic temperatures down to about 250 °C. The ease of sulfide mineral recrystallisation below magmatic temperatures ensures complete recrystallisation of the high-temperature minerals, which are never preserved in nature.

Most magmatic sulfide ores are dominated by a common assemblage of ore minerals – chalcopyrite ($CuFeS_2$), pentlandite $(Fe,Ni)_9S_8$, pyrrhotite ($Fe_{1-x}S$) and lesser magnetite (Fe_3O_4). Other Ni and Cu sulfide minerals may additionally be present as minor components. Chalcopyrite is for instance a product of recrystallisation of the mineral *iss* (intermediate solid solution) at temperatures of around 300 °C. The high-temperature Fe–Ni sulfide mineral is the mineral *mss* (monosulfide solid solution), which at high temperatures is a continuous solid solution between Fe- and Ni-end members. The mineral *mss* recrystallises to pentlandite and pyrrhotite at similarly low temperatures. Pyrrhotite is the dominant iron sulfide mineral rather than pyrite because there is always sufficient Fe in the sulfide melt to form this Fe-rich mineral. PGEs are rare elements even where concentrated in ores, and occur most commonly as assemblages of trace PGE sulfides, alloys, arsenides and related minerals.

Massive to disseminated Ni–Cu sulfide ores in intrusions formed from mafic magmas

There are numerous important ore bodies of this type in the Sudbury Igneous Complex of Ontario, Canada (Rousell *et al.*, 2003), which covers about 2000 km^2, and in medium-sized and smaller mafic or layered ultramafic–mafic intrusions down to outcrop areas of about 2 km^2 including at Noril'sk-Talnakh, Russian Federation (Spiridonov, 2010); Jinchuan, Gansu, China (Chai and Naldrett, 1992); Voisey's Bay, Newfoundland, Canada (Naldrett and Li, 2007); Nebo-Babel, Western Australia (Godel *et al.*, 2011); Madziwa, Zimbabwe (Birch and Buchanan, 1989). Mineralogically similar deposits in mafic bodies of uncertain origin in high-grade metamorphic belts such as at Selebi-Phikwe, Botswana (Maier *et al.*, 2008) likely also belong to this deposit type. Many ore-hosting intrusions are known to be members of major suites of mafic intrusion of regional extent that formed over short time periods over large areas of continental crust.

The majority of known resources of this style are in multiple ore bodies in two very large magmatic systems at Sudbury and at Noril'sk-Talnakh.

(a) The petrographic nature of Ni–Cu magmatic sulfide ores in intrusions

The ores at the known major deposits share many chemical and mineralogical commonalities. Nickel and copper have approximately equal concentrations of up to a few wt %, higher grades occurring in ores with higher sulfide content. PGEs are commonly by-products and are typically present at concentrations of order 1 ppm. Ore ranges texturally from **massive** sulfide ore with greater than 50% by volume sulfide in lenses up to a few metres thick, through net-textured or matrix ore, in which there are continuous aggregates of sulfide minerals enclosing silicate minerals, to **stringer** or **veinlet** ore in which there are small irregular segregations of sulfides, often resembling veins, to **disseminated** ore in which sulfides occur as xenomorphic segregations from about 1 mm to several centimetres diameter within a silicate matrix. In all ore types, the sulfides are intergrown mixtures of Ni and Cu minerals with pyrrhotite.

The **tenor** of the ore is the concentration of Ni or Cu in the sulfide fraction of the rock. The tenor ranges from a few percentage up to about 20% in these magmatic sulfide ores, hence rocks with a few percentage or more of sulfides constitute ore. Some intrusions have concentrations of sulfides that are spread out over relatively thick zones and are too weak to constitute economic ores. The intrusions of the large Duluth Complex of Minnesota, USA, contain a known resource of about 4000 Mt of rock with disseminated Cu–Ni sulfide minerals and about 0.2% Ni disseminated over tens of metres of thickness of the host intrusions. Although very large, these concentrations of magmatic sulfides do not constitute ore at the present time (cf. Figure 1.7).

(b) The nature of the host intrusions

Most of the host intrusions of massive Ni–Cu ores are broadly layered with lower more mafic layers and zonation upwards to less mafic compositions. However, they lack the marked cumulate layering on 10- to 100-metre scales that is characteristic of intrusions that host chromite ores (Section 2.2.2) and PGE ores (Section 2.2.4).

There are marked differences in the sizes, forms, and depths and styles of emplacement of the host intrusions:

- There are multiple, closely spaced small ore-hosting gabbro-norite to gabbro intrusions in the Noril'sk-Talnakh complex. Each intrusion is a flat-lying lens up to a few kilometres diameter in plan view (Figure 2.13). These intrusions formed in sedimentary rocks about 1 km below one of the eruptive centres of the extensive Permo-Triassic Siberian Traps flood basalts which contain about 10^6 km^3 of basalts and originally extended over an area at least 2000 km by 2000 km. The thicknesses of hornfelsic contact metamorphic aureoles around these intrusions indicate that magma flowed through the intrusions to feed the overlying basalt flows, rather than into the intrusions as a single pulse. The intrusions are thus interpreted to be parts of the upper-crustal feeder conduits that delivered magma to one of the world's largest flood-basalt provinces.
- The Mesoproterozoic Jinchuan intrusion is a small intrusion (~ 3 km^2), but has an elongate flared shape (canoe shape) above a relatively narrow underlying feeder dyke, similar to the Great Dyke of Zimbabwe (cf. Figure 2.6), suggesting that it is an erosional remnant of a larger igneous body and that the ores are hosted in the feeder zone to this intrusion.
- Two small ore-bearing intrusions of an area of a few square kilometres and that are connected by a feeder dyke are exposed at Voisey's Bay. These intrusions formed at several kilometres depth in gneisses, and are interpreted to be part of an originally much larger magmatic complex. At Voisey's Bay, the magma was originally picrite

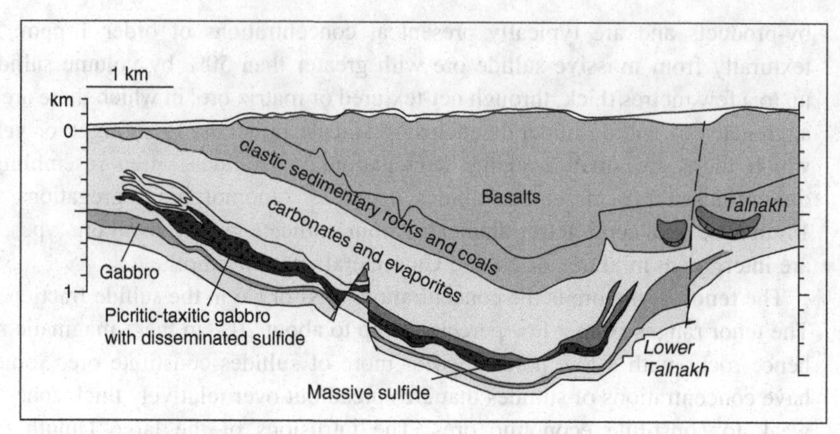

Figure 2.13 Cross section through the Permo-Triassic Talnakh deposits (Naldrett *et al.*, 1992), showing the typical position of massive sulfide ore bodies in the Noril'sk-Talnakh field at the base of relatively small lenticular to sill-like gabbroic intrusions in the foot-wall of the approximately 3-kilometre thick sequence of the Siberian Traps flood basalts.

(olivine-rich gabbro), but was compositionally modified as a result of assimilation of host-rocks and fractionation to troctolite (olivine–plagioclase).

- The Sudbury Igneous Complex is a several-kilometre-thick, elliptical bowl-shaped intrusion about 65 km by 25 km of Palaeoproterozoic age (Figure 2.14). It lacks an ultramafic cumulate basal sequence, and also lacks the cyclic and rhythmic layering of the similar-sized Bushveld Complex and Great Dyke intrusions. In contrast to these other intrusions it is relatively subtly layered from a norite (olivine–gabbro) base to a quartz–diorite top. Below the Main Mass of the intrusion there are widespread irregular masses of brecciated country rock (Sudbury Breccia). This igneous complex is uniquely interpreted as an astrobleme, a structure formed by impact of a meteorite, estimated to have been greater than 10 km in diameter. The energy of the impact caused wholesale melting of most of a bowl-shaped volume of rock that extended down to the base of the crust. The weakly differentiated gabbro to diorite is the product of crystallisation of this crustal melt.

(c) Settings of ore in the host intrusions

The sulfide ore bodies are commonly but not universally at or near the base of intrusions. They form complex lenticular shapes rather than a uniform layer in the intrusion, and are sited in different positions and in different host-rocks within the different intrusions. In the smaller host intrusions, we can recognise two common settings of ore:

- At Noril'sk-Talnakh the ore bodies form pods and lenses at the base, or rarely the top, of the relatively small host intrusions (Figure 2.15). At the base of the intrusions, the ores are typically massive sulfide with overlying stringer, veinlet and disseminated ore. These lenses of massive ore are typically overlain across a sharp contact by veinlet–disseminated or disseminated ore. Ores at the tops of some of the intrusions are

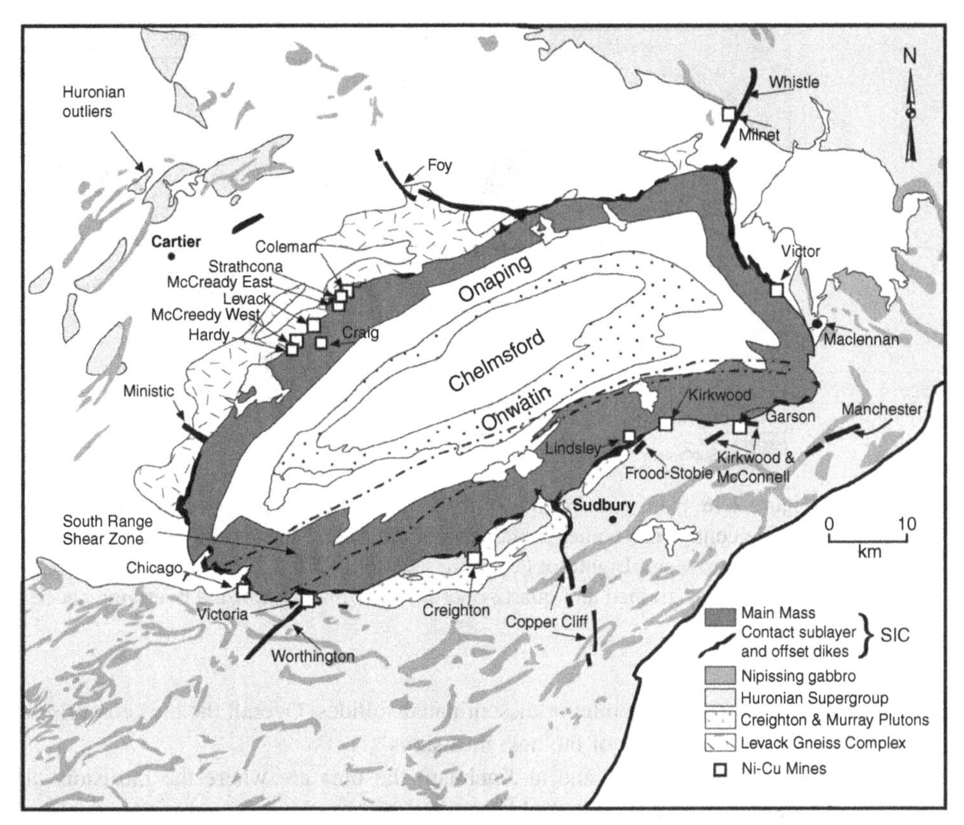

Figure 2.14 Map of the 1850 Ma Sudbury Igneous Complex (SIC), Ontario (Rousell *et al.*, 2003). The structure is bowl-shaped such that all sectors dip into the centre. The Ni–Cu–PGE ore deposits are located along the base of the Main Mass and in so-called 'offset dykes', which are intrusions that radiate outwards from the Main Mass of the intrusion into the foot-wall and which are up to about 200 m wide.

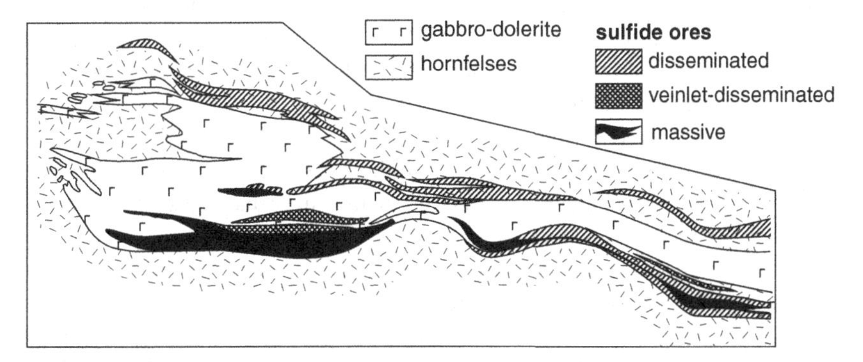

Figure 2.15 Example cross section through the Talnakh intrusion showing a typical distribution of sulfide ores in these relatively small gabbroic intrusions (Spiridonov, 2010). Massive ore forms irregular lenses at or just above the base of the intrusion, disseminated and veinlet–disseminated ore is above the massive ore and also as lenses at the top of the intrusion.

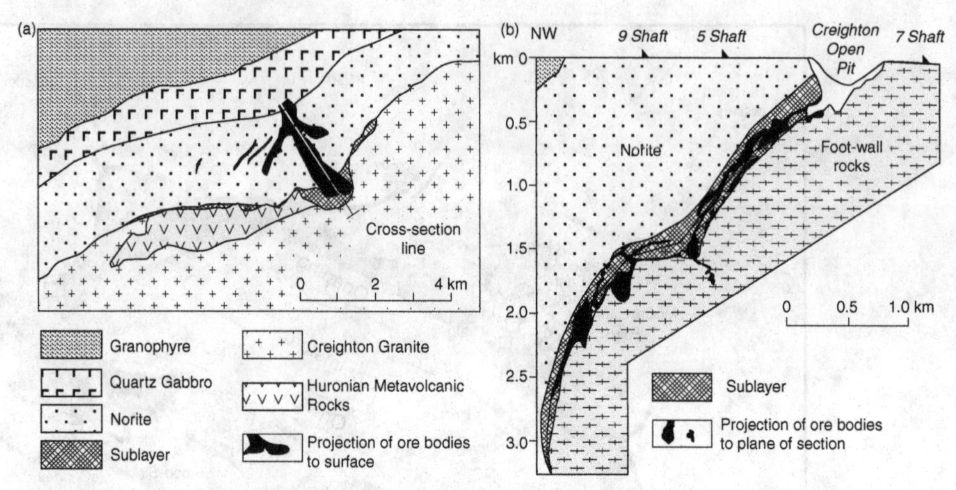

Figure 2.16 Plans and sections of the Creighton Deposit as an example of the many magmatic sulfide contact ore bodies at the base of the Main Mass of the Sudbury Igneous Complex. After Farrow and Lightfoot (2002), in Naldrett (2004). The 'sublayer' is a petrographically distinct finer-grained and quartz-poorer norite than that of the overlying succession.

comprised of veinlet or disseminated sulfides. Overall the ores comprise up to about 2–10% by volume of the host intrusions.

- At Voisey's Bay and at Jinchuan, the ores are where the intrusions flare, near and above where dykes would have fed magma into the intrusions. At Jinchuan the ores are almost entirely disseminated and net-textured rather than massive. The intrusion was formed from a mafic melt, but the sulfide ore bodies are unusually hosted in ultramafic cumulates (dunite, lherzolite and olivine pyroxenite) at the base of the intrusion.

The ore bodies in the large Sudbury Igneous Complex are irregular and complex lenses and ribbons of massive, breccia, veinlet and disseminated ore up to a few metres thick at about 70 localities along or close to the base of the main intrusion. Breccia ore contains clasts of sulfide and of wall-rock in a gabbro-norite matrix. Disseminated sulfides at concentrations below ore grade are present for many hundreds of metres above the ore bodies in the host norites.

There are different settings of the ore lenses in the Sudbury Complex. Some are 'contact deposits'. These may be within the basal 'sublayer', which is a petrographically distinct norite layer up to a few hundred metres thick that is discontinuously present along the base of the Main Mass of the intrusions (Figure 2.16). In some cases contact deposits are elongated along depressions of the basal contact of the main intrusion that plunge approximately down dip. Other ore bodies are in the Sudbury Breccia, which occurs as dykes and irregular masses of brecciated wall-rock up to a few hundred metres below the main basal contact of the intrusion. Ores also occur in 'offset dykes', or in apophyses and dyke-like projections of the intrusion into the **foot-wall** rocks below the main intrusion and in some cases several kilometres from the base (Figure 2.14). There are differences in the ore geochemistry in different ore bodies, especially with respect to Ni:Cu ratios. In general ore bodies below the Main Mass of the intrusion and in offset dykes have higher copper and also higher PGE contents.

Nickel sulfide ores in komatiites

Komatiites are ultramafic lavas and associated shallow intrusive units. They have greater than 18% (typically 30 to 40%) MgO and are petrologically peridotites (dunites, harzburgites and orthopyroxenites) and metamorphosed equivalents (serpentinites, talc-schists, tremolite-chlorite schists). They erupted at high temperature (up to ≈ 1600 °C), at low viscosity, and in many cases formed large-volume, very extensive flows that range in thickness from a few metres to locally several hundred metres thick. Individual flows are traced over tens of kilometres and are interpreted to have been originally up to hundreds of kilometres long. The flows are internally differentiated characteristically with dunitic cumulate basal units and more pyroxenitic flow tops. One petrographic characteristic of komatiites is spinifex texture, in which elongate skeletal olivine and orthopyroxene grains grew up to a few centimetres or as long as metres in length as a result of quenching of the tops of hot lavas on eruption. Komatiite units comprise repeated sub-aqueous lava flows and are a minor component in the now deformed and metamorphosed supracrustal volcanic and sedimentary sequences (**greenstone belts**) of Archaean and Palaeoproterozoic cratons. True komatiites are restricted to these early periods of Earth's history.

The magmatic sulfide ores in komatiites are Ni-rich sulfide ores (~ 1–5 wt % Ni, exceptionally to 20 wt % Ni) with only minor Cu (Ni:Cu ratios typically between 10 and 20) and with trace PGEs. Important examples include: Kambalda and elsewhere in the Yilgarn Craton of Western Australia (Gresham and Loftus-Hills, 1981; Cowden and Roberts, 1990); Zimbabwe Craton (Prendergast, 2003); Raglan Belt, Quebec, Canada (Lesher, 2007). Greenstone belts have characteristically undergone a multi-phase history of deformation, and komatiite ores along with their host lavas have thus been deformed, metamorphosed and recrystallised variously under sub-greenschist to upper-amphibolite-facies conditions.

(a) Nature and setting of sulfide ores in komatiites

The sulfide ores are dominated by pyrrhotite and pentlandite with minor chalcopyrite, and in some cases additional Ni sulfide minerals (violarite and millerite) and pyrite. The sulfide mineral assemblages in some ores are products of recrystallisation under metamorphic conditions, e.g. pyrite–milllerite (NiS).

The ores occur mainly in two characteristic settings in komatiite lava sequences:

- In relatively thin peridotitic lava flows that are up to a few tens of metres thick, there are bodies of massive ore up to a few metres thick (komatiitic peridotite-hosted ores) that are overlain with sharp boundaries by net-textured or matrix ores and disseminated ores (Figure 2.17). The ores most characteristically form ribbons of order 100 m wide and up to a few kilometres long above and extending locally just below the basal contact of a lava flow. These ore bodies are largest and most widespread in the basal flow of a sequence of flows (contact ores), but ores do occur at the base of higher flows (**hanging-wall** ores), and in many cases ore bodies are vertically stacked.
- Larger-tonnage ore bodies of low-grade disseminated sulfide ores up to a few tens of metres thick occur within bodies of olivine adcumulate that are themselves up to 500 m thick (komatiitic-dunite-hosted deposits). The host dunites are abnormally thick parts

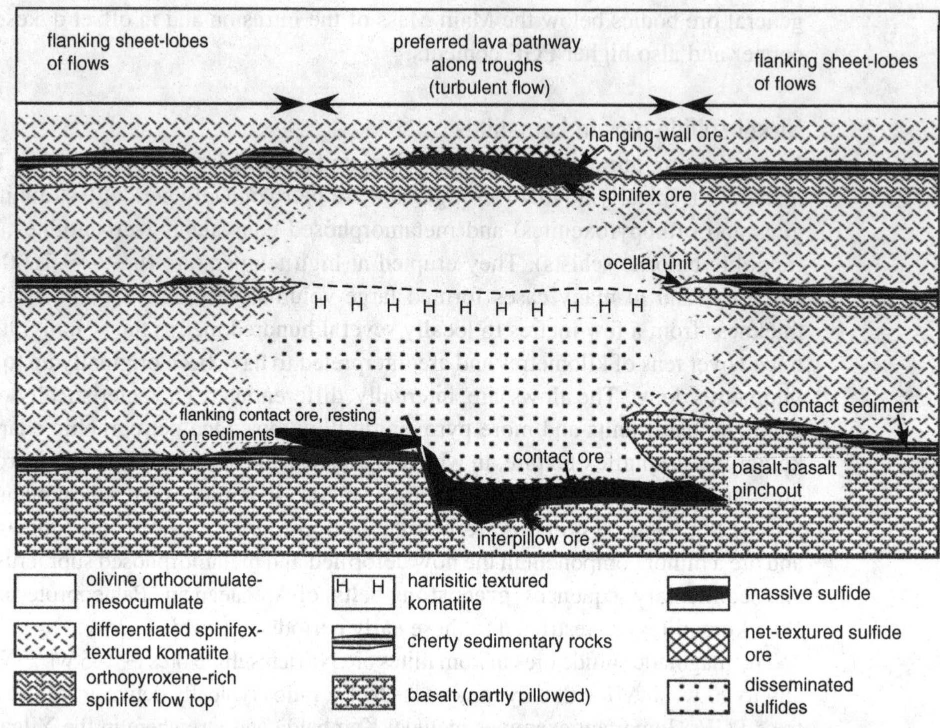

Figure 2.17 Section through a schematic sequence of three peridotitic komatiite lava flows showing the various possible settings and styles of Ni sulfide ores and different petrographic features of the host komatiite sequence (after Cowden and Roberts, 1990). Komatiite flows are up to about 20 m thick. The textures and petrography of each flow differs between the preferred lava pathway or channel facies in which flow is assumed to have been turbulent and the surrounding sheet-flow facies in which flow was laminar. Later and stratigraphically higher flows followed the same channel as the oldest flow, possibly because a depression on the surface remained after outflow of the channel facies. The largest ore body is hosted in a trough at the base of the basal flow and is layered with a few metres thickness of massive sulfide (contact ore) grading upwards into net-textured ore with $< 50\%$ sulfide in a silicate matrix, and disseminated ore in which isolated blebs of sulfide are enclosed between silicate grains.

of more extensive but generally thinner komatiite units. These ore bodies are also lenticular to ribbon-shaped in three dimensions.

Reconstruction of the exact setting and site of formation of komatiite-hosted ores is often difficult because original structural and physical relations have been disrupted by multiple deformation events. Faults and shear zones may be localised in massive sulfide ore because of its structural weakness. This deformation has led to the formation of sulfide-matrix breccias with wall-rock clasts in some ores. The ribbon-shaped **ore shoots** in both settings of ore are, however, reconstructed to occur in or above ‘troughs’ at the base of the flows and where the host flow is thickened. The troughs are lines along which the base of the flow is a few to several tens of metres stratigraphically lower than adjacent parts. They

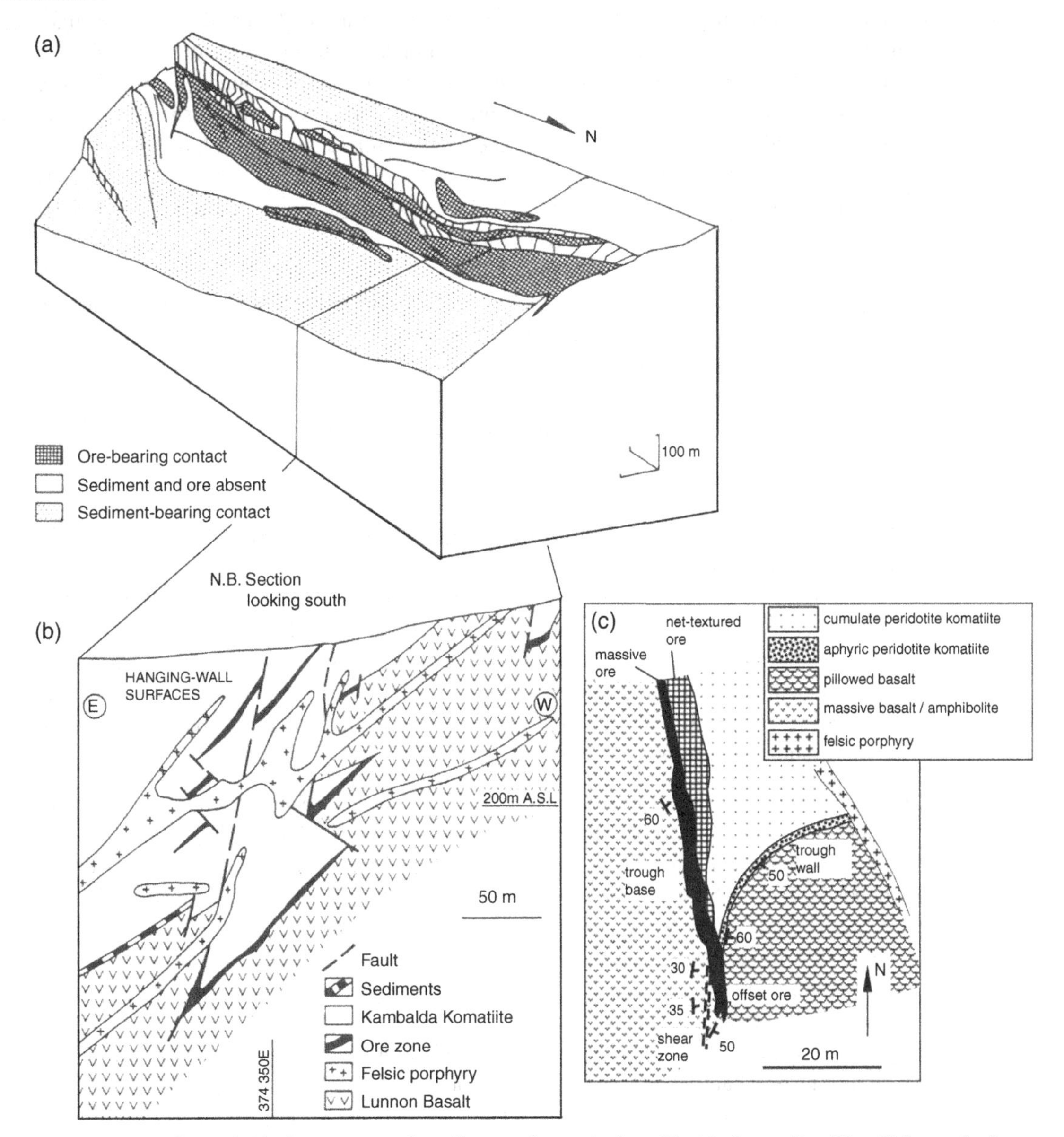

Figure 2.18 Geometry and attributes of a typical peridotitic komatiite Ni sulfide ore body. (a, b) Block diagram and cross section of the Lunnon deposit at Kambalda (after Gresham and Loftus-Hills, 1981). The shoot of massive ore is along the base of the basal komatiite flow of the sequence of flows and is mainly confined to a 20–30-m deep trough-like structure along the basal contact. Note that the basal sediments are missing along and adjacent to the trough. The felsic porphyries are later cross-cutting intrusions. (c) Plan showing detail of ore and rock types around a lateral limit of the ore zone at the base of a trough (after Lesher, 1989). The aphyric rim of the komatiite along the trough wall in combination with the different facies of foot-wall basalt beneath the trough and adjacent to the trough demonstrate that the trough is not primarily the result of deformation.

often have sharp, cuspate and overhanging lateral terminations (Figure 2.18). The troughs and thickened flows are interpreted to mark major lava channels (conduits), similar to those that form during some eruptions on contemporary basaltic shield volcanoes. The adjacent flanking parts of the flow are interpreted to be 'overbank' facies (sheet flow) with lower rates and volumes of lava flow. In the Raglan Belt, a single 1–2-km-wide strongly meandering channel is interpreted along a strike length of 20 km of exposure. At Kambalda, and elsewhere, there are multiple ore shoots within multiple sub-parallel troughs at the base of a single komatiite flow.

There is a common stratigraphic association of komatiites with thin black sulfide-rich cherty sedimentary rocks (Figure 2.17). These occur both immediately below a sequence of flows and between flows and are thus interpreted as background sediment deposited between eruptions. Best ore is often developed where a komatiite stratigraphically overlies such sedimentary rocks. However, the cherts are generally missing immediately below ore, especially in the troughs (Figures 2.17 and 2.18).

Genetic models for the formation of Ni–Cu magmatic sulfide deposits in intrusions and komatiites

The nature and characteristics of the two types of base-metal magmatic sulfide deposits match in general terms the history that was outlined on pages 42 to 44 for accumulation of sulfide from solution in a silicate magma. The massive or disseminated ore bodies are segregations of sulfide melt droplets that formed in a silicate melt and that partitioned the chalcophile elements and PGEs. The following aspects of genetic models for base-metal magmatic sulfide deposits are generally agreed:

- The Ni, Cu and PGEs in the ores were derived from the mafic or ultramafic magmas. Nickel and copper concentrations are 50–90% lower in much of the lowest kilometre of the approximately 3-km-thick sequence of flood basalts above the Noril'sk-Talnakh deposits than in higher lavas (Figure 2.19). These elements were largely partitioned from the mafic magmas into sulfide droplets before the magmas were erupted. PGEs are over 99% lower in the same lava units as Cu and Ni are reduced, hence were more strongly 'stripped', as is expected in view of the higher sulfide to silicate melt partition coefficients of these elements.
- The differences in Ni to Cu ratio between intrusion-hosted ores and komatiite ores are the result of significantly higher Ni concentrations in ultramafic melts. Calculations assuming batch melting models for magma generation indicate the komatiites formed through up to 40% partial melting of mantle peridotite and are Ni-rich compared to typical mafic magmas that would form through approximately 10% partial melting (Figure 2.12).
- The location of ore at the base of many intrusions and komatiite flows and the vertical sequence of ore types with net-textured and disseminated ores above massive ores is controlled by settling of dense sulfide liquid through actively crystallising silicate magma and by buoyancy effects (Figure 2.20). Net-textured ore is formed where cumulate minerals compact down into sulfide melt, and silicate melt that was interstitial to these early formed silicate crystals is displaced upwards into the flow or intrusion. Disseminated ore is formed of droplets that were trapped by crystallised interstitial silicate melt before they could settle.

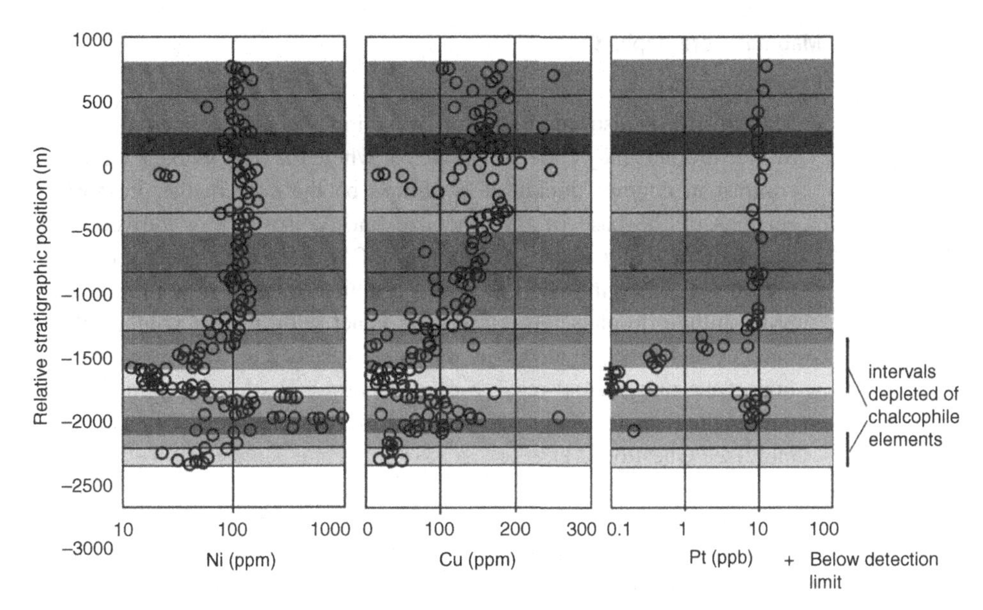

Figure 2.19 Concentrations or Ni, Cu and Pt through sections of the Siberian Traps flood basalts that were fed by the magmas that flowed through the Noril'sk-Talnakh intrusions (after Naldrett, 2010). The different shades are the different flow sequences. Note the relatively low concentrations of all these elements within two intervals towards the base of the sequence. These are interpreted to be the magmas from which the metals were stripped into sulfide melt droplets.

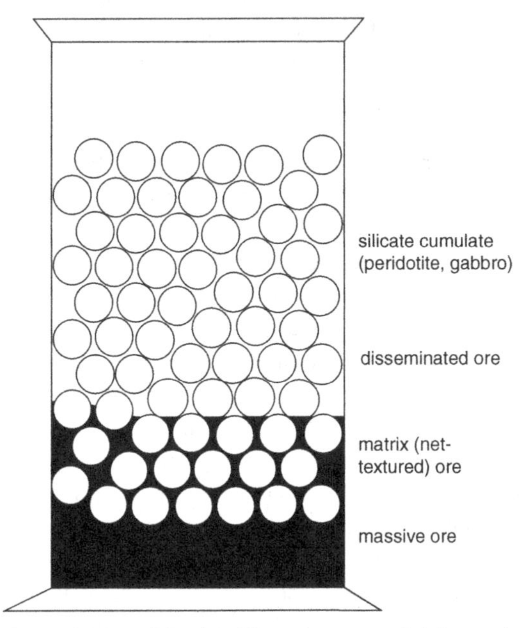

Figure 2.20 'Billiard-ball' analogue model for typical textures of magmatic sulfide deposits and their vertical distribution (after Naldrett, 1973). The open circles are billiard balls (= solid silicate crystals – intermediate density), the blank is water (= silicate melt – least dense) and the black is mercury (= sulfide melt – most dense). The upper boundary of the mercury is sharp and corresponds to the upper boundary of continuous sulfide melt and is controlled by the relative buoyancy of the solid silicates (billiard balls).

- Magma flow is also important for collecting and segregating sulfide droplets into ore. Sulfide droplets may collect in eddies or where the flow velocity decreases because of a contrast in conduit thickness, as would be the case in the ore-hosting flare-shaped sections of the base of the Jinchuan and Voisey's Bay intrusions. Where flowing magma effectively freezes before the sulfide droplets can settle, ore bodies are within the interior of an intrusion or flow. Turbulent magma flow may also be important to bring sulfide droplets into effective contact with a large volume of the host silicate magma, and hence to allow the droplets to 'scavenge' ore metals from a high proportion of the magma and reach high concentrations of Ni and Cu. The critical factor is termed the R-factor, and is the effective mass ratio of interacting silicate melt to sulfide melt (see Question 2.1).

The controls on sulfide saturation in mantle-derived silicate magma

Although we expect mantle-derived ultramafic and mafic magmas to carry sulfide in solution, magmatic sulfide ores are not known in all mafic intrusions of similar size that are derived through similar degrees of mantle melting, nor are they developed in all komatiite lava flows. One critical factor in the formation of magmatic sulfide ores is attainment of sulfide saturation in the silicate magma. For effective accumulation of massive sulfide ore, there must be saturation and separation of a sulfide melt phase before significant crystallisation of the host silicate magma. In the case of komatiites, for instance, saturation must occur before crystallisation and accumulation of large masses of olivine cumulates. The timing of sulfide saturation during cooling and crystallisation likely controls the nature and tenor of ore, the extent of accumulation of the droplets, and ore location.

Both komatiite magmas and primitive mafic magmas that formed intrusions such as Jinchuan and Noril'sk-Talnakh probably rose rapidly through the upper mantle and crust. Where magma rises rapidly through the upper mantle and crust, it follows an adiabatic cooling path, with only minor cooling on pressure drop. The solubility of sulfide in a silicate melt increases with increasing temperatures but decreases with increasing pressure (Figure 2.21). A silicate magma that was saturated with respect to sulfide in the mantle would thus become undersaturated on rising to near the Earth's surface and the sulfide should remain in solution until a large percentage of the magma has crystallised.

The relation between sulfide solubility and pressure in ultramafic and mafic magma shows that a mantle-derived magma can only reach sulfide saturation early in its crystallisation history at or near the Earth's surface if there is a change in magma chemistry. Sulfide saturation may be reached if there is a reduction in sulfur solubility, for instance, because of addition of silica into the magma. Alternatively, sulfide itself may be added to the magma. In either case, these changes are likely to be a result of assimilation of crustal rocks into the magma: in the first case, silica-rich rocks; in the second, sulfide-bearing rocks. Assimilation could be either from conduit wall-rocks as the magma intrudes through the crust (Figure 2.1), or from substrate foot-wall rocks in the case of komatiite lavas.

There are multiple lines of geological evidence for sulfide assimilation into the magma at or near the site of ore deposition of many magmatic sulfide deposits:

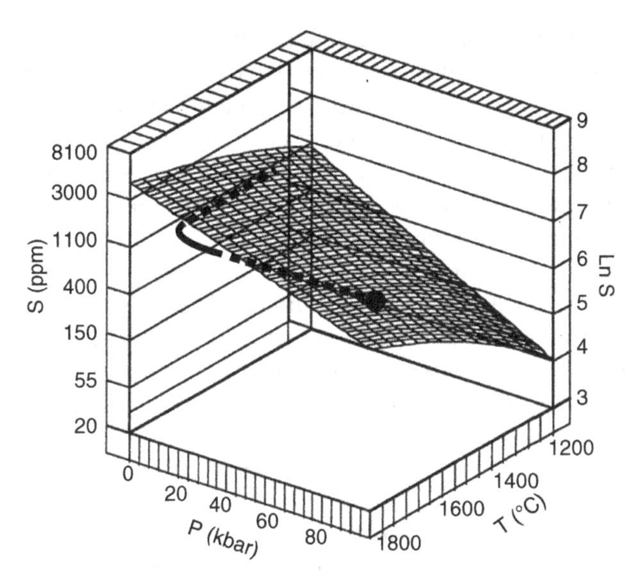

Figure 2.21 Experimentally determined solubility of sulfide in a basaltic magma as a function of pressure and temperature (after Mavrogenes and O'Neill, 1999). The thick solid and dashed line shows a possible decompression and cooling history for a magma that forms at 1600 °C and 75 kbar. If this magma is saturated with respect to sulfide melt (≈ 800 ppm) at the source, it will be undersaturated after adiabatic rise to the upper crust. It will only reach sulfide saturation in the upper crust after cooling to about 1200 °C and hence after significant amounts of crystallisation.

- The magmas of most major known Ni–Cu deposits have intruded through or extruded over sulfide- or sulfate-bearing country rock (e.g. gypsum–anhydrite bearing evaporites at Noril'sk-Talnakh, sulfide-bearing schists at Voisey's Bay, sulfidic sediments at Kambalda). This, however, may not be the case at all magmatic sulfide deposits. No sulfide-bearing unit has been identified adjacent to or near the ore-hosting Nebo-Babel intrusion, for instance.
- There are matches in the ratios of chemical elements and isotopes, for instance between the sulfur isotope ratios of massive sulfide ores and sulfides in the wall-rock or foot-wall. This evidence is discussed fully in Box 2.1.
- Units of the immediately underlying foot-wall are absent below ore-hosting troughs of komatiite lava flow. The likely rates of heat transfer into the substrate below turbulently flowing lava at ~ 1600 °C are such that the substrate would be melted by the lava in the main conduit and dissolved or incorporated into the magma. The cutting out of the foot-wall to form the troughs is thus interpreted as the result of ground melting. In some komatiite flows there is evidence of ground melting from localised concentrations of felsic ocelli (Figure 2.17), which are crystallised droplets of a felsic melt that were derived from sedimentary rocks and were incorporated into the ultramafic lava. Some troughs may, however, alternatively have formed where lava flowed preferentially along pre-existing valleys or topographic lows.

Working model for genesis of magmatic base-metal massive sulfide deposits

The different possible sequences of steps to form Ni–Cu magmatic sulfide deposits are shown in Figure 2.22. Massive or disseminated magmatic sulfide deposits may form in a large-volume mafic igneous system where sulfur from wall-rock of a conduit or foot-wall of a flow is assimilated into the magma. Deposits may also form where other changes in magma chemistry (for instance resulting from assimilation of silica-rich rock or, in the case of the Noril'sk-Talnakh intrusions, assimilation of coal) aid or induce early sulfide saturation and formation and separation of a sulfide melt phase.

The various sites of accumulate of the sulfide droplets into ore bodies are the result of physical processes during magma flow, including settling of droplets, or accumulation in eddies such as form at sites of changing thickness of channels or conduits of intrusions or lava flows. The most common setting for sulfide melt accumulation in intrusions appears to be in magma chambers which are localised thickenings along a conduit system. These would be sites at which rising magma slows down and can no longer carry droplets of dense sulfide melt in suspension. In komatiite lavas, sulfide melts collect at the base of steep-sided troughs (contact ores) in part because the majority of flow is along these channels. Accumulation of sulfide droplets may be aided by turbulent flow in the channels. Eddies in turbulent magma flow may also be important for the collection of the disseminated komatiite–dunite-hosted ores in the centres of thick lava channels.

Magmatic regimes of rocks hosting Ni–Cu sulfide ores

All known economic accumulations of magmatic Ni–Cu sulfide ores, except for those of the Sudbury Igneous Complex, occur where large volumes of mafic or ultramafic magmas derived through relatively high percentages of melting of mantle peridotite have intruded into and through continental crust. One reason for this relationship may be that high volumes of high-temperature melt are more likely to assimilate felsic rocks from the crust and hence become sulfide-saturated. Within the magmatic systems, ore occurs in most cases in sites along magma channelways along which there were large volumes of magma flow.

High-percentage melting of mantle peridotite is indicated by picritic olivine-rich and Mg-rich compositions of the most mafic rocks in the suites of intrusion that host ores. Very large melt volumes are preserved in the Permo-Triassic Siberian Traps, the flood basalt province that is centred on Noril'sk-Talnakh. This is a type example of an LIP, and as is typical of such provinces the large volumes of magma intruded into and through the crust over a time period of at most a few million years. The Nebo-Babel deposits in Western Australia are hosted in small intrusions that are part of the late Mesoproterozoic Giles Complex suite of layered mafic and ultramafic intrusions that is preserved over an area of about 50 km by 300 km, and this suite is interpreted as the remains of an LIP. The original extent of other ore-hosting Proterozoic and Archaean intrusive suites is not apparent in all cases because of incomplete preservation.

True komatiites are restricted to early periods of Earth history and were formed through high degrees of partial melting of mantle peridotite (apparent melt percentages of up to 40%) at high temperatures. It is interpreted that they extruded at temperatures much hotter than modern-day basalts (up to 1600 °C) and that the melts formed at much greater depth in the mantle (150–200 km). These high-temperature melts are almost certainly restricted to early Earth history because of the Earth's secular cooling.

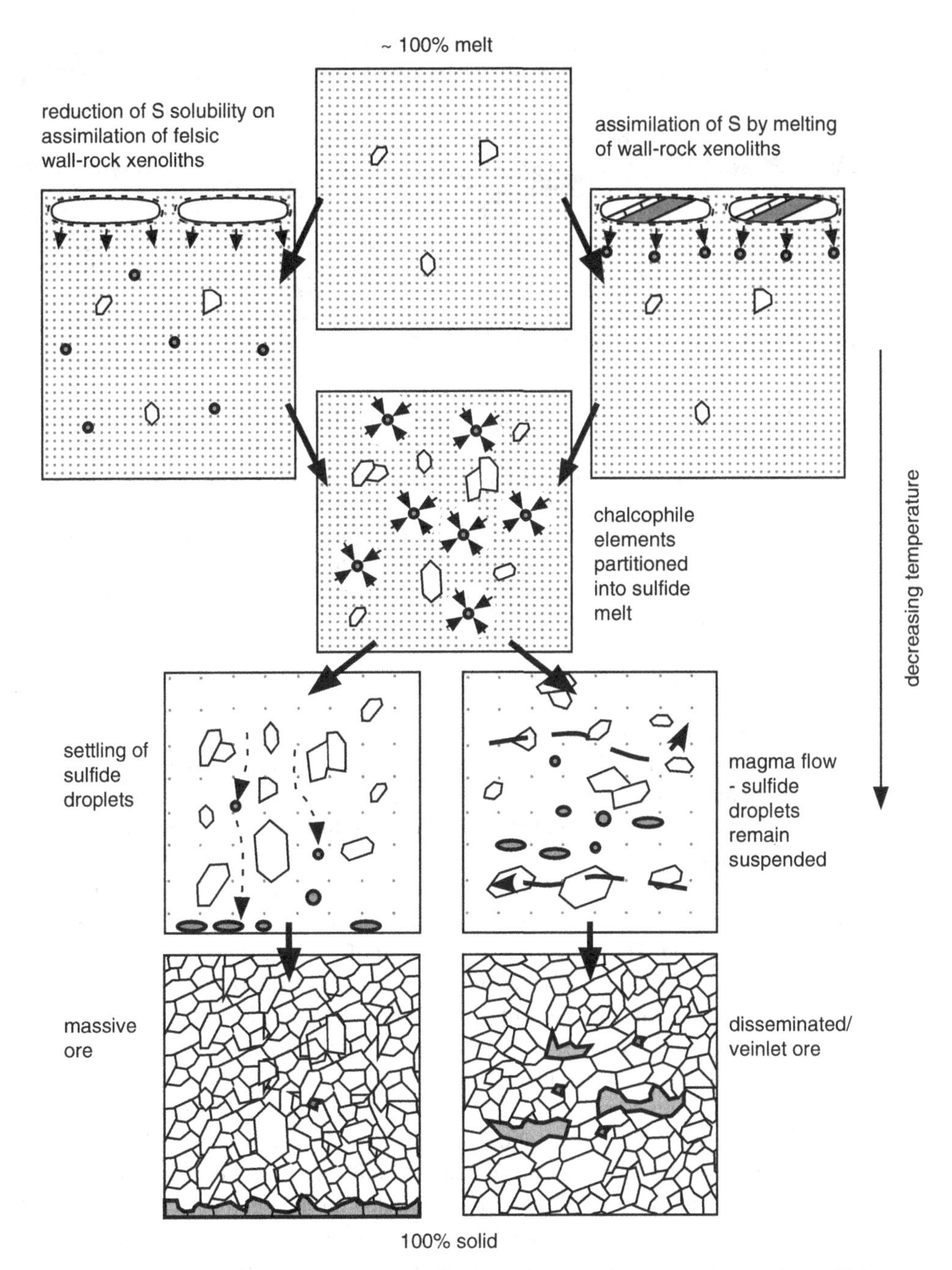

Figure 2.22 Schematic of pathways to formation of economic magmatic massive sulfide deposits in mafic intrusions and in komatiites. The stipple density represents the concentration of chalcophile and siderophile elements (Cu, Ni, PGEs). Sulfide saturation is reached by assimilation of either felsic wall-rock or S-bearing wall-rock. The ore elements are fractionated into the sulfide melt droplets after sulfide saturation. Droplets segregate due to settling to the base of the magma and/or due to irregular magma flow.

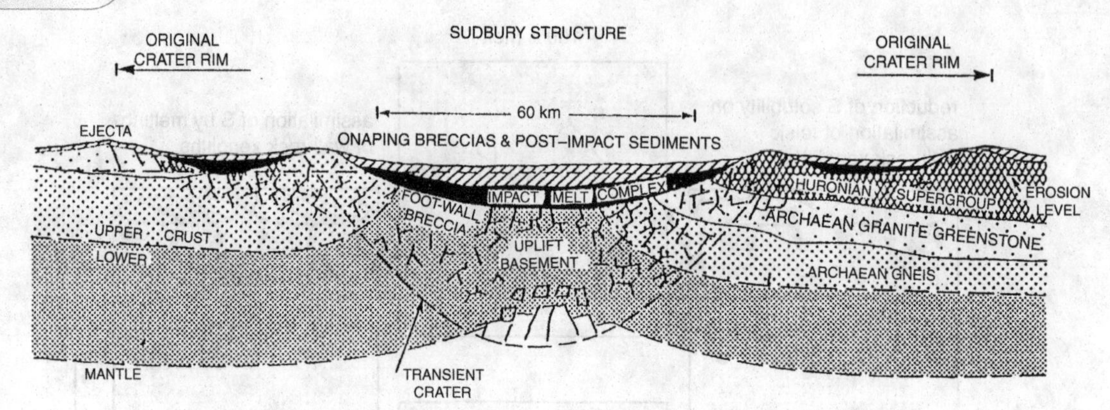

Figure 2.23 Interpretation of the cross-sectional form of the Sudbury structure immediately after genesis as a result of impact and impact melting (after Dietz, 1964; Rousell *et al.*, 2003). Numerical modelling of the formation of an impact crater shows that the sides of the crater are initially steeply dipping, but that the underlying crust and mantle rebounds to form the bowl shape. The original crater is interpreted to be much wider than the 25–60 km width of the preserved Sudbury Igneous Complex. Later tectonic activity, particularly compressional deformation during the later Proterozoic Grenville Orogeny has overprinted many original relationships and has caused the present-day elliptical shape of the complex in map view (Figure 2.14).

Komatiite lava flows generally occur in packages which probably extruded over time periods of at most a few million years. Individual lava flows have been interpreted from scattered preserved remnants to have been up to hundreds of kilometres long and they, similarly to more recent mafic LIPs, thus also formed in relatively short periods of intense magmatism.

The Sudbury Igneous Complex is unique as a host to Ni–Cu ores, not only with respect to its setting, but also in respect to many of the steps and processes of ore formation and accumulation. The intrusion was the result of wholesale melting of a volume of continental crust within a few minutes of impact of a meteorite (Figure 2.23). The composition of the bulk magma was dioritic, which corresponds closely to average crust, and thus contrasts with the picritic basalt or peridotite composition of the magmas that produced most other major accumulations of magmatic sulfide ores. The sulfides were incorporated from the melted crust into the magma. Melting and crystallisation at the surface means that pressure effects on sulfide saturation would not have played a major role in sulfide accumulation. The magma was likely initially strongly superheated above liquidus temperatures but may have been sulfur-saturated throughout once it had cooled sufficiently to start crystallising. Copper and nickel concentrations of the rocks in the upper levels of the main mass of the intrusion are relatively low and it is thus interpreted that the metals partitioned into sulfide droplets from all levels of the magma chamber.

The breccias along and below the main contact that host many of the ore bodies, and the 'offset dykes' that intrude outwards into the country rock, are interpreted to be the result of stress-induced fragmentation during the impact and during movement of the large volume of magma. The Sudbury Igneous Complex has the largest range of compositions of magmatic sulfide ores. The different Ni:Cu ratios of the ores reflects internal fractionation and repeated physical remobilisation of bodies of sulfide melt, and are such as to reflect evolution towards Cu-rich compositions. This is presumably a result of evolution of the intrusion geometry and migration of the sulfide melt during slow cooling. Ores furthest beneath the contact generally have lowest Ni:Cu ratios and the sulfide melt migrated to these sites after partial crystallisation.

Box 2.1 Isotope ratios as monitors of sources of elements in ore deposits. I. The case of sulfur in komatiite-hosted nickel deposits

The measurement and interpretation of concentration ratios of the isotopes of elements is a major part of current and recent research on the genesis of many ore deposit types. In the case of sulfur we are considering stable rather than radiogenic isotopes. There are four naturally occurring stable isotopes: ^{32}S, ^{33}S, ^{34}S and ^{36}S, with natural percentages of approximately 95, 0.75, 4.2 and 0.02% respectively. The isotope ratios are normally expressed in δ notation (δ = ‰ = parts per thousand) which expresses differences of abundance of a minor isotope relative to the most abundant isotope of the element, for instance:

$$\delta^{34}S = \left[\frac{\left(^{34}S/^{32}S\right)_{\text{sample}} - \left(^{34}S/^{32}S\right)_{\text{standard}}}{\left(^{34}S/^{32}S\right)_{\text{standard}}} \right] \times 1000 \tag{2.3}$$

where the standard for sulfur is the isotope ratio in an iron meteorite (Canyon Diablo).

Similarly to other relatively light elements with multiple stable isotopes (the most commonly measured in ore deposit geology are H, C, O and, to a lesser extent, B and N), the relative abundances can be accurately measured with mass spectrometry and vary naturally by up to a few per cent. The variations in abundance are the result of one or more of a number of geological and mineralogical processes. The strengths of atomic bonds of an element are dependent on the mass of the atoms bonded and variations in isotope abundance of up to a few per cent in relative terms develop because of equilibrium partitioning between coexisting phases (e.g. minerals and melts). Alternatively, partitioning can occur during physical and chemical processes such as diffusion and evaporation. A general rule is that partitioning is stronger in lower-temperature geological environments. For a fuller discussion of the geochemical behaviour of stable isotopes see, for instance, Sharp (2007) or Rollinson (1993).

In tracing the source of sulfur in komatiite Ni deposit we are interested in forensic-style matching of characteristic $\delta^{34}S$ of sulfide in the komatiites and in possible 'reservoirs' from which the sulfide may have been derived. We make the likely assumption that there was insignificant isotope partitioning either on separation of a sulfide melt from a silicate melt or on crystallisation of minerals from this melt.

Mantle-derived sulfur almost always has a $\delta^{34}S$ value within a narrow range of 0 ± 1‰. This uniformity is a result of small isotope partitioning between phases at high temperatures, the mantle being well mixed through continual convection in the solid state and percolation of melts, and homogenisation of isotope heterogeneities because

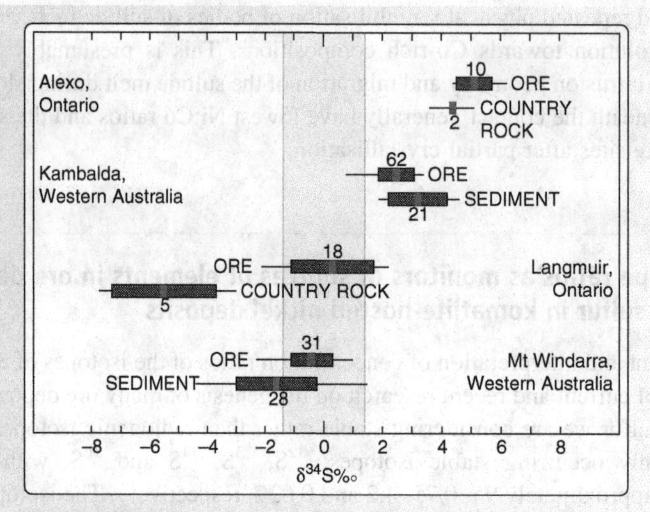

Figure 2.24 Sulfur isotope values of komatiite-hosted Ni sulfide ore compared to immediately underlying sediment or country rock (Lesher, 1989). The shaded band shows the range of mantle-derived sulfur.

of high rates of element diffusion at high temperatures. There are narrow ranges of $\delta^{34}S$ in sulfide minerals at many komatiite-hosted ores, but the values vary between deposits and some are 'heavier' (have a higher positive $\delta^{34}S$) than expected sulfur derived from the mantle. For many komatiites the $\delta^{34}S$ matches well with that of pyrite in the thin black, organic-rich cherty sedimentary rocks that underlie the lavas (Figure 2.24). These non-mantle 'signatures' of sedimentary rock pyrite are as expected if pyrite was precipitated from sulfur that had been cycled through oxidation and reduction reactions in near-surface environments, for instance during marine sedimentation or during diagenesis.

The data support field evidence for incorporation of sulfur into the komatiite by ground melting of stratigraphically underlying sedimentary rocks. From the perspective of ore genetic models, we thus infer that assimilation of sulfur at or near the Earth's surface is an important and possibly a required step in the formation of massive magmatic sulfide deposits. We can go one step further and estimate maximum possible percentages of mantle sulfide in the komatiite Ni ore, using the lever rule. About 15% of the sulfur in the Kambalda ores can be estimated to be mantle-derived based on the mean composition of the ores and assuming that sulfur in the mantle has $\delta^{34}S$ of 0 ‰ (Figure 2.24). The scatter in values at any locality can be the result of the combined effects of a number of processes and factors, both analytical and geological. Geological factors that may affect the isotope ratios include incipient weathering of sulfide minerals, incomplete mixing in the magma, and small amounts of partitioning as the minerals precipitated from the sulfide melt or recrystallised during cooling.

This interpretation of the source of sulfur in komatiite magmatic sulfide ores is based on a single isotope ratio and is thus potentially ambiguous and inconclusive. A match in ratios of a single isotope pair can only in rare cases be taken as proof of a relationship between the reservoirs. There are multiple possible reservoirs of sulfur and multiple processes that can fractionate isotopes, and measured isotope ratios can in many cases result from mixing between different combinations of two or more reservoirs in various proportions. We are further assuming that there were no processes that significantly partitioned sulfur isotopes between the time of incorporation into the magma and the time of crystallisation of the sulfide minerals. In general, isotope ratios can lend support to an ore genetic model, but cannot prove its validity.

2.2.4 Deposits formed from immiscible sulfide melt phases: PGE sulfide deposits

The six PGEs have similar geochemical behaviour as Ni and Cu in sulfide-bearing magmatic systems, but the richest ores of these elements are distinctly different from Ni–Cu sulfide ores. The elements all have highly siderophile behaviour and are the least abundant stable elements in the Earth's crust and mantle. Economic PGE ores need to contain 5–10 ppm combined PGEs, which is equivalent to concentration by enrichment factors of between a thousand and ten-thousand times crustal abundances.

Settings of PGE magmatic sulfide deposits

The most important primary deposits of PGEs are laterally extensive thin layers (= **reefs**), from less than 1 m thick to about 20 m thick in layered ultramafic to mafic large-volume and extensive intrusions in the upper crust. The reefs are broadly parallel to the cumulate layering, and are continuous along much of the extent of the exposed intrusions. There may be multiple reefs at different levels in one layered intrusion.

Excluding by- and co-product PGE resources of Ni–Cu sulfide deposits, almost all known economically significant PGE resources are in three exceptionally large intrusions more than about 4 km thick (> 10 000 km^3 volume), which have repeated multiple cycles of rhythmic cumulate layering and reversals of composition trends of cumulate minerals of layers as has been described in Section 2.2.2 for the Great Dyke of Zimbabwe. These three intrusions are hosted in Archaean cratons and are the Bushveld Complex, South Africa (e.g. Wagner, 1929; Kinnaird *et al.*, 2005; Schouwstra *et al.*, 2000; Figure 2.25), the Great Dyke of Zimbabwe (Prendergast and Wilson, 1989; Figure 2.6) and the Stillwater Complex, Montana, USA (e.g. Zientek *et al.*, 2002). Over 75% of the world's production and reserves are in the Bushveld Complex. The reefs in these larger intrusions generally have the highest PGE grades.

PGE-enriched layers have also been discovered and evaluated in a large number of layered ultramafic–mafic intrusions, most commonly in those that are more than a few kilometres thick. The majority are Archaean to Mesoproterozoic, but younger examples are known, for instance the early Tertiary Skaergaard Intrusion in Greenland. Many of the intrusions that host stratiform chromitite deposits also host PGE ores. The PGE reefs are at the base of some intrusions and in the interior in others and immediate host-rocks range from pyroxenite to gabbro. Resources in most of these intrusions have not been developed. This is partly a function of insufficient grade, but also because of the large investments that would be needed for processing plants to extract and separate the six metals.

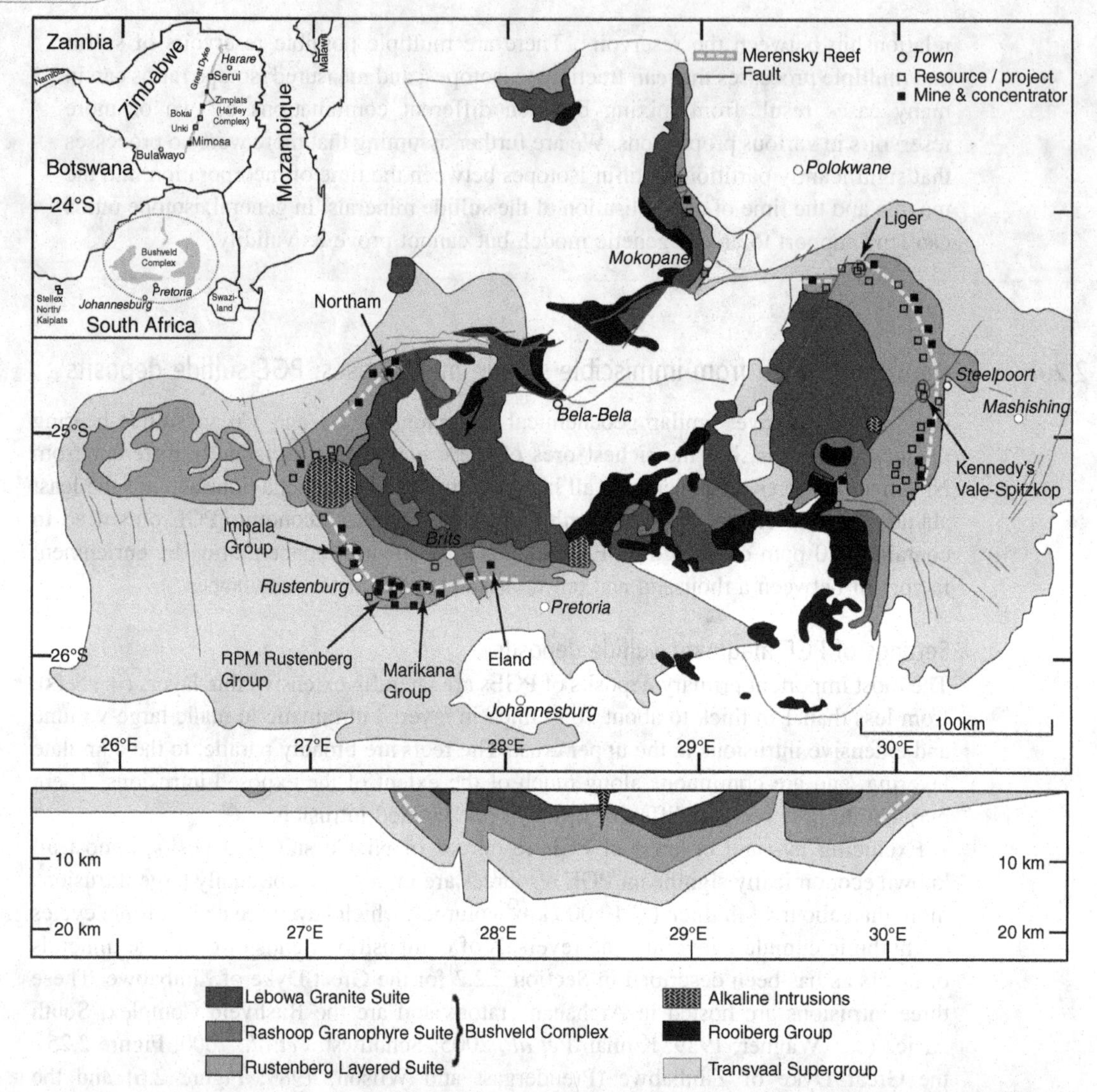

Figure 2.25 Geological map and cross section along 25° S interpreted from gravity and magnetic data of the 2050 Ma Bushveld Complex, South Africa (Webb *et al.*, 2004). The ultramafic and mafic layers of the complex form five layered, 5–10-km-thick, plate-shaped lobes, that dip overall towards the centre of the complex. The PGE ores in the Bushveld Complex are the Merensky Reef and two nearby parallel ≈ 1-m-thick layers that are parallel to the overall igneous layering. As shown by the distribution of mines and prospects, the reefs are mined over much of their outcrop length in the west and east lobes of the complex. They have been traced down dip to about 2 km depth.

The magmas that filled the intrusions that host PGE-enriched layers were high-magnesium basalts, tholeiites, or mildly alkalic basalts. The host intrusions of the economic reefs are layered intrusions formed from high-magnesium, low-titanium siliceous mafic magmas, variably classified as boninite–norite (BN) suite magmas,

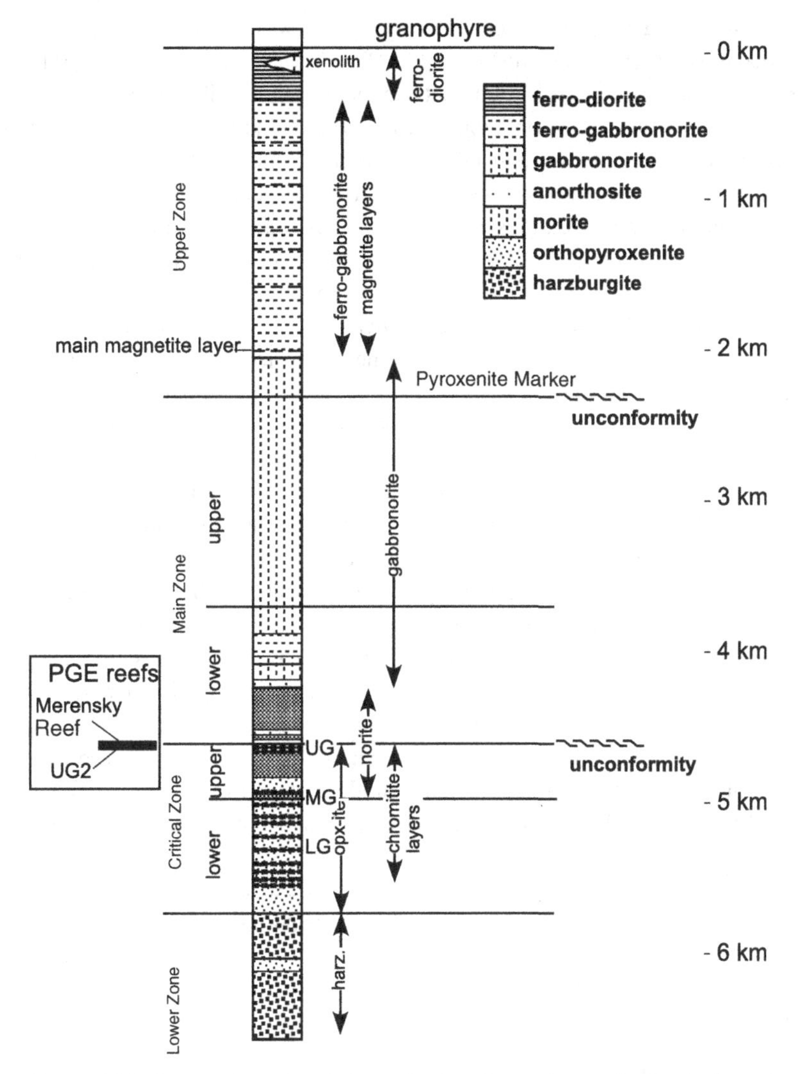

Figure 2.26 Succession of rock types in the ultramafic–mafic Rustenberg layered suite of the Bushveld Complex showing the positions of the main PGE reefs and chromitite and magnetite layers (after Kruger, 2005).

siliceous picritic magmas or siliceous high-magnesium basalts (SHMB). The high-magnesium contents are indicative of higher degrees of partial melting of mantle than would be the case for typical tholeiites. (Note that these boninite-like magmas are not closely related to similar magmas that occur as primitive magmas in oceanic island arcs.)

The Bushveld Complex (Figures 2.25 and 2.26) is Palaeoproterozoic (~ 2050 Ma) and intruded into the Archaean Kaapvaal craton and the overlying early Proterozoic sedimentary and volcanic sequences. It is the world's largest preserved intrusion. It extends over about 65 000 km^2, is up to 9 km thick, and is estimated to have intruded over a period of at most about 100 000 years. Its present geometry is bowl-like, outcropping along the rim, but at depths of up to 10 km in the centre. However, it was originally a sill-like intrusion that intruded most probably within about 4 km of the surface beneath a cover of felsic lavas and pyroclastic rocks (Rooiberg Group) that had formed a few million years earlier. Five lobes are distinguished which, based on geophysical evidence and the similarity of cumulate stratigraphy, are interpreted to be connected at depth under the centre of the intrusion. The area of the intrusion expanded through its growth such that higher levels onlap onto the foot-wall and there are internal unconformities in which lower cumulate layers are at least locally truncated.

Within the Bushveld Complex there are three major PGE-bearing horizons in the 'Critical Zone', which is a complex zone of rhythmic repetitions of rock type 1–2 km above the base (Figure 2.26). The Critical Zone is the transition from dominantly pyroxenites (below), to gabbros (above). The ore horizons are: (i) a pegmatitic pyroxenite layer with thin chromitite seams (Merensky Reef); (ii) the uppermost important chromitite layer (UG2) which is about 30–200 m below the Merensky Reef; (iii) a pyroxenite layer at the base of the northern lobe, where stratigraphically underlying layers are not developed (Plat Reef). The Plat Reef may be correlatable with the Merensky Reef. Both the UG2 and Merensky Reef have been traced along strike for 300–400 km and down dip to 2 km depth in both the Eastern and Western lobes of the Bushveld Complex. Small tonnages of PGEs have also been mined from dunite pipes that cross-cut the layered sequence.

The reefs in the late Archaean Stillwater Complex and in the Great Dyke are in very similar positions in the host intrusions as those in the Bushveld Complex. In each intrusion, ore occurs near the top of the layers with chromitite seams and at the boundary between the lower ultramafic layers of the intrusion and the upper mafic layers (Figure 2.27). In the Great Dyke, the major PGE reef is at a slightly lower relative position, near the top of the uppermost pyroxenite layer in the intrusion (Main Sulfide Zone, MSZ, Figure 2.6). A Lower Sulfide Zone (LSZ) occurs about 30 m lower in the sequence, but has lower PGE grades.

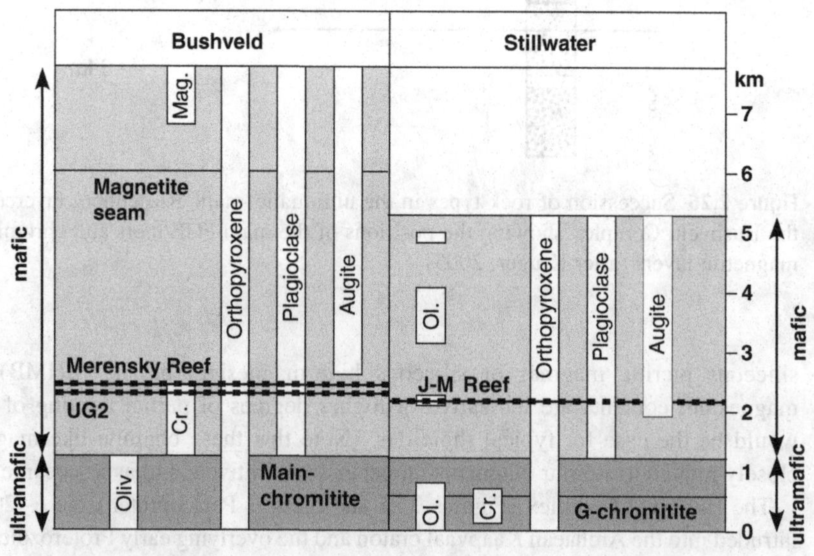

Figure 2.27 Comparison of positions of the main PGE-bearing reefs and main chromitite reefs relative to the distribution of the major cumulate minerals of the Bushveld and Stillwater complexes (Campbell *et al.*, 1983).

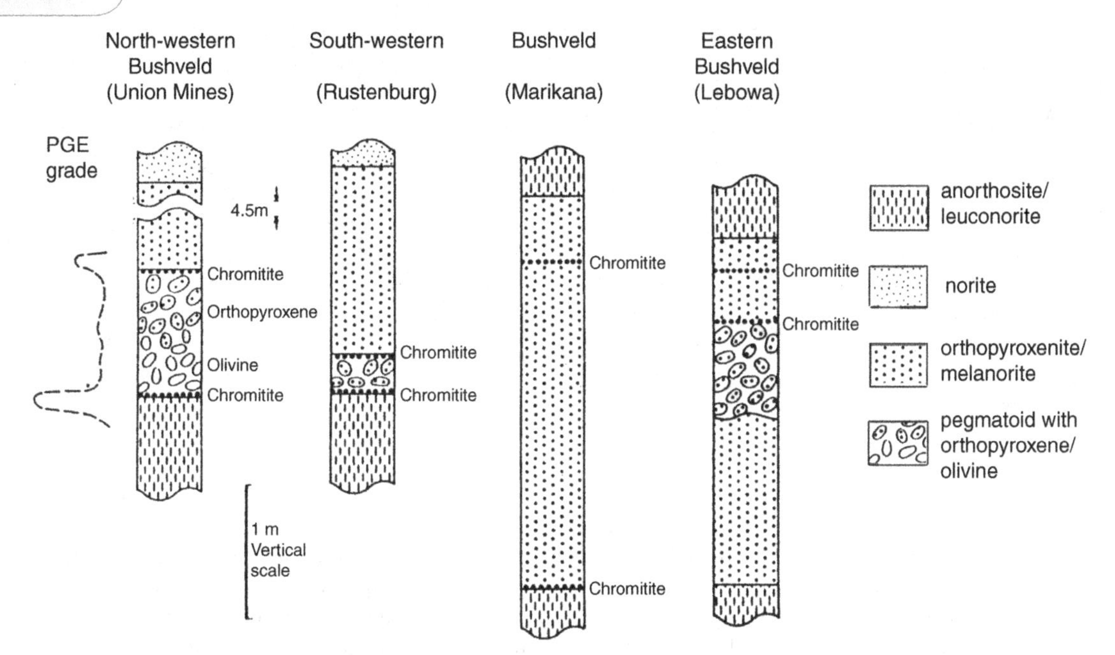

Figure 2.28 Variable nature of the Merensky Reef over distances of hundreds of kilometres at different mines around the Bushveld Complex. The distribution of PGEs is shown for the case of 'normal' reef at Union Mines (after Naldrett, 2004).

Nature of PGE ores

Most PGE reefs contain 0.5–5% disseminated base-metal sulfide minerals (dominantly pyrrhotite, pentlandite and chalcopyrite). These are significantly higher concentrations of sulfide minerals than elsewhere in the intrusions, but are low compared to magmatic sulfide base-metal deposits. PGE minerals (PGMs) are mixtures of trace concentrations of typically very fine grained (< 20 μm) and disseminated alloys (e.g. PtPd), sulfides, and related minerals (e.g. the arsenide sperrylite, $PtAs_2$). The PGM assemblage is very variable from reef to reef, and also within reefs, especially in the Merensky Reef. Like the sulfide assemblages in Ni–Cu magmatic sulfide deposits, these assemblages are not formed of minerals that crystallised from the melt, but are the results of low-temperature subsolidus recrystallisation of high-temperature minerals.

The Merensky Reef is typically less than 1 m thick (Figure 2.28). The exact make-up of the reef, the distribution of rock types and the position of the PGE ore vary somewhat along the strike length of the reef and down dip (Figure 2.28). 'Normal reef' is at the contact between anorthosite below and more mafic melanorite above and comprises a pegmatitic pyroxenite unit, about 40 cm thick, with unusually coarse (< 5 cm) grains of orthopyroxene in intercumulus plagioclase and with sharply defined thin chromitite (< 5 cm) seams at the base and top. In some sections the pegmatitic pyroxenite is not developed. In normal reef the PGEs have highest concentrations in the chromitite seams. However, in some sections there are considerable concentrations of PGEs up to about 1 m below the lower chromitite. The reef has a relatively high concentration of volatile-bearing minerals, in particular up to about 5% modal phlogopite and minor apatite.

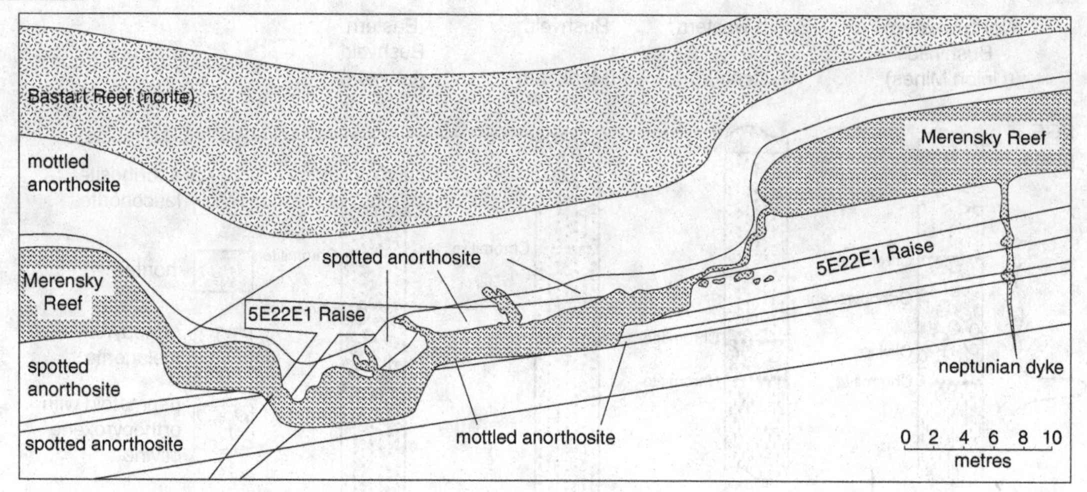

Figure 2.29 Cross section through a Merensky Reef pothole showing typical features (Carr *et al.*, 1994). The pothole is elliptical in plan view, up to a few tens of metres in diameter and is marked by apparent erosion of the cumulate layers up to a few metres below the Merensky Reef. The draping of the hanging-wall units shows that the potholes formed at the same time as cumulate crystallisation. Similar potholes occur on the main PGE-bearing J-M Reef of the Stillwater Complex. Carr *et al.* (1994) suggested that the potholes formed as a result of gravitational instability of largely crystallised foot-wall layers on the sloping floor of the magma chamber. They are alternatively the result of localised remelting of the underlying cumulates at the floor of the chamber after input of a pulse of hot magma.

The Merensky Reef is not a perfectly planar layer within the layered sequence, but is marked by 'potholes', which are sub-circular areas a few metres to hundreds of metres across where normal foot-wall to the reef is missing and the reef rocks sit a few metres lower in the cumulate sequence, and by some 'neptunian dykes' where the reef appears to intrude into the underlying cumulates (Figure 2.29). These irregularities influence PGE content of the reef. In fact, the Merensky Reef marks a major unconformity surface in the cumulate sequence of the Bushveld Complex: immediately underlying cumulate layers appear to have been 'eroded' beneath about half of the explored area of the reef in the western lobe of the complex.

The MSZ on the Great Dyke is directly beneath the lowest mafic units of the cumulate sequence and hosted in websterite (an orthopyroxene cumulate with intercumulus plagioclase) above orthopyroxenite. The zone is 2–3 m thick and can be divided into an upper base-metal subzone with the highest concentrations of Ni and Cu sulfide minerals, and a lower PGE subzone. The highest PGE concentrations are thus about 1 m below the level of highest sulfide contents. A similar separation of PGEs and base-metal sulfides occurs in other reefs, such as the sub-economic reef of the Munni Munni complex of Western Australia.

Ore genetic models for PGE reefs in layered intrusions

The processes of genesis of PGE reefs in layered intrusions is difficult to constrain from geochemical and petrological data. The following sections describe what has been generally inferred about ore genesis. The uncertainties in the ore genetic models and the reasons why there are uncertainties in the models are described further in Box 2.2.

(a) Magmatic evolution of the ore-hosting intrusions

The multiple cycles of rhythmic cumulate layering and reversals of composition trends of cumulate minerals indicate that there were multiple phases of magma input into the large magma chambers in which the PGE ores formed. As is discussed in Section 2.2.2, the chromitite layers in these intrusions are most easily explained as results of mixing between new magma and resident magma. Each new magma input into the chamber must have been after a time period short enough that the resident magma had only partially crystallised. The cumulates would have been 'soft' crystal mushes with interstitial melt and may have been deformed on intrusion of a new magma mass.

Internal unconformities and the onlapping of upper layers onto the underlying host-rocks also give evidence of multiple magma input into the chambers. The potholes (Figure 2.29) of the Merensky Reef are explained as results of deformation and possibly of remelting of partially crystallised cumulate on the chamber floor after a large mass of new magma filled the chamber. Isotope and trace-element ratios indicate that different levels in the cumulate sequence crystallised from magmas of slightly different compositions. This is clearest in the Bushveld Complex (see Box 2.3 for further discussion and explanation) but is also recognised in the Stillwater Complex. In the Bushveld Complex, all of the magmas were broadly mafic in composition and were genetically related, but cumulates of the lower series are interpreted to have crystallised from a mantle magma that had assimilated less continental crust than the magmas from which the Main and Upper zones crystallised.

(b) Processes of concentration of PGEs into reefs in layered intrusions

A number of features of the reefs indicate that input of new magma and mixing between this magma and magma resident in the chamber was critical for precipitation of the PGE minerals. The Merensky Reef marks a boundary between cumulate units and is at the major unconformity surface at the top of the Critical Zone, over which there is a marked change in some isotope ratios. Thin seams of chromitite are characteristic of the reef. The UG2 reef of the Bushveld Igneous Complex is hosted in the uppermost chromitite layer and lower chromitite seams in the Bushveld Complex are all enriched to some degree in PGEs.

The concentrations of PGEs in the host intrusions away from the reefs are uniform and are close to those of average crust (a few ppb). Enrichment of PGEs in ores is thus by about 1000-fold relative to average concentrations in the host intrusions. The elements must hence have been concentrated or 'scavenged' from a large thickness of magma into the thin continuous reefs. The presence of up to a few percentage sulfide in the reefs suggests that scavenging was a result of partitioning into a sulfide melt phase in a similar fashion as in magmatic Ni–Cu ores, although in contrast to Ni–Cu deposits, neither early saturation of a sulfide melt, nor an external source of sulfur are considered. Early formed sulfide droplets would have collected in a basal layer of the intrusion rather than in layers in the interior.

The model for the genesis of PGE accumulations in the intrusions therefore invokes formation of a relatively small concentration of immiscible sulfide droplets, and in some instances also chromite, in the magma chamber at a time of new input of a more primitive magma into evolved magma resident in the chamber. The sulfide droplets partition PGEs (and also Ni and Cu) from a large volume of magma to attain a high tenor of these highly siderophile elements. The sulfide droplets, and also chromite grains when formed, settle onto the melt–cumulate interface, or possibly flow as a component of density currents to cover the cumulate surface. In places, the sulfide droplets percolated as far as a couple of metres downward into the underlying partially molten cumulates (crystal mush) so as to separate the highest concentrations of base-metal sulfide and PGE minerals, for instance in the MSZ of the Great Dyke.

Textural and mineralogical recrystallisation during cooling means that the present-day reefs have very variable mineralogy. Sulfide minerals are absent in parts of the UG2 reef, and this may be the result of reaction between Fe-bearing sulfides and chromite such that the sulfide minerals were resorbed.

For magma mixing in a chamber to give rise to an immiscible sulfide melt and for the small volume of the sulfide melt to effectively concentrate PGEs by a factor of greater than 1000 from the silicate magma, the following must occur:

- Sulfur saturation must be overstepped to explain the strong concentration of sulfide in the reefs. Sulfur solubility in mafic magmas is most strongly controlled by temperature and magma FeO content. In a high-magnesium mafic magma it is calculated to drop strongly during early crystallisation of mafic minerals, but less rapidly once plagioclase is a liquidus phase (Figure 2.30). Sulfide saturation is potentially overstepped on

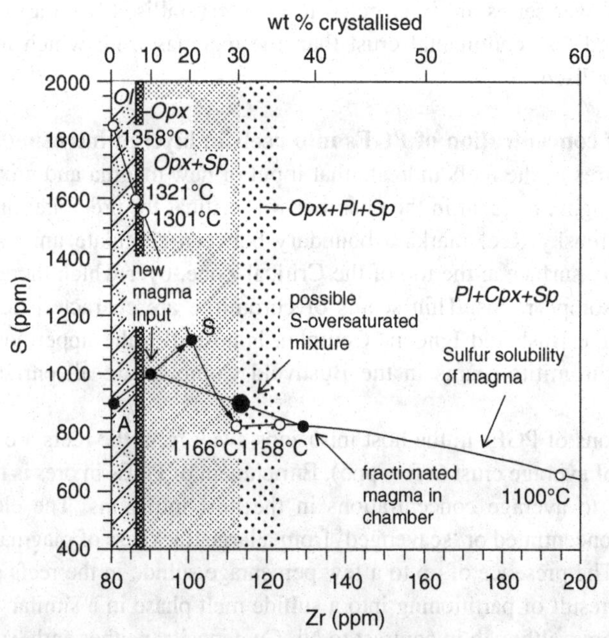

Figure 2.30 Schematic diagram illustrating estimated variations in the solubility of sulfide in a mafic silicate magma at low pressures and its variation in a cooling and crystallisation history (after Li *et al.*, 2001, based on a modelled composition for the input magma of the Bushveld Complex). If the input magma is sulfur-undersaturated, hot (~ 1300 °C) and has composition 'A' on simple cooling, it will reach sulfide saturation at point 'S'. Sulfide oversaturation could be induced on mixing with magma that has evolved through cooling and decreased FeO content beyond the onset of plagioclase (Pl) crystallisation, as shown, and will be most efficient if the magma has fractionated to a composition for which plagioclase + clinopyroxene (Cpx) are on the liquidus. Other abbreviations: Ol, olivine; Opx, orthopyroxene; Sp, spinel.

mixing between a range of compositional pairs of evolved and primitive magma, and appear to be most likely where the resident magma has fractionated to the point of having plagioclase + orthopyroxene on the liquidus, which is equivalent to approximately 50% fractional crystallisation of an input magma of the composition estimated for the Bushveld Complex. However, neither temperature nor FeO content is explicitly plotted in Figure 2.30 and the diagram cannot be used directly to determine sulfur solubility of mixtures. It has not been demonstrated that oversaturation will be attained on mixing pairs of magmas that may have been present in any of the intrusions. Other processes may be required to induce rapid precipitation of sulfide, such as assimilation of felsic wall-rock into the magma at the time of new input into the chamber.

- The PGEs need to be efficiently concentrated into the sulfide phase from a large volume of magma. If the magma had PGE concentrations typical of a mafic melt, the Merensky Reef could have formed if the elements were sequestered almost entirely from a 4-km magma column. Although the thickness of the Main Zone cumulates above the Merensky Reef is about 4 km, it is unknown whether this was the thickness of the magma chamber at the time of crystallisation of the reef. Similarly efficient sequestration is required for the underlying UG2. The efficiency of sequestration would likely be influenced by physical processes of intrusion. Mixing would be most efficient if the new magma rises into the interior of the chamber as a buoyant, turbulent plume, and neither hugs the floor nor the roof of the partially crystallised chamber (Figure 2.31). This will be most likely to occur where the new magma has a similar density as the resident magma. High-Mg mafic magmas become denser during an early stage of fractionation, because early formed cumulates are MgO rich, however, density begins to reduce on further fractionation, and magmas which are just starting to crystallise plagioclase have similar densities as primitive mafic magmas.

These two factors together imply that, given the estimated composition of the input magma into the intrusions, segregation of PGE sulfides would be most efficient on mixing a primitive magma with one that had fractionated to just beyond the point at which plagioclase starts to crystallise. These factors thus explain the common position of the best PGE reefs in the cumulate sequence of the different host intrusions (Figure 2.27).

(c) Melt regime of host magmatic systems

The percentage of melting in the mantle may affect the concentration of PGEs in large layered intrusions, and hence the grade of ores. As PGEs are strongly compatible in sulfide minerals and melts, they may remain in the mantle in unmelted, residual sulfides. PGE concentrations in the magma will be highest if the degree of partial melting in the mantle source is at or is slightly greater than the point at which all sulfide becomes dissolved into the magma (Figure 2.12). This point of sulfide melting is estimated to be a relatively high degree of partial melting, and to be that at which high-magnesium basalt magmas form from peridotite. Note also that in order to have a high concentration of PGEs, the melted peridotite must be ‘fertile’ and cannot have been depleted of sulfide because of an earlier partial melting event.

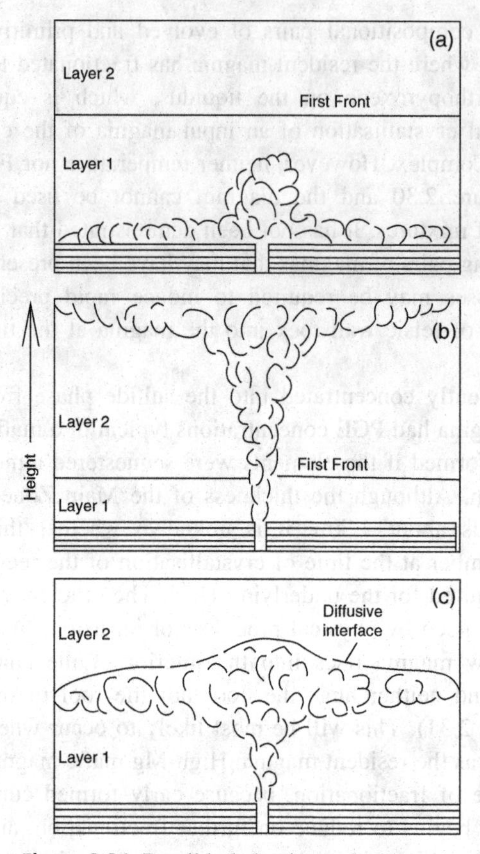

Figure 2.31 Possible behaviour of new magma input into a magma chamber. The magma chamber is assumed to be compositionally stratified ('First Front') with a denser lower layer and less dense upper layer. In (a) the new magma is densest, in (b) is least dense. In (c) it is more dense than the upper layer, but less dense than the lower. This last is the situation in which mixing between the input and resident magmas is most efficient (Campbell *et al.*, 1983).

Box 2.2 Debates over the genesis of PGE-bearing reefs in layered intrusions

The explanation for the formation of thin PGE-rich reefs in layered intrusions outlined in the text is an ore genetic model. It has been developed over a number of years through iterations of the thinking of researchers and exploration geologists. The ore genetic processes that are proposed have, however, neither been observed nor experimentally simulated, and many points in the model are inferences that are based on geological and geochemical reasoning. The model provides a rational explanation for many of the features of the deposits, although the fact that it provides explanation does not prove that the processes proposed actually took place. Further, the model may not be unique.

Although the ore genetic model is based on geological and petrological reasoning, a step in the reasoning may be flawed because we do not have a full understanding of what physical, chemical and mineralogical processes may take place in a large magma chamber that was multiply refilled, that remained molten for tens of thousands of years, and remained at near-solidus temperatures for a longer time period. At high temperatures and over long time spans, diffusion may occur over distances greater than the reef thicknesses, and interstitial melt may have percolated up and down through the partially molten crystal mush of the cumulate pile. Cumulates may have been deformed after deposition at the base of the magma chamber. The minerals that crystallised first from the melt may have been resorbed or recrystallised. The textural and geochemical record is imperfect and evidence may never be preserved for many mineralogical and physical steps in the formation of the reefs.

Additionally, it is not clear whether the model can explain sulfide saturation in the magma chamber. There are a number of features of the Merensky Reef and other reefs that the genetic model does not clearly explain. The mechanisms by which small PGE-bearing sulfide droplets were efficiently accumulated to the base of a magma, which is several kilometres in thickness, within the short time before overlying cumulate layers were deposited is, for instance, not clearly addressed. Other features that are incompletely explained include:

- the potholes on the reef and the controls on their localisation;
- the distribution of the PGEs within the Merensky Reef;
- the distinct pegmatitic texture of the reef pyroxenite.

Because of these and other incompletely explained features of the Merensky Reef and other PGE deposits, their genesis is continually debated. Some points of debate question the validity of the whole ore genetic model. It is argued that there are alternative processes that can equally well or better explain the genesis of the ores. Other debates concern modifications and additions to the model in attempts to better explain ore genesis and the nature of the reefs.

One debate that has been current since the first descriptions of the Merensky Reef (Wagner, 1929) is whether the PGMs crystallised from sulfide melt droplets that had settled from the overlying magma. The alternative is that the PGEs were transported upwards through the underlying cumulates and were precipitated in minerals in the reef when this level in the cumulates marked the interface between the cumulate pile and the magma chamber (Figure 2.32b, e.g. Boudreau *et al.*, 1986, Ballhaus and Stumpfl, 1986). Cumulates typically trap a small proportion of melt, and this trapped melt will release dissolved volatile components, including water and chloride on crystallisation (see Section 3.1 for further discussion of magmatic volatiles). That an aqueous solution played a role in the formation of PGE reefs was originally proposed because the Merensky Reef in particular contains a few modal percentage biotite. One suggestion is therefore that PGEs were dissolved in water that percolated upwards from the lower layers of the chamber over the time period of the last stages of crystallisation. The water and chloride in solution is suggested to have dissolved cumulate sulfide minerals and the PGEs

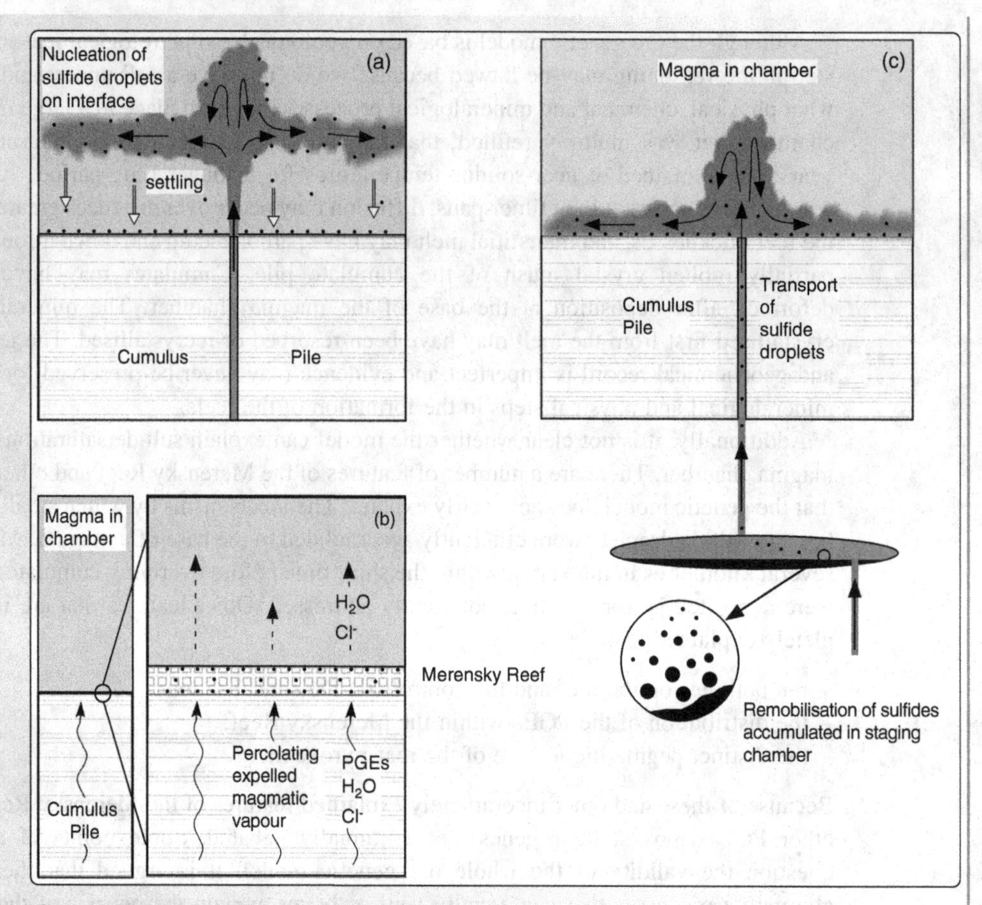

Figure 2.32 Cartoons illustrating some of the major genetic models proposed for the accumulation of PGEs in thin reefs in layered ultramafic–mafic intrusions. (a) The classic model presented in the text; (b) and (c) show an alternative and a variant. See text for discussion.

contained in them. The PGEs are proposed to precipitate as platinum group minerals (PGMs) at the cumulate–magma interface where much of the water and chloride redissolve into the overlying silicate melt, hence decreasing PGE solubility.

A different role that upward-percolating water could play in the development of the reef was investigated by Boudreau (2008). He proposed that the rocks in and immediately above and below the Merensky Reef became remelted and reconstituted because water had percolated in from below and had hence induced a reduction in the solidus temperature. Numerical simulation modelling of mineral growth and resorption, using likely values of a range of physical and chemical parameters of the linked processes that may occur around a reef in this situation, including temperature, the degree of melting, and diffusion of elements, reproduces some of the characteristic features of the reef, including spikes of PGEs at the base, and a pegmatitic texture. Cawthorn (2011) countered that there is no field evidence of melt remobilisation of the layer. However, irrespective of the counter arguments, and irrespective of whether PGEs were concentrated in cumulate sulfide or from rising aqueous solutions, Boudreau's (op. cit.) modelling has demonstrated that the distribution of metals and minerals on the scale of about a metre may have been modified by diffusion and recrystallisation of both silicate and sulfide minerals. There is little dispute that the ore mineralogy has been modified during cooling, but the modelling suggests that the distribution of ore elements may have also been modified on the scale of a mine face after accumulation.

An additional step in the generally accepted ore genetic model that is described in the text was proposed by Naldrett *et al.* (2009) to address the difficulties of reaching sulfur saturation and of physically sequestering small sulfide melt droplets from a large volume of magma into a thin reef (Figure 2.32c). Their proposal is based on analogy with the sulfide accumulation in staging chambers of magma pathways at Noril'sk-Talnakh and Voisey's Bay (Section 2.2.3). They proposed that sulfide melt first separated from silicate melt in a staging chamber below the main chamber of the Bushveld Complex and accumulated there. This sulfide melt was picked up and carried into the main chamber during intrusion of a later pulse of magma. Because the later magma would have physically picked up and transported (remobilised) sulfides, it would be sulfur-saturated and would have PGE contents much higher than has been interpreted for any of the input phases into the main chamber. The thin Merensky Reef could have been formed by settling of sulfide droplets from 250 m of enriched magma that intruded as a relatively thin layer above the cumulates, and it is thus not necessary to invoke settling from a 4-km-thick layer. There is also no need to invoke extremely efficient scavenging into a small volume of small sulfide droplets and rapid settling of these droplets. We do not know whether there was one or more staging chambers below the main chamber of the Bushveld Complex. The existence of staging chambers would however provide an explanation of two other features of the complex – the high-silicon content of the magma and its high $^{87}Sr/^{86}Sr$ ratio – both of which imply that crustal rocks were assimilated into the mafic magma, probably before it reached the main chamber, and that increasing proportion of crustal assimilation occurred through the phases of intrusion.

The proposal that there was a lower staging chamber and that this chamber played a central role in the formation of the PGE ores is speculative. One implication of the proposal is, however, that there may be a commonality between magmatic processes of formation of massive Ni–Cu sulfide ores and disseminated PGE sulfide ores. If this is correct, it implies that the complete geometry of magmatic systems is an important factor controlling which intrusions host economic PGE resources. An additional implication of the proposal is that there may be a style of PGE deposit that has not yet been discovered in small intrusions that formed in the deeper crust below now largely eroded major intrusions.

Box 2.3 Isotope ratios as monitors of sources of elements in ore deposits. II. Evidence from strontium isotopes of multiple magma inputs into the Bushveld magma chamber

The first conclusive evidence for multiple episodes of filling of the Bushveld magma chamber by slightly different magmas came from systematic measurement of the variation in ratios of isotopes of Sr, in particular the ratio $^{87}Sr/^{86}Sr$ in plagioclase and in whole rock (Hamilton, 1979; Kruger and Marsh, 1982; Kruger, 1994). Since then, other isotope and geochemical ratios have confirmed the interpretations (e.g. Maier *et al*., 2000).

Radiogenic ^{87}Sr is produced by β decay of ^{87}Rb, whereas ^{86}Sr is stable. The decay has a half life ($t_{½}$) of approximately 49 × 10^9 years, which is an order of magnitude longer than the history of the Earth. The $^{87}Sr/^{86}Sr$ ratios of rocks will not measurably change over the lifetime of a magmatic system, which in the case of the Bushveld Complex is of order 10^5 years. Nor will there be measurable fractionation of the isotopes between phases in a magma. The ratio will, however, increase slowly over much longer time spans: in the Earth as a whole the ratio has increased from 0.699 to about 0.704 since formation (Figure 2.33). The difference between the present $^{87}Sr/^{86}Sr$ ratio of any rock or reservoir and the initial ratio depends on the rock Rb/Sr ratio and time through the equation:

$$\frac{^{87}\mathrm{Sr}}{^{86}\mathrm{Sr}} - \left(\frac{^{87}\mathrm{Sr}}{^{86}\mathrm{Sr}}\right)_i = \frac{^{87}\mathrm{Rb}}{^{86}\mathrm{Sr}}\left(\mathrm{e}^{0.69t/t_{1/2}} - 1\right). \tag{2.4}$$

Rubidium is a more strongly incompatible element than strontium in most magmatic environments, and Rb/Sr ratios of melts and rocks derived from them tend to be higher than source rocks. At the present time, mantle rocks and melt derived from mantle have narrow ranges of $^{87}Sr/^{86}Sr$ from 0.703–0.704, whereas crustal rocks have a much broader range from 0.704 to 0.9 or higher, because of the time-intergrated effects of high Rb/Sr since formation. See for instance Rollinson (1993) for a summary of the history of Sr isotope ratios in different geological units and geological reservoirs in the Earth.

In the case of the Bushveld Complex, $^{87}Sr/^{86}Sr$ is back-calculated to the time of intrusion at 2.05 Ga and is expressed as $^{87}Sr/^{86}Sr_i$. $^{87}Sr/^{86}Sr_i$ varies systematically with height in the complex. It is generally relatively low (0.705), but with some 'excursions' to higher values in the Lower Zone and lower Critical Zone (pyroxene-dominated cumulates), but steps to 0.708 in the upper Critical Zone over a 200-m interval at and above the Merensky Reef, and is uniform above this for about 3 km of the Main Zone before dropping to approximately 0.707 in the Upper Zone (Figure 2.33). Some data have shown that the ratios may not be uniform in the chromitites and may thus give additional evidence of magma mixing in the formation of these layers (e.g. Kinnaird *et al*., 2002).

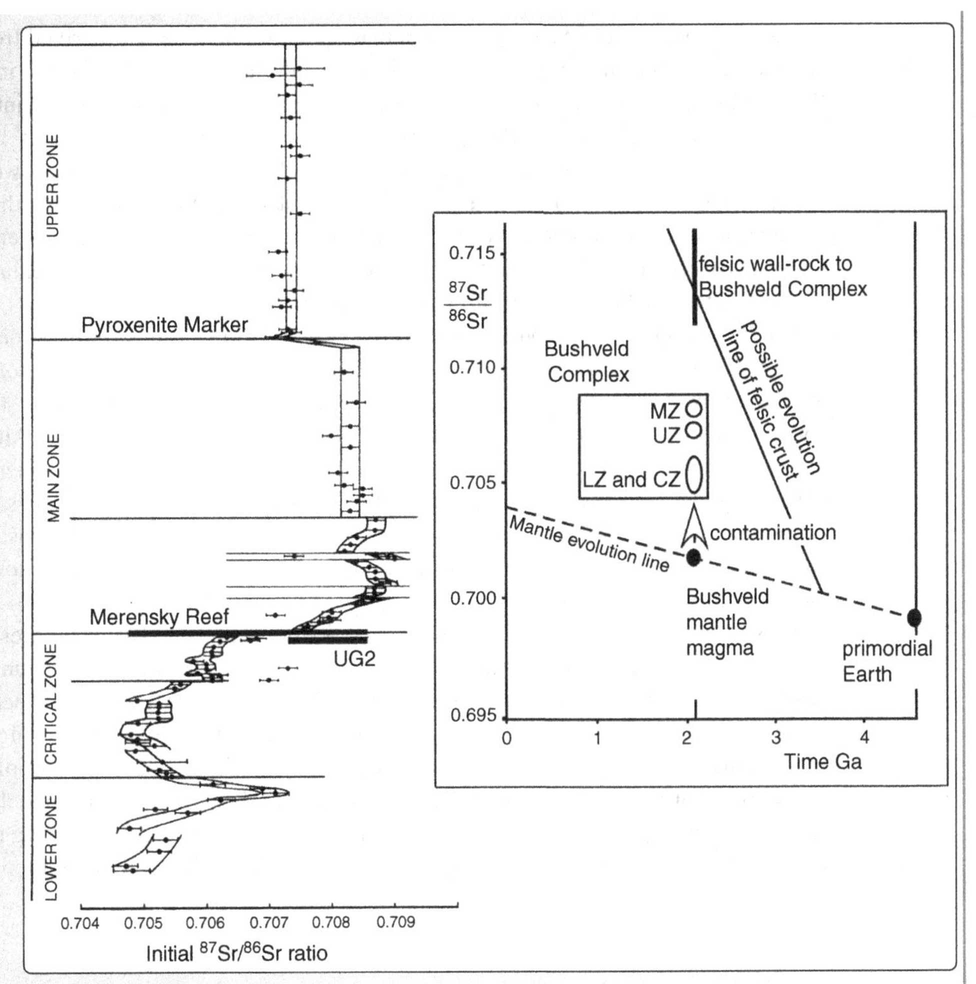

Figure 2.33 (Left) Whole-rock Sr isotope ratios calculated for the time of intrusions at 2.05 Ga in the layered sequence of the Bushveld Complex (after Kruger, 1994). The variations in isotope ratios show that multiple magma pulses entered and filled the chamber. Note the sharp step of ratios at the level of the PGE Merensky and UG2 reefs. (Right) Generalised evolution of the Sr isotope ratio of mantle and crust through time and interpretation of the Sr isotope data of the Bushveld Complex as the result of variable degrees of contamination with continental crust of the different pulses of mantle-derived magma.

The differences with height in the sequence imply that the different zones crystallised from melts derived from different source rocks or different mixtures of source rocks. There was in particular input of a large volume of (mafic) magma from a different mixture of sources at the time of accumulation of the Merensky Reef in the cumulate sequence.

What can we infer about the magma sources? The mafic composition of the magmas implies that they were derived from the mantle. However, even the Lower Zone has $^{87}Sr/^{86}Sr_i$ much higher than the mantle at 2.05 Ga (~ 0.702). The ratios indicate that up to a few tens per cent of Sr in the magma that filled the chamber was derived from old continental crust. This crust may have been assimilated before intrusion into the Bushveld magma chamber or from the walls of the chamber at the time of intrusion (see Kinnaird *et al.*, 2002). Because Sr is more abundant in crustal rocks than in mantle-derived mafic magmas, the percentage of rock assimilated would be lower than the percentage estimated of Sr. If assimilation took place before intrusion into the main chamber it would most likely have been in a deeper 'staging' magma chamber in the lower crust. We can also infer that the component of assimilated continental crust generally increased during the filling of the chamber.

Similar isotope data on other large, PGE-ore-hosting intrusions imply similar but slightly different evolution of the magmas in each case. Rhenium–osmium isotope data on the Stillwater Complex suggest that the magma inputs became less strongly contaminated by assimilated continental crust with time (Horan *et al.*, 2001). Although multiple magma input occurred into the Great Dyke chamber, there is no evidence for either large percentages of crustal assimilation or magmas of different sources. There is, for instance, no $^{87}Sr/^{86}Sr_i$ contrast across the MSZ in the Great Dyke (Mukasa *et al.*, 1998) and Os isotope ratios limit crustal assimilation into the magma to at most 25% by volume (Schoenberg *et al.*, 2003).

These data support other aspects of the genetic models for the Bushveld ores. They are consistent with the interpretation that the distinct siliceous high-magnesium composition of the mafic magmas that formed the Bushveld and other large Archaean and Palaeoproterozoic layered intrusions resulted from hot, high-magnesium, high-percentage mantle partial melts that assimilated continental crust to an amount of up to tens of per cent of the melt volume. The magma would thus have been formed through relatively high degrees of partial melt of the mantle ($> 10\%$ in a batch melting model, Figure 2.12) and would hence have the potential to have relatively high PGE contents.

2.2.5 Deposits formed through extreme fractionation of magma: rare-metal pegmatites

Pegmatites are coarse-grained igneous rocks. The word is normally used to describe a granitic rock. Granite pegmatites occur as dykes, lenses and smaller segregations, often as swarms within and around some granitic plutons. In ore deposit geology, the interest in pegmatites is in the so-called **rare-metal** or **rare-element** pegmatites. The latter name is unfortunate because of its closeness to 'rare-earth element', but indicates the presence of high concentrations of one or more metals and other elements that are present in trace concentrations (< 500 ppm) in average crustal rock and which are not extracted from other common deposit types. The elements enriched in pegmatites are mostly lithophile elements and include (see Figure 2.2) the LILEs (Li, Rb, Cs, Be), HFSEs (Ga, Sn, Hf, Nb, P, Ta, Y, U, Th, REEs), and elements that form compounds that are strongly soluble in aqueous solutions (B, F). Not all elements in this list are enriched in a single pegmatite or pegmatite swarm. There are, however, common patterns of co-enrichment, which are indicated by the abbreviations LCT (= Li, Cs, Ta) and NYF (= Nb, Y, F). Pegmatites with enrichments of mixtures of the elements of these types also occur.

As ores of metals, pegmatites are of greatest economic interest as ores of Ta, Sn, Cs, U and Rb. Ores of Li and Be were of recent interest, but alternative and cheaper sources are now available from evaporite brines and hydrothermal ores respectively. Economic interest is in the largest pegmatite bodies, and especially those that can provide multiple commodities, not only metals but also industrial minerals such as mica, feldspar, kaolinite clays and spodumene. Examples of currently or recently exploited pegmatites include: Rössing, Namibia for U (Berning *et al.*, 1976); Tanco, Manitoba, Canada for spodumene, Cs, Ta (Cerny *et al.*, 1996); Volta Grande, Brazil; Kenticha, Ethiopia; and Greenbushes and Wodgina, Western Australia for Sn, Ta (Partington, 1990).

Geology of rare-metal pegmatites

The largest pegmatites are typically irregular lenticular bodies up to 100 m by 1 km in cross section, or slightly larger (Figure 2.34). They are isolated bodies or are an abnormally large pegmatite within a swarm scattered over a few square kilometres, and are hosted in Archaean or Proterozoic metamorphic rocks, most typically in the upper-greenschist- or the lower-amphibolite-facies of the low-pressure baric type. Andalusite, staurolite, or more rarely sillimanite are index metamorphic minerals in the host terrains. The large pegmatites are detached from 'source' plutons by up to a few kilometres in map view, and some (e.g. Greenbushes) have no associated outcropping source pluton; the source is presumably at depth.

The pegmatites are compositionally granites, with quartz, K-feldspar and albite as major minerals but without either biotite or hornblende. Other minerals may include muscovite, rare-metal minerals and volatile-element-bearing minerals such as phosphates and tourmaline. The pegmatites are internally zoned with respect to major and minor minerals and also texture. Multiple, approximately concentrically arranged zones are typically mapped. A common pattern is for a relatively fine-grained granitic border zone, separate zones of albite and K-feldspar, and a near monomineralic quartz core. Ore is in specific zones, most typically in one or more internal zone, and each commodity mineral may be in different zones. At both Bikita (Figure 2.34) and Tanco there is an almost monomineralic small-volume pollucite (Cs-analcime) zone near the pegmatite core, whereas lithium was extracted from a more extensive spodumene- and petalite-rich intermediate zone.

Genesis of rare-metal pegmatites

The central process of rare-metal enrichment in pegmatites is melt fractionation (Figure 2.35). LCT pegmatites are some of the most strongly fractionated igneous rocks. Extreme fractionation is indicated by their bulk leucocratic granite composition with low concentrations of Ca, Fe and other elements that partition into the major minerals of granites, and with ratios of quartz, albite and K-feldspar that correspond closely to those of the 'minimum melt' or eutectic melt composition. Quantitative measures of the degree of fractionation are provided by concentration ratios of some pairs of elements that are strongly different from average crust, for instance, K/Rb in average crust ranges from about 200–500, but can be as low as 5 in pegmatites because of enrichment of

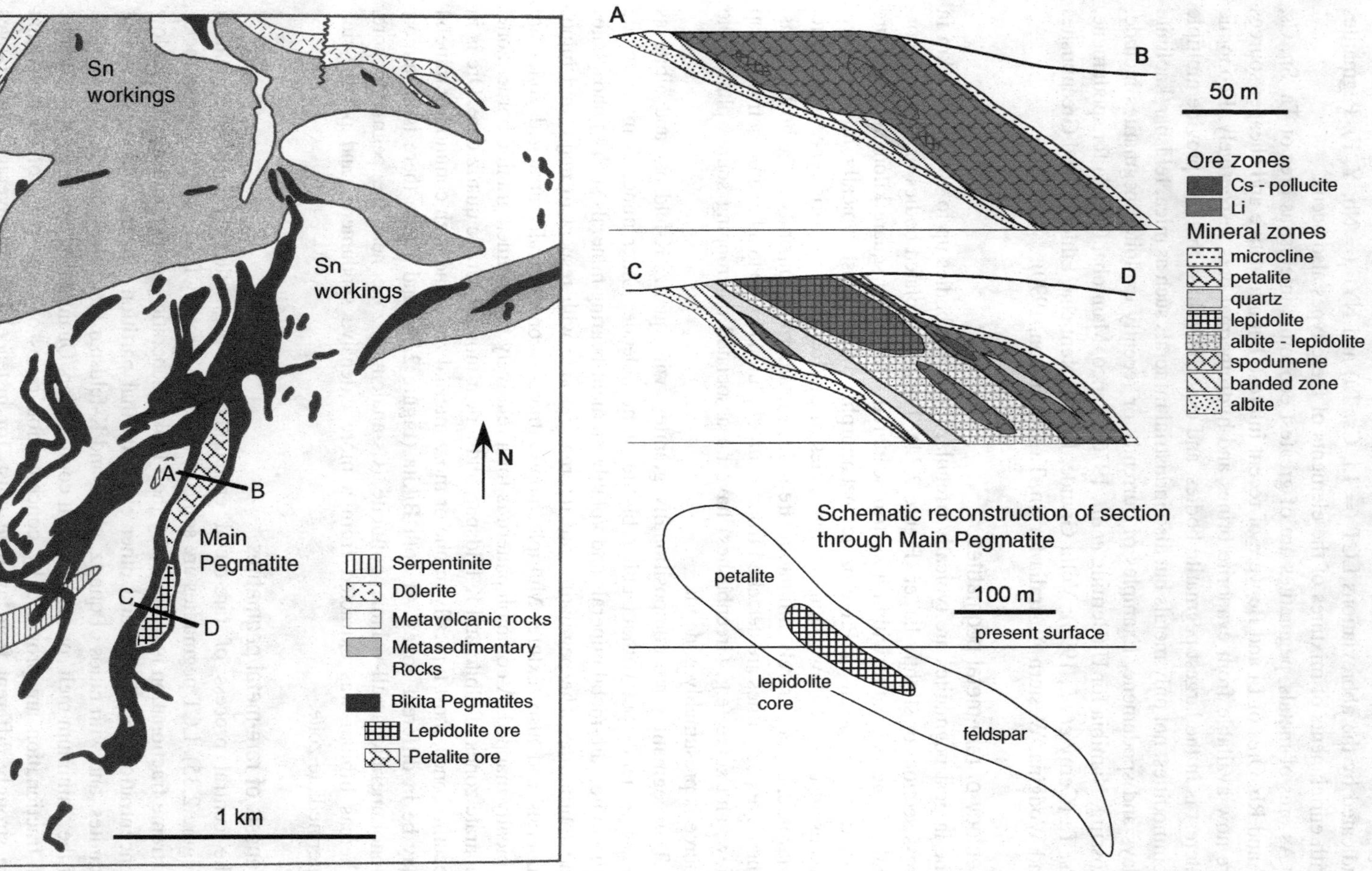

Figure 2.34 Geology and rare-metal ore distribution in the Bikita pegmatite, Zimbabwe, modified after Martin (1963). The Main Pegmatite is the largest member of a swarm of similar pegmatites hosted in deformed and metamorphosed rocks of a greenstone belt. The nearest outcropping granite is about 4 km distance. The Main Pegmatite is internally zoned as shown in sections A–B and C–D with separate lenses towards the pegmatite centre of lepidolite ore and petalite ore, both with up to about 4% Li_2O, and a smaller lens of pollucite ore of caesium. Minor cassiterite as an ore for Sn has been extracted from two of the smaller pegmatites in the swarm and minor tantalite as an ore for Ta has also been extracted.

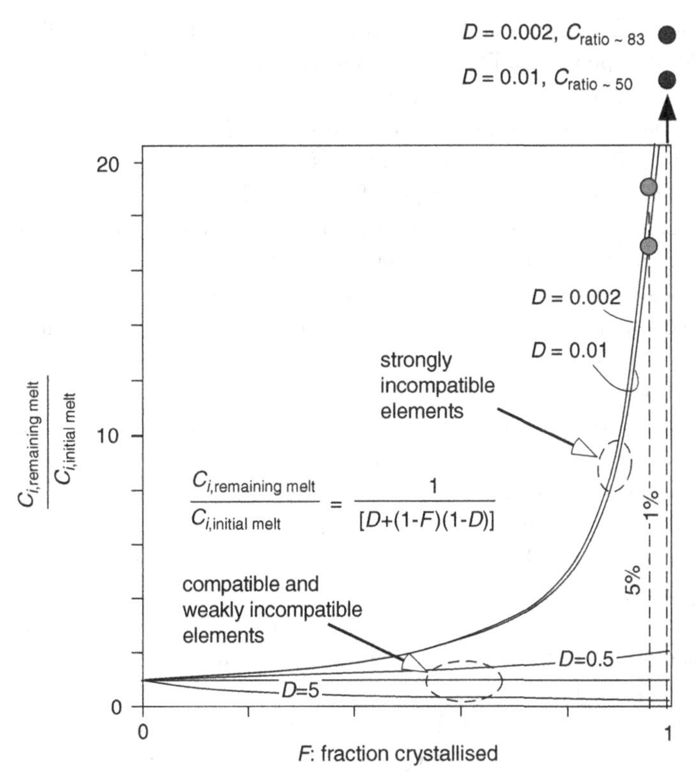

Figure 2.35 Concentration of trace elements in a melt during equilibrium fractional crystallisation, showing the effects of fraction of melt crystallised (F), and the bulk distribution coefficient of the element (D). Strongly incompatible elements are strongly enriched in the final melt fractions. Enrichment factors are shown for 5% and 1% remaining melt. The high concentration of an incompatible element may allow saturation with respect to a mineral in which this element is an essential component. Note also that concentration ratios of two elements with qualitatively similar behaviour but quantitatively different D values only diverge significantly at extremely low remaining melt fractions ($< 1\%$).

Rb, and K/Cs is reduced from 12 000 to as low as 20. Likewise, the ratios of some geochemically very similar elements (e.g. Al/Ga, Zr/Hf) vary little in most crustal rocks but can be abnormal in pegmatites because of small differences in partition coefficients between melt and minerals combined with large amounts of fractionation. Such ratios indicate that many pegmatite melts were the results of greater than 99% fractional crystallisation of typical granites.

Rare-metal pegmatites form at metamorphic pressures (3–5 kbar), broadly syn-metamorphically; the most typical timing being syn-orogenic to late-orogenic. Where it is possible to connect a pegmatite to a source pluton, this pluton comprises relatively strongly fractionated granite that crystallised at a similar depth as the pegmatite, and has relatively high concentrations of ore and volatile elements and is hence known as 'fertile'. The source fertile granite may be the upper layer of a larger stratified pluton. The reasons why some plutons develop a fertile phase are not fully understood. Some of the elements that are enriched in fertile granites, and also in pegmatites, can act as fluxes, including F as fluoride, P as phosphate PO_4^{3-}, and B as borate BO_3^{3-}. These fluxes, and additionally water, carbonate, and bicarbonate ions, disrupt the polymeric linkages of SiO_4 tetrahedral groups in silica-rich melt and hence reduce the solidus temperature. A high concentration of fluxes will also lower melt viscosity and hence facilitate segregation of the last small volumes of interstitial melt from crystals.

At some stage during crystallisation of the pluton, remaining melt intrudes pip-like or in sills or dykes into neighbouring metamorphic rock. The small volume of pegmatite melt rapidly loses heat and stalls where it becomes quenched. From the size of pegmatites and heat-conduction equations we can estimate that cooling through conduction would be within weeks to months.

Crystallisation is from the border inwards and progresses from complex multi-phase assemblages in the border zone to often monomineralic zones in the interior. The mineral assemblages indicate crystallisation at 650–450 °C, hence up to about 200 °C super-cooling below the solidus temperature of the melt and significantly below equilibrium temperatures of crystallisation. The 'depolymerisation' of the melt as a result of high concentration of fluxes reduces the rate of nucleation of silicate minerals. The combination of depolymerisation and super-cooling has two effects: mineral growth occurs on a small number of nuclei, hence promoting exceptionally coarse crystals; and feldspars and quartz crystallise episodically rather than simultaneously and thus form the internal zonation of the pegmatites.

A pegmatite melt is generally not saturated with respect to rare-metal minerals when it first leaves the source pluton, but progressive crystallisation of the pegmatite is itself a process of fractionation which concentrates many trace elements up to levels at which rare-element minerals are saturated. Experimental data indicate that 4–5 wt % Cs_2O is required to saturate a melt with respect to the caesium–aluminosilicate mineral pollucite, and approximately 2 wt % Li_2O to saturate spodumene.

The different classes of pegmatite can be related to different source granite compositions and hence to the settings of melting in the crust and to the geological make-up of the host terrain: LCT pegmatites, for instance, are associated with S-type or mixed S–I-type granites, which are peraluminous granites with Al concentrations in excess of those required to produce feldspars. These granites formed dominantly from crustal melts that incorporated partial melts of metasedimentary rocks, and hence are designated as S-type.

2.2.6 Ores formed through incorporation of a mineral from depth in the Earth into magma: diamond deposits in kimberlites and lamproites

Primary diamond ores occur in unique, rare and small-volume ultramafic rocks, which as a class are known as kimberlitic rocks. They occur as diatremes, dykes and small pipe-shaped intrusive bodies and rarely in extrusive bodies (tuffs etc.). Kimberlitic rocks form from alkaline, volatile-rich, potassic ultramafic magmas that are formed as small-degree partial melts of carbonate-bearing and hydrous-mineral-bearing mantle peridotite. Kimberlitic rocks are characterised by inequigranular textures with macrocrysts (0.5–10 mm), megacrysts and xenolith clasts in a fine-grained igneous matrix. The three types of diamondiferous kimberlitic rock are:

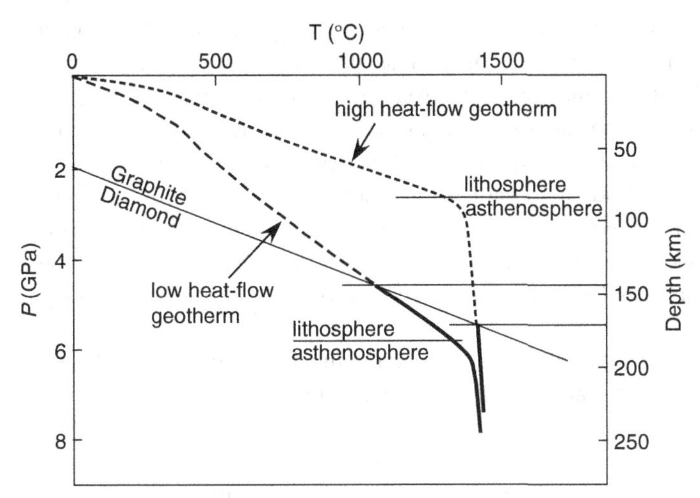

Figure 2.36 Pressure–temperature stability field of diamond, the depth of stability of diamond and the lithosphere–asthenosphere boundary along different geotherms and hence in different tectonic (thermal) settings. The diamond stability field is reached at between 140- and 180-km depth. The top of the diamond stability field is either in the lower lithosphere where there is a low geothermal gradient, or in the asthenosphere where there is a high geothermal gradient.

- Group I kimberlite – recognised worldwide – macrocrysts are dominated by olivine with lesser Mg-ilmenite, pyrope, diopside, phlogopite, enstiate and chromite in an olivine (or now serpentine)–carbonate matrix.
- Group II kimberlite (or orangeite) – only recognised in southern Africa – macrocrysts are dominately phlogopite with lesser olivine in an olivine–mica groundmass.
- Lamproite – recognised in Australia and India, and possibly elsewhere – major minerals are Ti-phlogopite, Ti–K-richterite, olivine, diopside, leucite and sanidine.

All kimberlitic rock types are susceptible to alteration during intrusion and cooling and after cooling at the Earth's surface, such that olivine is generally replaced by serpentine. Diamonds from these ores may be released by weathering and erosion, and selectively transported and deposited to form placer ores – see Section 5.2.

Formation and distribution of diamond

Diamond is the high-pressure polymorph of elemental carbon, and is stable relative to graphite at pressures of greater than about 4.0 GPa (40 kbar) in the Earth, and hence at depths of greater than about 150 km – the exact minimum depth limit of stability being dependent on the geothermal gradient (Figure 2.36). Its extreme hardness and extreme resistance to chemical breakdown is a result of its three-dimensional network of covalent C–C bonds.

Typical diamond grades in economic kimberlite and lamproite deposits are 10 to 100 carats per 100 tonnes (1 carat = 200 mg). This grade will include both gem-quality and industrial stones, and the value per carat can vary widely between deposits.

Diamonds in primary deposits are fragments of the mantle brought to the Earth's surface as xenocrysts or in xenoliths in the host magmas. There are a number of lines of evidence which show that diamonds are not cognate with the magmas (that is, they neither crystallised from the magma as phenocrysts nor were restite minerals): the source rock of the magma is hydrous and carbonate-bearing peridotite, but where diamonds are found in xenoliths in the kimberlitic rocks, they are in anhydrous peridotite or in eclogite; the ages of formation of diamonds within a single deposit are very variable and the diamonds are much older than the host intrusion; and different types of diamond with different suites of mineral inclusions and crystal form occur in one pipe. Diamonds are differentiated into P-type (derived from peridotite) and E-type (from eclogite) and different pipes have different proportions of E-type and P-type diamonds.

Based on diamond abundance in xenoliths in kimberlitic rocks we can estimate diamond grades in mantle in the diamond stability field to be 0.5–650 c per 100 t for peridotite and 17–34 000 c per 100 t for eclogite in the mantle. Diamonds are thus generally more abundant in the mantle than in kimberlitic ores. Diamond deposits fall outside the generality that element or mineral concentration is required to form ore.

Nature of diamond-bearing kimberlitic ores

There have been no recorded kimberlite eruptions. Figure 2.37 shows our view of what a typical funnel-shaped kimberlite diatreme would look like immediately after formation, built up from observations of kimberlites eroded to different depths. In reality the shapes are more complex, with multiple pipes feeding a diatreme. Some diatremes are bowl-shaped rather than the 'champagne glass' shape of the figure.

Most mines are exploiting a diatreme or its root. There is typically only a small volume of ore on the dykes or 'blows' along these dykes that were the pathways of magma below diatremes, but some have sufficiently high grades of diamond to be economically mined. Mines on diatremes are larger, with areal extents of 1–100 ha (10 000 m^2–1 km^2) and depth extents to about 800 m, whereas dykes are typically at most a few metres wide. Intrusions generally occur in clusters of 2–20 pipes or dyke swarms scattered over fields of up to 50 km across. These clusters mark single magmatic episodes in which multiple pipes intruded within time spans of at most a few million years. A few kimberlitic intrusions are apparently isolated rather than members of clusters.

Diatremes and pipes in particular, but also dykes, are internally complex with multiple kimberlite rock types, including volcaniclastic rocks, fall-back breccias and coherent intrusions. There can be multiple cross-cutting phases of each rock type (Figure 2.37) and these need to be distinguished by such features as xenolith clast type. The different phases within a pipe are probably all part of a single eruption and as such are estimated to have formed over time frames similar to volcanic eruptions in common igneous rock types of hours to months. Although differences between the phases are subtle, distinction of the phases and mapping of diatremes and pipes is important because diamond grade and value differ between phases, and also for instance with depth within a phase, and the differences thus control the economics of mining.

Diatremes form from the surface down to about 1-km depth. Their formation is either the result of phreatomagmatic processes or eruption processes. Phreatomagmatic diatremes form where kimberlite magma at 900–1100 °C heats near surface groundwaters. The heated groundwater flashes to steam and induces hydro-volcanic eruptions to form

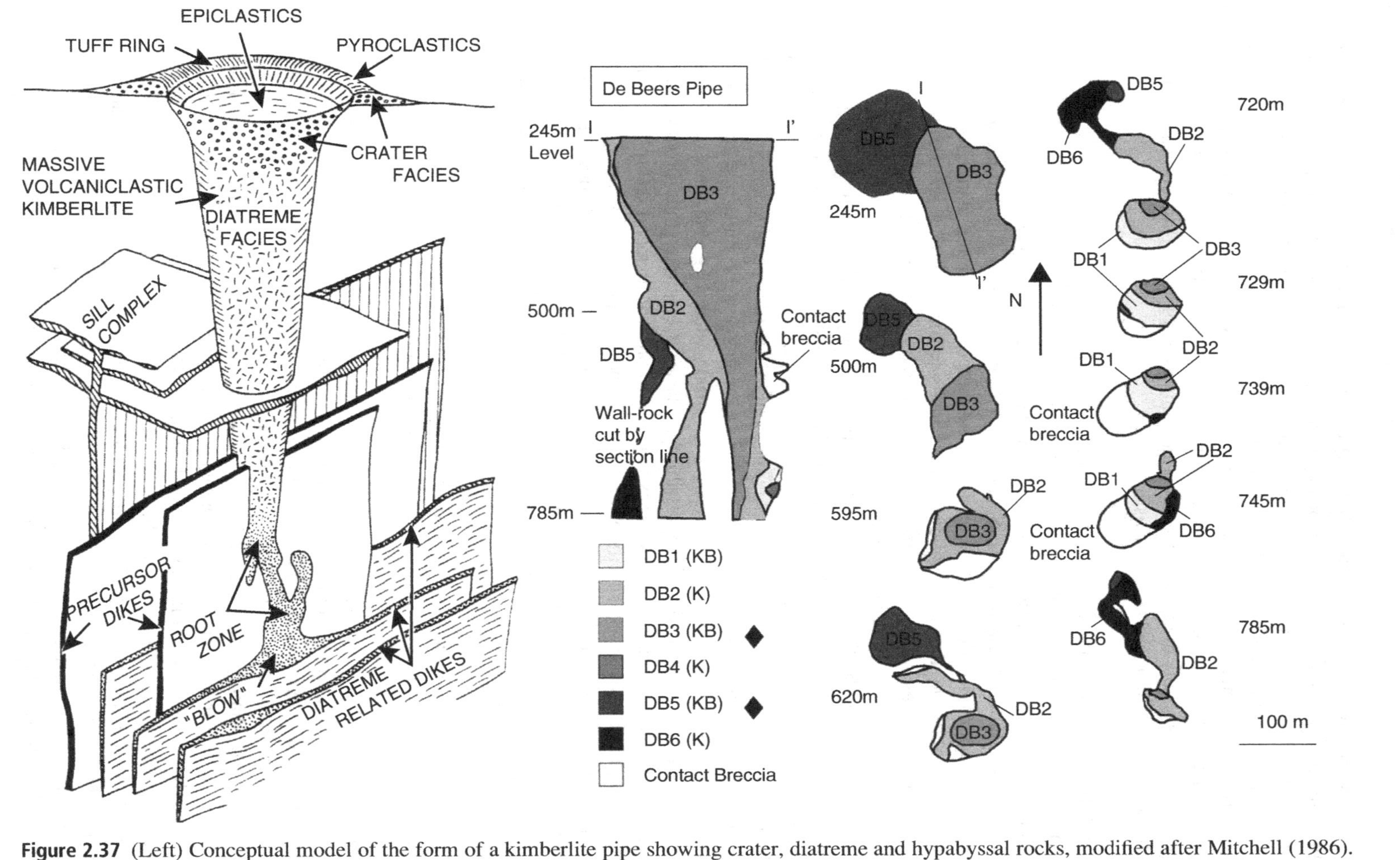

Figure 2.37 (Left) Conceptual model of the form of a kimberlite pipe showing crater, diatreme and hypabyssal rocks, modified after Mitchell (1986). (Right) Cross sections and plan views of the large De Beers diamondiferous kimberlite diatreme of South Africa showing the typical narrowing with depth, and the presence and distribution of multiple phases of kimberlite and kimberlite breccia within the pipe, modified from Field *et al.* (2008). K = massive kimberlite; KB = kimberlite breccia. Contact breccia is brecciated wall-rock with only a minor fraction of kimberlite in fractures and between clasts. The diamond symbols indicate which phases had high diamond grades.

breccias and excavate a maar crater. Magmatic processes can form a diatreme because of pressure fluctuations during eruption of a pyroclastic column into the atmosphere. Early in an eruption, magma pressure is high, but the pressure drops after eruption of much of the pyroclastic column, and this drop in pressure can induce inward collapse of the conduit wall to form magmatic breccia. Pressure fluctuations in the volatile-rich magma during ascent, induced for instance by obstacles to magma rise or constrictions along the magma pathway, are probably the cause of the internal complexity of kimberlitic bodies with their characteristic multiple pulses of intrusion.

Controls on the formation of diamondiferous kimberlitic rocks

(a) Processes of formation and ascent of kimberlitic magmas

Kimberlitic magmas form as a result of low degrees of partial melting of CO_2–H_2O-bearing peridotite (cf. carbonatites, Section 2.2.1). Low degrees of partial melting are a requirement for formation of these magmas, and temperatures in the mantle must therefore be only a little above the solidus. Partial melts formed at higher temperatures and greater percentages of partial melting would be basaltic. Low-percentage partial melts would have high CO_2 and H_2O concentrations ($>$ 5 wt % H_2O and 5 wt % CO_2), as is consistent with kimberlite chemistry and mineralogy. High volatile contents mean that the melts have a low viscosity and a low overall density if exsolved volatiles are retained in the melt as bubbles. Both of these characteristics would promote rapid magma rise. The buoyancy force for ascent may be self-reinforcing as volatiles become less soluble with decreasing pressure, exsolve and hence cause a decrease of bulk magma density. Reinforcement of buoyancy will be especially strong when large amounts of CO_2 are exsolved, as this has a high compressibility. These effects may thus promote rapid intrusion through the upper mantle, as is required to prevent diamond resorption during transport. The high compressibility of CO_2 may further promote unsteady magma ascent and hence pressure fluctuations in the magma, and may thus be a factor in promoting wall-rock brecciation and incorporation of abundant xenoliths and xenocrysts for instance through bubble nucleation. The xenoliths are typically derived from restricted depth intervals along the conduit.

(b) Which kimberlitic rocks contain diamonds?

About 10% of kimberlites and lamproites known worldwide are diamond-bearing and about 10% of these have sufficient concentrations of diamonds to have been developed as mines. A critical factor in the generation of a diamond deposit in ultramafic rocks is that diamonds are picked up during transport in xenoliths and xenocrysts. Most of the xenoliths and xenocrysts are picked up in the more brittle mantle lithosphere because this is where the wall-rock has been fragmented.

Kimberlitic diamond deposits and also major alluvial diamond deposits known before about 1970 are in restricted areas of the world – Southern Africa, Siberia, India, Brazil, West Africa. All diamondiferous kimberlites intruded into cratons – areas of ancient stable continental crust that has not been incorporated into or reworked in orogenic belts for at least 1600 million years. This observation is Clifford's Rule: 'significant diamondiferous kimberlites occur only in ancient shield regions, including Archaean cratons and Palaeoproterozoic mobile belts (orogenic belts) that border Archaean cratons, and were themselves undeformed since the end of the Palaeoproterozoic era' (see Figure 2.38).

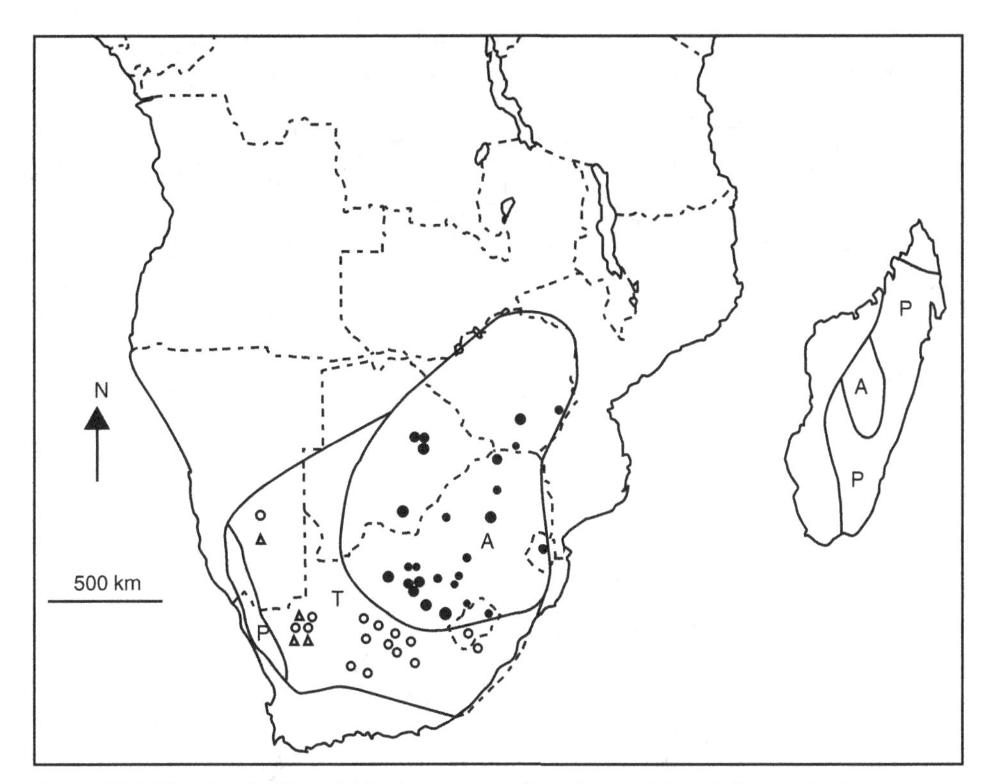

Figure 2.38 The distribution within the cratons of southern Africa of diamond mines on kimberlite pipes (filled circles), differentiating large and small mines. The open symbols show the locations of non-diamondiferous kimberlites and related ultramafic intrusions (open circles) and alkaline lamprophyres (open triangles). The tectonic provinces are after Janse (1994). A, P and T refer to a division of the cratons by time of latest tectonism: A = Archon, undeformed since 2.5 Ga and is the core of the Kaapvaal Craton, P = Proton, Proterozoic orogenic belts undeformed since 1.6 Ga, and T = Tecton, orogenic belts undeformed since 1.0 Ga. The terrains surrounding the craton are late-Proterozoic and early Phanerozoic orogenic belts. Note that all known diamondiferous pipes intruded through the Archaean craton and associated underlying lithospheric mantle.

Kimberlites that intrude younger orogenic belts are barren. The many diamond-deposit discoveries since the formulation of Clifford's Rule that have been made in other parts of the world (e.g. northern Canada, Australia, Finland) all fit the rule. The rule is a guide for exploration. Note, however, that although kimberlitic rocks intrude into ancient shields, the intrusions are much younger.

We can explain Clifford's Rule, and hence have confidence in its validity, through combining data on the thicknesses and make-up of the sub-continental lithosphere and on the geotherm (Figure 2.39). The thickness of the sub-continental lithosphere mantle (SCLM) has been mapped approximately using seismic waves and geothermometric petrological data to trace the origin of xenoliths brought up in kimberlites. In almost all cases the SCLM is significantly thicker (extending to 150–250 km depth) below Archaean cratons than below younger orogenic belts (60–150 km). Geothermal heat flow is

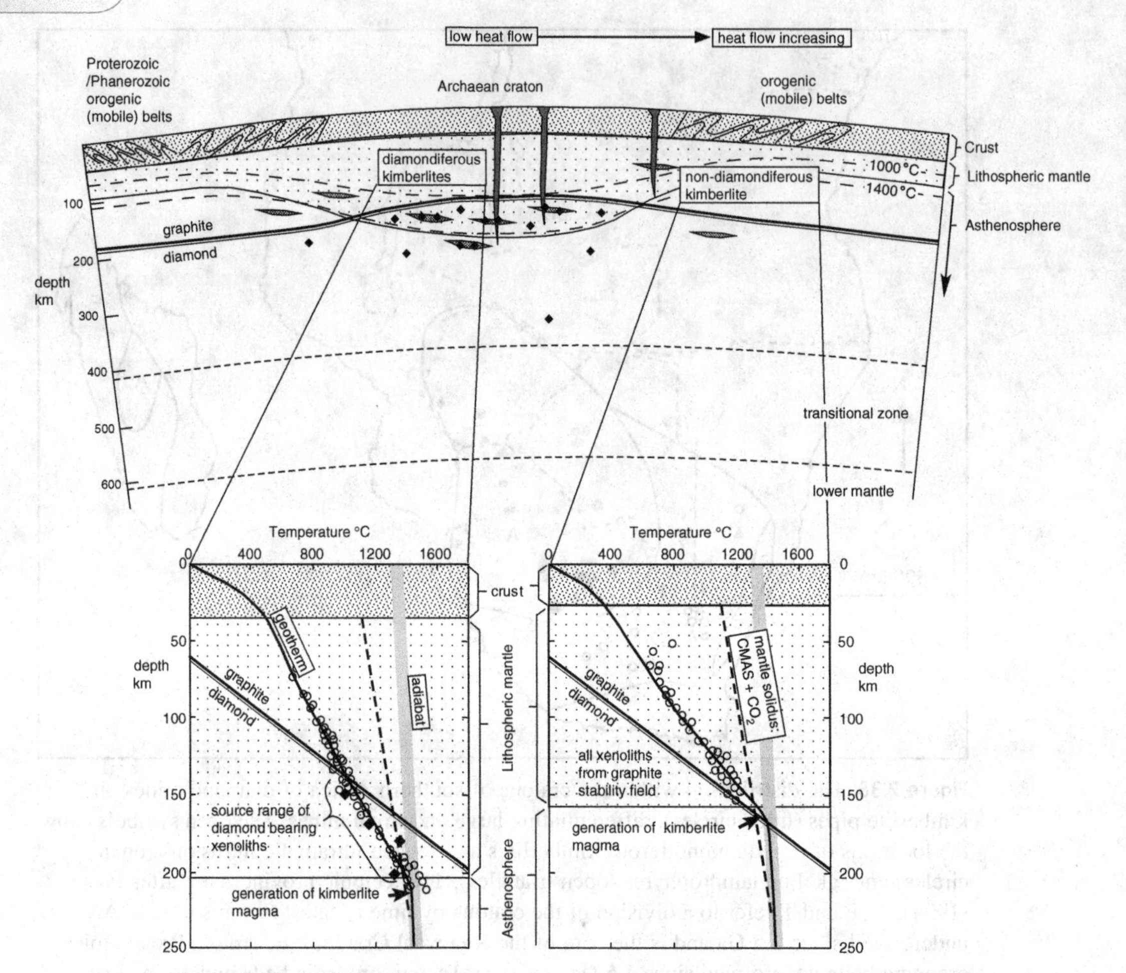

Figure 2.39 Geothermal and tectonic settings of diamondiferous and non-diamondiferous kimberlites in, respectively, thick sub-continental mantle lithosphere beneath Archaean cratons and thinner lithosphere beneath adjacent orogenic (mobile) belts, and hence explaining Clifford's Rule (see text). The geotherms are based on typical values of geothermal heat flow through the different types of terrain.

relatively low through Archaean cratons. Extrapolations of thermal gradients measured near surface (to 1–4 km depth) indicate that temperatures are lower at any depth in the upper mantle beneath Archaean and Palaeoproterozoic crust and higher beneath younger orogenic belts.

Because of the different geothermal gradients, the graphite–diamond transition lies within the lithosphere under some cratons, but in the asthenosphere under others and under all younger orogenic belts. Kimberlitic magmas may be generated either in the asthenosphere or in the lowermost lithosphere, but only those magmas that pass through lithosphere which is thick enough and cold enough that its base is in the diamond stability field can pick up diamonds as xenocrysts or in xenoliths of peridotite and eclogite. In addition, to preserve diamonds, the magma must ascend fast enough from the depths of diamond stability so that diamonds are not resorbed as they pass through the graphite stability field during ascent. Magma transport is expected to be faster as dyke-like intrusions through the brittle lithosphere than as percolations through the asthenosphere.

Concentrations of diamonds are presumably variable in the lithosphere mantle, and this will affect formation of economic diamondiferous kimberlites even in favourable thermal regimes. The oxidation state of the mantle (f_{O2}) can influence whether diamonds, rather than for instance carbonate minerals, are stable. Carbon in the mantle may be juvenile, and incorporated into the silicate fraction of the Earth at the time of primary aggregation, or be introduced as a result of subduction. Metasomatic addition of volatile elements and also incompatible elements to mantle peridotite, for instance by fluids released as a result of metamorphic reactions in down-going slabs in much older subduction zones (a process discussed further in Box 3.3), has been suggested to promote the formation of the unusual melts of kimberlites.

(c) Why and when do diamondiferous kimberlitic rocks intrude?

Clifford's Rule and the thermal structure of sub-continental lithosphere explain in what parts of the crust diamondiferous kimberlites are intruded, but does not explain when kimberlites intrude or what tectonic environments promote the formation of kimberlitic magmas.

The kimberlites shown in Figure 2.38 are, for instance, the result of multiple discrete phases of intrusions in the Kaapvaal Craton from 1600 Ma to 70 Ma. The youngest phase corresponds with the age of initial rifting of the South Atlantic, suggesting that a modest rise in mantle temperature at the margins of a zone of upwelling mantle may have induced kimberlitic magmatism in this area and at this time. Similarly subtle thermal events, in some cases at the peripheries of major tectonic events, are inferred elsewhere in the world. Flexure of the lithosphere due to loading by crustal thickening in an active orogenic belt, may be another cause, and may be an explanation of the intrusion of the Miocene Ellendale diamondiferous lamproite pipes in northern Western Australia.

A further requirement for formation of a primary diamond deposit is a limited amount of erosion (< 1 km) after intrusion. This requirement is not for formation of the kimberlite, but in order to preserve the diatreme and blow facies. This is thus an additional reason why economic diamondiferous kimberlites are hosted in stable Archaean and Palaeoproterozoic cratons.

Questions and exercises

Exercises

2.1 An example calculation of the mass balance of chalcophile element partitioning into immiscible sulfide melts.

Data:

(i) Concentration of Ni in the melt before sulfide separation = 1000 ppm

(ii) Partition coefficient $K_i^{sulf/sil} = 100$

Calculate the concentration of Ni in the silicate melt and in the sulfide melt for 0.01%, 0.1% and 1% sulfide melt. What trend do you determine for the wt % Ni in the sulfide melt phase?

The calculations you have completed are related to the R-factor that is discussed with respect to the tenor and hence economic value of magmatic sulfide deposits. R is the normalised effective mass of silicate melt with which the unit mass of sulfide melt equilibrates. R depends on both the S content of the starting magma and the efficiency of interaction the two melt phases after immiscible separation – sulfide droplets could for instance settle through the silicate magma too fast to collect diffusing chalcophile elements. Based on your calculations what factors and processes do you predict controls Ni (and also Cu and PGE) concentrations of magmatic sulfides?

2.2 Estimate the number of 1 carat diamonds per m^3 of diamond ore, assuming a grade of 100 c t^{-1}.

2.3 Using principles of chemical thermodynamics to estimate the depth of the field of diamond stability.

Principles and data:

If there are two polymorphic minerals with the same chemical composition:

- The stable mineral of polymorphs is that with the lowest chemical potential energy – Gibb's Free energy, G (J mol^{-1}). At surface pressure and temperatures conditions (25 °C, 1 bar), the G of diamond is 2.9 kJ mol^{-1} greater than that of graphite.
- The polymorph with the lower molar volume will become relatively more stable with increasing pressure. The change in energy with pressure is analogous to the relation 'work = force × distance', and in this case can be written as $\Delta G = V\Delta P$, where V is the molar volume of the mineral.

$V_{\mathrm{graphite}} = 5.2982\ \mathrm{cm}^3\ \mathrm{mol}^{-1}$, $V_{\mathrm{diamond}} = 3.4166\ \mathrm{cm}^3\ \mathrm{mol}^{-1}$.

Method:

(i) Recalculate the molar volumes in units of J $\mathrm{bar}^{-1}\ \mathrm{mol}^{-1}$.

(ii) Calculate the increase of G for each mineral per kilobar pressure.

We can see that G_{diamond} increases at a slower rate with increasing pressure than G_{graphite}.

(iii) Using the difference in G at surface conditions, determine the pressure at which $G_{\mathrm{diamond}} = G_{\mathrm{graphite}}$ at 25 °C.

(iv) Use typical densities of mantle rocks to estimate the depth that this pressure is attained in the Earth.

You have made a number of simplifying assumptions in the calculation. The best value based on experimental data is about 21 kbar (or 2.1 GPa). What simplifying assumptions may have affected the accuracy of the calculation?

2.4 Use the liquidus diagram for mafic rocks (Figure E2.1) to show that there is no liquid-line of descent which could give rise to a monomineralic layer of spinel (= chromite) in the interior of a differentiated layered intrusion.

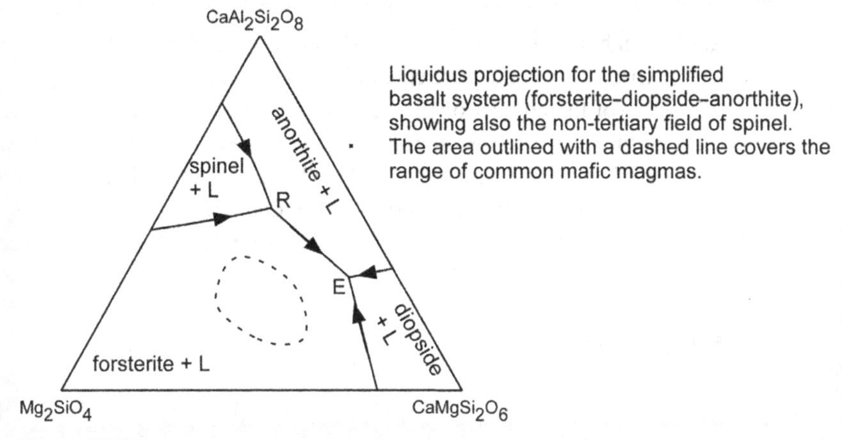
CaAl2Si2O8
spinel
+ L
R
anorthite + L
E
diopside
+ L
forsterite + L
Mg2SiO4
CaMgSi2O6
Liquidus projection for the simplified basalt system (forsterite-diopside-anorthite), showing also the non-tertiary field of spinel. The area outlined with a dashed line covers the range of common mafic magmas.

Figure E2.1.

2.5 (i) From reading or from your own observations in the field, give a summary description of an example of a rare-metal pegmatite. State three ore minerals that you might find in the pegmatite, stating the ore metal(s) or other economic commodities. Briefly describe the critical magmatic processes that give rise to its formation.

(ii) Some metals that are enriched in pegmatites are also enriched in carbonatites. Which metals are these? Explain the fundamental magmatic process of this phenomenon.

2.6 A number of different ore deposit types can be found in large layered ultramafic–mafic intrusions. Describe or give a labelled sketch of a schematic intrusion showing likely positions of the different ore types. Give a brief description (ore-body shape, whether ores are massive or disseminated, what major ore minerals would be present) of the nature of each ore type you have shown in your sketch.

2.7 Briefly describe and explain the critical magmatic processes involved in the formation of primary diamond deposits. Your description of primary deposits should include consideration of the lithospheric structure that promotes intrusion of diamondiferous rocks.

Discussion questions

2.8 There are distinct time distributions of the different types of magmatic sulfide deposits. Komatiites are dominantly Archaean, with rarer Palaeoproterozoic examples. The ages of large intrusive hosted Ni–Cu deposits range in age from ≈ 1850 Ma in the case of Sudbury to ≈ 250 Ma for Noril'sk. There are large earlier intrusions (pre ≈ 2000 Ma), e.g. the Bushveld Complex, but these seem to lack accumulations of massive sulfide. List processes and events in the Earth's history that may have contributed to the time spreads of the deposit types, giving a short explanation of why each process and event may affect the sulfide deposits.

2.9 Chromite has been suggested to be a marker of proximity to sulfide ore in komatiites. Based on your knowledge of the processes that may give rise to crystallisation of chromite in magmatic systems, and the processes that are proposed to give rise to 'contact' massive sulfide ores in komatiites, explain why this may be so.

2.10 A gabbro sill cuts the kimberlite diatreme of the Premier diamond mine in South Africa. The contact aureole is 30 m wide and diamond is replaced by graphite within 1 m of the contact of the 120-m-thick sill. Does this information allow an estimate of the rate of intrusion of kimberlite that is required to prevent diamond inverting to graphite on the way to the surface? We can probably assume that temperatures need to reach greater than 1000 °C for diamond to invert to graphite. This is thus a question of how long such high temperatures may have been imposed.

Further readings

Background to critical geological processes behind the formation of the ore deposit types described and discussed in this chapter is given in the following articles:

(1) Carbonatites and LREE ores – the environment of formation of carbonatites:

Bell, K. and Simonetti, A. (2010). Source of parental melts to carbonatites – critical isotopic constraints. *Mineralogy and Petrology* **98**, 77–89.

Dalton, J. A. and Presnall, D. C. (1998). Carbonatitic melts along the solidus of model lherzolite in the system CaO–MgO–Al_2O_3–SiO_2–CO_2 from 3 to 7 GPa. *Contributions to Mineralogy and Petrology* **131**, 123–135.

(2) Chromite ores – the formation of chromitites in ophiolites and in layered intrusions and the processes behind their unique textures in ophiolites:

Ballhaus, C. (1998). Origin of podiform chromite deposits by magma mingling. *Earth and Planetary Science Letters* **156**, 185–193.

Kelemen, P., Shizimu, N. and Salters, V. J. M. (1995). Extraction of mid-ocean-ridge basalt from the upwelling mantle by focused flow of melt in dunite channels: *Nature* **375**, 747–753.

Spandler, C., Mavrogenes, J. and Arculus, R. (2005). Origin of chromitites in layered intrusions: evidence from chromite-hosted melt inclusions form the Stillwater Complex. *Geology* **33**, 893–896.

Zhou, M-F., Malpas, J., Robinson, P. T., Sun, M. and Li, J-W. (2001). Crystallization of podiform chromitites from silicate magmas and the formation of nodular textures. *Resource Geology* **51**, 1–6.

(3) Magmatic sulfide ores – the processes of magma formation that produce large magmatic sulfide deposits and processes in magma chambers:

Kruger, F. J. (2005). Filling of the Bushveld Complex magma chamber: lateral expansion, roof and floor interaction, magmatic unconformities, and the formation of giant chromitite, PGE and Ti–V–magnetitite deposits. *Mineralium Deposita* **40**, 451–472.

Maier, W. D., Li, C. and de Waal, S. A. (2001). Why are there no major Ni–Cu sulfide deposits in large layered mafic–ultramafic intrusions? *Canadian Mineralogist* **39**, 547–556.

Maier, W. D. (2005). Platinum-group element (PGE) deposits and occurrences: Mineralization styles, genetic concepts and exploration criteria. *Journal of African Earth Sciences* **41**, 165–191.

Mungall, J. E. and Naldrett, A. J. (2008). Ore deposits of the platinum group elements. *Elements* **4**, 253–258.

Naldrett, A. J. (2004). *Magmatic Sulfide Deposits. Geology, Geochemistry and Exploration.* Berlin, Springer.

(4) Ores in pegmatites – contrasting views over the importance of water in the formation and development of pegmatites and their textures:

London, D. (2005). Granitic pegmatites: an assessment of current concepts and directions for the future. *Lithos* **80**, 281–303.

Thomas, R. and Davidson, P. (2012). Water in granite and pegmatite-forming melts. *Ore Geology Reviews* **46**, 32–46.

(5) Primary diamond deposits – the formation of kimberlites and the causes of their unique eruptive characteristics as igneous rocks:

Bailey, D. K. (1980). Volatile flux, geotherms, and the generation of the kimberlite–carbonatite–alkaline magma spectrum. *Mineralogical Magazine* **43**, 695–699.

Faure, S., Godey, S., Fallara, F. and Trepanier, S. (2011). Seismic architecture of the Archean North American mantle and its relationship to diamondiferous kimberlite fields. *Economic Geology* **106**, 223–240.

Pearson, D. G. and Wittig, N. (2008). Formation of Archaean continental lithosphere and its diamonds: the root of the problem. *Journal of the Geological Society of London* **165**, 895–914.

Sparks, R. S. J., Baker, L., Brown, R. J., Field, M., Schumacher, J., Stripp, G. and Walters, A. (2006). Dynamical constraints on kimberlite volcanism. *Journal of Volcanology and Geothermal Research* **155**, 18–48.

Wilson, L. and Head, J. W. III (2007). An integrated model of kimberlite ascent and eruption. *Nature* **447**, 53–57.

3 Hydrothermal ore deposits I: magmatic and orogenic environments

A hydrothermal ore deposit is one in which the ore minerals were precipitated from **aqueous** high-temperature **fluid** solutions, where:

- Aqueous implies that the solvent is water. The fluids are solutions, and can have salinities up to several times that of seawater. They are in some cases concentrated brines in which salts form more than half the solution by mass. Additionally the fluid may be a solution of water and dissolved gases (e.g. CO_2).
- High temperature can be from normal rock temperatures at a few kilometres depth ($\approx$ 100 °C) to magmatic temperatures ($\approx$ 800 °C).
- Fluid is a liquid or vapour or gas. These four words are used somewhat interchangeably. Most hydrothermal fluids at high temperature and pressure have a density intermediate between that of liquid water and that of water vapour at surface conditions. At temperatures greater than the critical temperature of pure water (376 °C), liquid-like pure water will not boil with either decreasing pressure or increasing temperature, but will steadily become less dense (Figure 3.1). The term **supercritical fluid** is used for these environments. Saline waters can boil at higher temperatures to produce a dense saline brine and a less-dense weakly saline fluid or vapour, and this fact affects the evolution of the highest temperature hydrothermal fluids that are associated with magmas.

Ore minerals may precipitate in the subsurface in void space in rocks or by **replacement** of other minerals, or at the interface between rocks and water bodies or the atmosphere (e.g. on the ocean floor).

Hydrothermal ore deposits are by value the most important class of deposit, and are the dominant deposit types for many metals of major economic interest (Cu, Au, Zn, Pb, U, Ag, Sn and Mo).

Types of hydrothermal ore deposits

Hydrothermal ore deposits are separated across two chapters in this book in the following types and sections:

(i) Hydrothermal deposits that have a close spatial and temporal association to magmatic activity (Sections 3.2.1 to 3.2.7).
(ii) Hydrothermal deposits which form during periods of regional magmatism and tectonism, but are not clustered close to or around magmatic centres. (Sections 3.3.1 to 3.3.3).
(iii) Hydrothermal deposits which form in sedimentary basins that lack widespread or temporally related magmatic activity (Chapter 4).

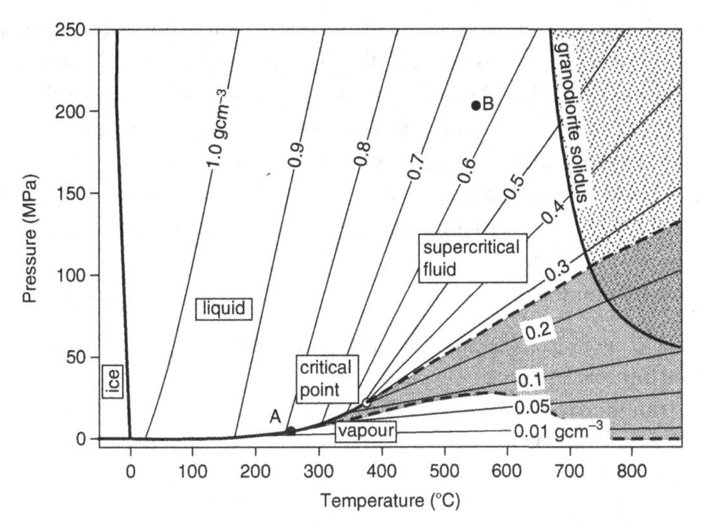

Figure 3.1 Pressure–temperature phase diagram of water over the range of conditions of hydrothermal ore deposit formation with contours of water density in g cm^{-3}. The light-shaded area shows the field of possible felsic melts. Water density varies from liquid-like at high pressure and relatively low temperature to vapour-like at low pressure. At conditions 'A', liquid water will boil to vapour if the pressure is reduced. At 'B' liquid water will not boil but will become steadily less dense on steadily decreasing pressure. This is the field of supercritical fluid. The dark-shaded area shows the field of possible saline liquid and less-saline vapour coexistence where NaCl and similar salts are in solution in the water.

3.1 Hydrothermal systems: generalities and background

Hydrothermal activity involves the transport of matter in aqueous solution in rocks. The term **hydrothermal system** is analogous to **petroleum system**. In the latter the source rocks of the hydrocarbons, the timings of maturation, migration pathways and traps of oil or natural gas are analysed. A hydrothermal system comprises the sources and sinks of the fluid, the sources and sinks of all chemical components transported in solution in the fluid or fluids, the complete pathways along which the hydrothermal fluids migrated through the crust and, additionally, the processes and events (tectonic, magmatic etc.) that caused fluid migration (the driving forces). Temperatures and pressures and their evolution through the lifetime of a system are also considered. Ore deposits extend over only small segments of any hydrothermal systems. The systems have dimensions of a minimum of a few kilometres vertically and laterally, where they develop around a magmatic centre, or as large as the scale of an orogenic belt or a sedimentary basin, and known lifetimes in some cases of millions of years.

3.1.1 What fluids form hydrothermal systems?

Waters from any source can in principle carry and precipitate ore minerals. We can divide hydrothermal fluids into those that rise from depth and those that are ultimately derived from the Earth's surface and flow downwards into rock. Fluids that are sourced from depth are sourced at pressures close to **lithostatic pressure** or rock pressure, and are thus **overpressured** with respect to most groundwaters. Where fluid pressures are approximately lithostatic pressures, the gradient in hydraulic head is such as to drive overall upward or oblique-upward fluid movement. Fluids derived from the surface have pressures close to **hydrostatic** pressures, that is, the pressure of the weight of the overlying column of water, and these fluids may migrate laterally or convect. Each of these two groupings can be subdivided to distinguish the following fluid types based on source.

Fluids derived from depth:

- Diagenetic and metamorphic fluids = fluids released from minerals through mineral **devolatilisation** reactions.
- Magmatic (**magmatic-hydrothermal**) fluids = fluids that were dissolved in silicate magma and is released from solution (**exsolved**) on decompression and/or crystallisation of the magma.

Fluids from the Earth's surface:

- Meteoric waters = groundwaters derived from the hydrosphere (rainfall, etc.) and heated on interaction with rock on percolation to depths of up to a few kilometres depth in the crust.
- Ocean waters – heated as a result of infiltration into the rocks of the ocean floor.
- Connate waters = waters buried with sediments in pore spaces.

Basinal and shield waters may also be differentiated. These are meteoric, ocean or connate waters that have circulated through and resided in, respectively, sedimentary or crystalline rocks such that their compositions have been modified by reaction with minerals. These are the fluids found at present at greater than about 1 km in many sedimentary basins, for instance in oil fields, and in crystalline rocks. Where these fluids are highly saline they are often termed basinal and shield brines.

More than one fluid may be involved in a hydrothermal system. The fluid type in a system may also evolve with time. Fluids other than aqueous fluids, most frequently hydrocarbon gases and oils, may also be present in hydrothermal systems and, although they are rarely considered to be carriers of ore metals, they may interact with aqueous fluids and influence their compositions.

3.1.2 Components of a mineralising hydrothermal system

We can distinguish five necessary components of a mineralising hydrothermal system (Figure 3.2).

(i) The fluid source and controls on fluid composition at source.

(ii) The source of the dissolved ore elements. In general, metals are both dissolved into the fluid at source and along the flow path through leaching and chemical reaction of rock through which the fluid flowed.

(iii) The pathway of fluid migration. This can be pore space in porous rocks. In crystalline rocks the pathways are more likely to be planar breaks (discontinuities) in the rock mass, that is, fractures, faults and shear zones. Fluid pathways are thus often

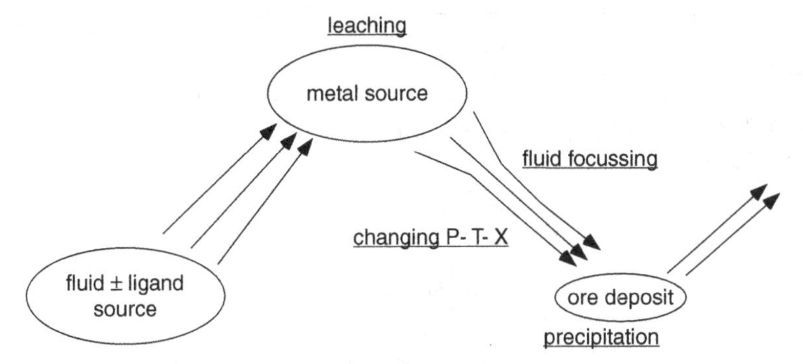

Figure 3.2 Schematic of a possible hydrothermal system. An alternative scheme would have the metal sourced and the fluid sourced from the same rock volume.

structures, and this fact gives rise to the concept of **structural control** on ore deposits (see Box 3.7).

(iv) The energy source to drive fluid migration. This may be thermal or mechanical, e.g. pressure variations caused for instance by deformation.

(v) The chemical driving forces for ore mineral precipitation. In general precipitation is a result of changing solubility of the ore mineral and hence one or more of changing temperature, pressure or fluid composition.

3.1.3 Mass-balance requirements for the formation of hydrothermal ore deposits

The hydrothermal system shown in Figure 3.2 incorporates two requirements for the formation of a hydrothermal deposit:

(i) Ore elements must first be dissolved into solution and later precipitated. Dissolution requires the fluid to be undersaturated with respect to the mineral, and precipitation the fluid to be oversaturated. At least one physical or chemical parameter of the fluid (P, T and X = composition) must change to cause precipitation of ore minerals. Changes in fluid composition may be the result of mixing of the fluid that is carrying the metal with a second fluid, or of chemical interaction with rocks. The hydrothermal fluid thus needs to move to form an ore deposit.

(ii) The fluids must be focussed from flow through a large volume of rock to flow through a small volume of rock at the ore deposit. Ore deposits are geochemical anomalies in which one or more ore element is present significantly above average crustal abundances. If ore metals were dissolved out of a rock, they must be dissolved from a much larger volume of rock, in many cases orders of magnitude larger, than the volume of rock at the ore deposit to obtain the required **Clarke of concentration** for economic ore (see Table 1.4).

3.1.4 Hydrothermal solution chemistry, aqueous complexes and mineral solubility

Most minerals are sparingly soluble in pure water at low geological temperatures. For instance, quartz has a solubility of a few ppm at 25 °C, and most silicate and sulfide minerals have solubilities in the ppb range. Common minerals that are exceptions to this

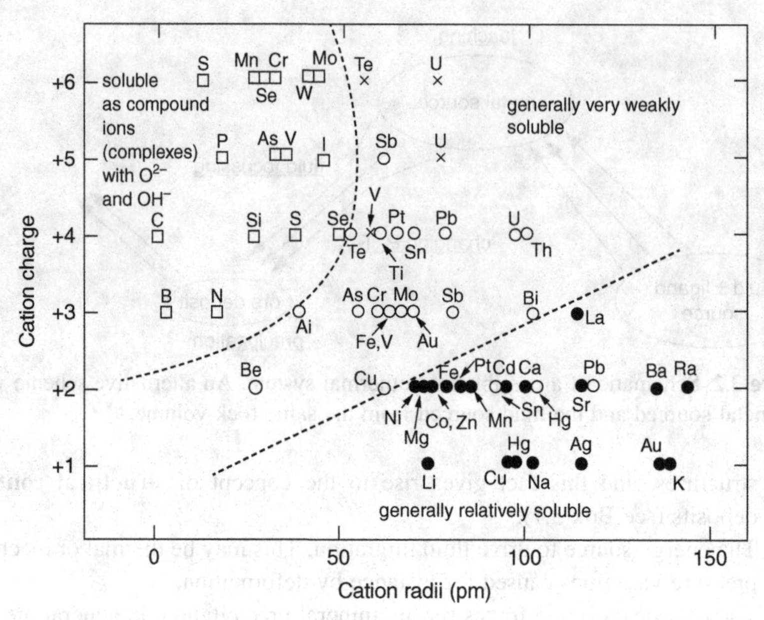

Figure 3.3 Cation charge vs. cation radii for ions in aqueous solution. Note the multiple valences of some elements. The symbols indicate the most common bonding of the ion by water in dilute aqueous solutions at near-ambient temperatures: filled circles, cations and aquocations; open circles, hydroxycations and hydroxyanions; crosses, oxycations; open squares, oxyanions. Three fields are distinguished. Ions that fall in the central field (mainly hydroxyions) are generally weakly soluble in near-neutral aqueous solutions, and hydrothermal ore deposits rarely form from these ions, although exceptions occur where hydrothermal fluids are of unusual chemical composition with high concentration of specific ligands. (Modified after Langmuir, 1997, which should be consulted for additional information.)

rule are the alkali halide salts (e.g. halite) and to a lesser extent carbonate minerals. Mineral solubilities tend to be higher at high temperature, and especially at high temperature combined with high pressure. For instance, quartz has a solubility of 1 g kg^{-1} or 1000 ppm at about 400 °C and 1 kbar, and higher solubility at higher temperatures and pressures.

Solubilities of most minerals are low because the component ions have low chemical stability in water. Many ions are too small and too highly charged to be electrostatically bonded to water molecules (Figure 3.3).

The solubility of many minerals is dramatically increased in waters that contain **ligands**. The metal ions bond with one or more ligands to form polyatomic ions or neutral polyatomic units (aqueous **complexes**) that are more stable in solution in water than the free ion. This effect is most important for the solubility of transition-row metals.

As an example, a simplified dissolution reaction of sphalerite, the most common ore mineral of Zn, in pure water can be written as:

$$ZnS + 2H^{+}_{(aq)} \leftrightarrow Zn^{2+}_{(aq)} + H_2S_{(aq)}. \qquad (3.1)$$

In water with a chloride salt in solution, the Zn would form chloride complexes, for instance a dichloro complex:

$$ZnS + 2H^+_{(aq)} + 2Cl^-_{(aq)} \leftrightarrow ZnCl_{2(aq)} + H_2S_{(aq)}. \tag{3.2}$$

Note that in both dissolution reactions, the sulfide ion bonds with protons. Many mineral dissolution reactions are protonation or deprotonation reactions (hydrolysis reactions) in which ions in solution compete with OH^- for bonding of protons and hence their presence influences the pH of the solution.

As metals are acids (cations), the ligands in a complex are bases and can be anions, negatively charged ion pairs, negatively charged molecular units, or molecular units with a negatively charged surface site. Common ligands in natural waters and hydrothermal fluids include:

$$F^-, Cl^-, Br^-, I^-, HS^-, SO_4{}^{2-}, HCO_3{}^-, NH_3, OH^-, HPO_4{}^{2-}, CN^-, CH_3CO_2{}^-\,(\text{acetate}), \text{etc.}$$

Example complexes are thus:

$$ZnCl_2,\ ZnCl_4{}^{2-},\ Au(HS)_2{}^-,\ Cu(NH_3)_4{}^{2+},\ Zn(CH_3CO_2)^+\ \text{etc.}$$

Molecular units with multiple metals such as

$$NaAuCl_2 \text{ or } NaAu(HS)_2,$$

are also possible.

As noted above, Cu, Au, Zn, Pb, U, Ag, Sn and Mo are dominantly extracted from hydrothermal ore deposits. A characteristic of all of these metals is that they can be present in aqueous solution in much higher concentrations as complexes than as free ions and are thus said to form stable complexes. Conversely, metals that do not form strongly soluble complexes with common ligands are rarely found in high concentrations in hydrothermal ore deposits. Examples include Ni, Ti, Sc and V. The different behaviours are related in part to charge and size of the metal ions (as illustrated in Figure 3.3).

Which complexes are important for which metals?

Which complexes carry the majority of a metal is a function of the ligands in solution and the relative abundances of ligands. There are incomplete experimental data for full prediction both of which metal–ligand bonds will form and the stability of complexes in natural hydrothermal solutions in which there are large numbers of solute species. As experiments on mineral solubility are difficult to perform at elevated temperatures and pressures, the limits to data are especially critical for these conditions. One general indication of which complexes are important for any metal is given by the hard–soft classification of metals and ligands (Table 3.1).

For instance, as chloride is the dominant anion in seawater and is known to be abundant also in most hydrothermal fluids (see Box 3.1), chloride complexes are generally predicted to be the ligand species of transition-metal complexes. Similarly, the soft metal ions Cu^+ and Au^+ are likely to be in solution as sulfide complexes.

Important general implications of aqueous complexation for the genesis of hydrothermal ore deposits are that:

(i) Metal solubility in hydrothermal fluids is critically dependent on fluid composition. For example, the concentration of the zinc dichloro complex $ZnCl_2$ in equilibrium with sphalerite (Equation (3.2)) increases in proportion to the square of the chloride content of the fluid, and hence the square of the salinity.

Table 3.1

Hard metals	Intermediate (borderline)	Soft metals
H^+, Li^+, Na^+, K^+, Be^{2+}, Ca^{2+}, Mg^{2+}, Sr^{2+}, Al^{3+}, Fe^{3+}, Cr^{3+}, La^{3+}	Divalent transition metals, including Zn^{2+}, Pb^{2+}, Bi^{3+}	Cu^+, Ag^+, Au^+, Hg^{2+}, Sn^{2+}, Tl^+, Ti^{3+}, Au^{3+}
Hard ligands		Soft ligands
F^-, **OH^-**, NH_3, NO_3^-, **HCO_3^-**, **CO_3^{2-}**, HSO_4^-, **SO_4^{2-}**, $H_2PO_4^-$, HPO_4^{2-}, $H_3SiO_4^-$	**Cl^-**, Br^-	I^-, **HS^-**, $S_2O_3^{2-}$, SCN^-, CN^-

Classification of metals and ligands with respect to hard–soft behaviour (as a measure of ionic vs. covalent behaviour). A metal will form a complex preferentially with a ligand of similar hardness (after Crerar *et al.*, 1985). Ligands in bold are typically the most abundant in geological environments, and are hence generally the most important complex formers.

(ii) Based on the concept of hard and soft acids and bases (HSAB) and preferences for complexation of similarly hard ions as summarised in Table 3.1, we expect elements that have similar solute characteristics to coexist in hydrothermal fluids and hence also in ore deposits (e.g. Pb and Zn). Conversely, elements with different solute characteristics (e.g. La and Au) are unlikely to be both enriched in the same hydrothermal system.

3.1.5 Products of hydrothermal fluid flow

Hydrothermal ores are only a small part of hydrothermal systems. In addition, the following features can form as results of hydrothermal fluid flow, and are hence markers of the passage of hydrothermal fluid through a rock pile.

Hydrothermal alteration

As hot aqueous solutions pass through rocks they induce mineralogical, chemical and textural changes in addition to those of ore mineral precipitation. Some gangue minerals will dissolve, others will precipitate. These changes are **metasomatism**. In ore deposit geology, metasomatism is generally referred to as **alteration**. Alteration may be pervasive, selectively pervasive (in which specific minerals or minerals in textural sites are replaced), or non-pervasive (for instance, only adjacent to fractures). In most hydrothermal systems, alteration affects a volume of rock many times that of the ore body. Alteration thus provides a much larger target for detection and mapping of a hydrothermal system than does a deposit and is commonly used as an indicator of proximity to ore in exploration.

The mineral assemblages in hydrothermally altered rock are controlled by a combination of temperature and pressure, fluid composition (concentrations of H^+, CO_2, Cl^-, B, F, Na^+, K^+, H_2S, etc. in solution), the host-rock composition before alteration, and also by the mass balance of rock and fluid (fluid/rock ratio). We define **alteration facies** on the basis of characteristic mineral assemblages in a similar fashion as metamorphic facies.

As discussed in the following sections, hydrothermal systems have characteristic fluid compositions and temperature and pressure environments. Many hydrothermal ore deposit types are thus correlated with specific styles, mineral assemblages, and hence alteration facies.

Veins

Veins occur both in hydrothermally altered rock and in ore. They are tabular bodies of minerals that have precipitated from hydrothermal fluids. In some older literature a distinction is made between **dilatant** veins and replacement veins, where the former implies mineral growth in open or fluid-filled space in the rock, and the latter implies a planar body of hydrothermally altered rock adjacent to a fracture or a channelway. Fractures can be present in the rock before hydrothermal fluid flow, can be formed as a result of tectonic stresses during fluid flow, or can be generated by the pressure of the fluid. **Hydrofracture** is the process of rock fracture by fluid that is at a higher pressure than the minimum principal rock stress (σ_3). This is the natural physical process equivalent to 'hydrofracking' that is employed by the hydrocarbon extraction industry to enhance rock permeability by inducing rock fracture.

There is a spectrum of modes of formation of veins from 'open systems' in which fluid is flowing through a rock, to 'closed systems' in which veins form as a result of local redistribution of fluids and solutes in a rock, for instance under a deviatoric stress. Where veins do mark pathways of fluid flow, the low concentration of minerals in solution means that the volume of fluid that migrated along the fracture is many orders of magnitude greater than the volume of minerals that are precipitated.

A wide variety of minerals are known as precipitates in veins. To a first approximation, the minerals that precipitate in a vein and their abundances are functions of (a) mineral solubility and (b) how solubility changes with the changes in physicochemical conditions of the hydrothermal fluid as it flows along the fracture, e.g. decreasing temperature. Quartz is an abundant vein mineral because SiO_2 has relatively high solubility at high P and T and its solubility decreases with decreasing fluid P and T under almost all circumstances. It hence precipitates in almost all environments in which a fluid that is saturated with respect to quartz is rising through the crust. Similarly, carbonate minerals are common vein minerals in veins formed in and around many hydrothermal ore deposit types because of their relatively high solubilities. The textures of minerals in veins are described further in Box 3.5.

Breccias

Breccias of different styles and textures are a common but generally minor rock type of many hydrothermal systems. They are, however, of particular importance as they are the hosts to ore in many ore bodies. Brecciation is a result of rock fragmentation in a pressure gradient. There are a number of different possible origins of a pressure gradient that will induce fragmentation. Common causes of fragmentation in rocks of hydrothermal systems are:

- Tectonic. The migration pathways of hydrothermal fluids are in many cases faults. Brittle faulting involves mechanical breakage of rocks, and fragmentation of the walls of faults may occur where faults develop with irregular fault planes. Rock fragmentation to form breccias that line and are adjacent to fault planes may be a result of abrasion during fault slip or of localised stress build-up at asperities along a fault plane.
- Collapse under gravity into open or fluid-filled space. Space in hydrothermal systems may be formed through subsurface migration of an underlying magma, or through migration of a hydrothermal fluid. Slip along a fault which has an irregular fault plane

may also generate space. Space may also be formed by corrosive leaching of rock into hydrothermal solution in the same fashion as karst caves in carbonate rocks.

- Explosive release of pressurised fluid. Hydrothermal fluids at depth may have a pressure higher than that of surrounding rock. If fluid overpressure is not released by slow fluid percolation, pressure can rise to a point of explosive release. A small decrease in fluid pressure can induce a large increase in fluid volume, especially where the drop in pressure causes fluid boiling through flashing to steam. This effect provides significant feedback to promote explosive release of fluid overpressure, most importantly in relatively shallow geological environments, and is in essence the cause of explosive volcanic eruptions. The release of pressurised fluid is in many cases to the Earth's surface (for instance the formation of hydrothermal eruptions and maars in **geothermal** and volcanic terrains), but it can also occur confined in a rock body at depth as a result of hydrofracturing.

Breccias that host ore bodies are commonly pipe-shaped bodies of rock, most typically sub-vertical, and less commonly tabular bodies, e.g. within a segment of a vein. Where breccias form in a hydrothermal system the matrix, or spaces between clasts, is commonly partly or completely filled with mineral precipitates similar to those that fill fracture veins in the hydrothermal system.

3.2 Hydrothermal deposits formed around magmatic centres

A large number of hydrothermal ore deposits of different metals are closely associated in space and time with magmatic centres of intermediate to felsic igneous rocks. Magmatic centres in the context of ore deposit geology are generally clusters of small intrusions such as dykes and stocks that mark the eroded roots of long-lived volcanoes. They are the sites of upward migration of magma at depths of a few kilometres in the crust. In many magmatic centres the small intrusions were derived from a larger pluton of which the top is exposed, or its presence at slightly greater depth can be inferred. In these cases, magmatic centres develop above dome-shaped protrusions (**cupolas**) or ridges on the roof of a pluton. In other cases there is no indication of an immediately underlying large pluton, and the small intrusions may have been sourced from significantly deeper in the crust. Multiple magmatic centres may form above large plutons.

At least two hydrothermal fluids of different sources will be present in and around magmatic centres:

(i) Magmatic-hydrothermal fluids: These are solutions of water and other **volatile** and soluble chemical components that were dissolved in the magma and that exsolved to form an immiscible aqueous fluid phase on decompression or on crystallisation. The exsolved aqueous fluid migrates or 'escapes' into the crystallised carapace of the intrusion and into and through overlying rock, in some cases reaching the atmosphere or hydrosphere. This process is sometimes known as 'magmatic degassing'.

(ii) Convecting groundwaters: A magma body within a few kilometres of the Earth's surface will act as a localised heat source and drive convection of groundwaters in surrounding crust. These waters may be meteoric, connate or, in the case of submarine volcanism, ocean waters.

Magmatic-hydrothermal fluids

Silicate melts can contain up to a few per cent by weight volatile components in solution. The term 'volatile' in the context of magmas describes those chemical components that emanate as vapours or gases from active volcanoes. Volatile components that are in solution in the magma and cannot be incorporated into volatile-bearing crystallising minerals (e.g. water in biotite) will be exsolved in a separate, immiscible, fluid phase at some stage in the rise and crystallisation of the magma. These components are mainly molecular compounds that form liquids or gases at surface conditions (~ 25 °C at 1 bar). Water is generally the dominant volatile component in a magmatic-hydrothermal fluid, but significant concentrations of CO_2, HCl, CH_4, H_2S and SO_2, amongst others, are also typically present.

The solubilities of water and most other volatile species in silicate melts decrease with decreasing pressure (Figure 3.4), and exsolution can thus occur as the magma rises through the crust (**first boiling**). At the same time, or after the magma has stopped rising, exsolution can occur due to increased concentration of the volatile elements in the residual silicate melt as the magma crystallises (**second boiling**).

At the time and place of exsolution the magmatic-hydrothermal fluid has the same temperature as the magma. The exsolved fluid may initially form dispersed immiscible bubbles in the magma, similar to those formed from immiscible sulfide phases. These bubbles will rise due to their lower density. If the fluid is released at or within about 1 km of the surface, it will have the density of a typical gas, as is observed in fumaroles and other hydrothermal features at active volcanoes. At higher pressures, it will have a density closer to that of liquid water. Under some conditions, a liquid-like- and a gas-like aqueous phase can simultaneously exsolve from a magma (Figure 3.1). An aqueous fluid phase may also coexist with an immiscible sulfide melt phase in the magma.

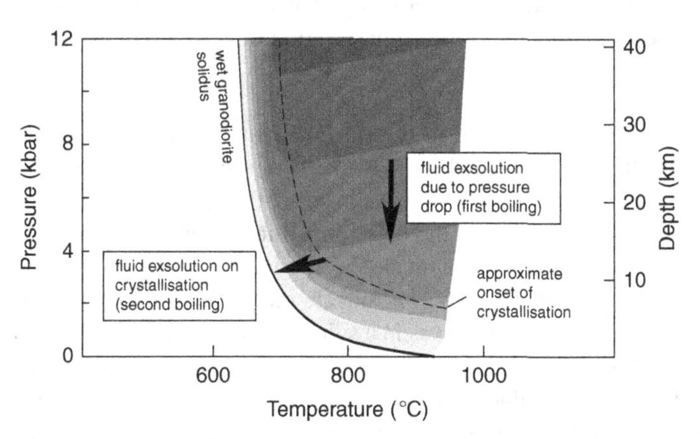

Figure 3.4 Qualitative representation of the variation of the solubility of water in a granitic magma with respect to temperature and pressure, shown by darkness of shading. Over the range of pressure and temperature that are shown the solubility will range from about 10 wt % at highest pressure to about 1% at the solidus. Solubility is taken to be the bulk solubility of water in the magma, hence both in the melt phase and in minerals that have crystallised from the melt. The onset of crystallisation as marked assumes water saturation in the melt.

Metal contents of magmatic-hydrothermal fluids

Elements will partition from a silicate magma into the magmatic-hydrothermal fluid depending on solubility and on equilibria between the solution, melt and minerals. Those elements that are both incompatible in crystallising phases and that have a high solubility in the fluid will be strongly partitioned into the fluid in a similar fashion as chalcophile elements are partitioned into sulfide melt bubbles. Solute species in fluids derived from typical intermediate to felsic magmas are dominated by soluble salts of metals, including chloride salts (NaCl, KCl, $FeCl_2$, etc.) and sulfate salts (Na_2SO_4, etc.). The total salt concentration and hence the salinity of the solution is largely controlled by the availability of anions, for instance, on the Cl^-:H_2O ratio in the magma.

Because of their high temperatures and high concentrations of multiple potential ligands, magmatic-hydrothermal fluids may dissolve greater concentrations of many ore elements than other geological fluids. Ores that form from magmatic-hydrothermal fluids have, however, very variable metal contents. The ores include, for instance, those which are mined for Au only and in which Sn is present at concentrations no higher than average crust, and ores that are mined dominantly for Sn with no enrichment in Au. Although in general we expect that there is a correlation between the metal contents of hydrothermal ores and the metal contents of the solutions, the fact that an ore body is mined, for instance, for Au and not for Sn does not necessarily indicate that the fluid at the deposit was 'Au-rich' and carried no Sn. An ore mineral precipitates from a migrating hydrothermal fluid where and when the mineral becomes saturated in the solution. Saturation can be reached either through changes in the physical conditions, for instance a decrease in temperature, or through changes in the fluid chemistry, for instance pH changes caused by acid–base reactions between the fluid and the rock through which it is migrating. The control of temperature on saturation means, for instance, that different minerals precipitate at different distances from magmatic centres as the fluid cools, giving rise to the ore metal **zonation** that is characteristic of many magmatic-hydrothermal systems.

Factors that potentially control the concentration of trace ore metals in a fluid that is exsolved from a magma include:

- The concentration of the metal in the magma at its source or sources. This will be controlled by the degree of melting and the composition and mineralogy of magma source rocks. The composition and hence history of the part of the mantle that is melted may be an important control on the ores formed in any area (see Box 3.2).
- The concentration of other chemical species in the melt, in particular of metal-complexing ligands (e.g. Cl^-, HS^-). A silicate melt is a chemically complex solution. Metals form coordination complexes in a silicate melt in a similar fashion as in aqueous fluids (see Section 3.1.2). Ligand concentrations and also parameters such as pH in the melt may thus strongly influence partitioning of metals both between minerals and melt at the site of melting in the source and between melt and fluid at the site of fluid exsolution.
- The degree and nature of fractionation in the magma at all stages along the magma pathway (Figure 2.1). The crystallisation of specific minerals can be important (Figure 3.5). Most of the metals of interest in ore deposit geology are trace components in magma, and the mathematical description of metal partitioning between melt, fluid and minerals is thus analogous to that discussed for melt–mineral fractionation in

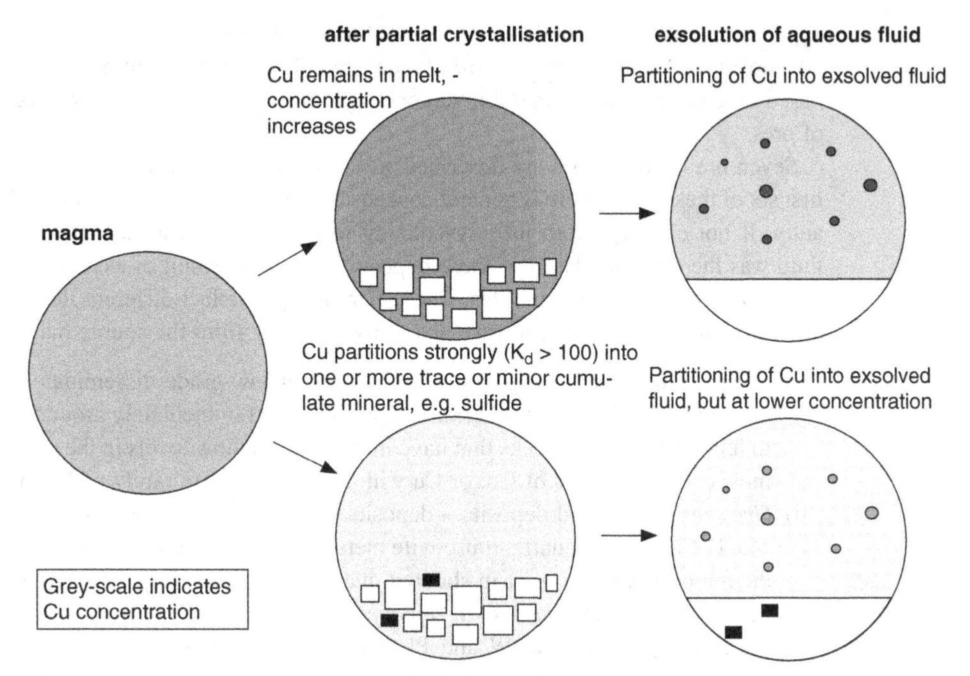

Figure 3.5 Schematic pathways of partitioning of Cu in a magma–fluid–mineral system. The upper line shows the case in which Cu is incompatible in all minerals crystallising in the magma, such that its concentration in the melt increases during fractional crystallisation. The metal's concentration in the exsolved magmatic-hydrothermal fluid will be high. In contrast, if one or more mineral with a high '$K_{d,Cu}$' crystallises from the magma, Cu will be sequestered, and Cu concentration in the remaining melt may decrease. The end effect will be for Cu to be dispersed through the whole pluton and the exsolved magmatic-hydrothermal fluid will have a relatively low Cu concentration.

Section 2.1. If the bulk solid to liquid partition coefficient of a trace element into a mineral multiplied by the modal percentage of the mineral in the crystallising assemblage is numerically greater than one (Equation 2.2), and the mineral is removed from contact with the remaining melt, e.g. into a cumulate pile at the base of a magma chamber, fractionation of this mineral will reduce the metal concentration in the remaining melt. Minerals which crystallise from the melt at all stages in the evolution and transport of the magma, including accessory minerals, are thus potentially important controls on the concentration of metals in magmatic-hydrothermal fluids.

- The history of fluid exsolution from the magma, including its timing relative to crystallisation and whether fluid release is continuous or takes place in batches. The timing of exsolution may be controlled by the depth of crystallisation and by the concentration of water and other volatiles in the magma. If fluid release is continuous, the fluid composition can change through the lifetime of the magma.

Empirical relations between magma chemistry and ore metal inventory in magmatic-hydrothermal systems are discussed further below in the sections on individual deposit types and in Box 3.3.

Types of hydrothermal ore deposit around magmatic centres

A number of common ore deposit types around magmatic centres are differentiated based on combinations of metal inventory, style of mineralisation, and geological setting of ores.

Seven ore deposit types are described and discussed in the following sections. For the first six of these types there is general consensus that the ore metals are sourced predominantly, if not entirely, from the crystallising magma and that a magmatic-hydrothermal fluid was the ore fluid. Each of these deposit types is the result of exsolution and escape (degassing) of magmatic volatiles. The different types reflect different fluid chemistries, and different styles of fluid release and transport away from the source magma.

(i) *Porphyry deposits* – large-tonnage deposits of low-grade disseminated ore associated with pervasive hydrothermal alteration in and immediately around intermediate to felsic porphyritic stocks that have intruded to shallow levels in the crust. These are most commonly ores of Cu, or Cu with Au or Mo, more rarely of Mo or of Sn–Ag.

(ii) *Greisens* and related deposits – deposits of predominantly Sn and W together with Mo, F, Li and B in quartz–muscovite metasomatically altered (greisenised) granite at the top of an intrusion, or in **sheeted** quartz veins in and adjacent to altered granite.

(iii) *Skarn deposits* and *carbonate-replacement deposits* – deposits of one or more of the metals Cu, Au, Fe, W and Pb–Zn in carbonate-bearing rocks that have been metasomatically replaced. We differentiate skarns in which the metasomatic replacement is by high-temperature calc-silicate gangue minerals and carbonate-replacement deposits in which the replacement is largely by sulfide minerals.

(iv) *Polymetallic veins*, including *vein fields* – swarms of veins centred on clusters of small intermediate or felsic intrusions, often extending over several kilometres from the intrusions, and often with a spatial zonation of metal content.

Epithermal deposits are a diverse set of deposits mainly mined as ores of Au, Ag and Cu that form at relatively low temperatures and shallow depths around volcanic centres and in geothermal fields. These are divided into:

(v) *High-sulfidation epithermal deposits* – these are closely associated with volcanic centres and are hosted in intensely altered rocks.

(vi) *Low-sulfidation epithermal deposits* – these have differing ore and alteration mineralogy, are more typically vein-hosted and generally more distal to volcanic centres.

Other rarer magmatic-hydrothermal ore deposit types that have been recognised include intrusion-related gold deposits (IRGD), also called reduced intrusion related gold deposits (RIRG) including Fort Knox, Alaska, USA, and Timbarra, New South Wales, Australia; Au deposits associated with strongly alkaline magmatic centres (Cripple Creek, Colorado, USA, Thompson *et al.*, 1985), and; Th–U–rare-earth element (REE) veins associated with alkaline granites such as at Capitan, New Mexico, USA.

The seventh important ore deposit type is one in which the volumetrically most important hydrothermal fluid is seawater that has convected through the crust of the sea floor driven by the heat of a shallow submarine intrusion. In these systems the metals are generally considered to be dominantly sourced by leaching from ocean-floor rocks into the convected ocean water and a magmatic-hydrothermal fluid plays at most a minor or secondary role in the hydrothermal system.

(vii) *Volcanic-hosted massive sulfide deposits* (VHMS, VMS deposits) – massive sulfide ores and associated vein ores formed at and just below the sea floor above a sub-sea-floor magma chamber.

We may find no ore deposits, one of the types listed below, or more rarely, more than one type around any magmatic centre.

3.2.1 Porphyry deposits

Porphyry deposits represent a repeated and distinct mode of hydrothermal fluid escape from large intrusions in the crust. The three most important commodities by value worldwide in these deposits are Cu, Mo and Au. They are the dominant world source of Cu (> 65% production) and Mo (> 95% of production) and account for significant proportions of production of Au, and also for Ag and Re as by-products. Porphyry deposits are large deposits with between 1 Mt and 10 Gt of ore in pervasively altered and veined rock in which ore minerals are uniformly disseminated at relatively low grades. Grades are commonly less than 1% Cu in a copper porphyry deposit, about 1 ppm Au where this is an economically important product, and about 0.1% Mo in a molybdenum porphyry deposit. The deposits are subdivided on the basis of metal content (Cu–Mo–Au), although most have above-crustal background concentrations of all three metals, and ore compositions form a spectrum between end members rather than clear clusters of types.

The common major characteristics of porphyry deposits are that:

- Mineralisation at porphyry deposits is hosted in many different rock types but is centred on and temporally related to small-volume intrusions, mostly of intermediate to felsic composition (52–77 wt % SiO_2). There may be multiple phases of minor intrusions in the deposit and in surrounding rock. At least one of these intrusions has a distinctly porphyritic texture.
- The large ore bodies are in the centres of even larger volumes of hydrothermally altered rock (> 10 km^3). Throughout the ore and much of the surrounding altered rock, pyrite is present at up to a few per cent of the rock as a prominent mineral. The hydrothermal systems have thus added a large amount of sulfide to crustal rocks in addition to ore metals.
- Ore and surrounding rock is characterised by closely spaced small veins and veinlets between which the rock is penetratively altered. There are commonly multiple generations of veins. These veins mark the pathways of hydrothermal fluid. Hydrothermal fluid flow is thus along dense networks of small fractures through large volumes of rock.
- Ore minerals occur both in veins and disseminated in altered rock. The important ore minerals are sulfides of Cu and Cu–Fe (chalcopyrite, bornite, chalcocite) and of Mo (molybdenite), native Au, and oxide minerals of W (scheelite, wolframite) and of Sn (cassiterite).
- The minor intrusions, related plutons and volcanic rocks are calc-alkaline to alkaline. Ore metal contents are somewhat related to tectonic setting and to the compositional characteristics of the host intrusive complex, although there is considerable overlap of the igneous rock compositions at deposits with different metals.

The most common porphyry deposits are those in which Cu is the most important commodity by value (Cu porphyries). Many of these deposits have concentrations of Mo of up to about 0.03% Mo and/or Au up to 1 ppm, either or both of which can constitute a secondary part of the value of the ore (Cu–Mo and Cu–Au porphyries). Typical Cu porphyries are associated with calc-alkaline diorite, quartz diorite, granodiorite intrusions of continental arcs and include the multiple examples in Chile and adjacent countries in the Andean chain (El Salvador, Gustafson and Hunt, 1975; Bajo de la Alumbrera, Ulrich and Heinrich, 2002; Proffett, 2003; Harris *et al.*, 2005), southwestern USA (Titley, 1982) and elsewhere (e.g. Oyu Tolgoi, Mongolia, Khashgerel *et al.*, 2006). Some porphyry Cu–Mo deposits are associated with calc-alkaline to quartz-poor alkaline monzonite, quartz monzonite, or granodiorite rocks in intra-ocean island arcs (e.g. Ok Tedi, Papua New Guinea).

Large Au-rich Cu–Au porphyries in continental arcs in which gold is of equal or greater value than copper are associated with igneous rocks of alkaline (sodic-alkaline or potassic) affinity, e.g. shoshonites (Cadia Hill, NSW, Australia, Holliday *et al.*, 2002). A relatively small number of Au-rich, Cu-poor porphyries occur. These are most commonly associated with relatively mafic intrusions such as diorite and quartz monzonites.

Molybdenum porphyries in which Mo is the main commodity by value are rarer. They can include co-product or by-product Cu, W and Sn. Two types are distinguished based on igneous rock association: quartz-monzonite-type deposits are associated with calc-alkaline to alkaline monzonite, quartz monzonite, granodiorite hosts in continental arcs (e.g. Endako, British Columbia, Canada); Climax-type Mo deposits are associated with alkaline high-silica granite and compositionally similar porphyritic intrusions in which F-bearing minerals fluorite and topaz are prominent (Henderson and Climax, Colorado, USA, Seedorff and Einaudi, 2004; Shannon *et al.*, 2004).

There are a small number of deposits which are very similar in style and setting to Cu, Au and Mo porphyries, but which are ores for Sn and Ag. These are associated with relatively peraluminous rhyodacitic porphyry intrusions in hinterland continental arcs (Cerro Rico/Potosi, Llallagua, Bolivian tin belt, Grant *et al.*, 1977).

The mineralogy and distribution of ore in many porphyry deposits has been strongly affected by processes after ore formation, in particular by deep weathering. These processes and their effects are described in detail in Chapter 6. The nature of the **primary ores** formed by high-temperature hydrothermal processes is described in this chapter.

Tectonic, spatial and temporal distribution of porphyry deposits

Almost all porphyry deposits form within volcanic arcs, both in continental arcs (e.g. the Cordillera of western North and South America) and in intra-oceanic island arcs (e.g. the 'Southwest Pacific' arcs of Papua New Guinea, Solomon Islands, etc.) (Figure 3.6). The deposits form either at the same time as active subduction, or immediately after cessation of subduction (post-subduction) in the arc. In the case of the Andes, for instance, a 10- to 20-km-wide belt of Eocene- to Oligocene-age porphyry deposits occurs along approximately 3000 km of strike length parallel to the continental margin from central Chile to northern Peru. A few porphyries, for instance Mo porphyry deposits associated with high-silica granitic magmas, form in more scattered areas of magmatism in 'back-arc' regions, as was the case during the early Cenozoic in interior western USA. Intra-continental back-arc rifting may have been a critical tectonic process for the formation of some of the porphyry deposits in this setting. Rarely, small

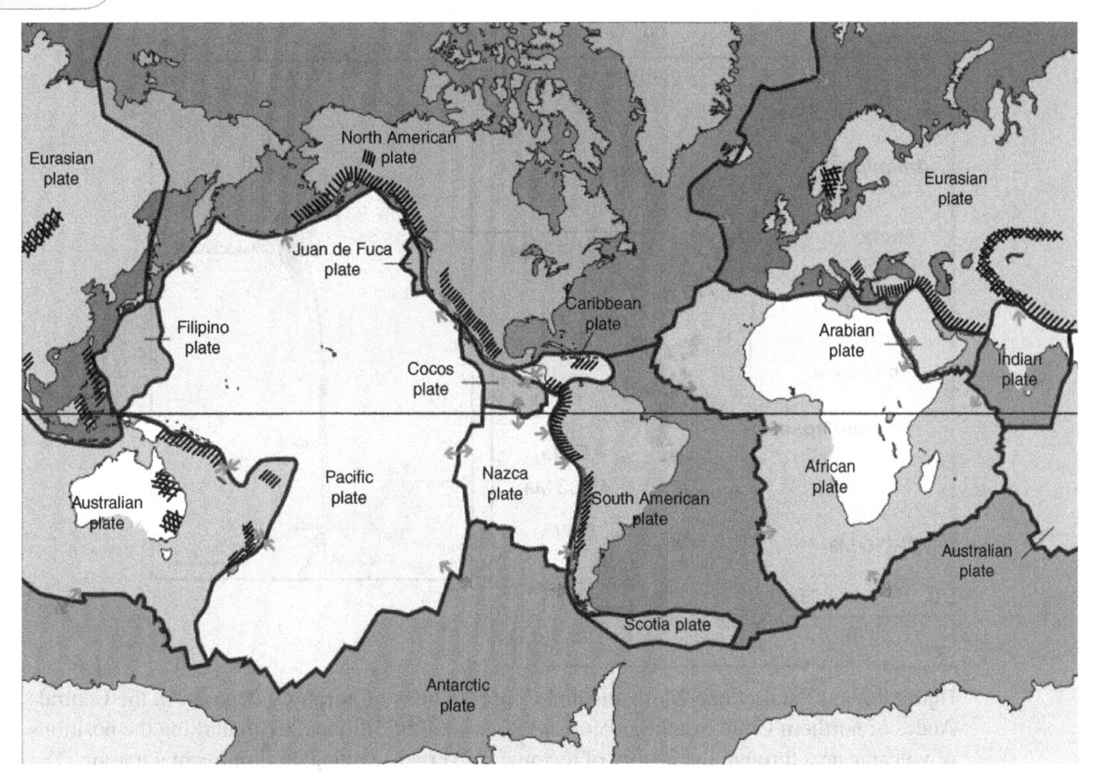

Figure 3.6 Distribution of Palaeozoic (cross-hatch) and Mesozoic–Cenozoic (diagonal shading) porphyry provinces and their relations to modern plate boundaries and plate motions.

porphyry deposits form in contrasting tectonic settings such as in the Palaeozoic continental rift of the Oslo graben.

The majority of porphyry deposits are relatively young (< 75 Ma). Extremely young porphyry deposits include Ok Tedi in Papua New Guinea and FSE in the Philippines, which both formed at approximately 1.2 Ma in currently active volcanic arcs. Older porphyry deposits formed in arcs along the currently active plate boundaries (e.g. circum-Pacific) or along inactive plate boundaries (e.g. central Asia orogenic collage of Kazakhstan and immediately adjacent lands).

In those parts of the Earth's crust such as the Andes which have had a long history of episodic subduction, sub-parallel belts of different ages occur, each corresponding to the position of the magmatic arc of that age (Figure 3.7). Within many belts there are often elongate or irregular clusters of deposits up to 30 km across, within which deposits can be as little as 1 km apart, and within which the deposits formed within a few millions of years (Figures 3.7 and 3.8). These clusters are interpreted to have formed above a single large, long-lived batholith-sized pluton, similar to those that are presently active in the volcanic arc of the Andes, for instance. This style of clustering is best developed in long-lived continental arcs in which large batholiths develop in the mid crust, such as in the Andes and in southwestern USA. Large batholith-sized intrusions are not recognised beneath porphyry deposits in island chains such as those of the southwestern Pacific Ocean, and clustering of deposits in either space or age is not recognised in these environments.

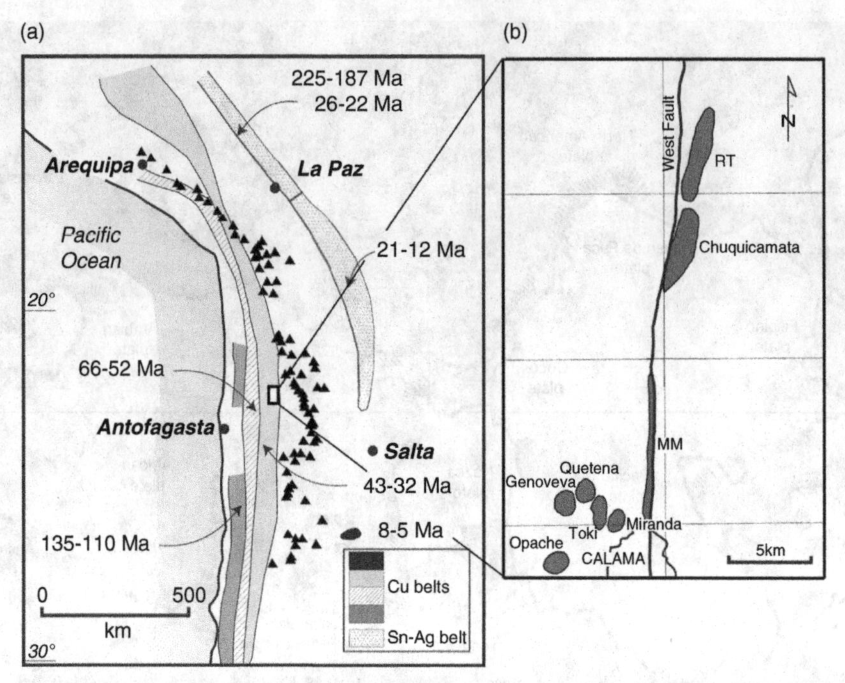

Figure 3.7 (a) Parallel late-Mesozoic and Cenozoic belts of porphyry deposits in the Central Andes of northern Chile, southern Peru and Bolivia (after Sillitoe, 2010) marking the positions of volcanic arcs through the history of tectonic convergence along this continental margin. The triangles show the position of the present active arc and distribution of active and recently active volcanic centres, which is presumed to be the belt of active porphyry deposit formation at a few kilometres depth. (b) Distribution of known porphyry deposits in a 10-km-wide segment of the Eocene belt in Chile showing the typical clustering of deposits within areas of this size, which are separated by a few tens of kilometres from neighbouring clusters.

Styles of magmatic centres that host porphyry deposits

The magmatic centres in which porphyry deposits occur cover areas of a few kilometres across and are typically marked by multiple phases of often nested small intrusions, stocks up to 1–2 km diameter, irregular dyke-like intrusions, and sub-vertical finger-like intrusions that can be as narrow as about 20 m (e.g. Figure 3.8). These intrusions can be divided into pre-mineralisation, syn-mineralisation (intra-mineral) and post-mineralisation based on the cross-cutting relations of veins and the intensity of hydrothermal alteration in the intrusions. Where the small intrusions have been radiometrically dated, they have been confirmed to have all formed within time periods of at most a few million years, and hence over the lifespan of the magmatic centre. They are petrogenetically related, although can vary somewhat in both chemistry and texture.

These magmatic centres are interpreted to be the eroded bases of stratovolcanoes (Figure 3.9) or clusters of volcanic domes. In some cases topographic remnants of the volcano are preserved. Reconstruction of the Eocene Bingham Canyon deposit (Utah, USA) suggests that the magmatic centre fed a volcanic edifice built up to about 2 km high above the surrounding terrain and about 30 km in diameter. However,

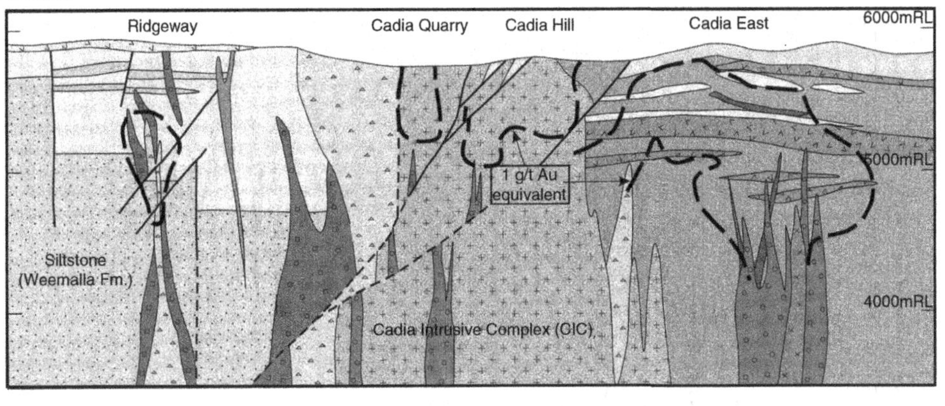

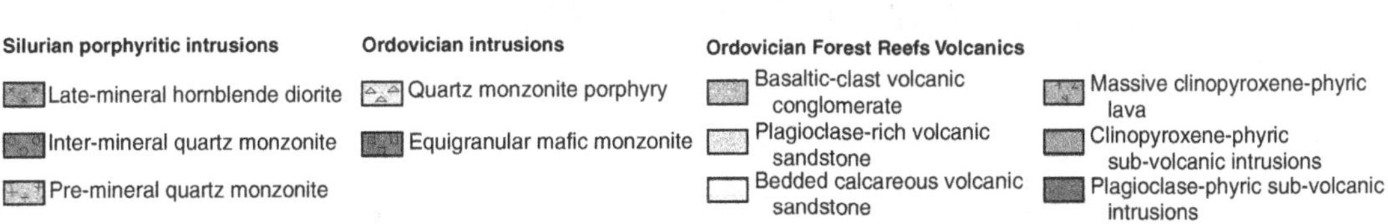

Figure 3.8 Cross section through a cluster of porphyry deposits, the example of the Cadia Cu–Au deposits in New South Wales, Australia, and the style and nature of associated multi-phase shallow-level intrusions. There are multiple stock- to finger-shaped intermediate to felsic intrusions spread over a few kilometres which intruded episodically over a time period of a few million years from the late Ordovician to early Silurian in older sedimentary and volcanic rocks. The dashed lines outline the ore bodies. The Cadia East and Ridgeway deposits are caps enveloping the tops of narrow finger-like intrusions which did not vent to the surface. The Cadia Hill deposit is also cap-shaped but is not closely associated with a specific intrusive body (Wilson *et al.*, 2007).

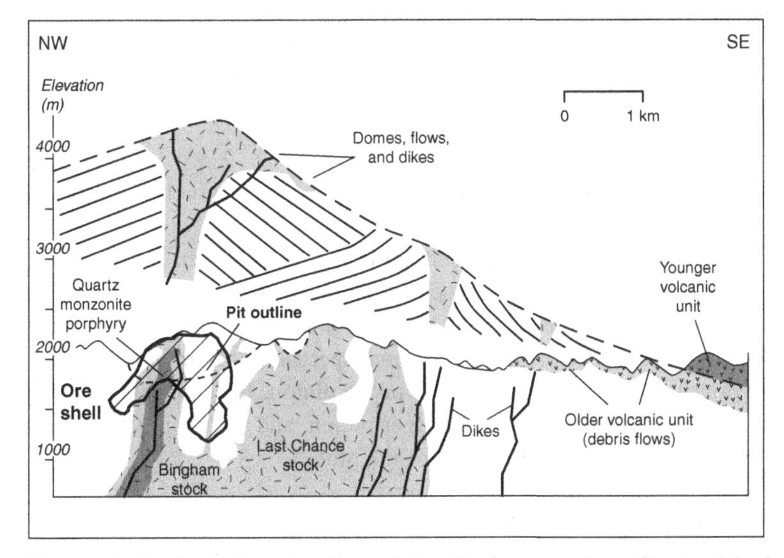

Figure 3.9 Geological section through the Bingham porphyry Cu–Au–Mo deposit (Utah, USA) showing the relation of the ore body (ore shell) to multiple phases of small intrusions in the host magmatic centre and interpreted reconstruction of a now eroded overlying stratovolcano. The deposit thus formed at about 2 km depth. After Waite *et al.* (1998).

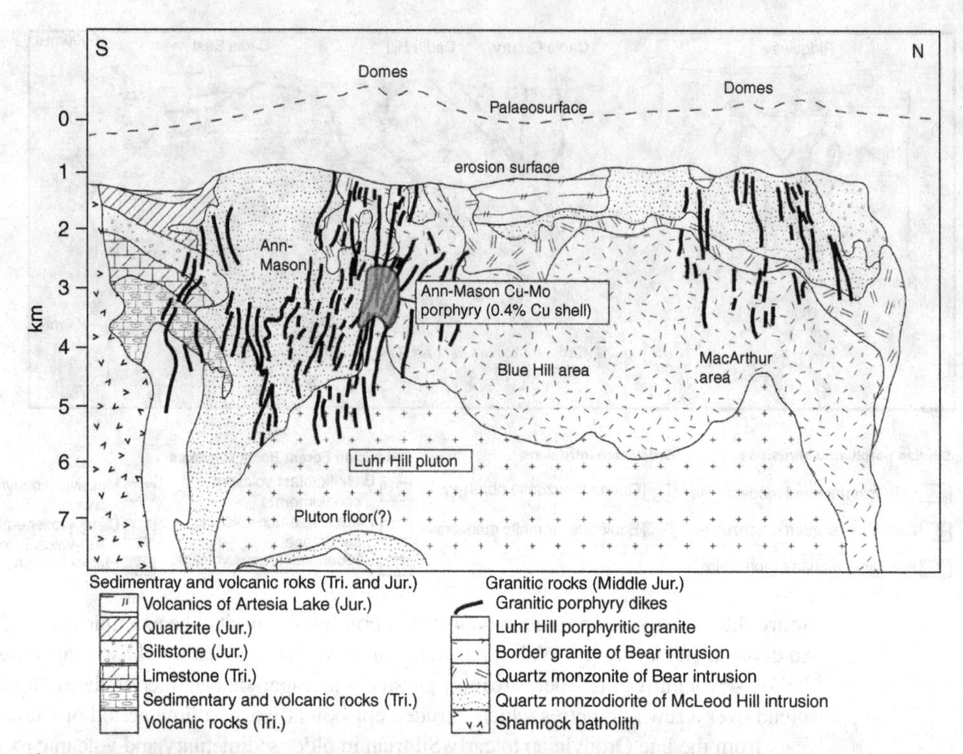

Figure 3.10 Siting and relations of the Ann-Mason porphyry Cu–Mo deposit, Nevada, USA to an underlying large granitic pluton, as reconstructed from the tilted crustal section in the area (after Dilles and Einaudi, 1992; Tosdal and Richards, 2001). The deposit is at the top of a cupola and is about 1 km above the top of the Luhr Hill pluton that is interpreted to be the large pluton from which the magmatic-hydrothermal ore fluid was derived. This pluton is interpreted from the area of exposure to have a minimum volume of about 65 km^3. Smaller amounts of fluid were also released from the pluton through the cupola in the MacArthur area.

many of the small intrusions in the magmatic centres pinch out upwards and hence were not feeders to volcanic rocks.

Rare examples of porphyry deposits in crust that has been tilted since ore formation show that the small intrusions are stocks or fingers that extend from a deeper, much larger parent intrusion of batholithic dimensions (Figure 3.10). The small intrusions have thus intruded from the top of the large intrusion, specifically from a cupola, or dome-shaped upward protrusion on the upper contact of the pluton. In some cases the porphyry stocks intrude into early, rim phases of the larger pluton, and the magma for these was presumably derived from the deeper interior of the pluton.

Siting and shapes of ore bodies

The deposits are estimated to have formed at depths of between about 1 and 10 km, at or below the base level of a volcanic edifice (Figure 3.9) and at or up to a few kilometres above the top of an underlying large pluton (Figure 3.10). Host-rocks to ore may include a rim phase of the pluton, one or more phases of porphyry intrusion and the host-rocks to

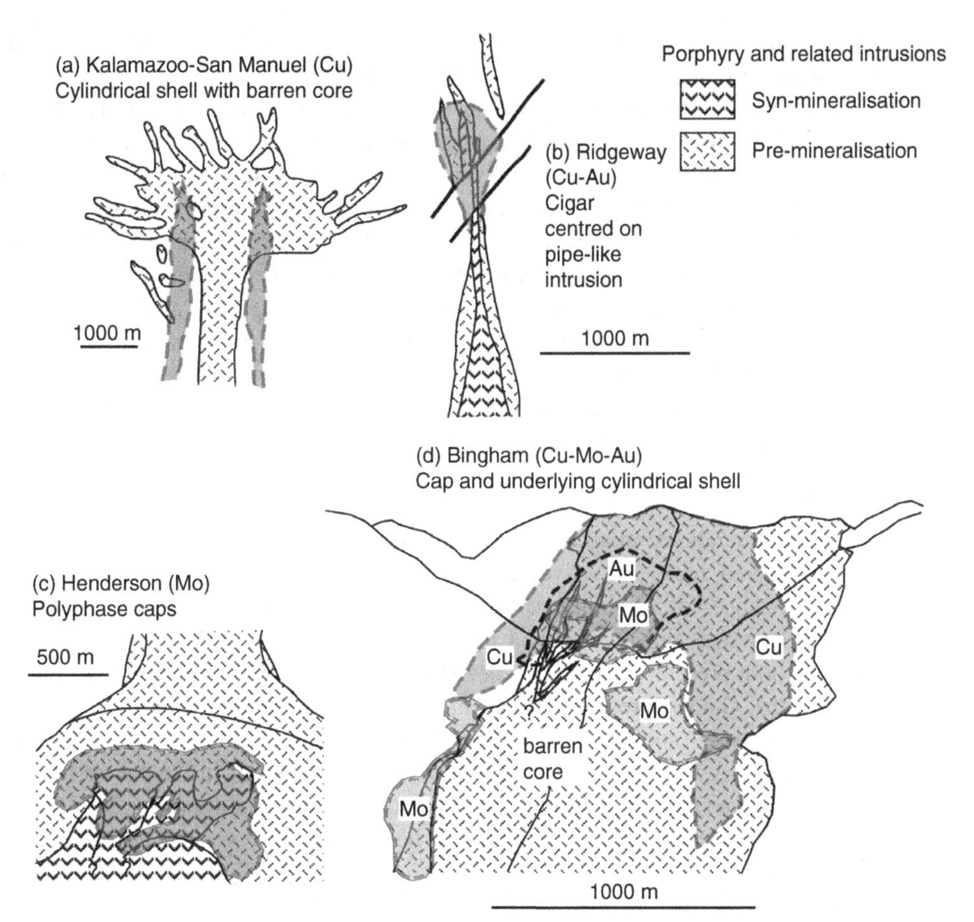

Figure 3.11 Variety of cross-sectional shapes of porphyry ore bodies. The shaded areas show the approximate outlines of the ore bodies, which in all cases are sub-circular in plan view. Note that exact outlines may change with changing cut-off grades of mining. The major commodities and ore body sizes are indicated in each case. (a) Kalamazoo–San Manuel, Arizona, USA, as reconstructed to original shape and orientation before tilting and truncation by a fault, by Lowell and Guilbert (1970); (b) Ridgeway, New South Wales, Australia (after Wilson *et al.*, 2007); (c) Henderson, Colorado, USA, with multiple overlapping caps to multiple cupolas of phases of porphyry intrusion (after Seedorff and Einaudi, 2004); (d) Bingham, Utah, USA, differentiating the shells of the three major commodities (Cu, Au and Mo) (after Gruen *et al.*, 2010).

the intrusions, which in different deposits can include older intrusive rocks, volcanic rocks of the same igneous complex, sedimentary rocks, and metamorphic basement.

The ore bodies are continuous bodies, hundreds of metres to a couple of kilometres in diameter and depth extent, centred on one or more of the small intrusions of the magmatic centre (Figure 3.11). In most economic porphyries, ore grade is uniform or varies gradually with position except where the body is cut by syn- or post-mineralisation intrusions.

The ore bodies have various approximately radially symmetric or elliptical shapes, which are often known as shells. Ore grade generally increases progressively and gradually from the peripheries to the centres of the bodies. The most common shape is probably the cap and underlying cylindrical shell, or inverted cone, with a barren core, as shown in Figure 3.11d. In a few cases, there are multiple nested and overlapping shells (Figure 3.11c) that formed within a short time period. Other ore bodies have most ore in a cylindrical shell and lack the cap (Figure 3.11a), and some are vertically orientated cigar-shaped bodies centred on narrow pipe-like intrusions (Figure 3.11b). Where Cu and Au occur together in the same ore body, the highest grades of Au and Cu are closely coincident. However, where Mo occurs with Cu and Au, it is in many cases differently distributed with concentrations either in a deeper-level shell below the Cu–Au ore body, and hence closer to the source pluton (Figure 3.11d), or in an annular concentration around the Cu–Au ore body (such as at Bajo de la Alumbrera, Argentina).

Ore minerals

Ore minerals are in all cases disseminated as millimetre- to sub-millimetre-sized grains in altered rock and in veinlets throughout the porphyry deposits. Maximum modal abundances of ore minerals are about 10%. The most important copper minerals are chalcopyrite and bornite, and more rarely chalcocite. Where bornite and chalcopyrite are both present, the ore body is zoned with the greatest abundance of bornite at greater depths. Gold is present as small inclusions of native gold with chalcopyrite or bornite and in solid solution in bornite. Molybdenite is the only primary molybdenum ore mineral. A few per cent by mode of pyrite is ubiquitous in the ore shell and also above and around the ore shell over distances of up to a few kilometres. It is generally absent form the barren core of a shell and below the ore body, where magnetite is present instead.

Veins and structures of hydrothermal fluid flow

Most ore and surrounding hydrothermally altered rock has a dense **stockwork** or **sheeting** of multiple generations of filled fractures, veinlets and centimetre-wide veins. Veins wider than a few centimetres are rare in porphyry deposits. In the most densely veined stockworks up to between 10 and 20% of the rock volume may be formed of veins. Veins are developed throughout the ore and immediately surrounding hydrothermally altered rock, although the cores of many ore shells have much lower densities of veins.

There are multiple generations of cross-cutting veins in the deposits. The mineralogies of the different generations of veins and their temporal evolution have been determined for a number of major deposits. Although the sequences and mineralogies of veins are different in detail, a consistent sequence of three generic types can be recognised in the ore zones at many deposits. The sequence at many porphyry Cu deposits is for instance: (i) early veinlets (e.g. EB-type = early biotite), which are thin and often wispy, with biotite together with, in some cases, magnetite lining the veins; (ii) the major phase of quartz-dominated veins and veinlets, often granular in texture, which contain disseminated chalcopyrite (A- and B-type veins); and (iii) a late widespread phase of D-type veins – quartz–pyrite ± chalcopyrite veins with sericite haloes, or chlorite–pyrite with chlorite haloes. Ore grade correlates with the density and abundance of A and B quartz-dominated veins.

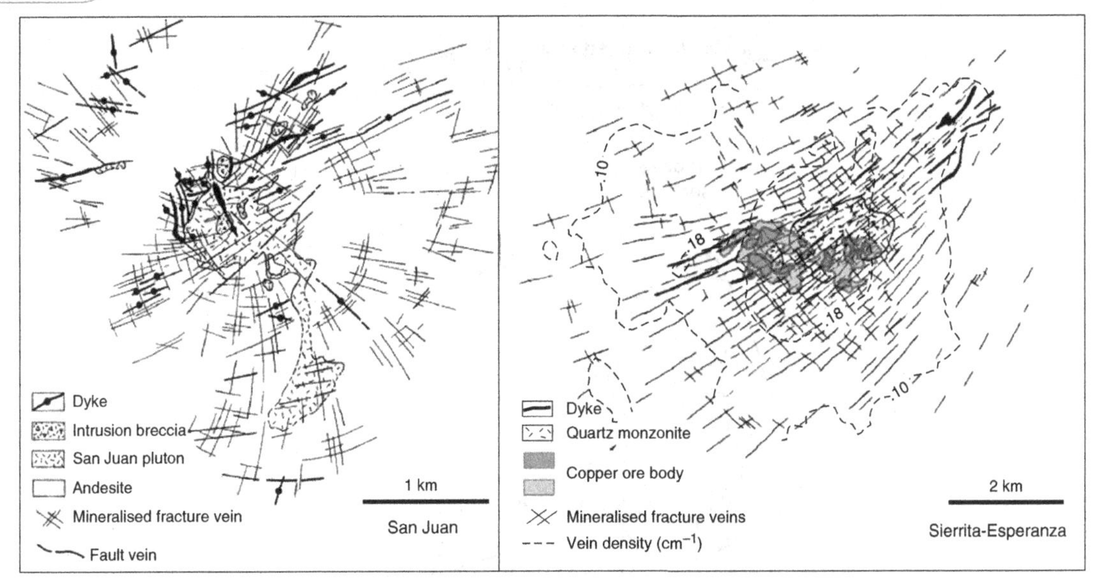

Figure 3.12 Distribution and orientations of major veins and fractures of the stockwork set that is related to ore at two porphyry deposits in Arizona, USA. The San Juan deposit is a prospect with the centre of the ore zone below the present level of outcrop. The pattern includes radial and concentric veins and a superimposed set with west-southwest–east-northeast strike. The Sierrita–Esperanza bodies are interpreted to be eroded to deeper levels and the vein orientations are interpreted to be related to regional tectonic stress. The orientations of dykes and elongation directions of sections of the ore bodies are approximately parallel to one or other of the vein orientations. Note that the veins are developed over a much wider area than the deposit but become less intense with greater distance from the magmatic centre. Modified after Heidrick and Titley (1982) and Titley (1993).

Veins form stockworks or are sheeted and are most commonly moderately to steeply dipping. Stockworks of any one type and phase of veins and veinlets are generally formed in a single phase with veins of different orientations formed essentially simultaneously such that there is no truncation of those of one orientation by those of another. Although stockworks may appear in outcrop to be composed of randomly oriented veins, structural measurements show systematic patterns of orientations, most commonly with either bimodal or girdle distributions. In some cases there are systematic variations in orientation across ore bodies (Figure 3.12).

A distinct and locally developed texture in the pluton cupolas below porphyry deposits and also in stock-like intrusions is 'brain rock' or 'unidirectional solidification texture' (UST). This has cuspate, bulbous, repeated irregular bands of generally coarse-grained to pegmatitic granite minerals with euhedral terminations where grains grew downwards into melt or fluid. These textures are interpreted to record mineral growth into pockets of hydrothermal fluid that had collected at the top of a crystallising body of magma.

Breccia **pipes** are present in some porphyry deposit. They may comprise part of the ore and include the same hydrothermally precipitated matrix minerals as in veins. Other breccia pipes are unmineralised: a large pipe at the El Teniente deposit (Chile) cut

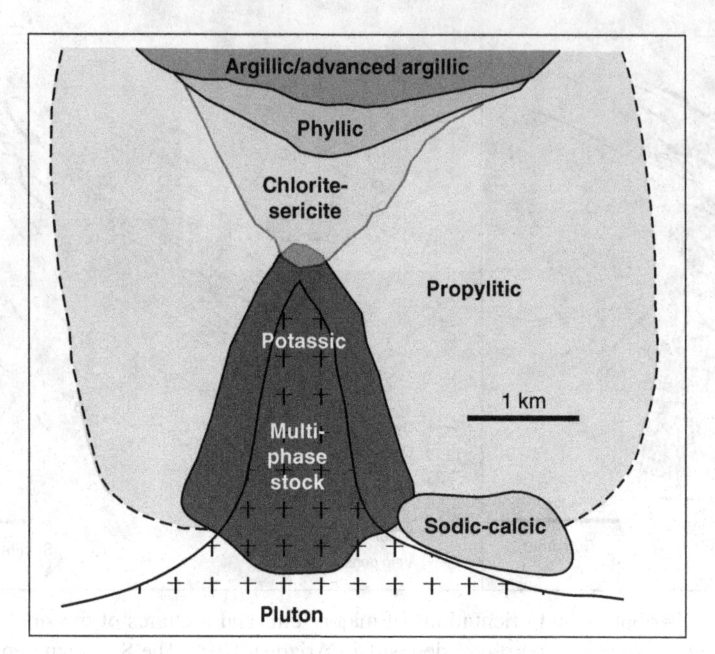

Figure 3.13 Schematic cross section through a porphyry deposit showing an idealised distribution of different alteration facies.

through the centre of the ore body after mineralisation. The breccia pipes are characteristically sub-vertical. Upward-flaring shapes of some pipes suggest that they formed in a similar fashion as diatremes in kimberlites, during venting of magmas or fluids to the surface. Others are collapse breccias which did not extend to the surface and in which clasts have fallen into void space in a similar fashion as roof clasts fill karst caves. Void space in a magmatic system may have developed on emptying of a magma or fluid-filled space at depth.

Hydrothermal alteration

Hydrothermal altered rock is a characteristic component of porphyry deposits. It is pervasive in the ore zones and in adjacent, overlying and underlying rock, in some cases extending kilometres beyond the limits of the ore body. There are multiple characteristic alteration mineral assemblages in both the intrusions and the surrounding country rock that formed at different temperatures and as the results of migration of different fluids. The assemblages and their patterns of distribution are different in detail at different deposits, but systematic and repeated patterns are recognised (Figure 3.13).

Up to six alteration types (**alteration facies**) are distinguished based on the presence of one or, in most cases, combinations of specific and characteristic minerals, and are repeatedly mapped in and around porphyry deposits within alteration **zones** in which one facies is dominant. The zones are each developed broadly concentrically around and above the porphyry intrusion and ore body. In many deposits, lower-temperature alteration assemblages partially overprint higher-temperature assemblages.

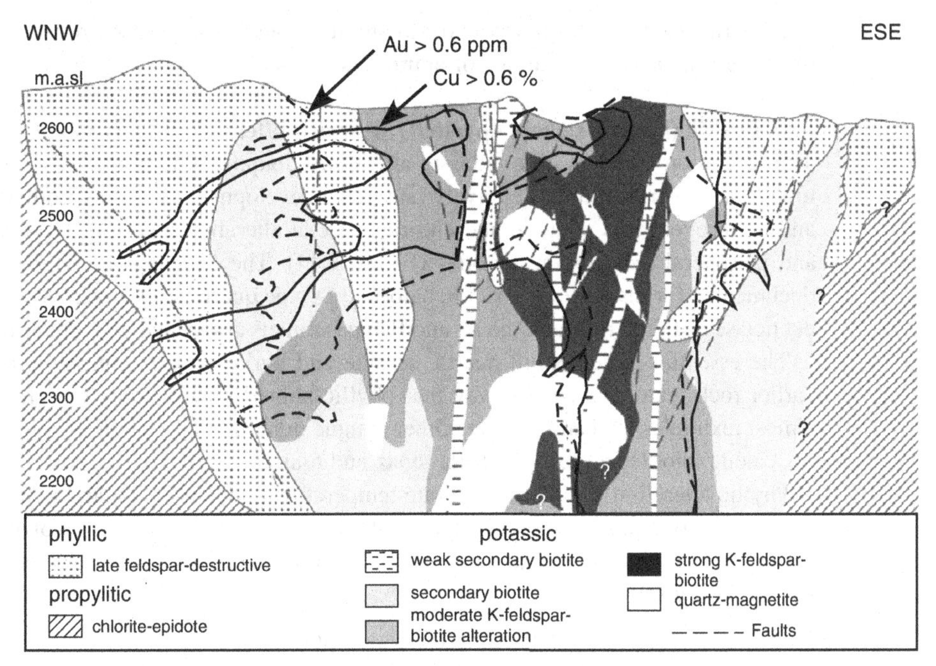

Figure 3.14 Relations between different facies of hydrothermal alteration and ore-grade mineralisation at the Miocene Bajo de la Alumbrera deposit, Argentina (after Ulrich and Heinrich, 2002). Ore deposition was synchronous with the main phase of potassic alteration. The ore zones of Cu and Au closely overlap and form a cap centred on the pipe of most intense potassic alteration. Phyllic alteration overprints the potassic alteration and is most strongly developed in tongues extending downwards from higher, now eroded levels of what is interpreted to have been a blanket of alteration, but does not significantly deplete or enrich ore grade.

(a) Potassic

This facies is ubiquitous and is characteristic of the ore zone and of deeper levels of most deposits, especially of the barren core beneath and in the interior of an ore shell.

Essential minerals of the potassic alteration facies are quartz and one or both of K-feldspar and biotite. These minerals are in many cases part of the primary assemblage of the rocks, especially of felsic rocks, but potassic alteration is recognised where a significant proportion of the K-feldspar or biotite has replaced earlier plagioclase and mafic minerals of, for instance, the host intrusion. Other alteration gangue minerals can include chlorite, albite, sericite, magnetite, anhydrite and pyrite. A high concentration of magnetite is a feature of potassic alteration zones in the core of the ore shell of many deposits. Sub-facies have been mapped in some deposits based on the relative abundance of biotite, K-feldspar and magnetite (Figure 3.14).

This facies of alteration is the result of metasomatic addition of potassium at high temperatures (450–600 °C). The fluid has a relatively high ratio of K^+/H^+.

(b) Sodic-calcic

This occurs in some deposits peripheral to the lower depth limits of the ore zones. Essential minerals are albite/oligoclase, actinolite and magnetite. The facies generally lacks pyrite and is almost never a host to ore. It is considered to be the result of infiltration of saline waters at temperatures of approximately 450 °C.

(c) Phyllic (quartz–sericite–pyrite/QSP/sericitic/feldspar destructive)

This alteration facies is, like potassic alteration, ubiquitous in porphyry deposits. Commonly, however, it formed relatively late in the development of the hydrothermal system and often overprints potassic and chlorite-sericite alteration zones, especially the upper and peripheral parts of these zones (Figure 3.14). The late timing of this alteration is documented for instance by sericitic haloes to D-type quartz–pyrite veinlets, which cross-cut potassic altered rock and the A- and B-type veinlets associated with potassic alteration.

The essential minerals are quartz, sericite and pyrite. These minerals grow over the earlier rock texture in such a way that phyllic altered rock appears bleached and often almost textureless in hand sample. Other gangue minerals can include K-feldspar, kaolinite, calcite, biotite, rutile, anhydrite, topaz and tourmaline.

Phyllic alteration occurs at moderate temperatures (200–450 °C) and is the result of a moderately to strongly acid fluid, such that the metasomatic reactions involve addition of H^+ and dissolution of K, Na, Ca, Mg, Ti, Fe from the rocks, e.g. in the replacement of K-feldspar by sericite:

$$3KAlSi_3O_8 + 2H^+ \rightarrow KAl_3Si_3O_{10}(OH)_2 + 6SiO_2 + 2K^+. \quad (3.3)$$

(d) Chlorite–sericite (css)

This alteration facies caps the ore zone at some deposits, although can include ore. It has a characteristic pale green colour. In addition to chlorite and sericite (or illite) other minerals include haematite, pyrite and, in some cases, smectite clays. It formed from similar fluids and at similar temperatures as the phyllic alteration zone, but of slightly lower acidity.

(e) Argillic

This may be locally developed at relatively shallow levels of the hydrothermal system above and peripheral to the phyllic zone. Essential minerals are clays (montmorillonite, kaolinite). Other minerals include biotite, illite, chlorite, pyrophyllite, diaspore, alunite, sulfides, quartz and andalusite.

Argillic alteration is the result of intense low-temperature metasomatism (100–300 °C) in which clay minerals are produced as a result of acid leaching of feldspars and mafic silicates.

(f) Propylitic

Propylitic alteration is very ubiquitous and is extensively developed around most porphyry deposits, and can extend up to several kilometres from the deposits with progressively decreasing intensity of development away from the ore body. Propylitic alteration assemblages closely resemble those of greenschist-facies metamorphism. Essential minerals are epidote, chlorite and calcite. Pyrite is normally present. Other minerals include iron oxides, sericite and apatite.

Propylitic alteration is the result of addition of H_2O, CO_2 and, in many cases, S^{2-} to the host-rocks without significant acid–base metasomatism or addition or leaching of metals at temperatures of about 250–400 °C.

Relations between ore and alteration zones

The highest grade ore is most commonly in the potassic alteration zone (Figure 3.14). A chlorite–sericite zone may also be host to ore. In porphyry Cu deposits, ore is localised near the boundary between the potassic and phyllic alteration zones in some deposits, or is largely within the phyllic zone. Where ore is in the phyllic zone, phyllic alteration is interpreted to have overprinted earlier ore-bearing potassic and css zones.

Compositions of hydrothermal fluids at porphyry deposits

Stable-isotope and fluid inclusion data show that a minimum of two fluids were involved in the hydrothermal systems of porphyry deposits, including one or more magmatic-hydrothermal fluid in the core of the hydrothermal system, which transported the ore metals, and convecting external fluids in the outer alteration zones.

The ore-forming magmatic-hydrothermal fluid is a chemically complex moderate- to low-salinity ('intermediate-salinity') fluid with metal chloride salts, minor CO_2, and significant concentrations of sulfur-bearing gases (SO_2, H_2S) in solution (see Box 3.1). The main salts in solution are NaCl, KCl, $FeCl_2$ and copper salts, but many other metals are present in solution, including metals that are not significantly enriched in the ore zones, for instance Pb, Zn and Mn. As the fluid rises, pressure loss means that the intermediate-salinity fluid generally undergoes phase separation (boiling) to form two fluids in the core of most deposits. These two fluids are a high-density, highly saline brine with greater than 40 wt % salts in solution, and a lower-density, lower-salinity aqueous-saline gas phase. Both the brine and the gas can have high concentrations of copper.

The hydrothermal fluids that caused the peripheral propylitic and sodic-calcic alteration zones around the deposits are interpreted to have been convecting surface-derived (meteoric) fluids. Where deposits formed in arid or semi-arid warm climates, the convecting groundwaters are in many cases highly saline.

Box 3.1 Fluid inclusions as sources of information on the conditions of hydrothermal ore deposits: the case of Cu transport in porphyry deposits

What is the composition of the magmatic-hydrothermal fluid that transports Cu, Au and Mo at a porphyry deposit? In what concentrations are the metals transported? What are the volumes of ore fluid and the efficiency of metal precipitation from the fluid? We can make some interpretations to partly answer these questions using the nature of hydrothermal alteration and its mineralogy, through for instance calculations of the temperature stability range of the mineral assemblages. However, fluid inclusions, or microscopic 'vacuoles' 1–50 μm in size of fluids and in some cases fluids and crystals trapped in growing crystals in veins or during healing of microfractures in crystals in hydrothermally altered rock, can be direct samples of the hydrothermal fluid in which the metals were in solution and from which they precipitated. Fluid inclusions are thus central sources of information about the conditions under which any hydrothermal ore deposit formed. They are thus a research tool for the development of ore genetic models of any hydrothermal deposit type.

Fluid inclusions are studied through a number of techniques; see for instance Roedder (1984), Shepherd *et al.* (1985) and Wilkinson (2001) amongst others. Fluid-inclusion petrography gives information about the timing of fluid flow relative to mineral growth, and on whether there were multiple flow events. A routine study of fluid inclusions involves microthermometry and the examination of phase transitions such as freezing and ice melting in an inclusion as it is heated and cooled. This analysis can give information about the overall chemical nature of the hydrothermal fluids – whether they are dilute or highly saline solutions and whether there are gases in solution. For highly saline fluids, it is possible to determine which salts are dominant in solution. Microthermometric data also give information on fluid density, and hence pressures and temperatures during fluid migration. More recent technical advances in chemical analysis, in particular proton beam probes using PIXE (proton-induced X-ray emission) and plasma mass spectrometry using LA-ICP-MS (laser-ablated inductively couple plasma mass spectrometry) now allow fluid inclusions to be analysed for concentrations of many elements (e.g. Pettke *et al.*, 2012). LA-ICP-MS analysis is fast, has relatively high precision and has allowed us to determine the relative concentrations of the major salts in many types of ore-forming solutions. Analyses at a range of ore deposits have shown that many ore fluids have unusually high concentrations of the elements that are in the ores (e.g. Audétat *et al.*, 2008). In addition, some constraints can be obtained on valence and speciation of ions in solution in fluid inclusions through X-ray analytical techniques such as XANES (X-ray analysis near-edge spectra) (e.g. Berry *et al.*, 2009).

What are Cu concentrations in porphyry ore fluids and what controls Cu precipitation and ore grades in porphyry Cu deposits? High-salinity chloride-rich fluids (30–60 wt % salts) are preserved in fluid inclusions almost ubiquitously in porphyry deposits, in addition to low-salinity, low-density fluids as inclusions in the veins of the deposits, as is consistent with gas–liquid immiscibility in saline aqueous solutions at the temperatures and pressures expected above intrusions in the shallow crust (e.g. 500 °C at 3–5 km depth, or about 10 MPa pressure) (see Figure 3.1). The characteristics, size and electron orbitals of the Cu(I) ion imply that it can form stable complexes in aqueous solution with both Cl^- and softer anions such as HS^-. Because chloride is the dominant ligand in most hydrothermal fluids, it was for many years assumed that copper was carried as copper chloride complexes in high-salinity fluids. It was further argued that high-grade porphyry Cu deposits may only form where high-salinity magmatic-hydrothermal fluids are available (e.g. Cline and Bodnar, 1991; Audétat and Pettke, 2003). The $Cl^-:H_2O$ ratio of magma may thus be an important control on where porphyry deposits form.

From LA-ICP-MS fluid-inclusion analyses the dominant salts have been determined to be sodium, potassium and iron chlorides (e.g. Landtwing *et al.*, 2005, Ulrich *et al.*, 1999) in ratios approximately as expected for equilibria with silicate minerals in the crystallising magma. Lead, zinc and manganese are at significant concentrations in the fluid but are not precipitated in porphyry ores. These metals remain in solution as the ore fluid passes through Cu and Mo ore bodies. Copper can be the most abundant cation in both brine and low-density, low-salinity vapours and can have Cu concentrations above 1 wt %, which for the low-salinity vapours are much higher than were expected from theoretical calculations based on laboratory measurement of copper solubility. These concentrations are, however, consistent with those assumed by Cline and Bodnar (1991) in their analysis of the source volume of magma required to produce a significant economic deposit. However, comparison of coexisting brine and low-density, low-salinity vapours formed by phase separation showed that Cu, together with some other elements (e.g. As), has higher concentrations in the low-salinity vapour than in the coexisting brine. This observation implies that Cu is most abundant in solution as a complex with a ligand species that partitions into the vapour on boiling, hence most likely a sulfide species (H_2S or HS^-) rather than Cl^-.

These data have thus led to a reassessment of whether there is a correlation between high-grade Cu porphyry deposits and magmas with high chloride contents, or a correlation to specific conditions of magma fractionation as was analysed for instance by Cline and Bodnar (1991). Further evidence that chloride content is not a factor in forming high-grade ore bodies came from investigation of drill core samples from holes that extend below the ore bodies (but above the source pluton) at Bingham Canyon and Butte, which are both major Cu–Mo porphyry deposits in the western USA. These showed that highly saline fluids are not present at these depths (Rusk *et al.*, 2008; Landtwing *et al.*, 2010). Rather, the only fluid in the veins below the ore bodies is a relatively low-salinity aqueous fluid (with ~ 6 wt % salts). As these samples are closer to the fluid source than the deposit we thus infer that this is the fluid exsolved from the pluton and the input fluid into the deposit. The high-salinity fluid in veins in the deposits condenses from this low-salinity fluid as it rises and loses pressure above the source pluton, and mass-balance calculations show that the highly saline fluid comprises only a minor part of the total fluid in the hydrothermal system.

The interpretation of the data on concentrations of copper and its partitioning between vapour and brine has, however, proven difficult. High-pressure–high-temperature laboratory experiments failed to reproduce the partitioning of Cu between brine and vapour that has been observed in the natural fluid inclusions and indicate, in contrast, a low degree of partitioning into the brine on phase separation (e.g. Simon *et al.*, 2006). The experimental partitioning data in combination with estimates of the masses of brine and vapour imply that both fluids contributed to transport of Cu once the fluid reaches the level at which phase separation takes place. Other experiments have, however, shown that small monovalent ions such as Cu(I) readily diffuse through quartz into and out of fluid inclusions at high temperatures (Zajacz *et al.*, 2009). The measured concentrations in the fluid inclusions may or may not be the concentrations at the time of hydrothermal fluid flow.

Irrespective of the uncertainties over the dominant complex of Cu in solution, Cu is precipitated in Cu–Fe sulfide minerals largely as a result of decreasing pressure and temperature of the magmatic-hydrothermal fluid as it rises above the source pluton. Although we cannot quantitatively constrain Cu concentrations, and although all magmatic-hydrothermal fluids carry all metals and there are probably relatively subtle differences of the metal contents of fluids derived from different magmas, the fluids of porphyry Cu deposits do appear to have relatively high concentrations of Cu.

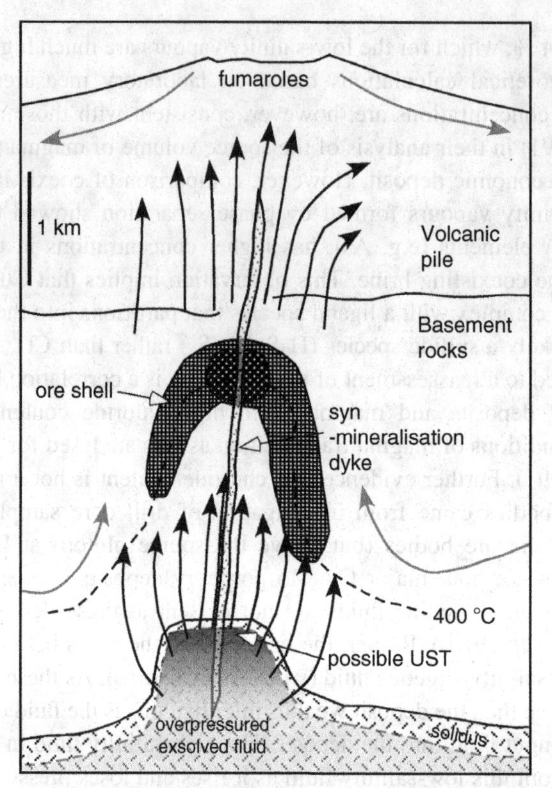

Figure 3.15 Conceptual section through a cylindrical magmatic-hydrothermal plume released from a cupola of a large crystallising pluton beneath a volcanic edifice and forming a porphyry deposit as it rises and cools (modified from Landtwing *et al.*, 2010). The centre of the plume remains warm to shallower depths and hence the ore shells follow an isotherm and have the characteristic 'cap' shape. Black arrows show the migration pathway of the plume, grey arrows show possible migration of meteoric waters and convecting groundwaters.

Interpretations of formation and development of porphyry ores and associated alteration

Ore minerals in porphyry deposits are precipitated from an approximately cylindrical plume of magmatic-hydrothermal fluid (a 'magmatic vapour plume') rising nearly vertically from an underlying magma chamber (Figure 3.15). The pathway of the plume is marked by potassic alteration and a dense network of A and B veinlets, and may be up to approximately 1 km in diameter. The high density of thin veins with uniform vein and alteration mineralogy shows that fluid migration was along a network of generally steeply dipping closely spaced fractures. In most cases these fractures formed at the time of fluid flow, although fluid migrates along earlier fractures and faults if these are present. Fluid pressure built up in the cupola and the magmatic-hydrothermal fluid formed its own pathways through the processes of hydrofracture.

Although the concentric distribution of ore around porphyry stocks or finger-like intrusions suggests that intrusions in the cores of the ore bodies are the sources of the hydrothermal fluids, the relationship between them is indirect. Both the mass of fluid and the mass of ore metals involved in formation of a porphyry deposit are far too large to be derived from either the porphyry stocks or even from the cupolas at the top of an underlying pluton. The concentration factor of Cu in ore is approximately 200 times relative to average igneous rocks (30 ppm to 0.6%), and the source volume of fluid that formed a 1 km^3 (~ 2 Gt) ore body must thus be a magma chamber of at least 200 km^3, and would need to be significantly larger if Cu is not efficiently partitioned from the magma into the hydrothermal fluid. It is thus inferred that the ore fluid collected from larger volumes of magma in the underlying pluton into the cupola before release upwards as the plume. The magma that intruded as porphyry stocks or fingers is also derived from the larger underlying pluton. A minor amount of convecting meteoric water may be incorporated into the plume and mix in with the magmatic-hydrothermal waters, especially late in its lifespan as fluid pressures in the vapour plume become reduced.

The source pluton does not outcrop at most porphyry deposits. Observations at the few deposits where plutons are exposed together with interpretations of geophysical data suggest that the distance of fluid migration from the source pluton to the site of ore precipitation is between a few hundred metres to a few kilometres. The plumes cool from magmatic temperatures as they rise, and will in general be cooler on their peripheries than in their centres. The ore body shapes reflect the geometries of these plumes and the distributions of temperatures and fluid flux within them. The cap of the common cap and shell shape of ore bodies marks the centre of the plume, with the shells forming where fluid flux was lower towards the margins and the fluid was cooler at any depth.

Sequence of events proposed in the formation of a typical porphyry deposit

A generalised common sequence of processes in the formation of a porphyry deposit is illustrated in Figure 3.16.

In detail the likely and common sequences of chemical and physical processes are:

(a) Intrusion of magma into mid- to upper-crustal chambers

(i) A pluton-sized body of magma that is crystallising (second boiling) or rising (first boiling) reaches saturation with respect to an aqueous fluid phase and starts exsolving hydrothermal fluid close to its roof, most probably as bubbles in the magma. This produces a volume of low density magma that intrudes buoyantly upwards above the roof of the pluton and crystallises as a porphyry stock.

(ii) Volatile components (HCl, CO_2, SO_2, H_2S, B species, etc.) and soluble elements are partitioned into the aqueous fluid. This separate phase may be transiently trapped as fluid pockets immediately under either the wall-rock or the crystallised carapace of the cupola. This will be the site of crystallisation of unidirectional solidification textures (UST), or brain rock. The release of this fluid may form early sets of veins in the surrounding wall-rock.

(b) Build-up and release of magmatic-hydrothermal fluids from magma

(i) Continued crystallisation of magma in the chamber allows accumulation of fluid at the top of a cupola. Magma from which dispersed water bubbles have separated is denser and may descend back into the main intrusion to be replaced in the cupola by new, bubble-rich magma. The presence of a cupola may be critical to allow the

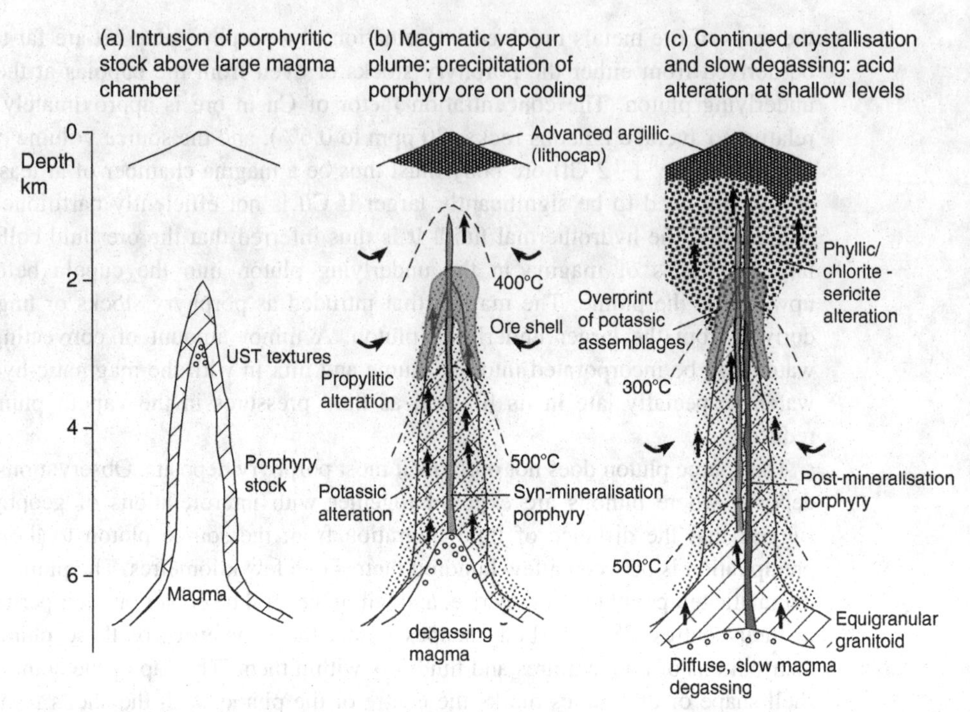

Figure 3.16 Schematic history of a magmatic centre above a large intermediate to felsic pluton in the upper crust leading to the formation of a porphyry deposit. See text for further discussion.

growth of a large fluid pocket at the top of a pluton and collection of fluid from relatively large volumes of magma.

(ii) Fluid pressure will likely be modulated by lithostatic or magma pressure at greater depth in the main body of the pluton, and hence can become higher in the cupola than rock pressure of immediately overlying wall-rock. A critical fluid pressure is reached and a network of stockwork veins or, in some cases, a breccia pipe, is formed by hydrofracturing at the top of the cupola, and propagates upwards to allow upwards escape of overpressured fluid. Fluid is released as a large-volume vapour plume that generates its own fracture pathway for several kilometres above the pluton roof through hydrofracturing. Because the fluid phase is strongly compressible, it becomes rapidly less dense as it rises and the plume is self-sustaining. The passage of this plume upwards above the cupola may be the main mineralising event because of the volume of fluid involved.

(iii) Fluid escape in a plume may be a relatively short-lived event (see Box 3.4). The three-dimensional patterns of orientation of the stockwork veins are related to hydrofracture formation in tectonic and magmatic stress fields.

(iv) On upward migration the fluid cools and loses pressure. Several chemical and physical processes occur in the plume fluid as it migrates away from the pluton, some of which may lead to ore mineral precipitation. The fluid may separate at high temperatures (> 450 °C) into highly saline brine and a dilute, low-density fluid. The precipitation of quartz-rich A- and B-type veins is one product of fluid cooling and loss of pressure. In the case of chalcopyrite, solubility in the fluid will be reduced on cooling, on loss of pressure, on increased concentration of sulfide due to reaction of magmatic sulfur dioxide and water through the reaction

$$4SO_2 + 4H_2O \rightarrow H_2S + 3H_2SO_4, \tag{3.4}$$

on dissolution of iron from minerals in the wall-rock, or on phase separation and the consequent changes in fluid chemistry. The positions and the shapes of the ore zones suggest that Cu is precipitated dominantly as a result of fluid cooling and loss of pressure. Gold is precipitated at a similar depth in an ore body that generally overlaps the Cu ore body. In porphyries in which there are concentrations of both Cu and Mo, the highest Mo grades are typically below those of Cu. The fluid thus reached saturation with respect to molybdenite at higher temperature than to Cu sulfide minerals. Lead and Zn are carried in the fluid, but do not precipitate, except in low concentrations peripherally around some deposits. The ore minerals precipitate both in the fluid channelways, i.e. in veinlets, and in altered rock as elements diffuse into the wall-rock from the closely spaced veinlets to form disseminated ore of effectively uniform grade.

The magmatic-hydrothermal fluid becomes more acid as it rises and cools as a result of increasing dissociation of dissolved acid volatile species as temperature is reduced through reactions such as $HCl \rightarrow H^+ + Cl^-$, and similarly for H_2CO_3, and of H_2S and H_2SO_4 formed through Equation (3.4). Increased acidity increases the solubility of most of the ore minerals, and hence acts to buffer the effect of the chemical processes that lead to ore mineral precipitation with decreasing temperature and pressure. However, despite the increase in acidity, the precipitation of Cu may be remarkably efficient, and up to 90% of the Cu in solution is estimated to be precipitated within approximately one kilometre vertical distance of fluid flow.

The potassic alteration zone forms from the magmatic-hydrothermal fluid as it cools from magmatic temperatures down to between about 500 and 450 °C. The chemical driving force of potassic alteration is the change in the position of equilibrium of reactions between aqueous fluid and feldspars with decreasing temperature. A fluid with concentrations of Na^+ and K^+ in equilibrium with Na-plagioclase and K-feldspar at high temperature becomes oversaturated with respect to K-feldspar and undersaturated with respect to albite as it cools.

(v) Sodic-calcic and propylitic alteration are formed where meteoric fluids convect through the surrounding rock driven by heat of the intrusion and possibly also the heat of the vapour plume, with sodic-calcic alteration formed where the convecting water is saline. While the magmatic centre is active, it will underlie the upwelling limbs of convection cells. Advanced-argillic alteration forms at shallow levels within about 1 km of the surface above the degassing magma.

(vi) Small intrusions may intrude from the cupola to form intermineral dykes and stocks, which have a lower density of A- and B-type veins than surrounding rock. The porphyritic texture of these small intrusions is most likely a result of pressure-induced quenching on the escape of the hydrothermal fluid and because of their rapid crystallisation after intrusion above the main magma chamber. The intrusions follow the same pathways as the magmatic-hydrothermal fluids and intrude within the period of flow of the magmatic-hydrothermal fluid. Pressure release in the magma chamber on escape of the fluid is a possible prompt for intrusion.

(c) Later and final crystallisation of the magma chamber

(i) Later release of fluid from non-convecting pockets of magma in the pluton lead to hydrothermal overprinting of ore and earlier phases of alteration and mineralisation at lower temperatures. This is because of sourcing of later fluid from greater depth in the pluton as it progressively crystallises from the chamber margins inwards, and hence greater cooling on upward migration. Post-mineral dykes may intrude through ore at this time, and are essentially barren.

(ii) Phyllic alteration, argillic alteration and chlorite–sericite alteration zones associated with D-type quartz–pyrite veins may overprint both ore and potassic alteration assemblages but do not appear to cause significant redissolution of ore minerals. The acidity associated with these alteration facies is a result of the processes mentioned above, most importantly HCl dissociation and disproportionation of dissolved SO_2.

The sequence of processes that has been outlined above may be repeated at a single magmatic centre while the underlying magma chamber is active. At the Henderson Mo porphyry deposit (Colorado, USA) there are multiple mineralisation events, each preserved in separate and overlapping ore shells in a cupola (Figure 3.11c). Early shells were truncated by later phases of intrusion, and the Mo that was contained in them possibly cannibalised and remobilised to be precipitated in later shells. At both Bingham Canyon (Utah, USA, Figure 3.11d) and Bajo de la Alumbrera (Argentina), the Mo ore is slightly separate from the Cu–Au ore shell, and the molybdenite-bearing veins cross-cut both syn-mineralisation dykes and the Cu-bearing veins. These veins may thus have formed either from a later vapour plume or from a late stage of the plume that caused the Cu mineralisation. Where there are clusters of porphyry deposits it appears that fluid exsolution took place repeatedly from a single large pluton from multiple closely spaced cupolas.

Development of magmatic centres of porphyry deposits

Porphyry deposits occur where there are clusters of porphyritic intrusions which can be interpreted to have formed above a long-lived, large, mid- to upper-crustal magma chamber. Some deposits (including a number in the island arcs of the southwest Pacific region) occur as isolated deposits, which are centred on a single pipe-like intrusion and lack evidence of localised prolonged magmatic activity from a sustained larger chamber. In these cases, the source pluton is presumably at greater depth.

The largest known active major mid- to upper-crustal intrusion in an arc is the Altiplano intrusion, which straddles the Chile–Argentina border along the Andean arc. This extends over 200 by 50 km and supplies multiple active and recently active andesitic to dacitic volcanoes. A number of comparisons of this intrusion with the igneous rocks associated with porphyry deposits suggest that it is analogous to those that underlie many porphyry hydrothermal systems. The intrusion has had a lifespan of about 10 million years, with individual volcanic centres being each active over up to a few million years. The small intrusions at the eroded magmatic centres of porphyry ore deposits are formed over similarly long time periods. The volcanic rocks are petrologically varied and have not evolved to felsic compositions over the lifespan of the intrusion, but have alternated through different intermediate rock types through its history. A similar evolution is seen in the small intrusions of the magmatic centres of porphyry deposits.

Large mid-crustal plutons such as the Altiplano intrusion are inferred to be continually replenished by hot magma input from a deeper level. This is required to explain the long lifetimes of the magmatic centres and the underlying upper-crustal magma chamber. Even a large pluton formed by a single phase of intrusion would freeze by loss of heat to the walls and would crystallise from the outside into the centre within at most a few hundred thousand years. The continual heat input to a porphyry-related pluton may either be of mafic magmas directly from the mantle, or of magma that has resided and evolved to more felsic compositions in a magma chamber deeper in the crust, possibly at the Moho. Many volcanic rocks in continental arcs have geochemical evidence that magma fractionation took place at pressures higher than those of an upper-crustal magma chamber, hence supporting the postulate of lower-crustal magma chambers. Whether magma input into upper-crustal magma chambers is directly or indirectly derived from the mantle, continual replenishment means that magmas in arc magmatic systems are dominated by mantle-derived rather than crustal-derived melt, as is indicated by isotope ratios. Although magma chambers may become compositionally stratified, most magma remains of intermediate composition throughout the evolution of a volcanic centre.

At most deposits, mineralisation is a short event towards the end of the lifespan of the pluton. It is not, however, synchronous with the last phase of magmatism, and post-mineralisation dykes cut most ore bodies. As has been discussed above, convection within the pluton may be important to ensure accumulation of large volumes of high-temperature fluid below a small area of the roof of the chamber before the formation of a deposit. We thus interpret that porphyry ores form while magma was convecting in the chamber. Convection requires a large continuous volume of magma with low crystal content. When convection is shut off once a critical degree of crystallisation is reached, the remaining magma crystallises inwards into the centre of the pluton, and the style of hydrothermal fluid release will be different, probably as a slow stream, through passive degassing. This evolution may be the cause of the commonly developed late phyllic overprint that is observed at many deposits. It is unclear whether the separate Cu–Au and Mo mineralising events at Bingham and other deposits imply repeated build-up of fluid pressure and release and, if so, what the causes of the different metal inventories are.

A volcanic centre may also evolve over the time period of the magmatic centre, for instance through erosion or collapse, and this may affect the timing, style and nature of mineralisation, for instance by inducing cooling of the rocks over time at the level of the deposit, and may be an additional or alternative cause of overprinting or telescoping of a deep (e.g. potassic) facies of alteration, with facies characteristic of shallower levels such as phyllic.

Why porphyry deposits are a products of volcanic arcs

The petrogenetic processes involved in formation of magma beneath volcanic arcs are critical for the formation of porphyry deposits. The calc-alkaline to mildly alkaline affinity of magmatism in arcs reflects a relatively high H_2O content of the magmas. A minimum of 4% by mass water in magma is estimated to be required to allow the magma to become saturated in water and exsolve an aqueous phase at a sufficiently early stage of crystallisation to generate the porphyry style of hydrothermal system from fluid collected in a cupola.

Much of the present-day mantle has been depleted in H_2O during previous episodes of partial melting (see Box 3.2). Calc-alkaline magmas are characteristic of convergent margin volcanic arcs because this is where water is re-added to the mantle of the supra-subduction wedge from subducted sediments and hydrothermally hydrated ocean crust, dominantly through dehydration reactions that occur as the subducted crust descends through depths of 75–150 km. Water addition reduces the temperature required for melting in the mantle, and primitive mantle melts in arcs are estimated to contain up to 6% water in solution. 'Wet melting' of peridotite is thus a critical process for the formation of volcanic arcs and their contained porphyry deposits.

The majority of active magmatic arcs and former arcs that are eroded down to the level of sub-volcanic intrusions are host to porphyry deposits, with the most strongly mineralised arcs having densities of on average about one per 30 km of strike length. However, no porphyry deposits are known within some arcs, notably the Japan arc. Other arcs do have porphyry deposits, but these are sub-economic (e.g. the late-Palaeozoic northern New England Orogen in southeastern Queensland, Australia). The deposits in any one area may have formed within narrow time intervals of the period of arc activity. The approximately 30 known economic Cu porphyry deposits in Arizona and adjacent areas formed in two short periods within the longer-lasting Laramide tectonism, at 76–70 Ma and 63–58 Ma. Specific magmatic or tectonic events within arcs thus appear to trigger the formation of porphyry deposits.

Critical factors of the chemistry and evolution of arcs controlling porphyry deposit formation

For a full understanding of porphyry deposit resources, we need to consider not only the petrological development of volcanic arcs, but also the geological, tectonic and geochemical factors or combination of factors that may lead to the formation of higher-grade and world-class ores of different metals, and to the abundance of economic deposits or the **prospectivity** of arcs and arc segments.

Some factors which might in principle affect arc prospectivity and the metal content and nature of porphyry ores include:

- The rates of subduction and hence rates of water addition to the overlying mantle wedge and the time span of subduction.
- The fertility of the mantle source and the maintenance of fertility, for instance through mantle flow and replenishment of the melting volume (Box 3.2).

- The geochemistry of both the mantle source rocks to the magma and also of wall-rocks to crustal magma chambers that may be assimilated are also potential factors.
- Sulfur concentrations, which may be one factor in view of the large mass of sulfide added to porphyry deposits and surrounding alteration haloes. The presence of trace sulfide minerals in intrusive rocks shows that the silicate magma was saturated with respect to a sulfide melt at some porphyry deposits.
- The settings and sizes of magma chambers at different levels in the crust. These are critical to allow build-up of concentrations of volatile and incompatible elements in the magma. The build-up of fluid pressure is critical to form a porphyry deposit and in this respect it is noted that sub-economic porphyry deposits typically do not 'hang together' but rather have a patchy distribution of ore within a larger volume of rock. They formed from a vapour plume but likely one that physically split and disintegrated on rising through the upper crust.
- The balance of mantle magma input to fractionation and assimilation of crustal wall-rock of a magma chamber, which may be critical. This will be controlled by heat advection and flow and may thus be controlled by such factors as the temperature of the crust surrounding a lower-crustal magma chamber.

Within a prospective arc, a porphyry ore body may not be developed at any individual magmatic centre because of a number of factors. A catastrophic eruption such as a major caldera-forming eruption may disperse accumulated fluids. The lifespan of an episodically replenished, fractionating magma chamber at one level in the crust may be one control on the development of porphyry deposits. Long chamber lifespans allow large degrees of fractionation and the build-up of concentrations of both volatiles, including water, and of incompatible ore elements in the magma.

The role of topographic evolution of an arc in the preservation of porphyry ores

The majority of porphyry deposits are geologically young. Porphyry deposits form at a few kilometres depth in active tectonic environments and will thus only be preserved into the geological record where there has been limited amounts of erosion. The majority of known porphyry deposits are hosted in arcs that are eroded down to the level of sub-volcanic intrusions. The prospectivity of an ancient arc is thus related to the tectonic evolution of the arc after deposit formation, and in particular the extent to which it becomes eroded. Many arcs or segments of arcs are deeply eroded and are now exposed as metamorphic belts, for instance the Cretaceous arcs of northwestern USA, which host no known porphyry deposits. A possible cause of deep erosion is crustal shortening at the plate boundary and consequent crustal thickening during or after active arc magmatism. A different factor appears to be important in the Andes. Here there is a correlation between distribution of late-Mesozoic and early-Cenozoic porphyry deposits and climate zones, presumably because climate is a major factor on rates of erosion. There are no known large deposits of these ages in the wet and glaciated southern Andes in which erosion rates are and have been orders of magnitude faster than in the desert climates further north, despite the high altitude along the whole length of the belt. Deposits may have formed in the south, but have been since eroded.

Box 3.2 The role of the mantle on metal contents of magmas and magmatic-hydrothermal fluids

Regional-scale belts of the Earth's crust can have characteristic ore metal inventories. Do compositional and petrological differences in the Earth's mantle have a role in controlling the distribution and metal content of ores? We would predict a role of mantle composition in particular for magmatic and magmatic-hydrothermal deposits. The primary division of the upper mantle at the depths at which magmas are sourced is into asthenosphere, and hence in general the convecting part of the mantle, and the subcontinental lithospheric mantle (SCLM). The SCLM can have long-term attachment to overlying buoyant continental crust and remain isolated from the convecting mantle for periods of billions of years. Arc magmas are dominantly mafic melts from hydrous asthenospheric mantle that fractionate to intermediate and felsic compositions in long-lived magma chambers at the base of the crust and in the lower or middle crust (e.g. Gill, 1981; Ulmer, 2001). The melts assimilate crustal rocks to variable proportions of their mass, especially where and if the magmas are resident in the magma chambers in the lower crust (e.g. Annen *et al.*, 2006).

Although the upper mantle is petrologically dominated by peridotite, there is evidence that it is not chemically homogeneous, even within the convecting part of the mantle where we might expect physical mixing during convection to smooth out any variability in chemistry. One line of evidence for chemical heterogeneity in the mantle is the small but significant variability of radiogenic and stable isotope ratios (Sr, Pb, Nd, Os, Hf, O) of basalts intruded through ocean crust, including ocean island basalts. Because oceanic basalts have intruded through thin oceanic crust, variability of isotope ratios and ratios of trace elements that are not influenced by fractionation must be inherited from mantle source regions. We compare the isotope ratios of the basalts with those of 'bulk silicate Earth' (BSE) and distinguish, for instance, depleted mantle (DM), which is the source of most mid-ocean ridge basalts and has had a long-term relative depletion of incompatible elements, and various types of enriched mantle (EMI, EMII), which have had long-term enrichment of incompatible elements. Basalts derived from EM also have higher incompatible trace-element contents and can be enriched in many of these elements by an order of magnitude or more compared to DM basalts of similar overall chemistry. For further discussion on the geochemical make-up of the mantle see geochemical texts such as Rollinson (1993) and Albarède (2003).

The geochemical differences in both the convecting mantle and the SCLM are reflecting geochemical processing and evolution over time in cycles of convection, episodes of partial melting, mixing in of material carried down from the Earth's surface in subduction zones, and percolation of melts and fluids, for instance fluids released in metamorphic reactions in subduction zones. As a generalisation we would expect mantle to become depleted in incompatible elements through time with each episode of partial melting. However, it becomes at least locally re-enriched (refertilised) where material is added by subduction. The added material is either precipitated from hydrous fluids percolated from subduction zones or comprises recycled and convected fragments of subducted crustal rock (e.g. Kogiso and Hirschmann, 2006). The eclogites that are an important source of diamonds in kimberlites (see Section 2.2.5) are one type of crustal rock recycled into mantle. In general, where mantle peridotite includes material from recycled crust it is likely to melt at lower temperatures, and hence first during upwelling or heating of a part of the mantle. The melts so derived are expected to be enriched in incompatible elements.

Do these and similar processes, and the differentiation of the mantle in general, have any effect on the formation and distribution of either magmatic or magmatic-hydrothermal ore deposits? There are few data on how ore elements behave in the mantle and we have imperfect information about which ore elements may be enriched in which type of enriched mantle, and hence which are enriched through which tectonic processes. It has, however, long been recognised that there are areas of the crust, characteristically belts up to about a thousand kilometres long, which have higher densities of ore deposits, often of multiple deposit types and origins, and in some cases in which mineralisation is of multiple ages. These belts have sometimes been referred to as metallotects or **metallogenic provinces.**

As an example, Sillitoe (2008) compiled Cu and Au deposit occurrence and size along the length of the American cordilleras and showed that segments of these convergent margins are 'Au-rich' with perhaps an order of magnitude higher concentrations of gold resources than neighbouring segments. To a lesser degree there are also segments that are 'Cu-rich'. The segment of the American cordilleras that has the most clearly marked concentration of gold in different deposit types of different ages is that of southwest and central-western USA (Figure 3.17). The ages of Au-rich ores in this segment extend back from the late Cenozoic to at least the Jurassic, and possibly to the Palaeoproterozoic. Although gold deposits are present outside of the segments in which they are concentrated, these segments appear much more prospective for hydrothermal gold deposits of all types, and particularly for large, world-class deposits.

Are these Au-rich provinces sourcing magmas from Au-rich volumes of underlying rock? In examining this question we should be cognisant that high Au contents of, for instance, magmatic-hydrothermal ores, do not necessarily result from high Au contents of magmas. Gold content of ores may be a reflection of other chemical parameters of the magmas, for instance concentrations of ligand elements which may affect element fractionation within the magmatic system and the degree of partitioning of Au into ore-forming hydrothermal fluids.

Solomon (1990) noted a correlation between gold-rich porphyry deposits and segments of island arcs in the southwest Pacific along which there had been reversal of arc polarity in the Cenozoic. The arc has remained in a constant position, but crust is now subducting from its other side. Papua New Guinea and the Solomon Islands are both arcs which have had polarity reversal and are hosts to major Au-rich porphyry deposits, whereas no large deposits are known in New Zealand. Solomon (1990) and later Richards (2009) proposed that the history of subduction zones in this part of the world may be a critical factor in the formation of these deposits. Gold is proposed to be concentrated in second-phase melts of the sub-arc mantle. The speculative explanation builds on interpretations of the behaviour of the sulfide component of the mantle. Some sulfide is proposed to remain in the mantle after a first phase of arc magmatism.

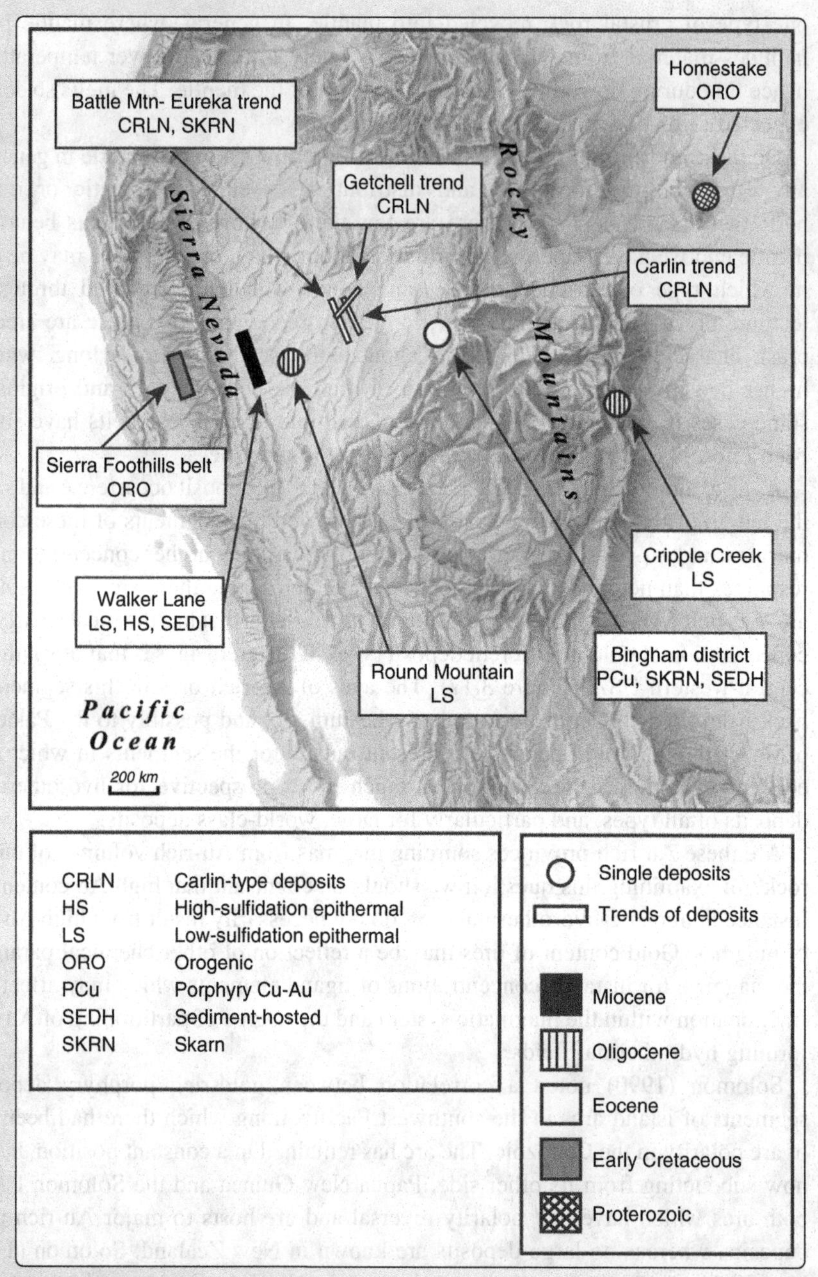

Figure 3.17 Major Au deposits (> 10-t-contained Au) of different types and ages in the western USA (after Sillitoe, 2008). An east-northeast–west-southwest trend is defined strongly oblique to the long-lived convergent margin in western North America. This belt parallels the trends of multiple middle-Proterozoic convergent margins along the southern edge of the older core of the North American plate (e.g. Whitmeyer and Karlstrom, 2007). Magmatic-hydrothermal ore deposits are abundant elsewhere within the belt of continental-margin convergent tectonics, for instance Cretaceous and early-Cenozoic porphyry copper deposits further south into northern Mexico, but do not contain large concentrations of Au.

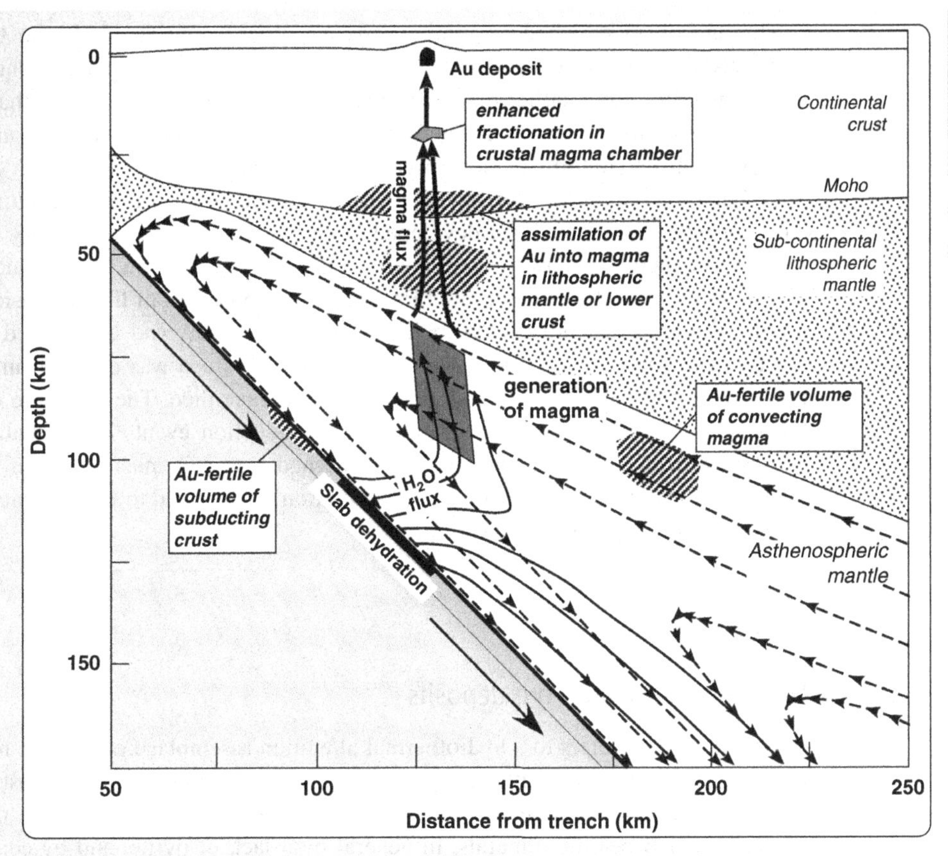

Figure 3.18 Model of dynamics of a mantle wedge and of generation and ascent of arc magmas showing possible Au-enriched reservoirs that could contribute to fertility of a magmatic centre in the shallow crust. Based on Sillitoe (2008) with mantle dynamics modified from Cagnioncle *et al.* (2007).

The Au in the mantle is strongly partitioned into the small amounts of remnant sulfide, and is incorporated into magmas when these sulfides are melted in the second phase of magmatism. This explanation of prospectivity is thus independent of mantle 'refertilisation' and does not invoke an Au-rich volume of the mantle.

Even if gold in prospective terrains is sourced from an Au-rich reservoir, there are different possible depths of enriched reservoirs which may provide an Au-rich component to arc magmas, including reservoirs in the crust (Figure 3.18). The differentiation between Sn belts and porphyry Cu belts in convergent margins is most likely a result of assimilation of lower continental crust, for instance (see Box 3.3). Titley (2001) mapped Ag:Au ratios of widespread late-Mesozoic and Cenozoic magmatic-hydrothermal ores throughout southwestern USA and showed a correlation between the ratio and the nature and age of deep continental crust, irrespective of the type of ore. Saunders and Brueseke (2012) showed a similar relationship in a neighbouring terrain based on Se and Te contents of ores. Titley (2001) argued that the lower crust is the 'reservoir' controlling ore composition, through derivation of melts at these depths. However, because crustal blocks are in most cases attached to blocks of underlying non-convecting mantle, these chemistries may alternatively be derived from the mantle.

The age of an enriched reservoir needs also to be considered. It can be either relatively recent and be related to the present configuration of plate boundaries or can be inherited from earlier plate margins. Hronsky *et al.* (2012) argued that refertilisation of the lithospheric mantle before or during the ore-forming subduction event is the important step to form a gold-prospective arc terrain. However, the lower crust and underlying mantle lithosphere appears in many arcs to be composed of fragments and slices of older 'basement' that developed in earlier periods of Earth history and may remain as coherent blocks through continental-margin tectonism (e.g. Grauch *et al.*, 2003). Pettke *et al.* (2010) measured Pb isotope compositions of fluid inclusions from porphyry ores in the prospective belt of the western USA and determined that the crustal or mantle reservoir from which the Pb was sourced was enriched in the mid Proterozoic and has remained chemically distinct since then. The signature could be the result of refertilisation of the mantle by subduction events in the Palaeo- and Mesoproterozoic. If this is the case, the enriched reservoir must reside in the non-convecting mantle (SCLM), and the refertilisation is unrelated to the present cycle of plate motions and sites of subduction.

3.2.2 Greisens and related ore deposits

The word greisen refers to a hydrothermal alteration assemblage of granitic rock, specifically quartz–muscovite with one or more F- and B-bearing mineral, most commonly fluorite, topaz or tourmaline. The alteration differs from phyllic alteration by the presence of the F- and B-bearing minerals, in general by a lack of pyrite, and by coarse-grained muscovite rather than the fine-grained sericite that is typical of phyllic alteration assemblages.

The deposit type is economically less important than porphyry deposits. Most deposits are relatively small, with maximum tonnages of tens to hundreds of megatonnes. The deposit type is, however, directly or indirectly the most important ore of Sn from the mineral cassiterite (SnO_2). Where it is indirectly the source of Sn, it is the primary source of the mineral mined from placer deposits (see Section 5.2). It is also an important source of W from the minerals scheelite ($CaWO_4$) and wolframite ($(Fe,Mn)WO_4$). Co-products and by-products of greisen deposits can be Cu, Zn, Bi and Mo, which are present in disseminated sulfide minerals, and fluorite. Deeply weathered greisens are additionally an important source of high-grade kaolin, for instance in southwest England. There are similarities in the chemistry and mineralogy of the ore and of associated magmatic rocks between greisens and the relatively rare Sn–Ag porphyries such as in Bolivia (see Section 3.2.1).

Greisen and related deposits were historically mined in many areas of the world, including in Devon and Cornwall, UK, the Czech Republic and the Palaeozoic orogenic belts of eastern Australia. For many years the world's major source of Sn was cassiterite placer deposits in Malaysia and neighbouring countries that were sourced from greisens. This source has been supplanted by mining of greisens, related deposits and cassiterite placers in four belts in the Mesozoic orogenic belts of southern China.

Settings of greisen deposits

Greisen deposits are those deposits in or temporally and spatially associated with greisenised granite. The term includes deposits that are spatially associated with greisen, but not in all cases hosted by greisen. These include skarn deposits (see Section 3.2.3) and vein deposits that extend from greisens into rocks that have different alteration assemblages. The deposits and their environments have been less thoroughly described than porphyry deposits. Although the ores are sited in and around cupolas and ridges on the upper surfaces of granitic plutons, they represent a style of magmatic-hydrothermal ore formation that contrasts with porphyry deposits both in terms of chemical nature and in terms of hydrothermal process.

The host intrusions to greisens are equigranular felsic granites rather than porphyritic minor intrusions and are the upper zones of large zoned plutons. There are generally no swarms of dykes intruding upwards through or upwards out of the cupola, but a few examples do exist, e.g. at Mt Bischoff, Tasmania, Australia, where greisen developed adjacent to a porphyritic dyke that is interpreted to have intruded about a kilometre above the source pluton. The host intrusions are generally strongly fractionated leucocratic granites with near-minimum melt mineralogies of quartz, K-feldspar, albitic plagioclase, biotite and muscovite. In the I- and S-classification the intrusions are S-type and are characteristically metaaluminous to peraluminous. In the ilmenite–magnetite classification the plutons are almost invariably ilmenite-bearing, and hence reducing.

Greisen ores and greisen alteration is typically strongest in and around a granite cupola (Figure 3.19). The alteration may be pervasive through a volume of the granite at the upper contact of the cupola, or in the interior of the granite. Multiple inverted cups of greisen may be present at a single cupola separated by unaltered granite. In many cases, granite underlying greisen is altered to albitite, i.e. albite-rich rock.

Alteration assemblages similar to those of greisen can extend into the wall-rock overlying the granite, where they are characteristically marked by the presence of abundant tourmaline. Dense stockworks of veins and veinlets of the style that are characteristic of porphyry deposits are not a part of the ore and surrounding alteration. Where veins are present they are typically sheeted, steeply dipping, large veins that are rooted in the greisenised granite and extend upwards into wall-rock. These veins are enveloped by greisen alteration haloes.

Nature of greisen ores

There is a variety of greisen ore styles. Many deposits are composed of lenticular bodies of disseminated, relatively low-grade ore (e.g. 0.25% Sn) with Sn in the mineral cassiterite in greisenised granite, typically forming sheet-like lenses sub-parallel to the upper contact of the granite. Adjacent unaltered host granite can have significantly higher concentrations of Sn (e.g. 60 ppm) than typical granites, and in some cases disseminations of ore minerals of sufficient grade to constitute ore can be present in unaltered granite. At Zaaiplaats (in the granite capping and in the centre of the Bushveld Complex, South Africa, see Section 2.2.4) there is disseminated cassiterite in epidote–chlorite–sericite altered sheet-like lenses of miarolitic granite a few tens of metres below the roof of the host intrusion.

More common and in many areas more important than disseminated granite-hosted greisen deposits are ores in sets of large sheeted quartz veins that extend above

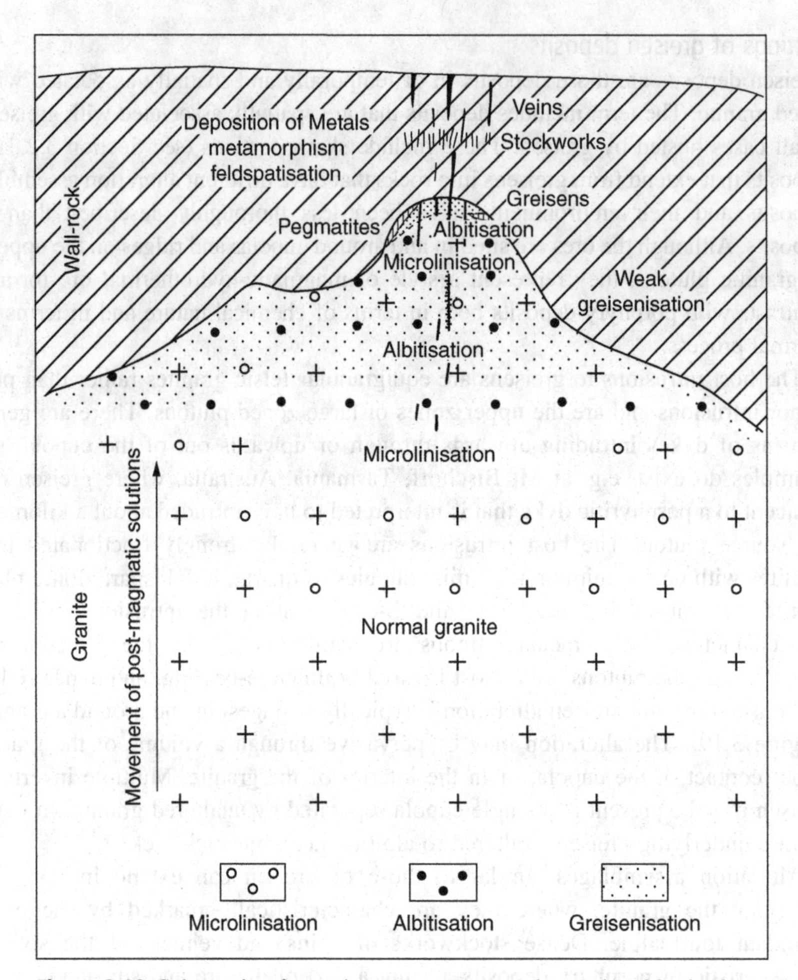

Figure 3.19 Schematic of a greisen ore system and associated hydrothermal alteration (after Scherba, 1970). Extensive and pervasive hydrothermal alteration is developed in the upper hundred metres to 1 km of a large granite pluton in and around a cupola and also in overlying wall-rock. Sheeted or stockwork veins are typically most abundant in the overlying wall-rock. Ore is variably in greisens in the cupola and in and adjacent to the veins in the wall-rock.

greisenised granite cupolas into overlying wall-rock. This is the setting at Panasquiera (Portugal) in which the veins are sub-horizontal, and also in most of the tin and copper mining districts of Cornwall (UK), in which the veins are typically sub-vertical. Both disseminated and vein ore may be present at a single intrusion, with veins overlying low-grade disseminated ore in some of the primary granite-hosted deposits of the Kinta Valley, Malaysia, for instance.

In some ore fields (e.g. Zaaiplaats) there are distinct sub-cylindrical pipes of extremely high-grade cassiterite ore with up to about 40% Sn. These are typically a few tens of centimetres across and extend, often meandering, across the sub-horizontal layers of the host granite. They are interpreted to represent channelways of either Sn-rich magmatic-hydrothermal fluid or magmas through the granite.

Interpretations of the conditions and settings of greisen deposit formation

Greisen deposits form from magmatic-hydrothermal fluids derived from large granite plutons. There are differences in the chemistry of both the ore-related intrusions (see Box 3.3 for further discussion) and of the ore fluids compared to porphyry deposits. The chemistry of ore and alteration indicate that fluorine and boron were relatively abundant in the fluid and that sulfur was in relatively low concentration. There are also differences in the mode and style of fluid migration.

The host granites are highly fractionated leucogranites intruded into continental crust. Many are of large areal extent of several tens of kilometres in diameter. Geophysical surveys have shown that many greisen-related granites are sheet-like intrusions with relatively restricted depth extent of at most a few kilometres. By analogy with granites associated with some pegmatite deposits (Section 2.2.5), it is possible that the exposed granites are the tops of vertically differentiated plutons. The rarity of porphyritic intrusions extending from the top of the ore-related plutons suggests that fluid pressure in the upper levels of the crystallising pluton was not episodically built up and released in the fashion that occurs in the plutons that feed hydrothermal fluid to porphyry deposits.

The ore-related plutons probably crystallised at a few kilometres depth in the crust, at pressures of about 1 kbar. Estimated temperatures of formation of typical greisens and associated ores are 250–450 °C. The temperatures are similar to those of phyllic alteration associated with porphyry deposits, but lower than those of potassic alteration and associated ore deposition. Albite alteration that is recorded below greisens (Figure 3.19) may be recording fluid migration at higher temperatures. Both highly saline and low-salinity fluid inclusions are present in ores, but it is not clear whether the two salinities are fluids from different sources or are the result of phase separation of a single fluid on upward migration.

The concentrations of cassiterite (and tungstates) in apparently unaltered granite implies that high concentrations of the ore elements were at least in some cases reached in the granite as a result of magmatic fractionation and independently of partitioning into a magmatic-hydrothermal fluid. However, in most cases it is apparent that the greisen ore and associated alteration formed from hydrothermal fluids that exsolved deeper in a crystallising pluton and that cooled as they percolated upwards into pluton roof zones and overlying rock. Tin is present in solution most probably as an Sn(II) complex such as $SnCl_2$. A schematic reaction for precipitation of cassiterite, in which Sn is present as Sn(IV), is thus:

$$SnCl_2 + 2H_2O \rightarrow SnO_2 + H_2 + 2H^+ + 2Cl^-. \qquad (3.5)$$

Causes of cassiterite precipitation from the hydrothermal fluid are thus either fluid oxidation or increasing fluid pH, for instance neutralisation of an acid fluid. Neutralisation of the fluid would take place as a result of the hydrolysis of feldspars to form muscovite as is characteristic of greisen style alteration, or of alteration of muscovite to topaz ($Al_2SiO_4(F,OH)_2$).

3.2.3 Skarn and carbonate-replacement deposits

Skarn is a historical Swedish mining term that has become to be used for the assemblages of Fe-bearing **calc-silicate** gangue minerals (garnet, diopside, wollastonite, etc.) that are associated with some sulfide ores. In the study of metamorphic petrology the word is used

Box 3.3 Regional controls on metal inventory of magmatic arcs

In eastern Australia there are metallogenic belts with different metal contents of ores, particularly of porphyry-style Cu–Au deposits and greisen-style Sn deposits (e.g. the Carboniferous arc of the southern New England Orogen), which trace multiple arcs at different positions (Figure 3.20). Some of these belts are marked by distinct metal inventories throughout their history and along their length, in others the inventory changes along strike of the belts. Each belt was developed over one or more periods of subduction and convergence along the Pacific plate margin between the Ordovician and the early Cretaceous. Similar but less pronounced variation of ore metal content is also apparent in the arcs developed over the shorter preserved history from the Jurassic to the present in the cordilleras of South America (Figure 3.7), and in the Canadian–Alaska segment of the cordilleras. Amongst the Cordilleran porphyry belts, there are belts of Au-rich porphyries, Cu–Au- and Mo-rich porphyries. There are also examples of arcs, or periods of arc development, in which neither porphyry nor greisen-style deposits occur. Neither deposit type is known from the Japan arc, for instance.

A magmatic-hydrothermal fluid derived from an intermediate to felsic pluton will contain multiple metals in solution, which may precipitate after different amounts of cooling and chemical evolution on transport away from its source. The importance of district-scale metal zonation is discussed in the text with respect to both polymetallic vein fields (Section 3.2.4) and within respect to skarns and carbonate-replacement deposits (Section 3.2.3). However, even after recognition and mapping of zonation, the magmatic belts on the scale of an arc or segments of an arc shown in Figure 3.20 are characterised by different deposit types and metal inventories. Magmatic events superimposed at different time periods within a belt have similar metal inventories.

Long-lived convergent plate margins undergo stepwise rearrangement over time of subduction-zone geometry, for instance the steepness of dip of the Benioff zone, and its position, and hence stepwise shifts in the position of the overlying magmatic arc. The question of what controls the metal inventory of a magmatic centre is therefore a question of what controls the inventory in an arc or a segment of an arc. Following the argument in Section 3.2, the differences are presumably related to the petrological fate or pathways of metals and also of ligands in the arc magmatic systems. They are thus expected to be correlated with differences in the chemistry of arc magmas. There are, however, multiple chemical parameters of magmatic suites (alkalinity, aluminosity, redox state, etc.) that may affect which minerals precipitate at which stage in differentiation, and one question is, therefore, which of the parameters are the dominant controls.

For granitic rocks we consider the following parameters of magma chemistry:

- Oxidation state of the magma. This can be expressed in terms of ilmenite versus magnetite series granites, or more finely divided based on the concentration ratio of redox-sensitive species of elements, particularly Fe(II) versus Fe(III), in igneous rocks.
- Alkalinity and aluminosity. These parameters are expressed by the classification of individual rock units as alkaline, metaaluminous and peraluminous, and of magmatic suites as tholeiitic, calc-alkaline, high-K calc-alkaline and alkaline.
- The degree of fractionation and compositional evolution.

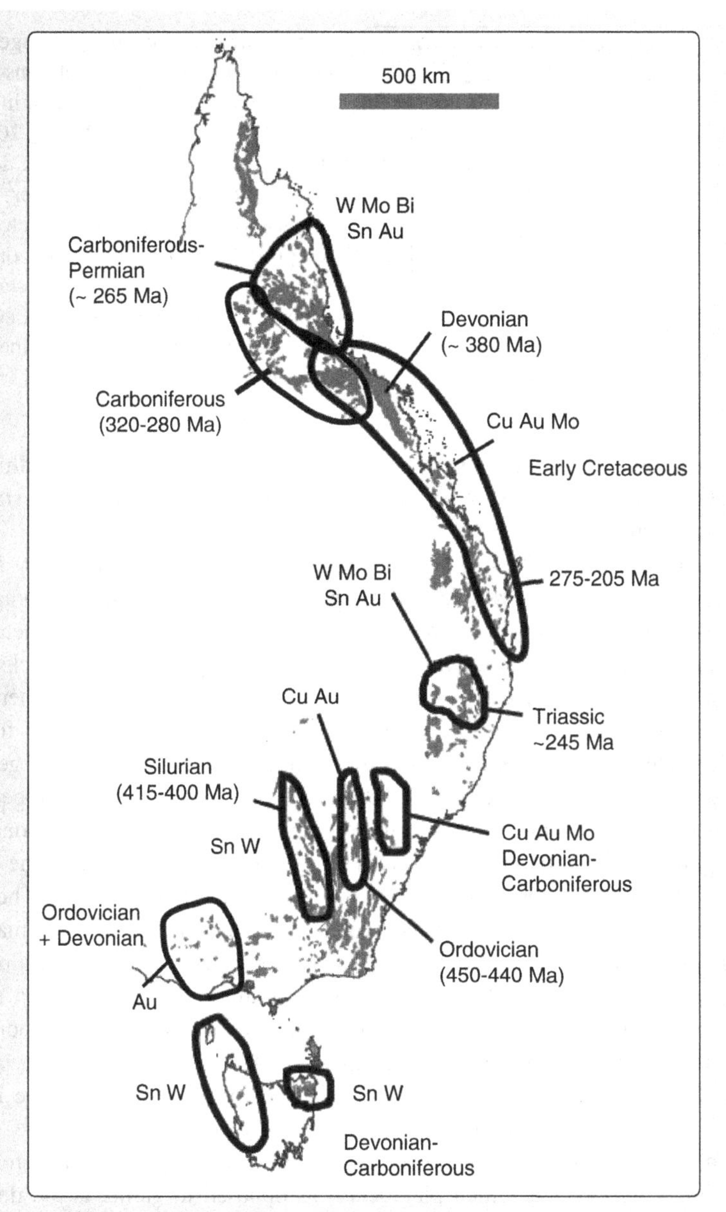

Figure 3.20 Metal associations and ages of granitoid-related ores in the Lachlan fold belt of eastern Australia (modified after Blevin, 2010). The fold belt and associated magmatism is the result of convergent-margin tectonism along an approximately north–south-trending plate margin located to the east largely continuously from early Ordovician to early Cretaceous times. The magmatic arcs and related areas of magmatism migrated over this time period and are now preserved as a series of north–south-trending belts of granitoids. The contrast between belts with Sn–W±Mo±Au and Cu–Au±Mo is preserved through multiple ages of magmatism in a belt of terrain and can therefore be interpreted to be likely related to source rocks or assimilation of rocks in the mantle lithosphere or lower crust.

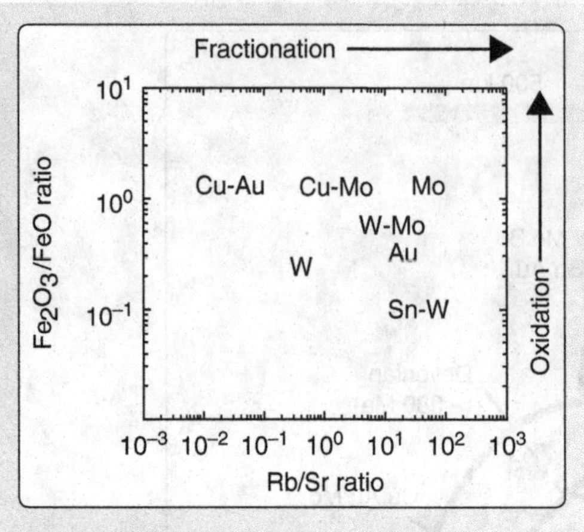

Figure 3.21 Compositional ranges and average compositions of plutons associated with different metal inventories of skarn deposit (after Blevin, 2010). The two measures effectively mark degree of magmatic differentiation. Other compositional parameters which are empirically important controls on ore metal inventory are the redox state (measured by the ratio Fe_2O_3:FeO) and whether the magmas are alkaline or calc-alkaline.

The classification of granites into I-type (igneous derived), S-type (sedimentary derived) and A-type (anorogenic) is based on a combination of oxidation state and alkalinity and is also a guide to the occurrence of ores of different metals.

A number of empirical relations are recognised between magma chemistry and ore metal inventory of magmatic-hydrothermal systems. Figure 3.21 illustrates a compilation of the degree of differentiation and oxidation state of magmas with different metals.

The oxidation state is reflected largely by the ratio of Fe^{3+}/Fe^{2+} in the rocks and is expected to control the balance of valence state of a number of other elements in a magma, perhaps most importantly of sulfur. The oxidation state is generally found to correlate with aluminosity, essentially because relatively reduced magmas are generally S-type magmas that have incorporated or been influenced by interaction with peraluminous sedimentary rocks, which also tend to be graphitic. The degree of fractionation is measured by the Rb/Sr ratio of the rocks. This ratio is expected to increase in the magma with fractionation as Sr is compatible in and is sequestered into plagioclase. The use of Rb/Sr assumes a simplistic model for the evolution through fractionation of magmatic systems in arcs. In reality, long-lived (< 10 million year) crustal magma chambers of magmatic arcs are kept active through episodic or continuous input of mafic, mantle-derived magma. The Rb/Sr ratios thus reflect a complex balance of fractionation in the chamber and refilling, and the balance of a combination of processes such as wall-rock melting, assimilation and magma hybridisation that is expected to take place in deep crustal magma chambers. The use of the parameter also assumes that the magma chamber in which the critical compositional evolution takes place is at a relatively shallow depth (< 20 km) at which plagioclase as opposed to garnet is the dominant Al-rich phase to be crystallised. A ratio that is proposed to measure the depth of a crustal magma chamber is Sr/Y as Y is strongly compatible in garnet (e.g. Richards, 2011).

One result, shown in Figure 3.21, is that oxidation state is a guide to whether magmatic-hydrothermal deposits are Cu ores. The relationship suggests that there is a mineral that crystallises in reduced magmas but not in oxidized magmas that concentrates Cu. From our knowledge of the geochemical behaviour of Cu, the most likely candidate is a sulfide mineral. Possible sulfide phases controlling Cu behaviour are pyrrhotite (Fe_7S_8), which typically contains 0.1 to 1% Cu where it occurs in magmatic rocks, and Cu-bearing sulfide minerals that are similar in composition to chalcopyrite but are only stable at high temperatures (intermediate solid solution, or *iss*).

The behaviour of Sn is best understood in this respect. Hydrothermal Sn deposits are almost invariably associated with reduced, strongly fractionated felsic plutons. In the S-, I-, A-classification of granites, the plutons that host Sn deposits are almost invariably S-type and are peraluminous to metaaluminous reduced, ilmenite series rather than oxidised, magnetite series. A common accessory mineral in I-type granites is titanite ($CaTiSiO_5$), in which Sn(IV) substitutes readily for Ti(IV) in the structure. Titanite does not crystallise in reduced granites: the most important Ti-bearing phase in these granites is ilmenite ($FeTiO_3$), and there is thus no sink for Sn in a mineral, its concentration increases as the magma fractionates, and it can hence be in high concentration in an exsolved fluid.

The oxidation state of magmas may be inherited from the convecting mantle (Box 3.2), but primitive arc basalts are almost invariably characterised by relatively high oxidation states (e.g. Rowe *et al.*, 2009), and the patterns of belts in time and space with common oxidation state suggest rather that the control is in the non-convecting lithospheric mantle or in the crust. S-type granites tend to form in continental rather than intra-oceanic island arcs, and are often present in the hinterland to the main line of the arc. The correlation between peraluminosity and low oxidation state in the ilmenite series of granites most likely reflects a control from incorporation of pelitic and graphite-bearing metasedimentary rocks, either directly at source or as assimilant picked up from the wall of a magma chamber or a magma conduit in the lower or middle crust. Belts of Sn granites may thus occur where thick sequences of turbiditic sedimentary rocks are present in the lower crust. This interpretation is consistent with the geological histories of belts of Sn ores.

The role of the degree of fractionation within a magmatic system on the metallogeny of hydrothermal ores is shown most strongly by the nature of magmatic suites that are host to Mo porphyry and Cu–Mo porphyry deposits. Molybdenum-bearing porphyry deposits are characteristic of continental magmatic arcs, and especially arcs developed over relatively thick continental crust. The ore-related porphyritic intrusions at the Climax-type Mo deposits are compositionally high silica rhyolite, and are hence the products of large degrees of fractionation of felsic magmas. Molybdenum is interpreted to be incompatible through all stages of magma fractionation, in contrast to Cu, which is sequestered in sulfide minerals, and it hence builds up to highest concentrations in the most fractionated phases (e.g. Audétat *et al.*, 2011).

to describe any **metasomatic** rocks with calc-silicate minerals. In ore deposit geology the use is more restrictive: a skarn deposit is an ore deposit in carbonate-bearing rocks that have been hydrothermally altered to assemblages of calc-silicate minerals together with, in some cases, magnetite or Mg-bearing gangue silicate minerals. The deposits are characteristically within about a kilometre distance of the contact of an igneous intrusion. Ore minerals occur in fractures or are disseminated through the altered rock.

Carbonate-replacement deposits are similar. These are deposits in metasomatically replaced carbonate rock that generally lack calc-silicate gangue minerals but in which ore is dominated by sulfide minerals. Carbonate-replacement deposits typically form at greater distances from igneous intrusions than do skarn deposits. There are some ore districts in which both types of ore are present in zonal distributions around an intrusion.

Both ore deposit types are ores for a variety of metals. Economically important types of skarn deposits as classified by contained metals (with examples) include:

- Cu, Cu–Au and Au skarns, such as at Ertsberg, West Papua, Indonesia (Mertig *et al.*, 1994) and Carr Fork area mines of the Bingham district, Utah, USA (Atkinson and Einaudi, 1978);
- W and W–Sn skarns such as at King Island, Tasmania, Australia (Kwak and Tan, 1981) and Cantung, Northwest Territories, Canada (Mathieson and Clark, 1984);
- Sn skarns, such as associated with greisens in Mesozoic magmatic belts of southern China (Chen *et al.*, 1992).

There are also occurrences of Zn–Pb skarns and Fe skarns, which are magnetitites. The latter have been sources of Fe, but neither Fe-skarns nor Zn–Pb skarns are of recent economic interest.

Carbonate-replacement deposits are ores for:

- Zn–Pb–Ag together with, in some cases, Au at for instance Leadville, Colorado, USA (Thompson and Arehart, 1990), Lark and US mines in the Bingham district of Utah, USA (Rubright and Hart, 1968) and Santa Eulalia, Mexico (Megaw *et al.*, 1988);
- Sn together with minor W at Renison Bell, Tasmania, Australia (Kitto *et al.*, 1997).

Both ore deposit types are formed in environments in which hydrothermal fluids infiltrate into and interact with carbonate-rich rocks within a few kilometres of magmatic centres. There are many examples in which one or both of these ore deposit types occur in the same ore district as porphyry deposits, or in the same district as greisens. In these districts the ores can be considered to form composite systems in which the same metal suite is present in all ore types, or in which metal inventories are systematically zoned with respect to distance from a magmatic centre. For instance, there are two neighbouring and similar-aged small-volume late-Pliocene porphyritic intrusions in the Grasberg region of Irian Jaya, Indonesia of which one is the host to the Grasberg porphyry Cu–Au deposit whereas the other is both surrounded by and encloses four Cu–Au skarn deposits (Figure 3.22). In contrast, at Bingham, Utah, USA, a single magmatic centre is host to and is the centre of zoned porphyry, skarn and carbonate replacement ores (Figure 3.23).

Characteristics of skarn deposits

Skarn deposits are relatively high-grade but low-tonnage deposits compared to typical porphyry deposits. For instance, the Grasberg porphyry deposit contains approximately 2000 Mt of ore at a Cu grade of 1.1%, whereas the nearby Ertsberg skarn had 25 Mt of ore at 2.6% Cu.

Skarn deposits form in carbonate-bearing rocks that are adjacent to or near to intermediate to felsic intrusions, including both large granite plutons and high-level porphyritic stocks. The position of the ores relative to the associated intrusions is variable: contact skarns line the contact of an intrusion, but other skarns extend within carbonate units up to about a kilometre away from contacts or form isolated bodies that do not abut an intrusion. In each situation, the ores are stratabound to the carbonate units of bedded sequences of rocks. In plan view the ore bodies may map out as irregular ribbons (e.g. Figure 3.23), known as **mantos** (**blankets**) where they occur in flat-lying units. The Ertsberg deposit shown in Figure 3.22 is unusual as it is a pervasively replaced roof

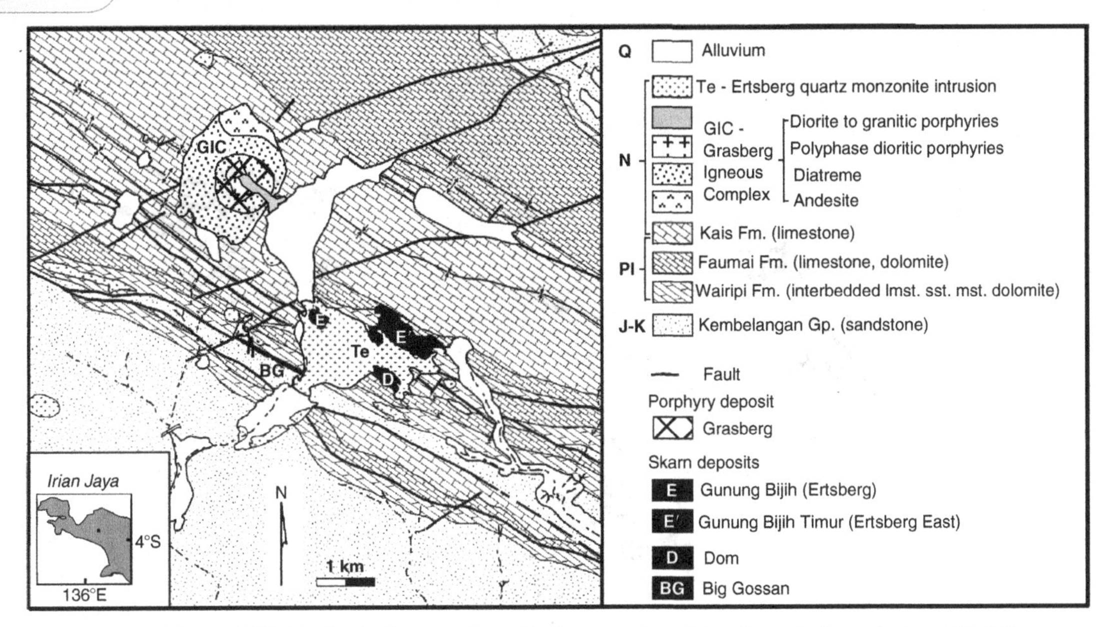

Figure 3.22 Geological map of the Ertsberg region, Irian Jaya, Indonesia (modified from Rubin and Kyle, 1997; Pollard and Taylor, 2002). Two similar pipe-like intrusions of late-Pliocene (4–2.1 Ma) age cut through a deformed carbonate-dominated sedimentary rock package and are associated with different styles of Cu–Au mineralisation: the Grasberg Igneous Complex (GIC) is a polyphase, largely intermediate volcanic and intrusive complex that is host to the major Grasberg Cu–Au porphyry deposit. The wall-rock to the ore-associated intrusions is an igneous diatreme breccia. Four major Cu–Au skarn deposits have formed by replacement of limestones in and around the neighbouring Ertsberg intrusion (Te). Two are along the contact of the intrusion; Big Gossan follows a lithological contact outwards from the wall of the intrusions and the Gunung Bijih skarn is a replacement of a large roof pendant of limestone in the intrusion (Mertig *et al.*, 1994).

pendant of carbonate within the host intrusion. Interleaved non-carbonate rock units may be altered and veined and include low concentrations of the ore metals but do not comprise ore (see Figure 3.24).

Economic ore typically forms bodies of coarse-grained disseminated ore within parts or zones of the replaced and recrystallised carbonate units. Veins are often not an important component of the ores. Ore mineralogies are very variable depending on the metals. A typical sulfide assemblage of a Cu skarn would be pyrite–chalcopyrite±bornite: pyrrhotite–arsenopyrite is the dominant sulfide assemblage in many Au skarns. Skarn deposits may be zoned with respect to ore metal content with distance from the ore-related intrusion.

In typical skarns, there is zonation of gangue mineralogy, textures and styles of metasomatic replacement of the host carbonate unit. A traverse towards the intrusion is thus from recrystallised carbonate (often marble or hornfels that formed as a result of isochemical contact metamorphism due to the heat of the intrusion), to marble with nodules and veins of skarn minerals, to massive skarn with almost total replacement of carbonate minerals by skarn minerals adjacent to the intrusion such that earlier textures

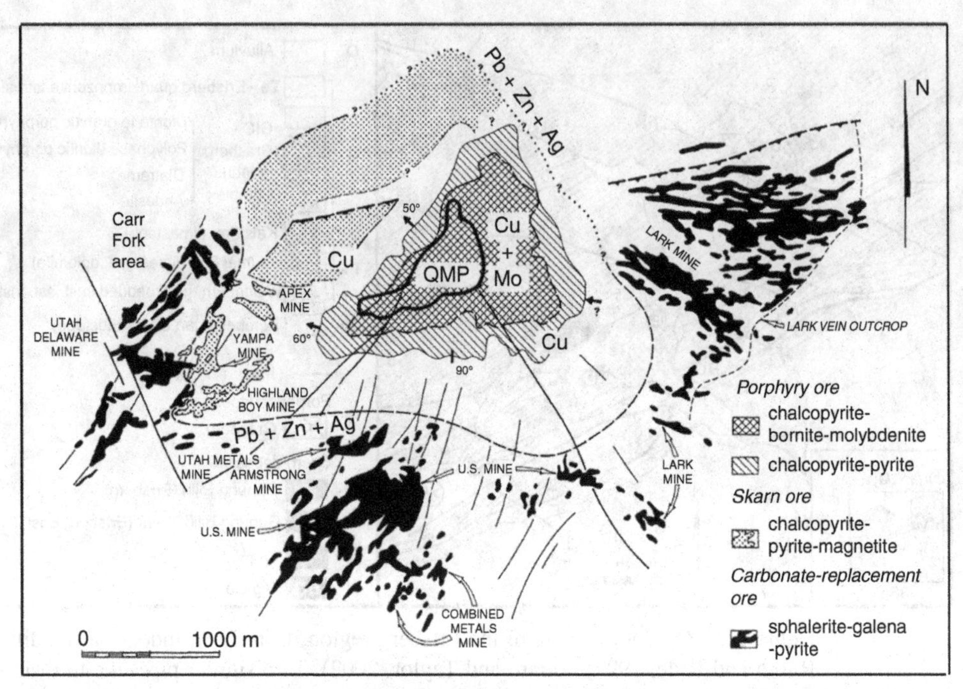

Figure 3.23 Ore deposits in the Bingham area of Utah, USA (after Atkinson and Einaudi, 1978). The major porphyry Cu–Au–Mo deposit is centred on a quartz–monzonite porphyry stock (QMP) – see Figures 3.9 and 3.11d. Historically important Cu skarn ores and Pb–Zn–Ag carbonate-replacement ores however surround the porphyry deposit in a zonal arrangement at up to about 2 km distance from the edge of the porphyry. Further out from the magmatic centre are minor Au–Ag pyrite vein ores.

are overprinted. The contacts between marble and massive skarn are characteristically sharp. They may however be irregular with bulbous and convex–concave shaped interfaces on scales down to a few centimetres. The different mineral assemblages within zoned skarns are likewise separated by irregular interfaces. In some cases there are also skarn assemblages in the intrusion close to its contact (*endoskarn*, as opposed to *exoskarn*).

Textures of skarns formed through hydrothermal activity are different to those formed through contact metamorphism; the skarns for instance are generally coarser grained. In many skarn deposits it is possible to recognise prograde stages and retrograde stages in their mineralogical and textural evolution, the former comprising assemblages of minerals that crystallised at the highest temperature reached during magmatic and hydrothermal activity, the latter comprising minerals crystallised during cooling. Retrograde assemblages partially or fully overprint and replace the prograde assemblages in many skarn deposits.

The gangue mineral assemblages of skarns are similar to metamorphic assemblages in metamorphosed impure limestone or dolomite. Common prograde minerals include diopside, garnet, wollastonite, vesuvianite, idocrase, olivine and magnetite. Many of these minerals are stable only at relatively high temperatures of above about 450 °C. We

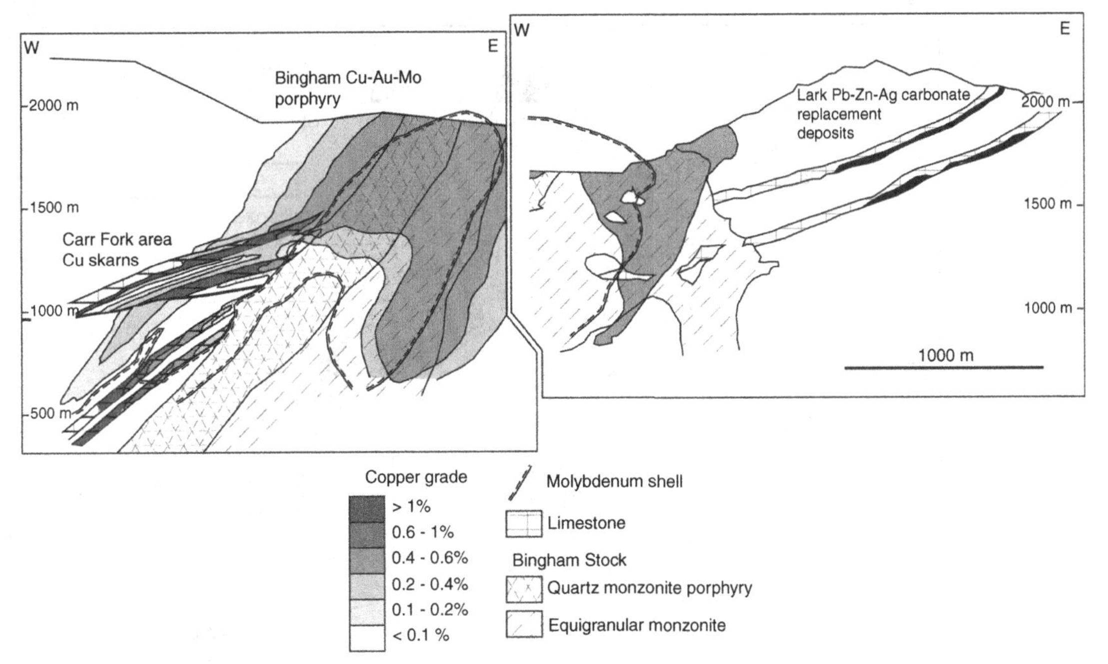

Figure 3.24 Staggered cross section through the Bingham magmatic centre (see Figure 3.23) showing relations between porphyry, skarn and carbonate-replacement ore bodies (after Rubright and Hart, 1968; Atkinson and Einaudi, 1978; Gruen *et al.*, 2010). No precise information is available on the shape of ore bodies in the Lark deposits and these are thus shown schematically. The skarn ores in the Carr Fork area are within the low-grade limb of the porphyry ore shell and are interpreted from petrographic relations to have formed at the same time as the porphyry ore, whereas the carbonate-replacement deposits at Lark are outside the explored shell and may be related to later quartz–sericite–pyrite veins and associated alteration (Einaudi, 1982).

distinguish skarn types based on the dominant minerals in this list, e.g. diopside skarns or garnet skarns. Magnetite skarns are a relatively common type and their development presumably reflects the high concentration of iron in many high-temperature magmatic-hydrothermal fluids. Skarns formed in dolomite in contrast to limestone have significant concentrations of Mg-bearing minerals, e.g. olivine. Common retrograde minerals include amphibole, epidote, chlorite, serpentine and talc, hence hydrous minerals that form at temperatures down to approximately 300 °C in environments in which there is ample water.

Characteristics of carbonate-replacement deposits

Carbonate-replacement deposits, as do skarns, typically form as a result of localised pervasive stratabound replacement of carbonate units within bedded sedimentary rock sequences. Like skarns, they can be relatively high grade compared to porphyry and greisen deposits. Carbonate-replacement deposits are generally within about 2 to at most 4 km of shallow intrusions (Figure 3.23). There may, however, be multiple intrusions of different age within a few kilometres distance of some

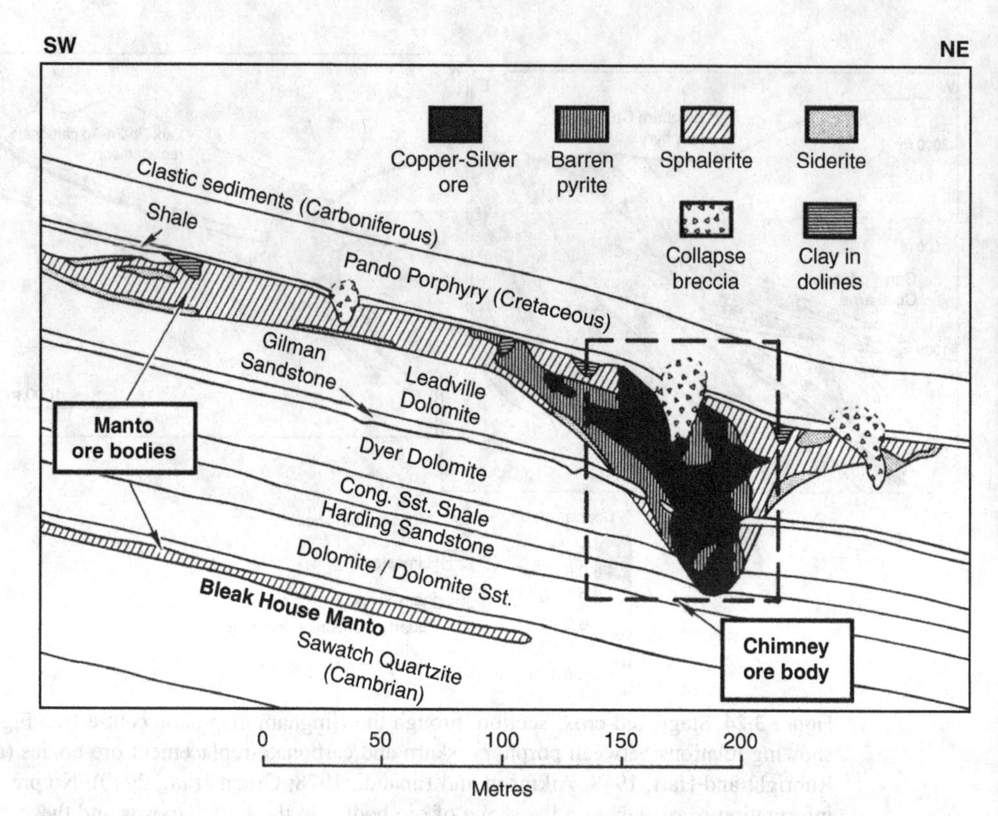

Figure 3.25 Cross sections through the Gilman Pb–Zn–Ag carbonate-replacement deposit, Colorado, USA (after Titley, 1993). Manto carbonate-replacement ore bodies occur at the top of the Carboniferous dolomitised limestone where it is overlain by shale and to a lesser extent in Cambrian quartzite. The discordant chimney ore body presumably follows a fault or fracture that cuts the carbonate sequence. The ore assemblage with Cu sulfide minerals in the chimney can be interpreted to have formed at higher temperature than the assemblages of the manto bodies and to have fed hydrothermal fluid into the mantos up dip.

carbonate-replacement deposits, and it is thus not in all cases possible to link the ore bodies with a specific intrusion.

Carbonate-replacement deposits are also similar to skarns in that they have irregular shapes within host beds. Ore bodies can likewise often be described as mantos, but can also include **chimneys** in which ore bodies transgress the stratigraphy (Figures 3.25 and 3.26). Replacement is often preferentially within certain horizons of the host carbonate unit, for instance at the top or bottom of limestone beds. Pods or lenses of replacement may be adjacent to faults or to dykes that cross-cut the units, and the radial ore shoots at for instance deposits in the Bingham district (Figure 3.23) reflect sets of radial fractures and faults around the central intrusion. The overall shapes of many ore bodies have strong resemblances to the shapes of cave systems in karst terrains (Figure 3.26).

Many ores are essentially bodies of massive sulfides, intermixed with unreacted carbonate host-rock and some quartz. The ore assemblage in Pb–Zn–Ag carbonate-replacement deposits within 1–2 km of the Bingham Cu–Au–Mo porphyry deposit is

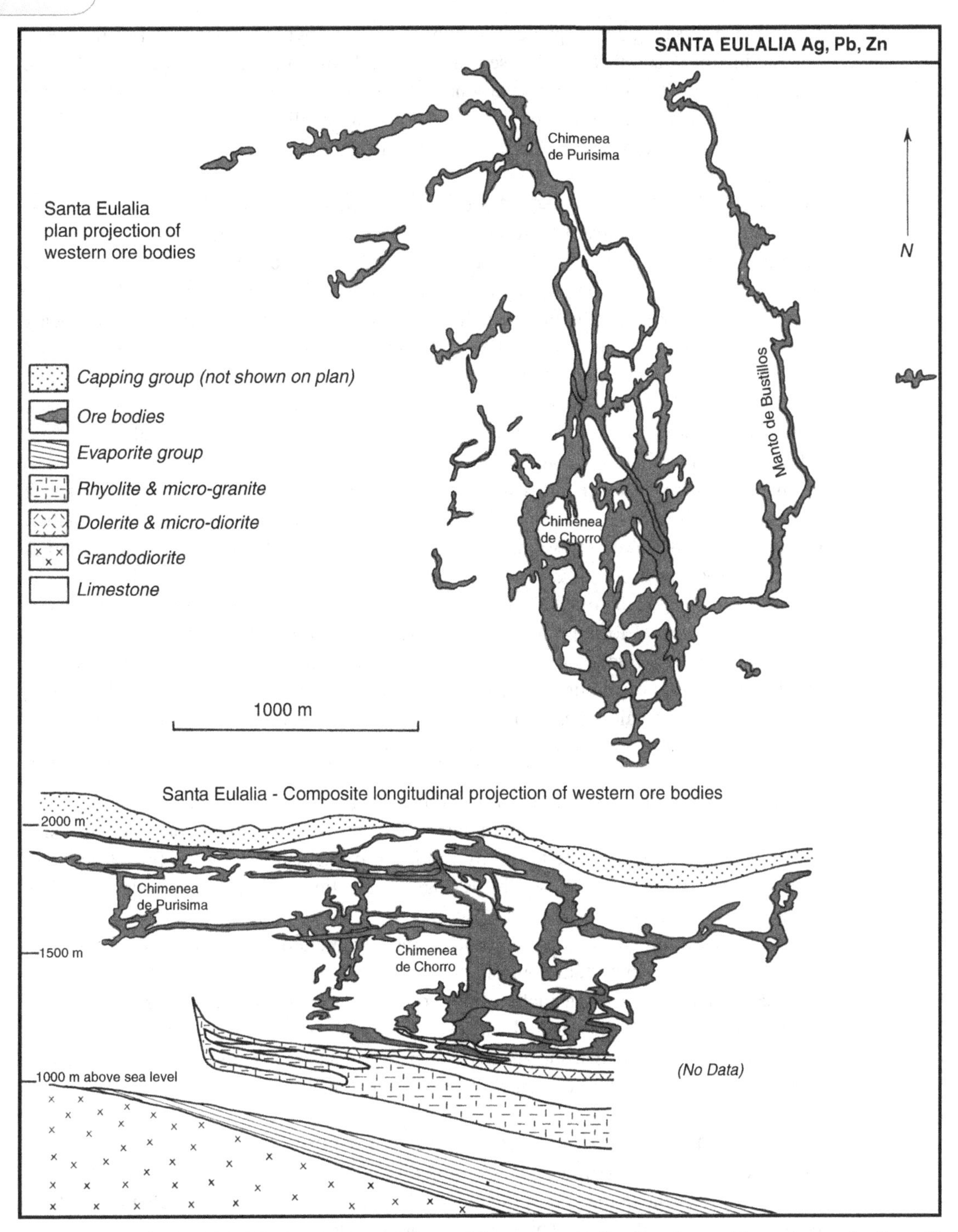

Figure 3.26 Maps and cross sections of the geometrically complex and irregular ore bodies with mantos and chimneys (chimeneas) at the Santa Eulalia carbonate-replacement deposits, Mexico (after Hewitt, 1968). There is a close resemblance between the ore-body geometry and karst cave systems. The ores are mineralogically mainly either sulfide ores with pyrrhotite (or pyrite)–galena–sphalerite, or **oxide ores** with a Pb and Zn carbonate and sulfate minerals and Fe and Mn oxide minerals.

for instance pyrite–galena–sphalerite, with up to 70% pyrite. One characteristic set of non-sulfide ore minerals that is present in some carbonate-replacement ores include zinc oxide (e.g. zincite, ZnO), zinc carbonate (smithsonite, $ZnCO_3$) and lead carbonate (cerrusite, $PbCO_3$) minerals, for instance at Santa Eulalia. Zinc-rich ores with these minerals form one type of so-called oxide ores of zinc.

Chemical and physical processes in the formation of skarn and carbonate replacement

The unique characteristics of these two deposit types result from the unique chemistry of carbonate minerals and rocks. Carbonates are strongly soluble in high-temperature hydrothermal fluids that have low concentrations of dissolved CO_2 and low ratios of Ca^{2+}/H^+. Importantly, carbonate minerals can dissolve orders of magnitude faster than silicate minerals at similar temperatures. The characteristic sharp contacts between unaltered and altered rocks are 'reaction fronts' and the bulbous shapes of the fronts are typical of 'reaction-infiltration instabilities' that develop where there is positive feedback between chemical reaction, in this case dissolution of carbonate minerals, and fluid infiltration, hence focussing further reaction.

There are many instances in which there is a reaction front separating calcite marble from skarn with an assemblage of diopside together with possibly garnet and magnetite. Comparison of the mineral assemblages and textures across these reaction fronts shows that the carbonate unit was effectively completely dissolved, and was replaced by the skarn assemblage with little change in rock volume (isovolumetric replacement). The chemical reactions to form prograde skarn assemblages are complex and can be inferred, for instance in simplified and schematic form for a clinopyroxene skarn in limestone to involve dissolution of calcite, addition of iron and silica from the hydrothermal solution, precipitation of the calc-silicate minerals, and liberation of CO_2, in schematic form:

$$\text{calcite}(CaCO_3) + Fe_{(aq)} + SiO_{2(aq)} \rightarrow \text{clinopyroxene}(CaFeSi_2O_6) + Ca_{(aq)} + CO_2. \quad (3.6)$$

Some calcium will be released into solution if the reaction involves isovolumetric replacement of calcite by clinopyroxene. Similarly complex reactions, for instance involving replacement of calcite by pyrrhotite or pyrite, can be written for alteration at carbonate-replacement deposits. The retrograde reactions in skarns are approximately isochemical except for the addition of water.

There are multiple factors that control alteration mineralogy in a hydrothermal system (see Section 3.1.3). Important controls on whether garnet, diopside or wollastonite is the dominant mineral in a skarn include: the composition of the country rock (particularly whether dolomite or limestone and whether a clay-rich carbonate unit); the fluid temperature, and hence distance from an associated intrusion or position within a magmatic vapour plume; and the hydrothermal fluid composition.

Ore genesis at skarn and carbonate-replacement deposits

The assemblages of prograde gangue minerals in skarns form at approximately 450–600 °C, which is the upper end of the range of fluid temperatures that are implied by fluid inclusions in these deposits. The complex paragenetic histories of many skarn deposits with prograde and retrograde stages reflect heating and cooling of the rock volume around high-level intrusions. Temperatures of formation of carbonate-replacement deposits have been estimated to be around 300 °C to locally about 400 °C (for instance in the Cu–Ag chimney at Gilman, Figure 3.25) based on fluid-inclusion data and on isotope fractionation between coexisting minerals. These deposits thus formed at significantly lower temperatures than skarn deposits and the differences in the gangue minerals are related dominantly to which minerals are stable at each range of temperature.

The common metal content of skarn and carbonate-replacement ores and associated porphyry and greisen ores implies a common magmatic-hydrothermal source of the metals. The gangue minerals that are paragenetically associated with ore are thus also likely formed as the result of infiltration of magmatic-hydrothermal fluids. In contrast to the prograde reactions, the retrograde reactions are approximately isochemical except for the addition of water, and may thus be the result of convection of groundwaters during cooling of the intrusion.

The relatively high grades and polymetallic nature of both skarns and carbonate-replacement deposits can be explained as results of the unique chemical nature of carbonate rocks. As described in the section on porphyry deposits, magmatic-hydrothermal fluids become acidic on cooling as a result of disassociation of acid aqueous species such as HCl in solution. Dissolution of carbonate minerals, however, buffers fluid pH, through reactions such as:

$$CaCO_3 + 2H^+ \rightarrow Ca^{2+} + CO_2 + H_2O. \tag{3.7}$$

Most ore minerals are more soluble in acidic than neutral solutions. The solubility, for instance, of sphalerite in a fluid in which Zn is transported as a chloride complex can be expressed as:

$$ZnCl_4{}^{2-} + H_2S \rightarrow ZnS + 2H^+ + 4Cl^-. \tag{3.8}$$

Neutralisation of an acidic fluid by dissolution of carbonates thus induces a decrease in ore mineral solubility. The relatively high ore grades of skarn and carbonate-replacement ores compared to adjacent and related porphyry or greisen deposits are the result of efficient chemical precipitation of ore minerals as acid is neutralised.

Patterns of fluid flow and sources of hydrothermal fluids at skarn and carbonate-replacement deposits

The settings and distributions of skarn and carbonate-replacement deposits around magmatic centres and their forms are variable.

The Bingham Cu–Au–Mo porphyry deposit is surrounded by a partial halo of Cu skarn deposits and by Pb–Zn–Ag carbonate-replacement deposits (Figure 3.23). The Cu skarn deposits lie within the low-grade Cu halo of the porphyry deposit. For instance, the Carr Fork deposits, which had grades of about 2–3% Cu, are hosted in carbonate units that are interbedded between quartzite units and are adjacent to one flank of the Bingham Stock and within the low-grade halo to the lower levels of the shell of the major porphyry ore body (Figure 3.24). Both the interbedded quartzite and adjacent stock have sub-economic Cu grades of between 0.1 and 0.3%. Temperatures of skarn ore formation are within the range of those estimated for the potassic alteration zone in the enveloping low-grade halo of the porphyry deposit and the paragenetic timing of ore mineral precipitation is the same as that of potassic alteration. The skarn deposits can thus be interpreted to have formed in the peripheral parts of the magmatic vapour plume from which the porphyry deposit was formed (cf. Figure 3.15). The metals and sulfur are thus most likely derived from the pluton that fed the small intrusions, including the Bingham Stock, and not directly from the Bingham Stock that is adjacent to the deposit. The 10- to 20-times higher grades of the skarns than adjacent low-grade porphyry ore are presumably the result of the efficiency of copper sulfide precipitation on acid neutralisation, although they may have been enhanced by channelisation of fluid flow along the carbonate units as porosity is developed on dissolution of calcite.

The Pb–Zn–Ag carbonate-replacement deposits in the Bingham district are 0.5 to 1.5 km from the edge of the Cu shell of the porphyry deposit (Figure 3.24). They likely formed at temperatures of approximately 300 °C, as would be expected at the outer edges of a rising cylindrical plume of magmatic 'vapour'. The concentrations of lead and zinc in the input fluid to the porphyry deposit are similar to those of Cu (Box 3.1). Galena and sphalerite are present in the outer parts of the porphyry ore shell, but at abundances much less than 1% and too low for Zn and Pb to be extracted as by-products. These peripheral deposits may thus also be formed at the margins of the same vapour plume that formed the porphyry deposit with efficient metal precipitation on acid neutralisation. The paragenetic timing of ore mineral precipitation suggests rather that the ores formed during infiltration of the later, lower-temperature fluids of dominantly magmatic origin, which formed the D-type quartz–pyrite veinlets with sericitic haloes that are abundant through both the porphyry ore and immediately surrounding rock. The radial distribution of ore shoots suggests that fractures were an important control on fluid flow in this peripheral zone of the hydrothermal system.

The Gilman (Figure 3.25) and Santa Eulalia (Figure 3.26) deposits are different in that they are not spatially associated with known magmatic centres. Both are in the neighbourhood of intermediate to felsic sills, which formed broadly during the same tectonic event, although the overlying sill at Gilman is earlier than the ore. Isotope compositions of ore and gangue minerals imply, however, that the ore fluids were dominated by magmatic-hydrothermal fluids and hydrothermal temperatures of up to about 400 °C imply a relatively proximal, but unexposed, magmatic fluid source. Ore fluids appear to have been very strongly channelised along highly permeable channelways in order to retain a high temperature. In both areas, and also elsewhere, earlier karst was likely a major control on ease and patterns of hydrothermal fluid flow. Clay-filled doline depressions at the top of the Gilman ore body indicate some karstification before deposition of the overlying clastic sedimentary rocks and hence more than 100 million years before the early-Cenozoic mineralisation. Karst cavities would have been themselves controlled by faults and fractures and hence these would also control the geometry of the ores.

3.2.4 Polymetallic veins and vein fields associated with magmatic centres

There are a large number of small, mainly historic mines in and around magmatic centres in which multiple individual veins were or are **selectively** mined in contrast to **bulk** mining of large volumes of veined and mineralised rock at porphyry deposits. Many **vein fields** are well-defined clusters of veins spread over several kilometres diameter, each vein separated by altered but unmineralised rock or by unaltered rock. It is in many cases a question of economics as to whether mining is bulk-rock or selective. These veins mark a hydrogeologically different style of fluid migration away from magmatic centres than porphyry deposits in particular.

Veins of vein fields include economic grades of one or more of the metals Sn, W, Mo, Bi, U, Au, Ag, Cu and Zn. Many are polymetallic and contain ore extracted for multiple commodities.

Vein fields in some ore camps overprint porphyry ores but may be significantly more extensive, such as at Main Stage Veins at Butte, Montana, USA (Meyer *et al.*, 1968) and at Morococha, Peru (Catchpole *et al.*, 2011). Other examples are peripheral to either greisens or porphyry deposits (tin mining districts of Cornwall, UK) or occur around magmatic centres which lack known porphyry or greisen deposits (Idaho Springs – Central City district of Colorado, USA, Rice *et al.*, 1985).

Common characteristics of veins around magmatic centres include the following:

- The veins are quartz- and calcite-dominated, with disseminated sulfide or small pods of massive sulfide in the vein and in adjacent altered wall-rock.
- Ore veins can range from a few centimetres to about 10 m in width and can be spaced at up to about 1 km distance.
- There are typically one or multiple sets of veins by orientation, with multiple sets by orientation and gradual changes in vein orientation similar to those of the veinlets at porphyry deposits (Figure 3.12).
- The size of vein fields and the nature of the associated magmatic centres vary. Tin-bearing systems are typically the most widespread and many of these are associated with equigranular granites with large areas of surface exposure that are much wider than deep and that intruded to a few kilometres depth in the crust. A vein field extends over 30 km from the outcrop centre of the Mole Granite of the New England Orogen, New South Wales, Australia (Figure 3.27). The most peripheral veins of the field occur up to about 10 km beyond the lateral extent of the pluton, which is interpreted from geophysical data to be disc-shaped and not more than about 5 km thick. In contrast, the Pb–Zn–Ag–Au system such as at Central City is centred on a magmatic centre formed of multiple phases of pipe-shaped, small porphyritic intrusions similar to those in and around porphyry deposits. The area of the vein field is about 5 km across and hence much smaller than at the Mole Granite.
- Ore fields are zoned with respect to metal content of the veins and also to the alteration assemblages in the vein haloes. In the case of the Mole Granite, there are zones from the centre of the granite outwards with vein ore of (Figure 3.27):
- W–Bi (wolframite–bismuthinite)
- Sn (cassiterite)
- Sn–Cu–Pb–Zn–As (cassiterite–arsenopyrite–chalcopyrite)
- Pb–Ag–Zn–Cu (sphalerite–galena).

The W deposits in the centre of the field are associated with silicification and albite alteration, whereas greisen-style alteration (muscovite–quartz) is dominant around the Sn veins, and chlorite–sericite alteration around veins at greatest distance from the pluton. There is thus similarity of both the metal content and alteration mineralogies of the ores with those of greisen deposits.

Isotope data show that the hydrothermal fluid in these vein fields is in most cases dominantly a magmatic-hydrothermal fluid, although with varying proportions of

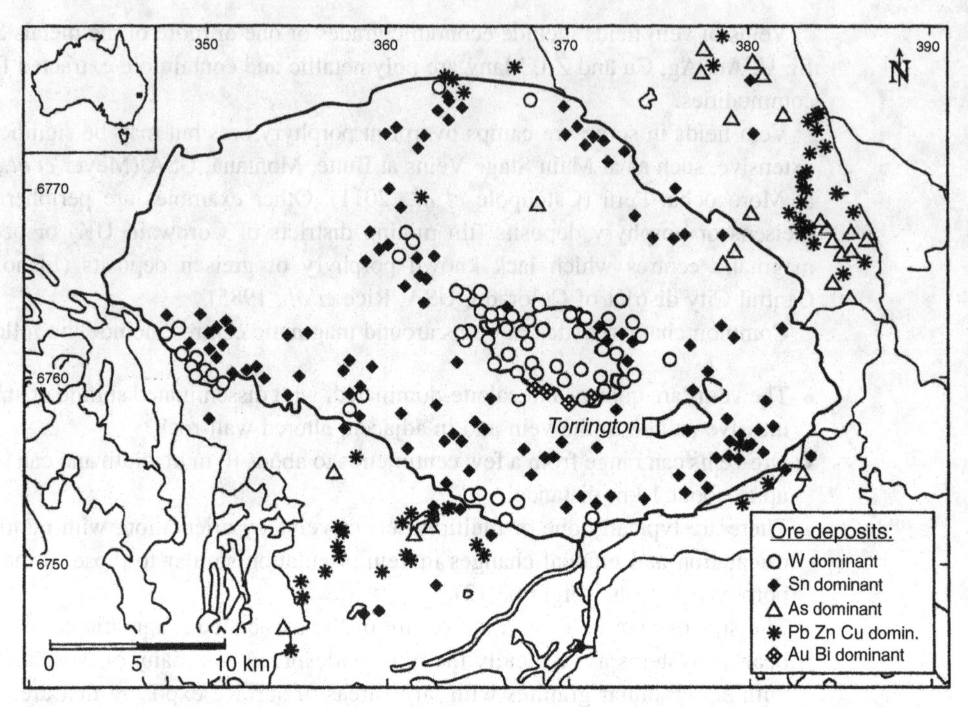

Figure 3.27 Distribution of metalliferous veins around the Permian Mole granite, New South Wales, Australia, showing the zonation with respect to metal content (Audétat *et al.*, 2000, after the Geological Survey of New South Wales). The outcrop limit of the granite is indicated by the thicker solid line.

in-mixed meteoric waters, especially in the peripheries of the fields and late in the paragenetic histories. The large-scale metal zonation in vein fields reflects saturation of the fluid with respect to different ore minerals at different positions within the hydrothermal system, due to fluid cooling and evolution of the fluid composition because of interaction with wall-rock, and mixing of the metal-carrying fluid with other fluids. In the case of the Mole Granite (Figure 3.27), the W deposits are interpreted to have formed at 500–600 °C, and the Pb–Zn–Cu deposits at about 250 °C. Fluid-phase separation is recorded in a number of fields, but in the same fashion as in porphyry deposits it may not be a critical factor promoting ore mineral precipitation. Mixing with groundwaters will in general be most important at the outer edges of the vein field and during the later infill stages of the veins.

Different vein fields appear to have formed from different styles of fluid migration. The smaller fields such as at Central City are of similar dimensions as lithocaps and of phyllic caps above some porphyry deposits (e.g. Figure 3.14). Fluid exsolution to form these veins fields was thus less focussed at the source pluton than that of typical magmatic vapour plumes. In the case of the broader fields such as at the Mole Granite dispersed fluid exsolution seems inferred, as does lateral migration of fluids away from the source pluton, or alternatively it seems that the fluids in the peripheral veins were derived from deeper buried larger intrusions of the same suite as the outcropping pluton.

3.2.5 High-sulfidation epithermal Au–Ag deposits

Two types of epithermal deposits are recognised. This section describes and discusses high-sulfidation Au–Ag epithermal deposits, and the next section low-sulfidation epithermal deposits. The distinction between the two is discussed in the following introductory section

General introduction to epithermal deposits

Epithermal, a word meaning low temperature, was originally applied to ores based on mineral textures such as euhedrally terminated crystals and open spaces (**vughs**) in veins, and on the presence of specific ore minerals (e.g. **sulfosalts**) and gangue minerals (e.g. clay minerals), which indicate hydrothermal mineral precipitation and alteration at lower temperatures than most other types of ore deposit that are associated with magmatic centres.

The word has taken a more restricted meaning in ore deposit geology:

> Low-temperature ($< \approx 300\,°C$), precious- or base-metal deposits with close temporal and spatial association with volcanic centres. Precious metals (Au and Ag in various ratios) are the major products of epithermal deposits, although some produce by-product Hg and Sb, and by- or co-product base metals Pb, Cu and Zn.

The two contrasting types of epithermal deposit, high-sulfidation- and low-sulfidation-, form in similar tectonic settings as porphyry deposits (Figure 3.28) in continental volcanic arcs, intra-oceanic island volcanic arcs and in areas of diffuse volcanism in continental back-arc regions. As epithermal deposits are formed at shallow depths (< 2 km), many probably become eroded relatively rapidly after formation. The majority of known deposits are thus geologically recent (Cenozoic), and occur in areas of active or recent arc volcanism (circum-Pacific volcanic belts – Chile, western USA, Japan, New Guinea, New Zealand). The geologically youngest mined epithermal deposit (Ladolam on Lihir Island, Papua New Guinea) formed at less than 0.4 Ma, and is hosted in the crater of a recently extinct volcano that is still geothermally active, and which is possibly also still an active ore-forming system. A few pre-Cenozoic deposits have been preserved from erosion, for instance, in Palaeozoic magmatic arcs of eastern Australia, the Appalachians of eastern USA and in central Asia. Some deposits occur in deformed and metamorphosed Proterozoic volcanic terrains such as at Enåsen, Sweden.

Sulfidation state and the division of epithermal deposits

Epithermal deposits have very variable ore mineralogy, alteration mineralogy, metal content and grade, and also ore and structural style, including whether the ore is in veins or is disseminated in altered rock. The first-order rationalisation of this variability is into a two-fold end-member division of high-sulfidation and low-sulfidation epithermal deposits.

This division is formally based on **sulfidation state**. This is analogous to oxidation state and is a measure of the fugacity, or gas pressure, of molecular sulfur (f_{S2}) in the hydrothermal fluid. Sulfidation state is effectively a measure of the ratio of sulfur to chalcophile elements in ore minerals. It controls the valence states of chalcophile elements, and hence which sulfide and oxide ore minerals form. Note, however, that the term does not necessarily indicate the amount of sulfur present.

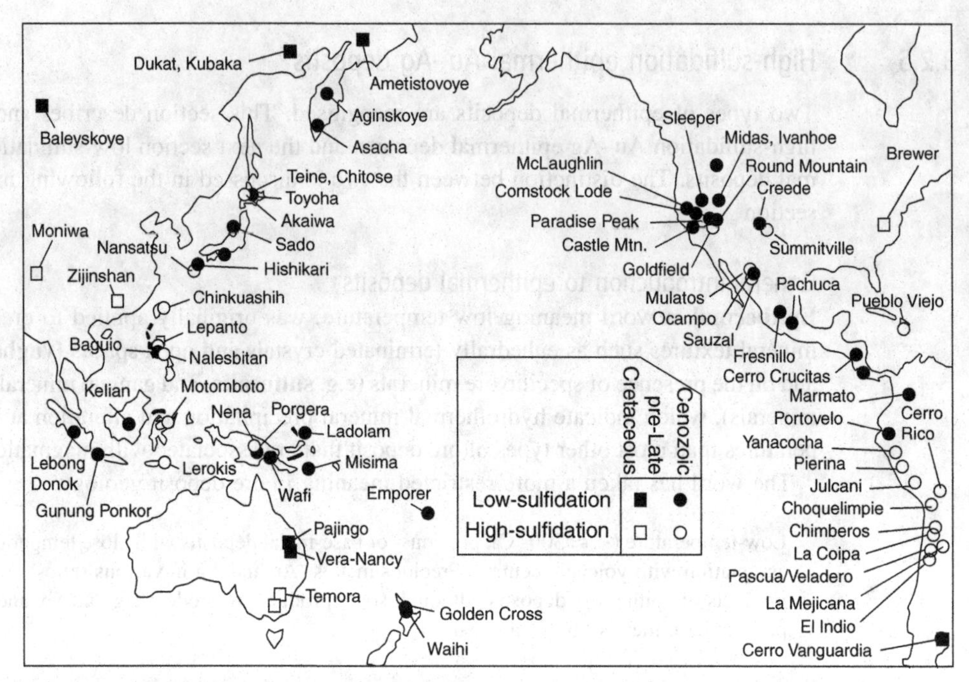

Figure 3.28 Distribution of epithermal deposits in the circum-Pacific region (modified after Cooke and Simmons, 2000). The division by age distinguishes deposits that formed along convergent plate boundaries that are active at the present day from those related to boundaries in the geological record. Low- and high-sulfidation classes of deposit are discussed in the text. Note the close correspondence of the distribution to that of porphyry deposits shown in Figure 3.6.

The divisions between sulfidation states are marked by mineral reactions that are balanced with addition or subtraction of molecular sulfur. Some of the more important of these mineral reactions are, with the higher sulfidation state assemblages to the right:

$$\underset{\text{pyrrhotite}}{Fe_7S_8} + 3S_2 \rightleftharpoons \underset{\text{pyrite}}{7FeS_2} \tag{3.9}$$

$$\underset{\text{chalcopyrite}}{5CuFeS_2} + S_2 = \underset{\text{bornite}}{Cu_5FeS_4} + \underset{\text{pyrite}}{4FeS_2} \tag{3.10}$$

$$\underset{\text{tennantite}}{0.67Cu_{12}As_4S_{13}} + S_2 = \underset{\text{enargite}}{2.67Cu_3AsS_4}. \tag{3.11}$$

Based on these and other sulfidation reactions we can divide the spectrum of sulfidation state of ores into:

- low-sulfidation ores, with arsenopyrite (FeAsS), chalcopyrite ($CuFeS_2$) and pyrrhotite (Fe_7S_8);
- intermediate-sulfidation ores, with pyrite (FeS_2) in place of pyrrhotite and with tennantite ($Cu_{12}As_4S_{13}$) replacing arsenopyrite;

- high-sulfidation ores, in which pyrite is important, enargite is the main arsenic-bearing mineral (Cu_3AsS_4), and either bornite (Cu_5FeS_4) or covellite (CuS) is typically the major Cu ore mineral.

The sulfidation state of epithermal ores generally correlates with both their oxidation state and their acidity:

- The ores of low-sulfidation deposits are reduced, such that sulfur is present dominantly as reduced S(II) in sulfide minerals. Gangue assemblages of these deposits indicate a near-neutral hydrothermal fluid. Alternative names of this deposit type are, adularia–sericite, based on the common presence in veins of sericite and adularia, a low-temperature potassium feldspar, or low-sulfur deposits.
- High-sulfidation deposits are oxidised, such that the sulfur is present in part as oxidised S(VI) in sulfate minerals such as barite, anhydrite and alunite or as native sulfur (S(0)). Altered rock in and around ore is strongly leached of many elements. The leaching indicates interaction with a strongly acidic hydrothermal fluid. These deposits are thus sometimes known as acid-sulfate, or high-sulfur deposits.

Some care is needed in classifying an epithermal ore deposit based on these characteristics because rocks altered to acid-sulfate assemblages are spatially associated with some low-sulfidation deposits.

Low-sulfidation epithermal deposits and high-sulfidation deposits form in different environments through different processes and the distinction between the types is generally clear when their settings and all geological features and the distribution of these features are observed. The two end-member types and their modes of formation are thus described and discussed separately. Deposits with intermediate-sulfidation ore assemblages are similar to those with low-sulfidation assemblages and have been generally combined into the low-sulfidation class. As shown in Figure 3.28, both low- and high-sulfidation deposits can form in volcanic arcs, but there are some segments of active volcanic arcs in which one of the types is widespread and the other largely absent. A brief investigation of the reasons for the differences between the deposit types is given in Box 3.6.

High-sulfidation epithermal deposits

High-sulfidation epithermal deposits are Au or Au–Cu deposits with by- or co-product Ag, which are hosted in intensely altered rock, in most cases in intermediate to felsic volcanic or high-level intrusive rocks, including lavas and pyroclastic rocks. Well-studied and described examples include Yanacocha, Peru (Teal and Benavides, 2010); Summitville, Colorado, USA (Gray and Coolbaugh, 1994; Bethke *et al.*, 2005); Pascua, Chile–Argentina (Chouinard *et al.*, 2005); and Nansatsu, Japan (Hedenquist *et al.*, 1994).

Geological settings of high-sulfidation ores

The environment of formation of most of the geologically young deposits of this type can be interpreted to have been within about 1.5 km of the Earth's surface and less than about 2 km distance from a volcanic centre that was active over essentially the same time period as ore formation. Some typical settings within and around volcanic centres are within

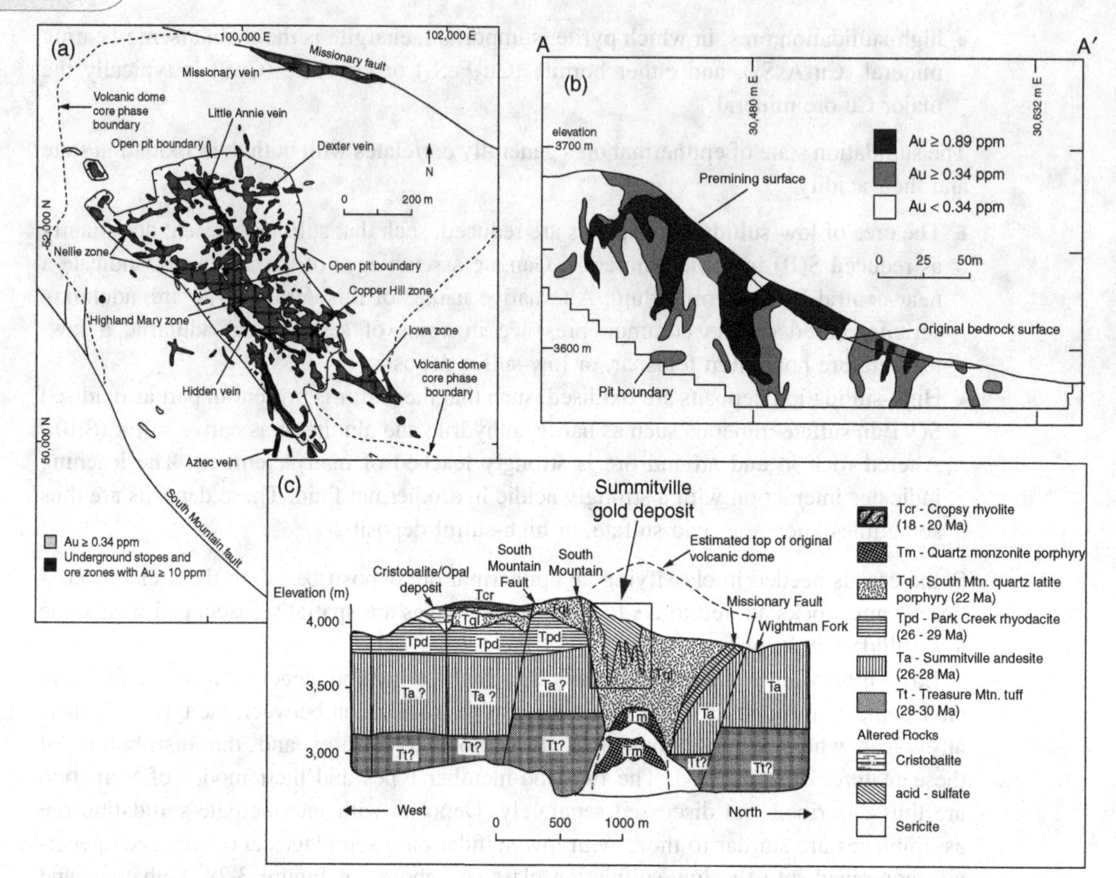

Figure 3.29 Ore geology and setting of the mid-Cenozoic high-sulfidation epithermal Au–Ag deposit at Summitville, Colorado, USA (after Gray and Coolbaugh, 1994). (a) Map showing the trellis-like distribution of ore with multiple orientations of ore zones which formed along and adjacent to fractures of different sets. (b) Cross section A–A′ showing a section through partly eroded multiple ore shoots, each of which has approximately inverted tear-drop shapes with heads at about the same altitude. (c) Interpreted setting of the ore body with the host volcanic sequence. The deposit is within a porphyritic intrusion that was the feeder pipe to a lava dome that extruded above the cupola of a larger intrusion about 1 km beneath the surface. The faults are ring faults of the host caldera. The cristobalite–opal deposits that are shown are presumed to have formed at the same time as the ore at the palaeosurface.

calderas, especially below lava domes that have built up within a caldera (Figure 3.29), or beneath lava domes and central craters in fields of intermediate to felsic volcanoes. In other cases the deposits formed under the flanks of large stratovolcanoes rather than immediately under the positions of craters (Figure 3.30). Phreatic breccia pipes which splay upwards are a common host to ore or parts of an ore body.

The larger deposits are often composed of effectively closely spaced clusters of individual ore bodies scattered over areas of about 2 km^2 in the case of Summitville, Colorado, USA (Figure 3.29) or Pascua, Chile. In each of these deposits, the ore is restricted essentially to within a relatively narrow altitude range of 200–300 m and forms an overall blanket. The large Yanacocha deposit in Peru is exceptional in that multiple ore

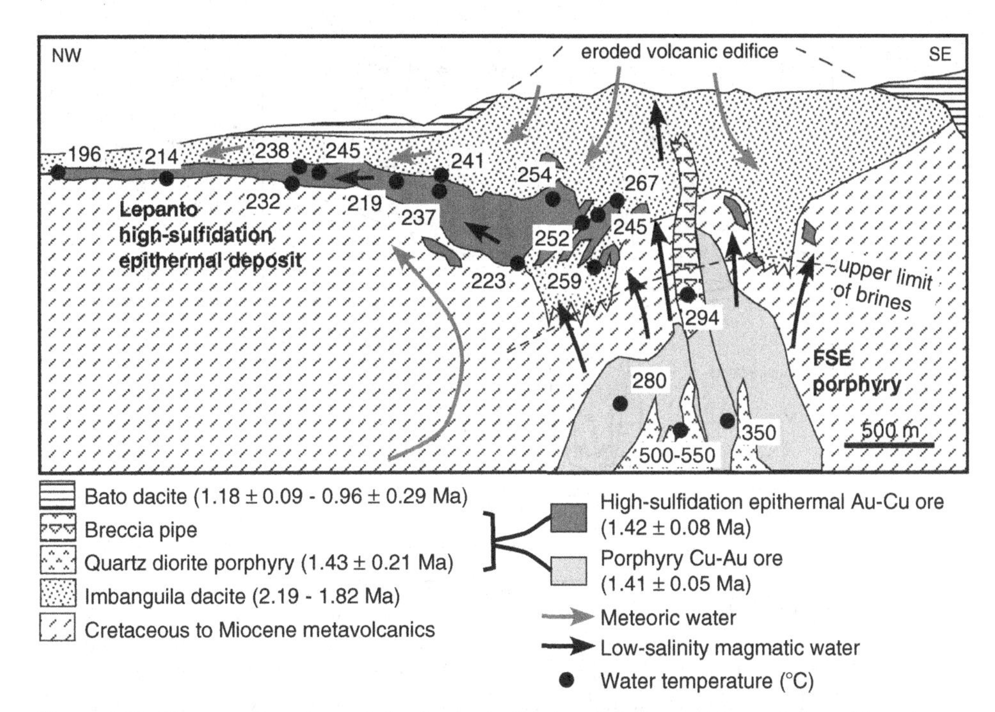

Figure 3.30 Cross section through the Lepanto (high-sulfidation epithermal deposit) and adjacent FSE (porphyry deposit), Philippines. The Lepanto deposit straddles the basement–lava-pile contact under a flank of the central volcano, whereas the porphyry deposit mantles a quartz–diorite porphyritic stock at about 1 km depth (Arribas *et al.*, 1995). The two deposits give indistinguishable radiometric ages with uncertainties constraining the time of formation to within less than 0.12 million years. Interpreted fluid flow paths are shown. All fluid inclusions contain low-salinity fluids above the 'upper limit of brines' whereas both high-salinity and low-salinity inclusions are present below this level. Water temperatures (in °C) are measured from fluid inclusions and document progressive cooling of the magmatic-hydrothermal fluid as it migrates away from the source and mixes with groundwater.

bodies are scattered over about 100 km^2 within an extensive volcanic field of lave domes. In this deposit the different ore bodies formed over a time period of about 5 million years, probably beneath lava domes and craters that were active at slightly different times.

The term **lithocap** describes a sub-horizontal body of erosionally resistant, strongly silicified, intensely altered rock that is generally barren of ore and that forms at and just below the surface at many volcanic centres in volcanic arcs. A lithocap is in most cases the host to high-sulfidation ores where they are developed.

Nature of the ores

Ore minerals are disseminated through irregular ore bodies that are hosted in many cases in much larger volumes of intensely altered rock. Ore may be concentrated along and adjacent to fractures in the host-rocks, to give a trellis-like pattern in map view. Veins are generally not a significant host to ore, except in the deepest levels of a few deposits. An inverted tear-drop shape is characteristic of many isolated ore bodies. These are

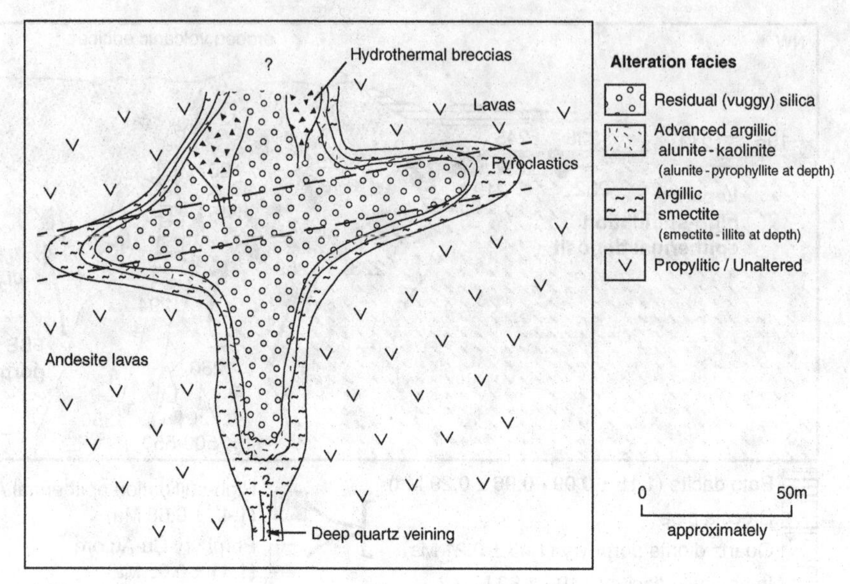

Figure 3.31 Schematic cross section through a typical high-sulfidation epithermal Au deposit – based on Iwato deposits in the Nansatsu district, Kyushu, Japan (after Hedenquist *et al.*, 1994). Gold ore is hosted in residual silica, and is associated with pyrite, enargite, covellite and native sulfur. Hydrothermal fluid migrated upwards through the volcanic pile, but high permeability of pyroclastic beds allowed lateral fluid infiltration. The hydrothermal breccias bottom out at about the same level as the ore zone and may be the result of boiling of groundwater where it was heated by magmatic vapour.

narrow, vein-like ore zones at depth that splay upwards into wider 'mushroom heads' of ore. The larger deposits are often composed of effectively closely spaced clusters of such inverted tear-drop-shaped ore bodies or irregular-shaped high-grade ore centres (Figure 3.29b).

There is characteristic zonation of alteration facies around the ore bodies (Figure 3.31). The central ore zone of most deposits consists of porous to massive quartz-rich (up to 95% quartz) rock. This style and mineralogy of alteration is known as *vuggy silica*, or residual silica. In essence, all major chemical components of the rock other than silica have been leached by the hydrothermal fluids. Primary volcanic quartz phenocrysts may be distinguishable from fine-grained quartz that crystallised from the silica component of leached matrix minerals, for instance feldspars. In general, however, primary rock textures are not preserved. Millimetre- to centimetre-sized **vughs** are scattered throughout the altered rock and comprise up to a few per cent of the rock. These formed during alteration as a result of overall leaching of matter from the rock.

The immediately surrounding alteration zone shows *advanced-argillic* alteration. The essential minerals of this facies are quartz, alunite ($KAl_3(SO_4)_2(OH)_6$), and a clay of the kandite series, most commonly kaolinite or dickite. Anhydrite ($CaSO_4$) is common in this alteration facies. Pyrophyllite is an important mineral at deeper levels with advanced-argillic alteration and may replace alunite. In rare cases andalusite is present. The alunite may occur as pseudomorphs after original feldspar. This assemblage marks slightly lower degrees of leaching than vuggy silica, such that K and Al that were originally in the rock remain in addition to silica.

The surrounding alteration zone is *argillic* facies in which one or more of kaolinite, illite, smectite or interlayered illite–smectite is a major mineral. This zone may be pervasively developed between ore zones in larger deposits, but around isolated small deposits, such as shown in Figure 3.31, the argillic alteration zone is surrounded by a propylitic alteration zone with assemblages including calcite, sericite and epidote similar to those at porphyry deposits.

Ore minerals are typically disseminated through the most strongly altered rock, most particularly in the vuggy-silica zone. Ore mineral concentrations are up to a few per cent by mode. The minerals are either intergrown with quartz or have grown as euhedral grains into the vughs that formed as a result of the alteration. Characteristic ore minerals are pyrite, chalcocite (Cu_2S), covellite (CuS), and multiple Cu–As sulfosalts such as enargite–luzonite (Cu_3AsS_4). The sulfosalts may occur together in complex intergrowths. Gold is present as both native gold and as solid solution in sulfide minerals. Native sulfur is present locally with sulfide minerals in some ore bodies.

Interpretation of the environment of formation of high-sulfidation epithermal deposits

The geological settings of high-sulfidation epithermal deposits show that they formed in magmatic centres during active magmatism. The ore and alteration assemblages and other geochemical and mineralogical data such as differences in ratios of the stable isotopes of S between sulfate and sulfide minerals indicate ore formation at 150–350 °C, most commonly at about 250 °C. However, there is evidence from for instance fluid-inclusion data, and the presence of fine intergrowths of sulfosalt minerals that grew from droplets of sulfide melt, that hydrothermal fluid temperatures were higher, up to about 600 °C, at least transiently in some high-sulfidation deposits. Temperatures may thus not have been held at around 250 °C through all stages of ore formation, and may have varied through ore bodies during their formation.

The almost complete leaching of all major cations other than Si from vuggy-silica alteration zones indicates that the core of these deposits formed from extremely acidic fluids (pH of 1–2). Acidity decreased outwards through the enveloping alteration zones. Strongly acidic conditions are also inferred for the formation of the lithocaps that form as a widespread blanket over degassing magma chambers. These strongly acid conditions can be produced through chemical reactions similar to those that produce acid rain, that is, the condensation of acid gas species (SO_2, H_2S, CO_2, HCl) into water. Sulfur dioxide is probably the most critical component for producing pH of 2 or lower at these deposits through the reaction:

$$H_2O + SO_2 + \tfrac{1}{2}O_2 \rightarrow H_2SO_4 \rightarrow H^+ + HSO_4^-. \quad (3.12)$$

The textural setting of many ore minerals as euhedral crystals that have grown into vughs implies that they grew either after the intense acid alteration or at a late stage in the development of the alteration assemblages.

The ratios of stable isotopes of H, O and S in alteration and ore minerals that have precipitated from the hydrothermal fluid, e.g. alunite, show that the water and the sulfur in solution were dominantly sourced from magma. Fluid inclusions preserve multiple hydrothermal fluids of contrasting composition and density similar to those in the upper levels of porphyry deposits (Box 3.1). Three types are repeatedly recognised: low-density vapour-like fluids, relatively low-salinity (3–10 wt % salts), liquid-like fluids, and rarer high-salinity brines (up to greater than 40 wt % salts). A low-salinity magmatic-hydrothermal fluid that is exsolved from magma at greater than 700 °C at depths of less than about 2 km will have vapour-like density at source (see Figure 3.1). The different fluid-inclusion types can be interpreted as having formed from such a fluid as it rises and loses pressure above a magma chamber. Vapours may have been little modified from source or may have expanded as pressure is reduced with little loss of temperature: vapours that cool as they rise may contract to low-salinity liquid-like fluids; brines may have condensed from the vapour as it decompressed and cooled.

Volcanic degassing and metal flux through volcanoes

High-sulfidation epithermal deposits are formed at a shallow level in active magmatic centres by rising hydrothermal vapours, some of which may have degassed to the atmosphere. In view of the settings of ore formation, we can use observations and measurements on fluids and vapours at active arc volcanoes in order to understand the hydrothermal processes that contribute to the formation of these deposits. The available data include those that have been collected, for instance, for understanding volcanic hazards and styles of volcanic eruption.

Many active volcanoes emit gases between eruptions at temperatures of between about 200 °C and magmatic temperatures of up to about 900 °C. These emissions occur even when cooling magma is not present at the surface, and historical records show that they can either be short-lived, during and after an eruption, or be more continuous, in some cases over time spans of at least a few hundred years. Where emissions are continuous, the temperature and flux of gases can be very variable over time spans of a few years.

The emissions are variably diffuse through craters or lava domes, or through discrete hydrothermal vents. Fumaroles are vents of high-temperature gases; solfataras are fumaroles that emit sulfuric gases. Where gas temperatures are high (> 600 °C), the gases can be inferred to be sourced essentially directly from magma and have risen as a plume upwards above the magma with little interaction with lower-temperature rock or with surface-derived waters on their rise. The source of the gases is presumably partially molten magma at shallow depth in a chamber such as a feeder pipe to the volcano, possibly at a depth of less than about 1 km. Lower-temperature emissions indicate greater cooling between the magma and the surface. Fluid cooling may be because of a deeper source, escape through less-permeable rock and hence greater heat exchange with rock, mixing with surface-derived waters, for instance in a perched water table on or under the flanks of the volcano, or a combination of these processes.

At active calc-alkaline volcanoes, the emitted gases are dominantly water vapour, but include a few per cent of other gas species, especially SO_2 and CO_2. Higher concentrations of SO_2 in particular are measured emitting from some volcanoes with alkaline magmas. These gases carry metals in solution. At many volcanoes, ore mineral sublimates line fumarole vents, for instance the hydrated molybdenum oxide ilsemanite around vents emitting gases at around 800 °C at the andesitic Merapi volcano in Indonesia. Direct sampling of the gases allows detection of and measurement of the concentrations of the metals, with the most commonly recorded at arc volcanoes being As, Cd, Cu, Au, Fe, Pb, Mo, W and Zn. Although the metal concentrations are low, as would be expected in solution in these extremely

Table 3.2

	White Island, New Zealand	Satsuma Iwojima, Japan	Augustine, Alaska, USA	Mt Etna, Italy	Ladolam, Papua New Guinea	Rotokawa, New Zealand
Style of discharge	Eruption	Fumarole	Fumarole	Eruption	Geothermal	Geothermal
Temp. (°C)	> 850	877	870	900	260	320
Flux (Mt a^{-1})						
H_2O	1.9	5.2	0.03	50	1.5	4.5
CO_2	0.5	0.04	0.003	13	0.014	0.3
Cl	0.04	0.06	0.005	0.1–0.5	0.0032	0.0023
S	0.06	0.09	0.005	0.2–0.75	0.0048	0.0012
Cu (t a^{-1})	110	0.16	0.011	480–850	71	ND
Au (kg a^{-1})	> 36	0.02	ND	80–1200	24	37–109

Fluxes of gas components and metals in solution in emitted gases at example volcanoes. The fluxes are those emitted into the atmosphere from lavas at the surface during eruptions, those emitted from fumaroles during quiescent periods, or the upward flux through the shallow subsurface in geothermal fields above magma chambers (Hedenquist and Lowenstern, 1994; Simmons and Brown, 2006; 2007; Calabrese *et al.*, 2011). ND – not determined.

low-density gases, they are significant and the concentrations are such that, given the measured gas flux, up to hundreds of tonnes of Cu and hundreds of kilogrammes of Au can be emitted per year into the atmosphere from a single volcano (Table 3.2).

Active volcanoes also give evidence of the presence and nature of hydrothermal alteration within a few hundred metres of the surface through, for instance, blocks ejected during eruptions. Strong hydrothermal alteration is occurring under craters of many active volcanoes, especially those volcanoes which are also emitting high-temperature gases. In many cases, ejected blocks show high-sulfidation styles of alteration. Blocks of tuff and andesitic lava ejected during recent explosive eruptions at the White Island volcano (New Zealand), for instance, contain cristobalite, alunite, anhydrite and pyrite, i.e. minerals typical of advanced-argillic zones of high-sulfidation deposits (Figure 3.32). This and other arc volcanoes have also been a source of native sulfur mined from precipitates in the pore space of loose volcaniclastic rubble in craters.

Genesis of high-sulfidation epithermal deposits

The typical lithological settings, the overall geochemical characteristics (acidity, oxidation state) of the ore fluids together with geochemical source tracers (e.g. isotope ratios) and the comparisons with observations at active volcanoes allow us to interpret the typical settings and processes of formation of high-sulfidation epithermal deposits.

The deposits form above a crystallising and degassing magma chamber; in most cases the chamber is at depths of slightly greater than 1.5 km below the surface (Figure 3.33). The deposits may be formed within the edifice of a stratovolcano, or perhaps more commonly either beneath relatively small lava domes or at the base of magmatic-hydrothermal breccia vents.

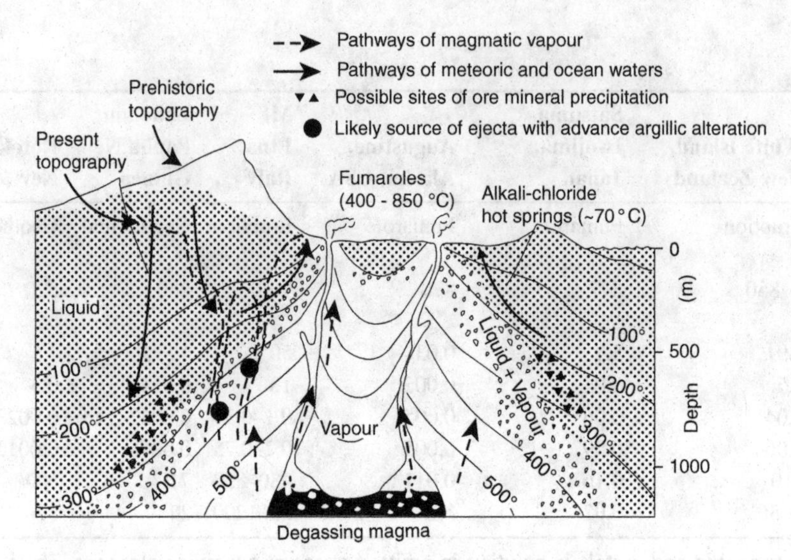

Figure 3.32 Interpretative section through the active andesitic arc volcano of White Island, New Zealand showing the possible distribution of temperature and flow paths of magmatic gases and groundwater and the likely locations of active high-sulfidation epithermal ore mineral precipitation and alteration of the type sampled in blocks ejected in explosive phreatic eruptions (after Hedenquist *et al*., 1993).

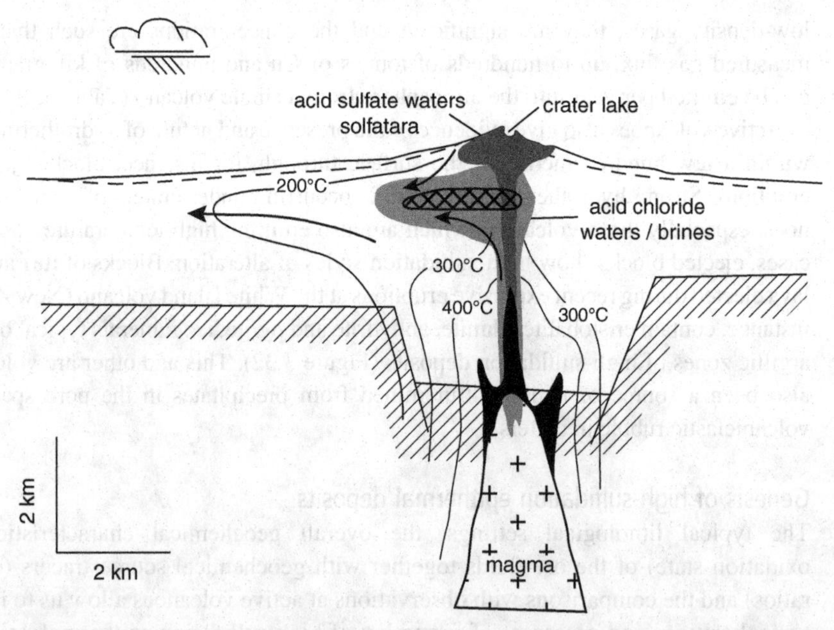

Figure 3.33 Interpretation of the setting of acid alteration and high-sulfidation epithermal ore (the oval-shaped hatched zone) in a volcanic edifice (Cooke and Simmons, 2000). Ore is interpreted to form within the edifice of a volcano, dominantly at the interface between meteoric water descending through the volcanic rocks and acid, gas-rich fluids rising from the underlying magma chamber.

The magmatic gases are water-dominated but contain up to a few weight per cent of SO_2 and other acidic species in solution, and are similar in composition to those collected from degassing calc-alkaline volcanoes. The gases percolate upwards from the magma chamber, in many cases as a near steady stream of 'passive' degassing between eruptions. Gas migration is controlled by permeability and may be focussed along sets of fractures or migrate through more permeable rock units such as unwelded pyroclastic units (as illustrated in Figure 3.31).

The fluid at source is at magmatic temperatures. It is neither strongly oxidising nor acidic at source and is in equilibrium with minerals of 'normal' felsic rock, including accessory sulfide minerals if these are present, and contains both oxidised (SO_2) and reduced (H_2S) sulfur species in solution. Because of short transport distances, high permeability of vuggy-silica alteration zones, and the limited chemical buffering capacity of altered rock that is composed predominantly of silica, the fluids of high-sulfidation deposits are 'gas-buffered'. A high oxidation state develops as a result of changing position of equilibrium between SO_2 and H_2S in the fluid as it cools. The extreme acidity is the result of disproportionation of SO_2 in aqueous solution to H_2S and H_2SO_4 as the vapour-like fluids cool and decompress and, additionally, in many cases, the result of mixing with and condensation into surface-derived groundwaters.

Depending on the volcanic topography and the pathways of the magmatic vapour and the position of surface-derived meteoric waters, condensation of magmatic vapours into groundwater may occur within a narrow depth band within a volcanic edifice above the chamber of an active volcano, around the periphery of a hot, 'magmatic vapour plume' as shown in Figure 3.32, or under the flanks of the volcanic edifice (Figure 3.30). Formation under a volcanic flank is expected where there is a topographic head driving radial movement of groundwater and underlying magmatic vapours (Figure 3.34). The inverted tear-drop shape

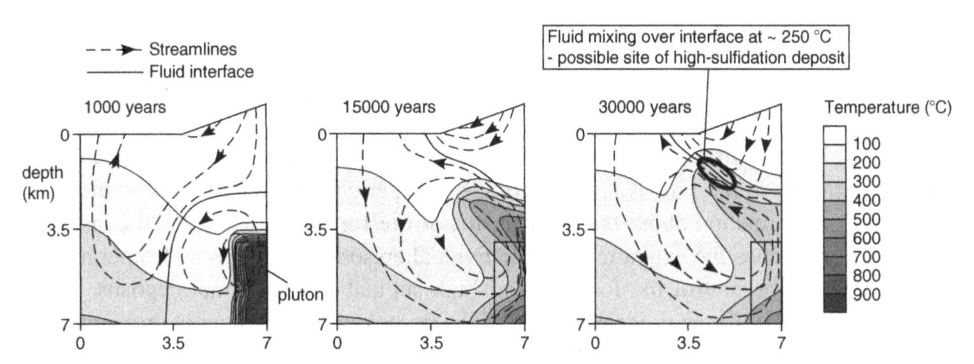

Figure 3.34 Simulation of groundwater flow beneath the topographic edifice of a volcano during cooling of the magma chamber. Two fluid flow cells are developed – high-level, topographically driven flow of surface-derived waters and free convection driven by the magmatic heat. Cooling of the deep convecting fluid to epithermal temperatures (< 300 °C), mixing with groundwater along the hydrogeological interface, and hence also the formation of high-sulfidation alteration assemblages and Au precipitation, are predicted to take place predominantly beneath the flanks of the volcano in this scenario (Forster and Smith, 1990). This simulation is run with the assumption that there is no fluid production in the magma. Fluid production would increase the time period of high-temperature fluid convection in the lower cell but would not substantially change the geometry of the convection cells.

of many zones of vuggy-silica alteration may be a result of downward percolation of relatively dense acidic condensates. Lithocaps form above many shallow-level magma chambers, and document the widespread formation of acid condensates during degassing of shallow-depth magma chambers.

The metal contents of vapours sampled from gas vents of active volcanoes demonstrate that ore metals may be transported in solution in low-density and high-temperature magmatic sulfuric vapours. However, the presence of metals in the fumarolic gases demonstrates that at least a proportion of these metals are not precipitated below the surface. Ore precipitation in the subsurface above shallow degassing magma may be relatively inefficient at many volcanoes and much of the metal in solution vented into the atmosphere. Specific chemical and physical conditions may be required for effective ore mineral precipitation in the subsurface in these environments.

Where ore minerals have grown in vughs, precipitation can be interpreted to have occurred after intense vuggy-silica alteration. If the metals are precipitated at this later stage we can explain the siting of ore in vuggy-silica altered rock as a result of fluid migration through permeability that was formed during acid alteration. Some precipitation may be through direct sublimation at high temperatures from solution in the gas phase as pressure and vapour density are reduced where the fluid rises without significant cooling. Alternatively, ore precipitation may be from condensed brines which have cooled as they migrate through altered rock. Historical records of the nature and temperature of degassing show further that active volcanoes are very dynamic environments with conditions of hydrothermal fluid flow changing often on yearly or decadal time scales. The very variable temperature of mineral precipitation in the ores is likely a result of variable amounts of cooling of the gases over the lifetimes of degassing and is a reflection of the dynamic nature of the volcanic environment and, for instance, feedbacks between permeability and progressively developing alteration of the rocks.

Box 3.4 How long does it take for a hydrothermal ore deposit to form?

Are hydrothermal events effectively instantaneous on a geological time scale or are they drawn out? Knowledge of the time periods of ore formation would give important information about the mechanisms of ore formation if we could constraint them, for instance the extent to which individual episodes of degassing from a magma chamber might control the formation of high-sulfidation epithermal deposits. However, even order-of-magnitude estimates of the times required to form hydrothermal ore deposits have been very difficult to obtain.

Conventional mineral geochronology such as the zircon U–Pb method provides ages of crystallisation or of cooling rather than time periods of growth of the mineral. Time periods determined from geochronology are generally maximum periods between the ages of bracketing events, such as crytsallisation of pre- and of post-ore intrusives (e.g. Henry *et al.*, 1997). The most precise radiometric mineral ages are from relatively recent rocks, especially from the Neogene, and can be in general obtained for specific geological units, such as rapidly cooled small intrusions or lavas. Precisions of order of 10 000–20 000 years have been provided from U–Pb ages of zircons (e.g. von Quadt *et al.*, 2011). Ar–Ar ages of potassium-bearing hydrothermal mineral can be almost as precise. As Ar–Ar dates from minerals are expected to record cooling through specific temperatures, times of cooling of rocks can in some circumstances be constrained and used as input into bracketing.

Precision of ages is a measure of analytical reproducibility, and is not necessarily a measure of accuracy, that is, whether the age is the true age that the event occurred. Mid-crustal magma chambers that supply small porphyritic intrusions in porphyry deposits, for instance, have long lifetimes of potentially millions of years (e.g. Grunder *et al.*, 2008). Even small shallow-level magma chambers that feed volcanoes have lifespans of tens of thousands of years. Data from zircons from historic eruptions show that they grew over the time periods of the life of shallow magma chambers and that the erupted crystals have not in all cases grown immediately prior to eruption (e.g. Schmitt *et al.*, 2010; Strom *et al.*, 2012). There may thus be limits to the accuracy of dating of events at a magmatic centre and even the best-precision geochronology may be insufficient to constrain periods of hydrothermal activity.

There are, however, a number of types of observation and data that can provide constraints on time periods of hydrothermal activity complementary to those of geochronology. These constraints indicate that the time periods required for ore formation may vary widely between different ore deposit types. Some constraints on three types of ore deposits include:

(1) Hydrothermal ores in sedimentary basins such as at MVT deposits (Section 4.2.1) may form over relatively protracted time periods, especially if fluid flow is the result of long-lived driving forces such as a topographic gradient. Waters in the downstream end of the topographically driven flow paths of the active Great Artesian Basin of Australia have been dated at up to about 2 Ma (Bentley *et al.*, 1986) and tufa of mound springs formed by discharge of the water has formed over the last 700 000 years (Prescott and Habermehl, 2008). These ages can be considered minimum time periods of groundwater flow in the Great Artesian Basin, and by analogy in similar geological environments in the past.

Migrating fluids can carry heat, and rising hydrothermal fluids heat surrounding rock. To a first approximation there is a relatively simple mathematical relationship between the lateral gradient of temperature around a channelway, which is a measure of the diffusion of heat away from the channelway, and the flux of fluid and the temperature gradient along the channelway. This is a central relationship used to assess the results of numerical modelling of coupled fluid flow and heat flow at MVT deposits (see Box 4.1), and for these deposits suggests time periods of fluid flow of order of 1 million years.

The complex paragenesis of the ores shows, however, that ore mineral precipitation may not have occurred throughout the period of fluid flow and, if this is the case, the time period of ore precipitation may be shorter. Conversely, Tompkins *et al.* (1997) used interpretations of a change in the structural setting from extensional to compressional tectonics between two phases of ore precipitation, in consort with the tectonic history of the host terrain to the MVT deposit at

Cadjebut, Western Australia, to imply that the deposit formed over two periods separated by as long as 35 million years.

(2) Order-of-magnitude constraints for the time period of formation of magmatic-hydrothermal deposits such as high-sulfidation epithermal (Section 3.2.5) and low-sulfidation epithermal deposits (Section 3.2.6) around or above high-level intrusions are provided by measurements of the fluxes of ore metals in emissions from volcanoes and in subsurface fluids in geothermal fields (see Table 3.2). Emissions of Cu and Au in SO_2-bearing gases through fumaroles at the active andesitic volcano of White Island, New Zealand are sufficient to form a deposit with about 1 Mt of Cu and 45 t of Au in about 10 000 years (Hedenquist *et al.*, 1993), and similarly for the andesitic Galeras volcano in the Andes of Colombia, where a 200-t Au deposit could form over this period of time (Goff *et al.*, 1994). More recent direct measurements of the Au content of fluid and the flux of geothermal fluid at about 1 km depth below the Pleistocene Ladolam epithermal Au deposit, Lihir Island, Papua New Guinea, give a similar time period, with sufficient gold flux to deposit 1300 t of Au in about 50 000 years (Simmons and Brown, 2006). In all cases these measurements provide a limit for the minimum time required to form a deposit. The estimates based on gas-emission measurements have the additional uncertainty that they are of emitted metal rather than metal deposited below the surface. The order-of-magnitude estimates of ore formation over 10 000 years are however consistent with independent estimates of the time period of cooling of small, shallow-level intrusions in the crust (e.g. Cathles, 1977). Similar time periods of activity have been determined directly for VHMS mounds on the ocean floor (e.g. Lalou *et al.*, 1998, and see Section 3.2.7).

(3) Porphyry deposits (Section 3.2.1) may constitute extremely rapid ore formation. These deposits are interpreted to result from upwards release of overpressured fluid from crystallising plutons. Multiple intersecting vein sets in the ores opened and filled simultaneously in a single phase of relatively short-lived fluid flow. Chemical diffusion into rock from channelways of hydrothermal fluid flow is mathematically analogous to diffusion of heat around a fluid channelway, and the widths and lengths of potassic alteration haloes around veins in major porphyry Cu deposits were used by Cathles and Shannon (2007) to estimate the period of hydrothermal fluid flow in example deposits. The results suggest formation over time periods in the two cases of about 100 and about 700 years. The estimates are dependent on relatively poorly known rates of chemical diffusion in rock, but suggest that these large deposits formed rapidly through 'controlled explosions' of high-pressure fluid at a few kilometres depth in the crust.

3.2.6 Low-sulfidation epithermal deposits

Low-sulfidation epithermal deposits are mined predominantly for Au and Ag, although many have concentrations of Pb, Zn and Cu in sulfide minerals. The two precious metals have differing relative importance in deposits around the world. Some are markedly

Ag-rich, in particular the deposits in the Eocene and Miocene magmatic belts of central and northern Mexico.

Low-sulfidation epithermal deposits are hosted by a greater variety of rock types and are typically at greater distances (≈ 2–10 km) from volcanic centres than high-sulfidation deposits. These deposits are hosted in sub-aerial volcanic rocks within a few kilometres of central volcanoes, and the typical host sequences are bedded sequences of intermediate- to felsic-composition lavas and pyroclastic rocks. The host sequences generally lack abundant cross-cutting shallow intrusive rocks such as are characteristic of magmatic centres. One characteristic siting of these deposits is just outside of a caldera rim fault. Well-studied examples include Creede, Colorado, USA (Bethke, 1988); Waihi and other deposits in the Coromandel Peninsula (Hauraki goldfield), New Zealand (Simpson *et al.*, 2001; Simpson and Mauk, 2007); Hishikari, Japan (Ibaraki and Suzuki, 1993).

The style and setting of mineralisation of low-sulfidation epithermal deposits is itself very variable. However, some of the variability of the deposits that have been classified as low-sulfidation epithermal is probably due to lumping of deposits into the category on the basis of alteration and ore mineralogy, even though the deposits may have formed through different processes. Many deposits in and around alkaline magmatic centres, for example, especially centres with silica-undersaturated rocks, do not share all the characteristics described below, e.g. Cripple Creek, Colorado, USA (Thompson *et al.*, 1985) and Ladolam, Papua New Guinea (Carman *et al.*, 2003).

Geological settings of low-sulfidation ores

The most characteristic style and setting of low-sulfidation epithermal deposits is in interconnected networks or swarms of steeply dipping small to large (up to 10 m thick) veins, in which the ore is in the veins and in immediately adjacent, hydrothermally altered wall-rock. A **structural control** (see Box 3.7) on the orientation and siting of veins is common – multiple veins commonly have sub-parallel orientations in a deposit (Figure 3.35) and are, for instance, hosted along segments of faults in fault sets that are more extensive than the host volcanic centres.

The veins and ore may be continuous along strike lengths of up to a couple of kilometres, but although veins may be traceable to more than 1 km depth, economic ore is only developed over restricted vertical intervals of about 500 m, and in some cases as little as 200 m. Ore may thus form a near-horizontal ribbon along the strike length of the vein. The depths of mineralisation is estimated from fluid-inclusion data to be typically between about 200 and 700 m below the water table, or rarely to about 1.5 km depth. Ore can also occur in breccia pipes, with the ore minerals in either the clasts or the breccia matrix.

Nature of the ores and associated hydrothermally altered rocks

Figure 3.36 shows components of an idealised low-sulfidation epithermal deposit. Ore veins are quartz-dominated with adularia (low-temperature K-feldspar) and variably calcite, chlorite and other gangue minerals. Many veins are zoned vertically with for instance calcite and adularia only within restricted depth intervals. Ore minerals are disseminated in the veins or occur in massive pods and may also be disseminated in strongly altered rock adjacent to veins. The sulfide mineral assemblage in the veins is very variable from deposit to deposit. Some of this variability is related to sulfidation state, with pyrite or marcasite being the dominant sulfide in intermediate-sulfidation-state ores and pyrrhotite together

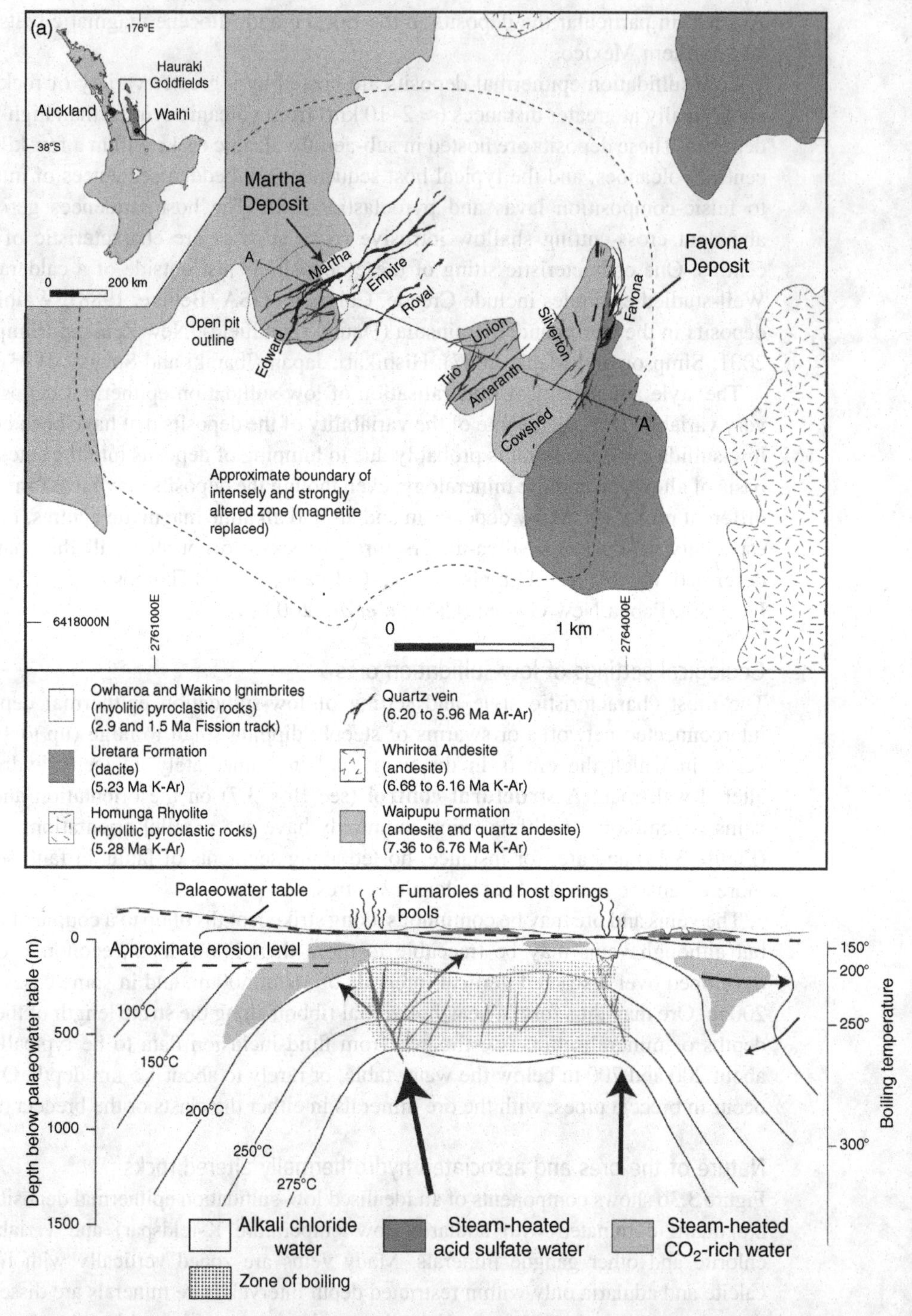

Figure 3.35 The Au–Ag deposits at Waihi, New Zealand as an example of typical vein-style low-sulfidation epithermal deposits (after Simpson and Mauk, 2007). (a) Surface geological map of the Waihi area, showing a projection of the veins that comprise the Martha and Favona deposits. Gold occurs with chalcopyrite and Ag phases in the steeply dipping zoned veins with

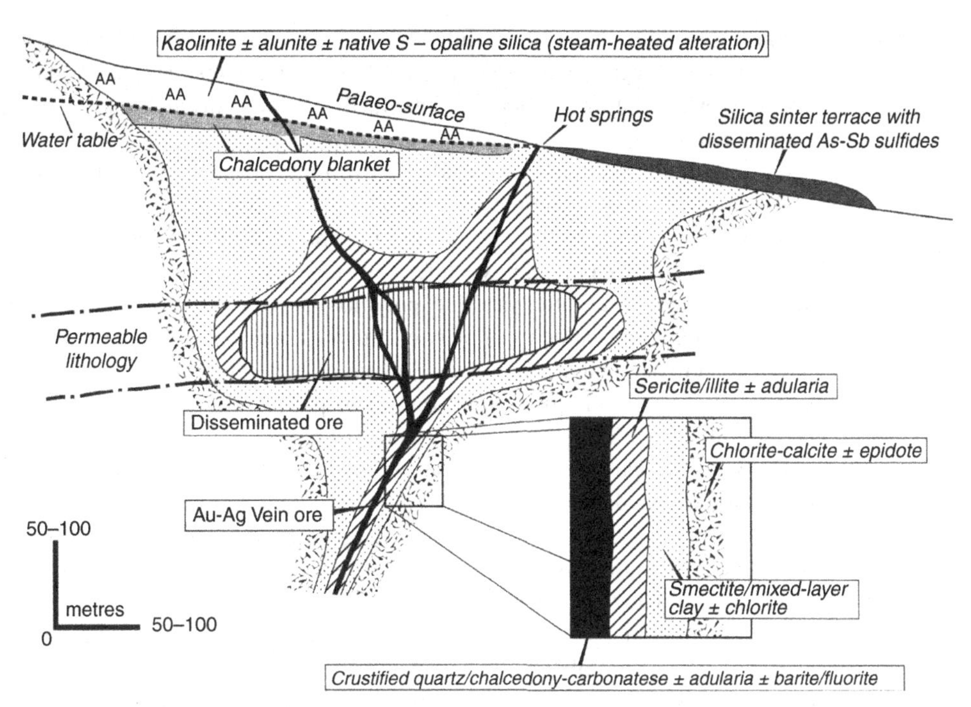

Figure 3.36 Schematic cross section through a low-sulfidation epithermal ore system showing a hypothetical composite system containing different styles of ore: vein ore, disseminated ore and hot-spring ore, together with the distribution of alteration assemblages (Hedenquist *et al.*, 2000).

with arsenopyrite (FeAsS) in low-sulfidation-state ores. The base-metal sulfide minerals galena, sphalerite and chalcopyrite are common, as are Mn minerals including the manganese carbonate rhodochrosite. Stibnite (Sb_2S_3) is an important mineral in some deposits. Native Au or electrum (Au–Ag alloy) is the main host of Au. Silver is hosted in electrum and the silver sulfide acanthite (Ag_2S) and silver sulfosalts such as proustite (Ag_3AsS_3) and pyrargyrite (Ag_3SbS_3). Native silver is a common **secondary** mineral formed after these Au sulfide minerals in incipiently weathered levels of deposits. Gold and silver telluride minerals are important hosts for the precious metals in some deposits. Extremely high-grade **bonanza** ore zones (with greater than 30 ppm Au, and locally greater than 1000 ppm) occur locally along and within the depth range of some veins.

Caption for Figure 3.35 (*cont.*) quartz adularia along walls and quartz–clay in the centres. Note that the ore veins are a network of two structural sets. The ore veins are in the centre of a larger zone of intensely and strongly altered rock that is marked by low magnetic susceptibility as a result of hydrothermal replacement of magnetite. (b) Northwest–southeast section through the deposits showing an interpretation of the ore fluid (geothermal fluid) temperature at the time of formation and generalised hydrothermal fluid flow paths (based on fluid inclusions and alteration mineralogy and analogy with the structure of the Broadlands–Ohaaki geothermal field). The deposits are interpreted to have formed at the top of an up-flowing cell of geothermal waters, which boiled as they rose. Gases from boiling caused the formation of CO_2-rich and acid-sulfate waters below the surface above the ores.

Many ore veins have approximately symmetrically developed bands with different mineral assemblages on a scale of order 1–10 centimetres. The bands are indicative of evolving or oscillating physical and chemical conditions during progressive vein fill from wall to centre. The vein minerals are also characterised by a number of distinct mineral textures (Box 3.5). Vughs and euhedral growth into space are indicative of open 'fissure' space in the vein at the time of mineral growth. Bands of fine-scale crustiform and colloform growth are indicative of rapid precipitation of minerals. Where the colloform bands involve a silica mineral, we infer that silica was initially precipitated from the fluid as fine-grained or cryptocrystalline polymorphs such as chalcedony or opal. Quartz pseudomorphic replacements of platy calcite are common and likewise indicate changing physical and chemical conditions in the vein over the time of mineral precipitation such that calcite first rapidly grew from solution as platy grains and the solution later became undersaturated with respect to calcite and was replaced by quartz.

Hot-spring deposits are hosted in sub-horizontal masses of porous silica sinter or recrystallised banded silica. The ore minerals include very fine-grained sulfides of a few micrometre grain size of As, Sb and Hg sulfides, for instance cinnabar, realgar and orpiment, which are disseminated in silica sinters and give the silica characteristic grey colouration in hand specimens.

Low-sulfidation epithermal veins have haloes up to a few metres wide of argillic alteration facies with illite and silicification and are hosted in volumes of altered rock that extend up to several kilometres laterally. This regionally developed alteration is propylitic, but generally with clay minerals, especially smectite, as essential members of the mineral assemblages. The alteration assemblages thus formed at slightly lower temperature (200 to 275 °C) than the chlorite-bearing propylitic assemblages around porphyry deposits. Subtle lateral and vertical zonation of the intensity and mineral content of the regional propylitic alteration may be apparent. With respect to clay mineral content, for instance, smectite, interlayered smectite–illite and illite zones may be recognised. This zonation is interpreted to be due to temperature variations from about 275 to 150 °C: ore veins are typically in the highest temperature zones. A strongly developed vertical zonation from propylitic to argillic to low-temperature styles of advanced-argillic alteration occur above many deposits, for instance Hishikari, Japan (Figure 3.37). Chalcedony or opal rather than quartz is the dominant silica mineral in these types of alteration. The zonation is evidence of increasing acidity of the hydrothermal fluid upwards above the ore zones.

Ore fluids and the environments of formation of low-sulfidation epithermal deposits

The hydrothermal fluid in the adularia-bearing veins that are surrounded by a zone with smectite-bearing propylitic alteration assemblages can be interpreted to be alkali chloride, near-neutral, relatively reducing waters at 200–300 °C with very low to low (0.5–5 wt %) salinity and in some cases up to about 1 wt % gases in solution, especially CO_2. Similar temperatures are implied for formation of the alteration assemblages in rock surrounding the veins, and lower temperatures down to about 150 °C a few hundred metres distant from the veins (Figure 3.35). Unlike porphyry and high-sulfidation epithermal deposits, saline brines with greater than about 25 wt % salts are not recorded in these deposits. However, the presence of low-density, vapour-rich and higher-density, more-saline fluid inclusions in growth zones in the vein quartz indicate that the fluid was at least transiently boiling in the veins.

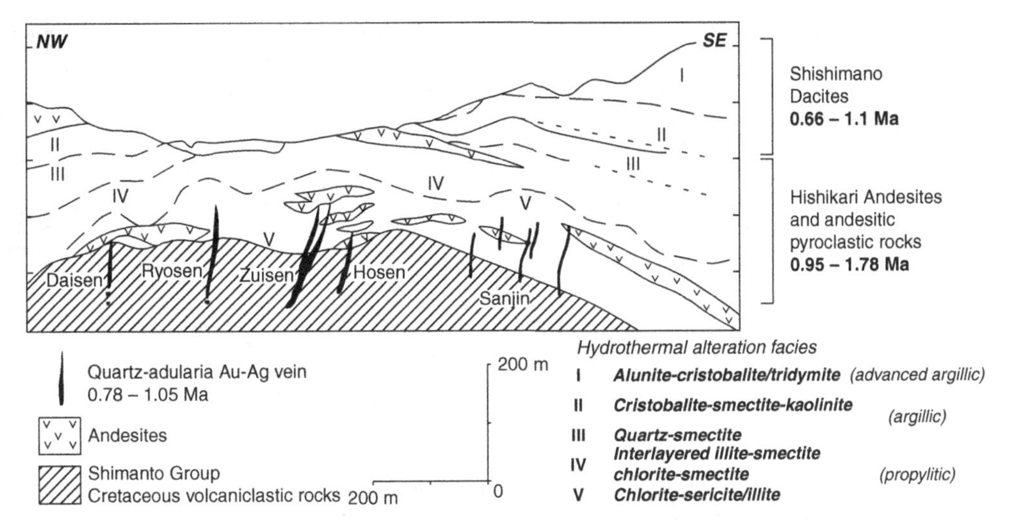

Figure 3.37 Cross section through the low-sulfidation Au–Ag epithermal veins of the Hishikari Deposit, Japan (after Ibaraki and Suzuki, 1993). The veins are hosted in Pleistocene volcanic and volcaniclastic rocks and formed at a few hundred metres below the palaeosurface. Ore is over a restricted vertical extent of the sub-parallel veins in the set. There are sub-horizontal zones of alteration around and above the veins with facies indicative of cool (~ 100 °C) acid-sulfate fluids close to the palaeosurface.

Depths of ore formation of less than about 1 km are implied by the geological settings of vein formation and fluid-inclusion densities. The ore veins were thus formed in broad zones of unusually high temperature at depths of a few hundred metres. The depths are such that water-rich solutions would boil at temperatures of 200–300 °C. The chemistry of an aqueous solution, including its pH and oxidation state, will evolve during progressive boiling because volatile gases such as CO_2 and H_2S partition into the gas phase. The vertical zonation of ore minerals in the veins can be explained as the result of progressively increasing degrees of boiling of a chemically complex ore solution upwards in a fissure such that different minerals reach saturation after different degrees of boiling (Figure 3.38). Gold is likely carried dominantly as a bisulfide complex in solution. The partitioning of H_2S into the gas phase on boiling will thus reduce Au solubility, but changes in fluid pH may counteract this effect. High-grade bonanza zones in veins may form where changes to the various chemical parameters that control Au solubility during boiling reinforce each other to give a strong decrease in solubility.

Argillic and advanced-argillic (acid-sulfate) alteration above the ore zones of the veins indicate that more acidic fluids were present above the ore zones. The acidity is a result of the same chemical processes of condensation into water of acid gases CO_2, SO_2 and H_2S that operate at high-sulfidation epithermal deposits, e.g. as a result of the reaction:

$$H_2O + CO_2 \rightarrow H_2CO_3 \rightarrow H^+ + HCO_3^-. \quad (3.13)$$

In the low-sulfidation geothermal environment, however, the acid gases are sourced through boiling of gas-bearing, circulating, near-neutral geothermal fluids. The resulting alteration is distinguished from advanced-argillic alteration of high-sulfidation epithermal deposits mainly by the presence of cryptocrystalline polymorphs of silica, which is most

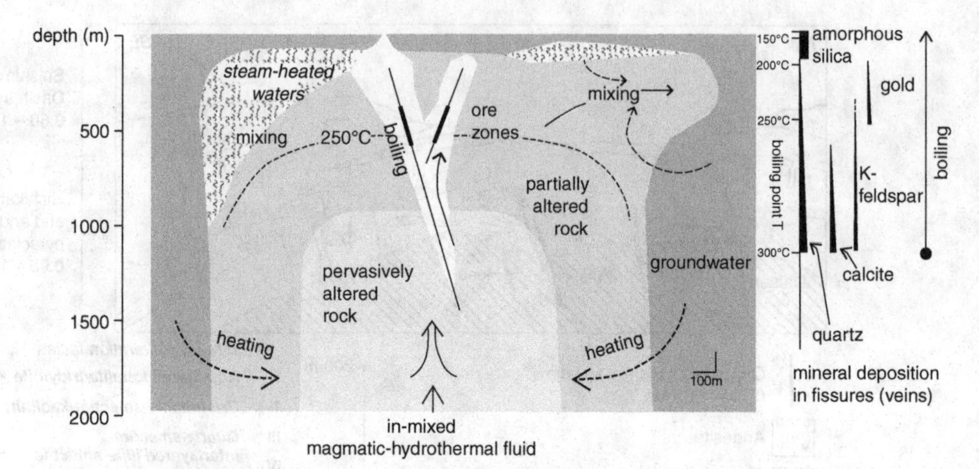

Figure 3.38 Interpretation of processes during Au–Ag mineralisation in the upper 2 km of the low-sulfidation epithermal deposits such as at Waihi. The ore deposit is interpreted to be in an up-flow zone of a regionally extensive geothermal field in which the fluid has pervasively interacted with the reservoir rocks and in which a magmatic-hydrothermal fluid was fed in from depth. Fluid boiling occurred on up-flow at depths shallower than about 1 km, and was restricted to sub-vertical open conduits or fissures. Boiling gave rise to mineral deposition in veins, with different minerals precipitating at different depths and Au precipitated only after a certain degree of boiling. Mixing between gases exsolved from the deep fluid on boiling and groundwater gave rise to steam-heated acid waters on the periphery of the up-flow zones (after Simmons and Browne, 2000).

likely to form and be preserved where temperatures are low and probably of order 100 °C. These acidic waters are known as acid-sulfate and steam-heated waters, and the associated alteration is known as steam-heated alteration (Figure 3.38).

Silica sinters are porous silica masses that precipitated in hot-spring pools as a result of evaporation and cooling of the water that is fed into the pool from below. Although no case is known of hot-spring ore overlying vein-style epithermal ore, the known hot-spring ores overlie hydrothermally altered rock with advanced-argillic assemblages similar to those above vein ores, and low-grade veins are common in underlying rock. Hot-spring Au–Ag–Hg ores are thus interpreted to form at the surface above sites where low-sulfidation epithermal veins might form (Figure 3.36).

Geothermal settings of low-sulfidation epithermal deposits

The geological setting of epithermal deposits in the vicinity of generally sub-aerial volcanic centres, the depths and temperatures of ore formation, and the extent of alteration around ore-hosting veins, are all comparable to high-temperature **geothermal systems** that occur within magmatically active belts on continental crust. These analogies have led to the interpretation that many if not all low-sulfidation epithermal deposits formed in geothermal fields. The active geothermal fields of the Taupo Volcanic Zone of New Zealand, for instance, are sited between and around late Pleistocene small volume dacite lava domes (Figure 3.39). We have good knowledge of the behaviour and development of many geothermal systems as a result of extensive drilling and investigations for geothermal energy potential.

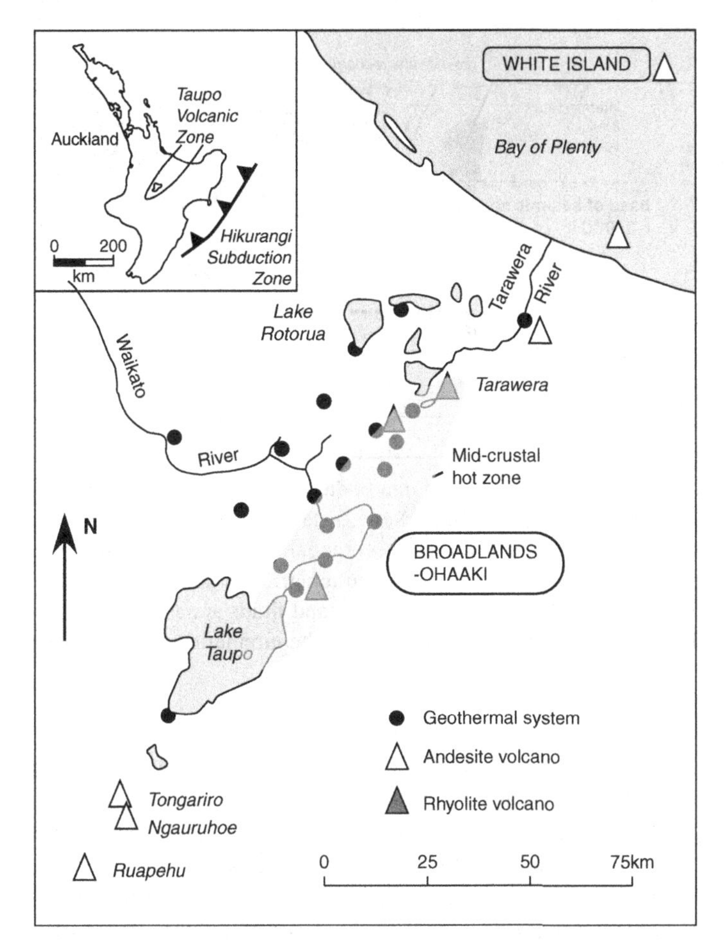

Figure 3.39 Map of the Taupo Volcanic Zone, New Zealand, showing the distribution of active arc volcanoes and known geothermal systems in the central rhyolite-dominated segments of the zone (after Simpson *et al.*, 2001). The mid-crustal hot zone is as shown in sectional view in Figure 3.40 and is the probable position of active magma chambers at about 8–20 km depth.

Critical observations on active geothermal fields that demonstrate that the environments of low-sulfidation epithermal deposits are analogous include:

(i) Geothermal fields such as in the Taupo Volcanic Zone are regionally extensive areas of elevated ground temperature and abnormally steep geothermal gradients within the upper few kilometres of the crust. The geothermal fields of the Taupo Volcanic Zone are sited in a rift zone about 40-km wide along a 100-km segment of the active volcanic arc (Figure 3.39). Porous rocks at up to a few kilometres depths in these fields are the geothermal reservoirs in which waters in pore-space and in fractures convect and from which the reservoir water can be extracted to provide energy. The fields are characterised by up-flow zones of up to a few kilometres across in which temperatures are highest at depths of less than about 1.5 km. The up-flow zones are separated by

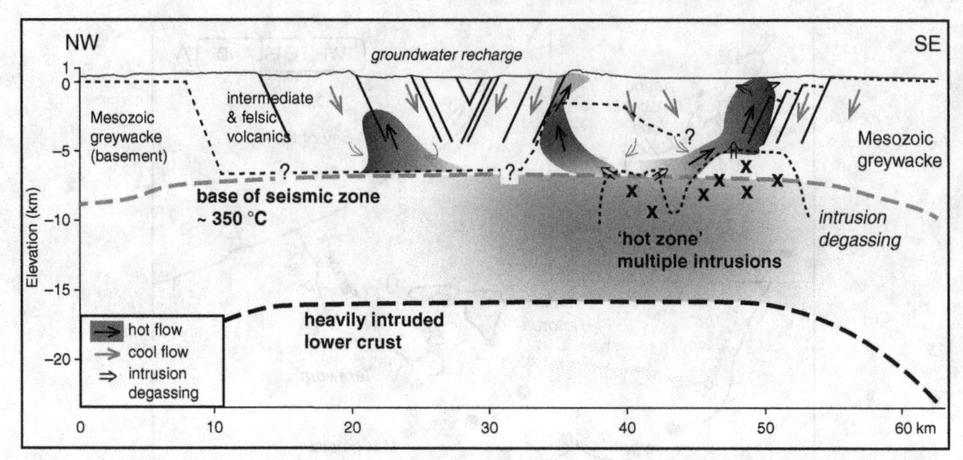

Figure 3.40 Model for the settings of magma bodies and fluid circulation in a geothermal field, based on the Taupo Volcanic Zone, New Zealand (after Rowland and Simmons, 2012). The grey zones show uprising limbs of convection cells of geothermal waters in which temperatures will be up to about 350 °C. The volumetrically dominant waters are meteoric infiltrating from surface recharge; however, gases and fluids are assumed to be released from magma chambers in the middle crust and mix into the circulating meteoric waters.

more extensive zones in which flow is downwards or laterally. It is the up-flow zones that are the settings of the characteristic geothermal features such as hot-spring pools, fumaroles, hydrothermal eruption craters and mud pools.

The geothermal fields are interpreted to overlie active or recently active and still hot magma chambers. The top of the magma chambers are inferred to be at least 4 km beneath the Lardarello geothermal fields of Italy, for instance, based on geothermal temperatures at the base of exploratory drill holes, and at slightly greater depths in the Taupo Volcanic Zone based on seismic imaging (Figure 3.40).

(ii) Waters in the reservoirs of geothermal fields, i.e. at depths of between about 0.5 and 3 km, are chloride waters with low salinity, a near-neutral pH and are relatively reducing. They are thus similar to the ore fluids of low-sulfidation deposits.

(iii) Large volumes of rock and reservoir waters in geothermal fields have temperatures of around 250–300 °C at depths of a few hundred metres. Temperatures are only rarely higher because they are buffered by the latent heat of boiling at the boiling temperature unless the water boils off completely to steam.

(iv) Many geothermal fields have areas of advanced-argillic alteration assemblages in the near subsurface, especially around hydrothermal eruption vents and fumaroles. Acid fluids to form these alteration assemblages are products of condensation of gases boiled off near-neutral geothermal fluids into near-surface-derived meteoric waters. Rocks drilled at 1–2 km depth below the depth of boiling have most commonly propylitic or, more rarely, argillic or sericitic alteration. Other facies of alteration occur at deeper levels and include assemblages with amphibole and clinopyroxene.

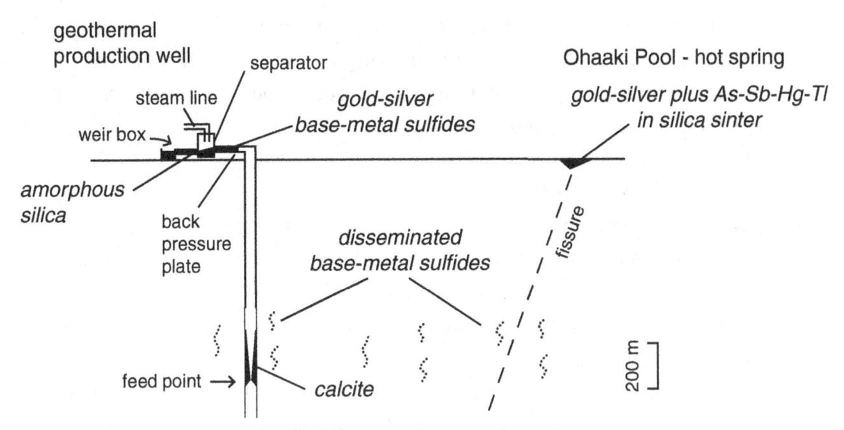

Figure 3.41 Schematic diagram showing the settings of mineral precipitation in a production well at the Broadlands–Ohaaki geothermal field and in the rocks of the geothermal field and the silica sinter of the Ohaaki Pool, which contains average concentrations of 0.2 ppm Au. High concentrations of Au and other minerals have been precipitated at various points in the pipe-work of the geothermal power station. Fluid pressures are reduced from about 40 to 10 bars across the back-pressure plate. Sulfide precipitates in pipes on both sides of this plate are dominated by chalcopyrite and contain a few wt % Au and Ag (Brown, 1986).

Additional observations show that many active geothermal fields are active ore-forming systems:

(v) Exploration by drilling in geothermal fields where reservoir temperatures are between 200 and 300 °C has intersected thin veins of quartz with pyrite and that contain a few ppm of Ag, and also disseminations of base-metal sulfides sphalerite and galena in altered rock where reservoir temperatures are between 200 and 300 °C, for instance in the Taupo Volcanic Zone and in the Salton Sea geothermal field of California, USA.

(vi) Silica sinters that are forming in some active geothermal pools (e.g. Champagne Pool, Waiotopu, and Ohaaki Pool, Broadlands, New Zealand) contain extremely high concentrations of up to 1–2% of the relatively low-boiling-point semi-metals As, Sb, and lower but strongly elevated concentrations of Hg and Tl (Figure 3.41). These elements are present in the form of fine-grained amorphous sulfide minerals (for instance Sb_2S_3 and As_2S_3) intergrown with amorphous porous silica. The sinters also contain concentrations of up to a few ppm of 'invisible' Au and Ag, which are in the crystal lattices or are adsorbed onto the surface of the fine-grained sulfide particles.

(vii) Although waters extracted for geothermal energy production generally do not contain concentrations of metals high enough to precipitate ore minerals, the low concentrations are in at least some cases the result of mineral precipitation as fluid pressure is reduced between the reservoir at depth and the power station. Precipitates of metal sulfide minerals and precious metals at pressure-release valves in feeder pipes show that the waters at depth and under pressure do have metals in sufficiently high concentrations in solution to form ore minerals. These precipitates were first observed at the Broadlands geothermal power station, New Zealand, where assemblages of Cu–Fe sulfide minerals with extremely high concentrations of up to 4 wt % Au, Ag and other metals had collected inside pipes between the well-head and the outlet between installation and maintenance a few years later (Figure 3.41). The sulfide minerals and the Au precipitate as a result of pressure release and changes to the chemistry of the water that are a consequence partial boiling and loss of gas from the fluid as it passes through the back-pressure plate. It is estimated that about 99% of the gold in solution is precipitated as the water partially boils between the reservoir and the power station.

(viii) High but variable Au and Ag contents have been measured in direct samples of hydrothermal fluids from active geothermal reservoirs below the depth of boiling where temperatures are between 200 and 320 °C, e.g. Broadlands, New Zealand, and below the very young, less than 0.4 Ma, Ladolam gold deposit, Papua New Guinea (Table 3.2). The Au concentrations in waters of both of these fields are below saturation with respect to native Au, but the solutions would reach saturation if they cooled or lost pressure and boiled. The measured concentration of up to 20 ppb Au at Ladolam is sufficient that, given the measured current flux of geothermal water to the surface, a large (1000 t Au) deposit could form within 50 000 years (see Box 3.4).

(ix) Gold concentrations have also been measured in the waters of hot-spring pools in geothermal fields. The concentrations are lower than in geothermal waters, as is to be expected at lower water temperatures: the highest measurements are of the order 100 ppt, which are, like the reservoir waters, close to saturation with respect to native Au.

Processes in the formation of low-sulfidation epithermal deposits

Quartz–adularia-vein-style low-sulfidation epithermal deposits form at shallow depths of a few hundred metres in geothermal fields which have geothermal reservoirs with water temperatures greater than about 250 °C. These high-temperature geothermal fields are spatially and genetically associated with magmatism, most commonly at convergent margins. We use observations and data from active geothermal fields in these settings to interpret the formation of deposits in the geological record.

(a) Origin of the ore fluids and constraints on their compositions

The ore fluids before boiling are low-salinity fluids with a few weight per cent salts in solution and low but finite contents of gases in solution, in particular CO_2 and H_2S. Geochemical fingerprinting using isotopes of O and H shows that geothermal waters in the Taupo Volcanic Zone and other geothermal fields associated with arc volcanoes are dominantly of meteoric origin. The fluids are thus waters that have infiltrated into the reservoir rocks through groundwater recharge (Figure 3.40). However, concentration ratios of gases in solution in the waters, most notably high ratios of N_2/Ar and He/Ar, show that these gases are mainly derived from magma. It is thus inferred that many geothermal waters in magmatically active terrains include a component of in-mixed fluid from a degassing underlying magma chamber, which may be at several kilometres depth. We can thus infer that magma was supplying fluids through magmatic degassing into the geothermal reservoir and that magmatic water is mixed into the ore fluid, even if only as a minor fraction. Isotope data allow that of the order of 10% by volume of the geothermal reservoir fluids may be of magmatic origin.

Although there is a magmatic component in the geothermal waters, the ore fluid has a different composition than that of high-sulfidation epithermal deposits or porphyry deposits. It is neutral or weakly alkaline and is not highly oxidised at and below the level at which it boils as it rises.

The widespread hydrothermal alteration around low-sulfidation epithermal deposits shows that ore fluids likely interacted chemically with large volumes of rock. Although the fluid has a magmatic component the composition of the fluid becomes rock-buffered. Acidity produced during cooling of gas-bearing magmatic waters is neutralised by reaction with rock, most importantly by the production of sericite and clay minerals from feldspar, e.g. through reactions that are characteristic of both phyllic and propylitic alteration such as:

$$\underset{\text{K-feldspar}}{3KAlSi_3O_8} + 2H^+ \rightarrow \underset{\text{muscovite/sericite}}{KAl_3Si_3O_{10}(OH)_2} + 2K^+ + \underset{\text{quartz}}{6SiO_2}. \tag{3.14}$$

Oxidation state is buffered to relatively low values in rock-buffered systems by conversion of Fe(II) to Fe(III) in rock-forming minerals, for instance the formation of epidote in propylitically altered lavas.

(b) Sources of metals and sulfur in ore minerals

Low-sulfidation epithermal ores are in most cases polymetallic with significant concentrations of Au, Ag, Cu, Pb and Zn. They are also sulfidic. Metal and sulfide contents are at concentrations close to ore mineral saturation in waters in geothermal fields in which ore mineral deposition is known to be occurring. Ore metals and sulfide may in principle be in solution in both the magmatic-hydrothermal fluid and in the convecting meteoric waters. A magmatic-hydrothermal fluid derived from degassing arc magma would likely have similar metal contents as fluids of high-sulfidation epithermal deposits and porphyry deposits, but cannot be directly sampled in a geothermal system. Metals could also have been leached from minerals in parts of the large volume of rock that the heated geothermal waters interact with during convection in the reservoir. This latter process is further discussed in Section 3.2.7. The different metals in polymetallic ores may be derived in different proportions from the two fluids.

(c) Controls on the location and development of ores

Low-sulfidation epithermal deposits are formed in the up-flow limbs of convection cells that extend down to a few kilometres in geothermal fields. Up-flow zones may be above localised heat sources such as ridges or domes on the top of an underlying crystallising intrusion, or their positions may be controlled by the distribution of permeability, such as faults or facies variations of strata in the interior of the reservoir. Within the up-flow zones, fluid flow is predominantly along open fractures, which are progressively filled by mineral precipitates to form the ore veins. Breccias and vein textures show that veins were in some cases active faults during the period of ore formation. Most active geothermal fields are sites of frequent seismic activity, in part because of their setting within arcs, but also because of mass and stress redistribution during fluid and magma flow in the upper crust, and we thus expect fractures and faults to form and to be periodically opened and re-opened to form void space as a result of seismic activity in these environments.

The available analyses of geothermal reservoir waters show that they have elevated concentrations of ore metals, but are undersaturated with respect to ore minerals in the reservoirs. Boiling of the geothermal water as it rises to within a few hundred metres of the surface would cause ore mineral saturation and precipitation (Figure 3.38). The importance of boiling is indicated most strongly by precipitation of ore minerals in the pipework of geothermal power stations. Where ores are of limited vertical extent, this may mark the range of depths over which boiling took place in the rising hot fluid.

The repeated bands of different mineral assemblages in the veins, including thin bands with high grades of Au and Ag, indicate that there are repeated episodes of ore precipitation in the vein. These are likely reflections of repeated and oscillatory changes to physical and chemical conditions in the veins. Hydrothermal eruptions and the consequent changes in fluid pressure and compositions, including partial boiling of the water in the fractures, or other similar disturbances to the hydrological regime, may be an important cause of these oscillations. Groundwater tables and hydraulic head throughout a geothermal field respond rapidly to hydrothermal eruptions anywhere in the field, and we can infer that the eruptions need not be immediately above a vein to induce an episode of mineralisation.

Figures 3.36 and 3.41 are drafted to imply a direct link between hot-spring Au–Ag–Hg deposits and low-sulfidation epithermal veins. The hot-spring deposits are formed where geothermal waters, which have retained sufficient concentrations of metals on cooling, rise and reach the surface at temperatures of typically between about 70 and 95 °C. Ore mineral saturation at the surface is a result of fluid cooling and evaporation in mud pools or hot-spring pools. The primary minerals to crystallise in these pools are cryptocrystalline silica and As and Sb sulfide minerals. These minerals recrystallise to the fine-grained silica sinter and sulfides that are characteristic of hot-spring deposits in the geological record.

Box 3.5 Mineral paragenesis and mineral textures in ore deposits

The determination of ore paragenesis was historically a major step in the scientific study of an ore deposit, especially of a hydrothermal deposit.

Paragenesis refers to associations and co-occurrences of minerals in an ore deposit, which can be inferred to have grown or have been present contemporaneously.

Paragenetic sequence refers to the time sequence of minerals. It refers in particular to the relative timing of growth of different minerals and mineral assemblages as determined by interpretation of mineral textures.

A number of texts have been published in which typical ore parageneses and their interpretation are discussed in detail in combination with methods of petrographic determination of ore minerals, including those of Ramdohr (1980) and Craig and Vaughan (1994).

We expect the paragenetic histories of many hydrothermal ore deposits and associated hydrothermal systems to be complex. By their nature, hydrothermal systems evolve. A hydrothermal deposit precipitated from fluids rising through the crust evolves from ambient temperatures, through a phase of heating, and finally through a phase of cooling. The chemistry of a hydrothermal fluid may similarly evolve with time, for instance as aliquots of inflowing fluid progressively dissolve minerals and hence overcome mineral buffers (e.g. Pettke *et al.*, 2000), or as minerals are precipitated along the walls of veins (e.g. Cathles, 1991). Magma chambers evolve over time with progressive and repeated infill, cooling and crystallisation. In addition, the channelways of ore-forming hydrothermal fluid, be they permeable rock units or structures, are likely to be the preferential paths of infiltration and flow of fluids over long periods of geological history, including before and after an ore-forming event, and during weathering after exposure at the Earth's surface.

Ore paragenetic studies are thus critical steps for understanding the history of mineral development of many ore bodies. Studies are preparatory steps for sampling for geochronology and geochemistry, for instance if the aim is to determine the isotope ratio of ore fluids or to use mineral geothermometry to determine the temperatures of ore mineral precipitation. They further allow the establishment of visual guides to ore grade and effective day-to-day mining of an ore body that developed through a complex hydrothermal history.

In many respects, the techniques and observations that are used to determine ore paragenesis and paragenetic sequences are similar to those of igneous and metamorphic petrography in, for instance, hand-specimen and microscopic determinations of crystallisation histories of magmatic rocks or of prograde and retrograde metamorphic histories. Mineral textures that can indicate relative timing thus include pseudomorphs, intergrowths, inclusions of different origin (exsolution, relicts, chance) and growth zonation, among others. As in igneous and metamorphic petrography, many textures should be interpreted with care:

- Reactions involving many types of ore minerals, in particular sulfide minerals, can be orders of magnitude faster than reactions involving silicate minerals at the same temperature (e.g. Barton, 1991). This fact has a number of implications: many ore assemblages have partially re-equilibrated on cooling, and many have been completely re-equilibrated during cooling where they have been held at high temperatures for prolonged periods of time. It was noted, for instance in Sections 2.2.3 and 2.2.4 that many magmatic sulfide ores have no relicts of primary mineral assemblages: the primary minerals have been completely overprinted during cooling of the ores.
- Veins and breccia infill provide a unique environment of mineral growth in which sequences of mineral growth into 'open' space are preserved from zoned growths and cross-cutting relationships.

Specific information is also searched for in mineral textures of hydrothermal ores, including textures in veins that are indicative of the physical and chemical environment of crystallisation, and whether growth was into open space. One conclusion from hydrothermal ore textures is that dramatic overstepping of equilibrium to reach high degrees of supersaturation of a solution with respect to one or more minerals during their growth is common in many types of hydrothermal system. Overstepping can develop as a result of rapid infiltration of fluid into a rock that it is not in equilibrium with, rapid fluid migration along a temperature gradient, or 'instantaneous' changes in fluid pressure in a seismic event. We should not assume equilibrium crystallisation of minerals. Chalcedony, the various polymorphs of opal, and marcasite are phases that have no equilibrium stability field and can only grow where there is significant overstepping of saturation for growth of the stable polymorphs, which are in these cases quartz or pyrite. These minerals are common in a number of different types of hydrothermal ore deposits

The degree of overstepping of equilibrium conditions is reflected by the form of a mineral, as has been shown experimentally (e.g. Okamoto *et al.*, 2010). Different forms reflect different rates of mineral growth and the differing balance between the rates of the multiple steps in the reaction pathways on the atomic scale of grain

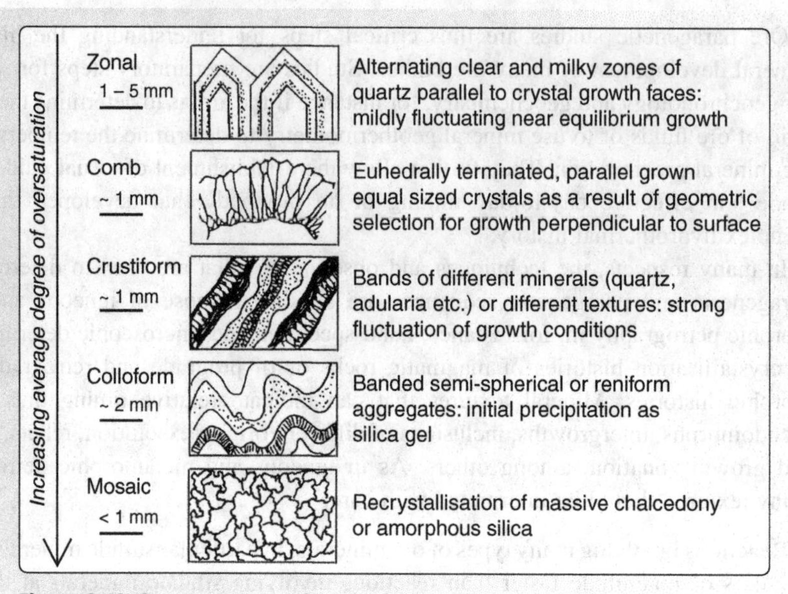

Figure 3.42 Some textures of quartz in hydrothermal veins and their interpretation (after Dong *et al.*, 1995). The scales shown are relative and approximate only and are given to indicate typical scales of grains and bands.

nucleation and growth. In the case of vein quartz, Dong *et al*. (1995) put forward a classification of common textures, largely with textural terms of traditional usage, and a number of these are ranked according to the degree of supersaturation with respect to quartz in Figure 3.42. Some of the distinct textures are recrystallisation textures which develop on inversion of one or more metastable form of silica to quartz, and hence record initial precipitation of silica as a metastable form. Silica gel formed from colloidal silica appears to be one common 'first step' crystallisation product from hydrothermal fluids, especially at the relatively low temperatures of epithermal quartz veins (Fournier, 1985; Saunders, 1990), but similar crystallisation products are also indicated by quartz textures in 'mesothermal' temperatures (e.g. Herrington and Wilkinson, 1993).

Box 3.6 Relations between the classes of magmatic-hydrothermal ore deposits

The types of ore deposit discussed in Sections 3.2.1 to 3.2.6 all involve hydrothermal fluids that exsolve from calc-alkaline to mildly alkaline intermediate to felsic magmas in an intrusion in the upper approximately 10 km of the crust, most commonly in magmatic arcs. In all cases the deposits are formed within a few kilometres of the source pluton (Figure 3.43). The ore metals are considered to be derived from the magmas and transported in magmatic-hydrothermal fluids, although this may not necessarily be the case for the complete metal inventory of low-sulfidation epithermal deposits. Chemical and mineralogical reasons for the division by metal content into Cu–Au±Ag±Mo ores that are characteristic of most porphyry deposits and Sn–W±Ag ores of greisens are discussed in Box 3.3. However, multiple ore types with similar metal inventories may be formed in the vicinity of an intrusion. Deposits of Cu–Au±Ag±Mo are considered specifically here.

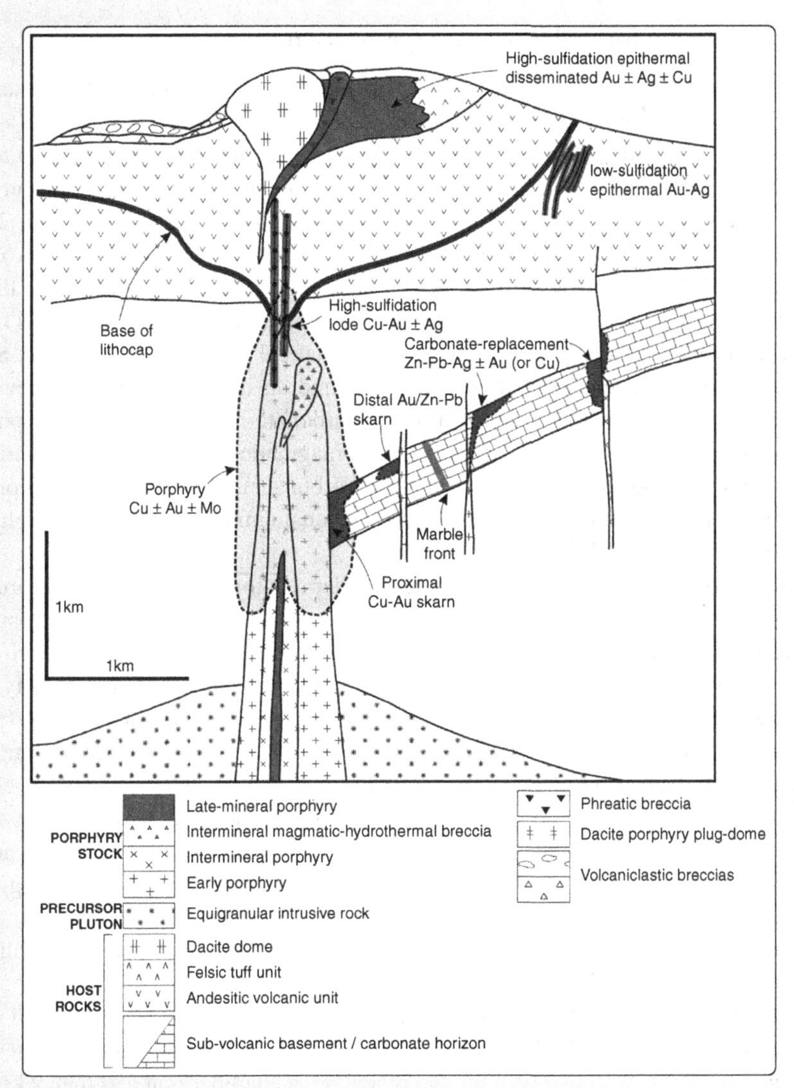

Figure 3.43 Composite of porphyry, skarn and carbonate-replacement deposits and low- and high-sulfidation epithermal deposits around a schematic volcanic centre in a magmatic arc (modified from Sillitoe, 2010). The figure is schematic, and there is no case in which all the deposit types shown are known around a single magmatic centre. It illustrates the potential combinations and the relative positions of deposits. Additionally, late and zoned polymetallic vein sets may be present, centred on and above the porphyry deposit and extending to the limits of the figure.

Two questions of interest both for our understanding of the origins of these ores and in exploration programmes are:

(1) What factors control where and when each deposit type may form?
(2) Can multiple deposit types form around a single magmatic centre, and if so under what circumstances?

(1) Which deposit type will form at a magmatic centre?

One factor differentiating the different deposit types is the mode of fluid escape and transport away from the source intrusion.

Porphyry deposits form from a vapour plume that migrates upwards from a cupola or ridge at the top of a large intrusion. The release and transport of these vapour plumes have been likened to slow or controlled explosions (Cathles and Shannon, 2007) in which sufficiently high fluid pressures are built up to result in rock hydrofracture within a relatively narrow column above the source intrusion. Vein fields are the result of flow dominantly along relatively widely spaced planes of weakness such as fractures and faults and are generally marked by greater lateral dispersion of the fluid. Skarn or carbonate-replacement deposits can form around an intrusive centre wherever magmatic centres intrude into or are overlain by carbonate-bearing strata. The deposits can be interpreted in some cases as the result of a magmatic vapour plume interacting with one or more carbonate units, even where the carbonate unit is not within the core of the plume. Reactivity and rapid dissolution of carbonate minerals promotes a replacement rather than a vein style of mineralisation.

The two end-member types of epithermal deposit form at similar temperatures and depths, and have a similar metal inventory. In contrast to porphyry deposits, high fluid pressures appear not to build up during the formation of either type of epithermal deposit and the styles and nature of veins in these deposits generally do not indicate formation through hydrofracture. Fluid release appears to be 'passive' streaming and, as discussed in Box 3.4, may take place over two orders longer time spans. Both deposit types are associated with intense acid alteration, and the chemical reasons for the acidity and the chemical processes leading to acidity in the two deposit settings are similar: in the case of high-sulfidation ores acidity is caused by condensation of acid gases (SO_2, CO_2) derived directly from a degassing magma chamber, in the case of low-sulfidation ores it is the result of condensation of these gases where they are boiled off from geothermal reservoir waters but ultimately derived from degassing magma.

The differences between the types of epithermal deposits can be interpreted in terms of setting and sites of fluid release from magma at different magmatic centres. High-sulfidation deposits form where the top of a magma chamber is at less than 2 km depth. In this environment, the rock above the chamber may be dominated by a magmatic-hydrothermal fluid, with at most only minor influx of meteoric waters into ore zones. Magma chambers under geothermal fields that host low-sulfidation epithermal deposits are most likely centred beneath up-flow zones of geothermal convection cells and their tops are interpreted to be at depths of at least 4 to 8 km. The low-sulfidation ore bodies precipitated at a few hundred metres depth are thus several kilometres distance from magmatic sources.

Although there are significant differences in ore fluid sulfidation state (f_{S2}), pH and oxidation state at the two types of epithermal deposit, these differences may not be reflecting fluid composition at source in the magma. Rather, the different compositions are end-member results of differing degrees of interaction between fluids of the same composition at source with groundwaters and wall-rocks as the fluids cool on transport away from a source magma: fluid chemistry at high-sulfidation deposits is gas-buffered and that of low-sulfidation ore fluids is rock-buffered. For additional discussion of the chemical pathways of fluid compositional evolution in these environments see Giggenbach (1992) and Einaudi *et al.* (2003).

An underlying control of the differences between high-sulfidation and low-sulfidation epithermal deposits is thus the depth of the degassing magma chamber (Figure 3.44). The differences in ore style are consequences of the pathways of fluid flow at the sites of the deposits: irregular diffuse ore bodies of high-sulfidation deposits form in secondary porosity that itself formed as a result of leaching into the acidic fluids; fluid flow in low-sulfidation deposits is dominantly along open or transiently open fractures.

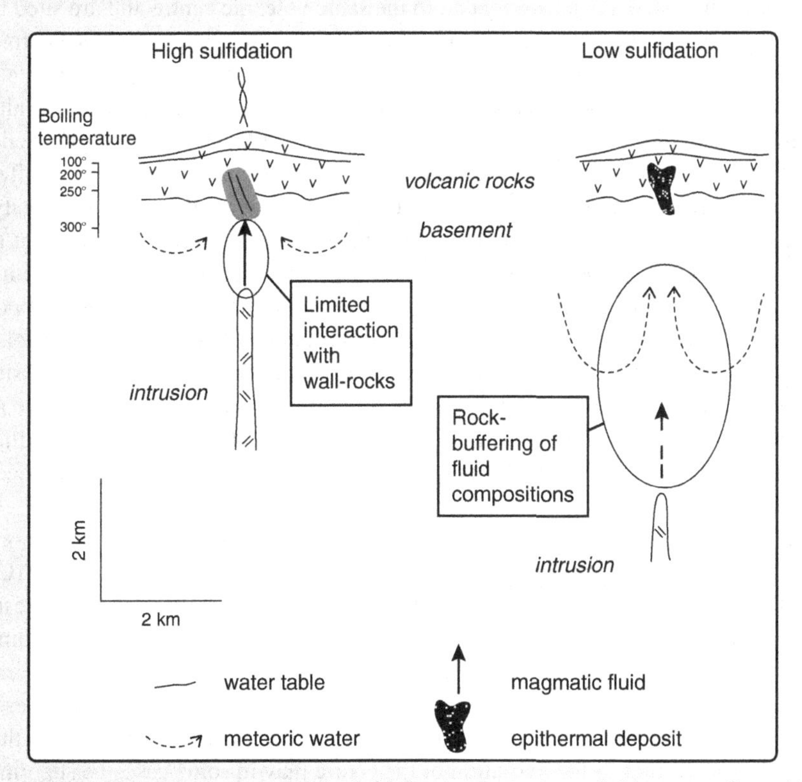

Figure 3.44 Schematic possible fluid flow paths and ore fluid sources for epithermal deposits in geothermal fields (after Simmons *et al.*, 2005).

(2) Multiple deposit classes around single magmatic centres

The case of skarn or carbonate-replacement deposits forming around the same intrusive centre as porphyry deposits with similar metal inventory is relatively common in any environment in which magmatic centres intrude into carbonate-bearing strata (e.g. Bingham, Utah, USA; Cadia, New South Wales, Australia; Potrerillos, Chile). Although synchronous formation of the different ore types has not been demonstrated, skarns and carbonate-replacement deposits can be interpreted in some cases to be the result of the traverse of a magmatic vapour plume that formed a porphyry deposit through one or more carbonate units (e.g. Figure 3.24).

The possible coexistence of porphyry and both types of epithermal deposit is of greater interest. The two classes of deposit form at different depths around a centre (Figure 3.43) such that, for instance, a porphyry deposit may be unexposed at depth where there are epithermal deposits at the surface.

There are magmatic centres where both deposit styles have been discovered and explored:

- The Pleistocene Lepanto high-sulfidation epithermal Cu–Au deposit is 1–2 km distant from the FSE porphyry Cu–Au deposit in northern Philippines (Figure 3.30). The two deposits are related to the same volcanic centre and are sited beneath and within a single extensive lithocap, with the porphyry deposit centred on a porphyry stock, and the epithermal deposit shallower and largely hosted by volcanic rocks under the eroded flanks of the volcano. Radiometric dates of the alteration minerals in the two ore types are indistinguishable (Arribas *et al.*, 1995; Hedenquist *et al.*, 1998) and the two ore bodies are interpreted to have formed effectively simultaneously after fluid phase separation of a high-salinity, high-density brine and a low-salinity vapour, and consequent separate migration pathways of the two fluids above the source intrusion. This interpretation is in contrast with recent realisation that highly-saline fluids are not required for the precipitation of porphyry ores. As the uncertainty on the relative ages is about 10^5 years, and hence of similar or greater magnitude as the likely time periods of ore formation, it is possible that the two ore bodies formed sequentially with an intervening time gap. There are also later, low-sulfidation epithermal veins elsewhere in the same volcanic edifice that formed within a couple of hundred thousand years of the porphyry and high-sulfidation ores (Chang *et al.*, 2011).
- The Ladolam (Papua New Guinea), low-sulfidation Au–Ag ore overprints weak Cu–Au–Mo porphyry-style mineralisation, and both are less than 0.4 Ma (Carman, 2003). The rapid overprinting of porphyry-style mineralisation that presumably formed at a 1–2 kilometres depth is by mineralisation that formed within a few hundred metres of the surface. In this case, this is considered to be the result of mass-wastage collapse of the volcanic island, an example of so-called 'telescoping' of ore styles. A consequence of mass wastage is that deposits of two styles that form at different times in the evolution of the centre may in some cases be superimposed. Volcanic centres are dynamic and may build up and be eroded such that material may be buried or exhumed.

- There are a number of cases in which vein fields overlap porphyry deposits, including Butte, Montana, USA (Meyer *et al.*, 1968; Rusk *et al.*, 2008) and Morococha, Peru (Catchpole *et al.*, 2011). Where there is geographical overlap, the veins overprint the porphyry ores, and in the case of the Butte may have formed up to 4 million years later (Rusk *et al.*, 2008). Based on these and other examples, low-sulfidation polymetallic Au–Te veins at Central City in Colorado, USA were proposed to be veins formed distally to an unexposed and as yet undiscovered porphyry Mo deposit (Rice *et al.*, 1985).

These examples show that in most cases where there are multiple deposit types at a single magmatic centre, the different deposits may have formed at different times within the evolution of the centre rather than simultaneously and hence through different episodes of fluid release. It was noted in the discussion of porphyry deposits that many magmatic centres have lifespans of a few million years. The chemistry and geometry of magma chambers in the crust are dynamic over these time spans (e.g. Grunder *et al.*, 2008).

3.2.7 Volcanic-hosted massive sulfide (VHMS) deposits

Volcanic-hosted massive sulfide (VHMS) or volcanogenic massive sulfide (VMS) deposits are stratabound and sometimes stratiform bodies of massive sulfide hydrothermal ore and associated sulfidic ores (sulfidic sediments, disseminated replacement ores, stockwork sulfide-bearing veins) that formed at or just below the sea floor near active magmatic centres in relatively deep marine environments. The host-rocks to these ores are submarine volcanic rocks, or more rarely turbidites or other deep-sea sedimentary rocks intercalated with volcanic rocks. Deposits in the latter settings are sometimes classified as SHMS – sediment-hosted massive sulfides. VHMS ores are polymetallic and contain variable combinations of sulfides of Cu, Zn and Pb together with Au and Ag: some are mined for pyrite as a source of sulfur or for production of sulfuric acid. They have variable metal contents, with division for instance into Zn–Pb and Cu–Zn ores.

Sulfide deposits forming on the ocean floor

Actively forming massive sulfide bodies on the sea floor are laboratories in which the processes of ore formation can be studied and are potential ores if suitable technology for collection and mining is developed. Many of these bodies contain high grades of a number of metals, with in many cases up to several per cent Cu and several ppm Au, in addition to Zn, Pb and Ag. The deposits forming at relatively shallow depths of less than about 1500 m in the oceans are more likely to be exploited in the first instance.

Manned and unmanned ocean submersible programmes have located and investigated through observations, and through collection of rock and fluid samples, at least 150 sulfide deposits that are actively forming where hydrothermal fluid vents from ocean crust into ocean water. These sites range in depth from a few tens of metres below the surface to greater than 5000 m depth. A small number of these sites have been drilled in the Ocean

Drilling Program in order to obtain information about the thicknesses of sulfide bodies and about the rocks and hydrothermal alteration in the subsurface.

(a) Tectonic and geological settings of sea-floor sulfide deposits

The first discovery of active sea-floor hydrothermal activity in the 1960s was of pools of relatively low-temperature hydrothermal waters (60–70 °C) at depths of over 2000 m near the spreading axis of the Red Sea (Atlantis II Deeps) (Miller *et al.*, 1966). Heated hydrothermal water vents through sediment. The fluid salinity is about 20 wt %, which is interpreted to be derived by dissolution of evaporites on the flanks of the rift ocean. The vented saline waters pond in the lowest 170 m of a basin a few kilometres wide on the ocean floor because they are denser than ambient seawater. The ore deposit interest in this site is that a few metres of sediment composed dominantly of iron and lesser manganese oxide and hydroxide minerals overlie the basalt substrate of these basins. Importantly, this sediment includes layers with up to a few per cent base-metal sulfide minerals, particularly sphalerite and chalcopyrite. The brine ponded in the basin has orders of magnitude higher concentrations of Cu, Fe and Zn than normal seawater. The minerals in the sediments are thus chemical precipitates that grow in the brine pools as the hydrothermal fluid cools and mixes with overlying normal ocean water.

Geothermal heat-flow data indicate that hydrothermal activity is also occurring on and adjacent to mid-ocean ridges (MORs), but the first direct evidence of such activity was the detection of elevated water temperatures and coincident abnormally high concentrations of the isotope ^{3}He – which is likely derived from the mantle – in near-bottom seawater over the East Pacific Rise. Detection of these anomalies led to the discovery in 1979 of high-temperature hydrothermal vents where fluids at temperatures up to 400 °C emanate from sulfidic chimneys and mounds in the narrow axial graben of the East Pacific Rise at 21° N at depths of about 2600 m (Francheteau *et al.*, 1979). The numerous more recent discoveries have been in a wide range of tectonic and geological settings, on all major mid-ocean-ridge spreading centres, including a few sites up to a few kilometres away from an axis of spreading, along spreading centres in marginal basins and in back-arc basins, on volcanic seamounts, and on submarine volcanoes along intra-oceanic volcanic arcs (Figure 3.45). Sub-sea-floor hydrothermal activity and sea-floor sulfide deposits occur in any tectonic setting where there is sea-floor volcanism.

Actively forming deposits on submarine volcanoes in arcs, in marginal basins and back-arc basins are closer analogues of deposits in the geological record than those on MORs. The majority of sulfide deposits on arc volcanoes are forming at depths of between a few tens of metres to about 1600 m, and in back-arc basins from about 400 to 3500 m, in contrast to greater than 2200 m depth along MORs. The rocks of volcanic arcs and back-arc basins are more likely to be tectonically emplaced onto continental crust, for instance between two segments of continental crust during compressional or transcurrent tectonism, and are thus more likely to be preserved in the geological record than MORs, which are mostly entirely subducted.

The hydrothermal vents are in most cases at distinct topographic sites. The first discoveries on the East Pacific Rise were on the edge of a narrow fault-bounded graben nested within the axial graben of the ridge. Many others are on topographic highs, particularly on ridges (Figure 3.46a). Vents on active sea-floor volcanic edifices

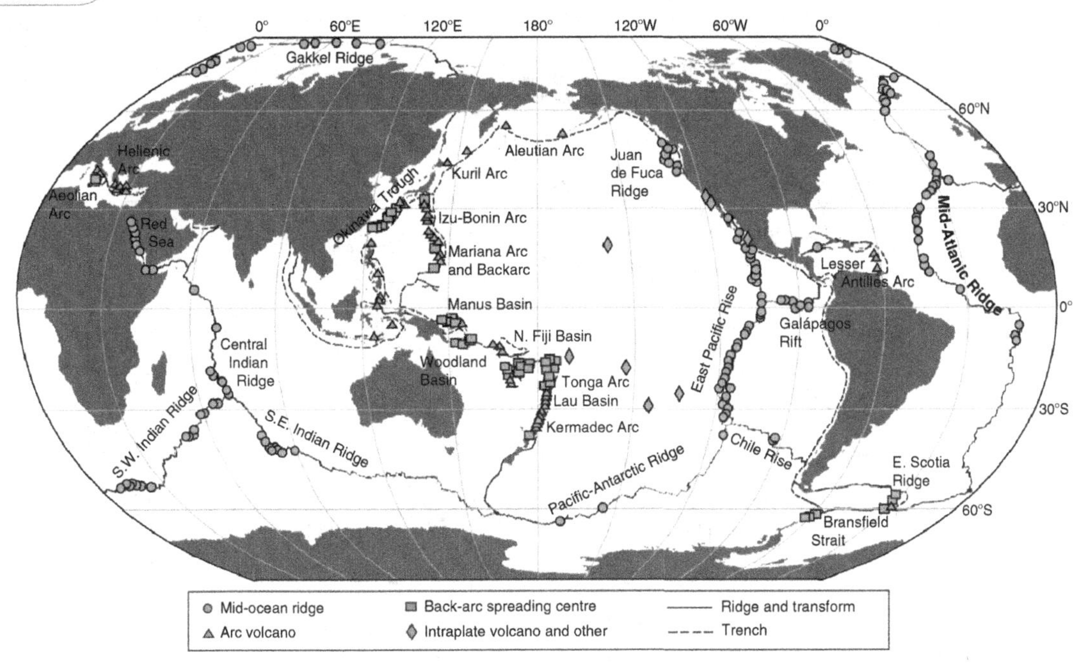

Figure 3.45 Locations of known active and recent sea-floor hydrothermal venting and accumulation of polymetallic sulfide deposits or Fe–Mn–Si chemical sediments (after Hannington *et al.*, 2011). Note the range of tectonic settings of hydrothermal activity – mid-ocean ridges (e.g. Mid-Atlantic Ridge), spreading centres of back-arc basins (e.g. Manus Basin, southwest Pacific), submarine volcanoes of volcanic arcs (Izu Bonin arc, Japan; Kermadec Arc, New Zealand), and intraplate volcanoes of seamounts (central Pacific, including Loihi Seamount, Hawaii).

are often near the top of volcanic cones or at the base of the crater walls within a crater.

(b) The nature of hydrothermal activity on the sea floor

The active sulfide-precipitating hydrothermal systems are marked by fields of continuous or transient exhalations of high-temperature acid aqueous solutions, most typically between 200 and 350 °C with a pH of between 3 and 4 through the ocean floor. There is a relationship between the maximum temperature of fluid vented in any one field and the depth of the field, which is a result of the buffering of fluid temperature by boiling, the temperature of which increases with increasing pressure (cf. Figure 3.1). Vents with fluid at greater than 300 °C, for instance, can thus only form at water depths of greater than about 900 m. Many sea-floor hydrothermal fields also include vents of low-temperature fluids (as low as 4 °C), which are hence at only slightly higher temperatures than ambient ocean bottom water (~ 2 °C). At many sites, multiple vents occur in hydrothermal fields covering a few hundred metres across and in which venting is very heterogeneous with different vents of different fluid temperatures (Figure 3.46b).

The vents may be 'black smokers' (exhalations with fine dark sulfide grains in suspension), more rarely 'white smokers' (exhalations with sulfate minerals and silica in suspension), or transparent shimmering vents that are not emitting significant amounts of particulate matter.

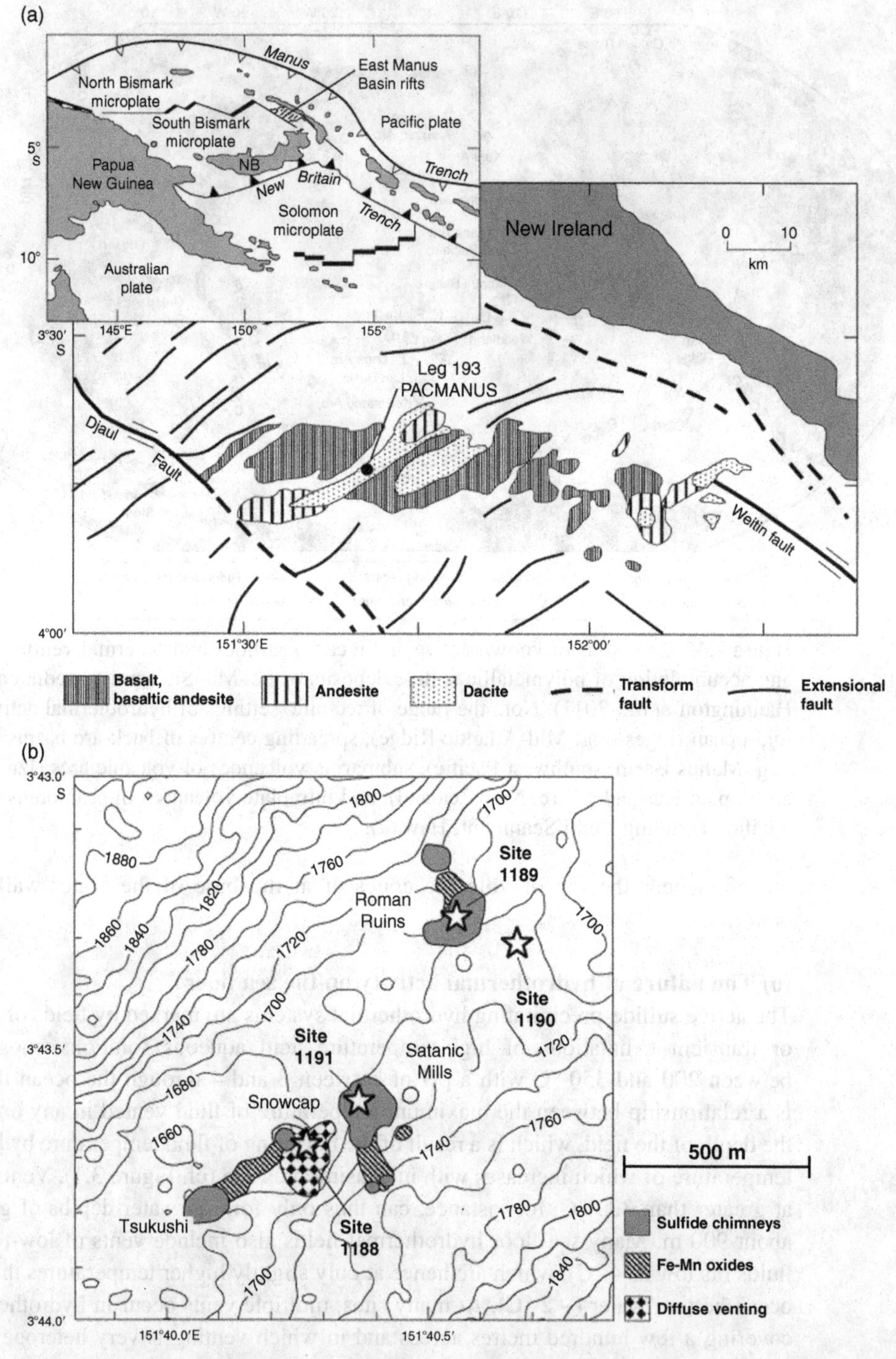

Figure 3.46 Setting and distribution of the Pacmanus sea-floor hydrothermal field in the Manus back-arc basin, Papua New Guinea (Binns *et al.*, 2007). (a) The setting of the field on the segmented spreading centre of the back-arc Manus Basin, and the range of volcanic rock types along the active and Quaternary ridges. The site of hydrothermal spreading is at about 1700 m water depth along an evolved, dacitic segment of the ridge. (b) Distribution of different vent types and precipitates within the hydrothermal field near and on the ridge crest.

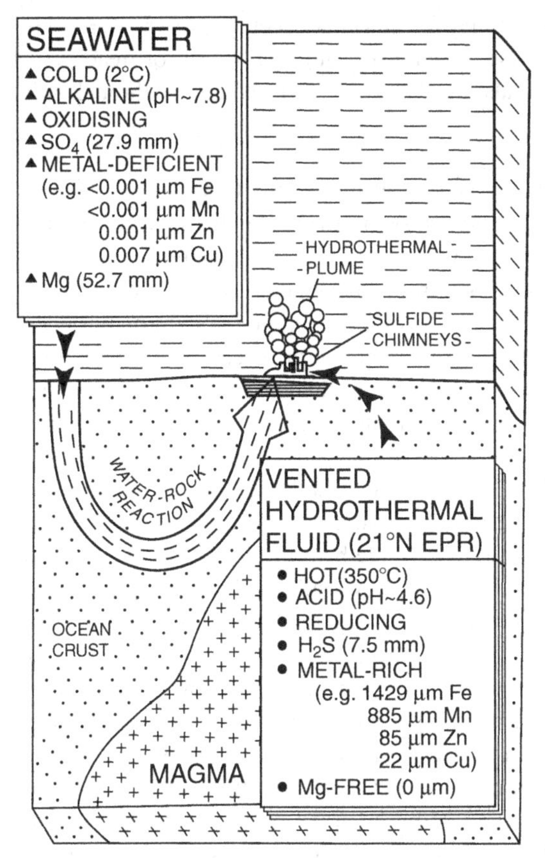

Figure 3.47 Chemical evolution of circulating seawater in a sea-floor hydrothermal system through reaction with basalt at temperatures of up to 350 °C, based on analysis of fluids vented at one of the vents of the 21° N East Pacific Rise hydrothermal field (after Scott, 1997).

In the open ocean, the high-temperature vented hydrothermal fluids rise up or spread out into the ocean column as a hydrothermal plume distinguished from surrounding seawater by slightly higher temperature, higher concentrations of gases such as methane, and elevated light scattering (turbidity) because of a high suspended load of minerals, mainly iron and manganese oxides and hydroxides.

The venting hydrothermal fluid has been sampled at a number of localities (Figure 3.47). The fluid has salinity similar to, but generally not identical to, that of seawater (= 3.5 wt % salts), and is thus interpreted to be dominated by heated seawater. A magmatic component to the fluid is, however, indicated by some aspects of water chemistry at active vents. The small variations in salinity relative to seawater can be explained as results of boiling of the solution as it rises below the ocean floor, hence forming a residual liquid with slightly enhanced salinity and a vapour with lower salinity.

There are, however, significant differences between the solute content of the vented fluids and seawater (Figure 3.47). Sulfate is absent and is replaced by sulfide at similar but slightly lower molar concentrations. The solutions have orders of magnitude higher concentrations of Fe, Mn, Zn and Cu and other metals than seawater, but much lower concentrations of Mg. These differences are the result of reactions between the convecting fluid and minerals in the rocks and sediments through which it passes (see p. 194).

(c) Nature of hydrothermal precipitates on the sea floor

Sub-sea-floor active venting is often through narrow, approximately cylindrical, **chimneys** up to a few metres high by a few tens of centimetres diameter that are composed of massive sulfide that has precipitated from the vent fluid. Elsewhere, venting is through low mounds of sulfide precipitates a couple of metres high, through small 'finger' chimneys of sulfide growing above sulfide-rich sediments, or diffusely through sulfidic or oxide crusts on sea-floor rocks. Vents of low-temperature fluids (as low as 4 °C) are typically diffuse and produce encrustations of iron and manganese oxide and hydroxide minerals. A hydrothermal field may be formed of multiple active and inactive chimneys scattered over areas of between about 100 m^2 to tens of thousand square metres interspersed with sulfide crusts and mounds (Figure 3.46). Some extinct or dormant fields have been discovered that are composed of mounds of massive sulfide beneath thin crusts of iron oxide.

The large chimneys are tubes of massive sulfide that have grown around single or multiple orifices which are the conduits for the hydrothermal fluid. They are composed of multiple intergrown sulfide minerals, mainly chalcopyrite, sphalerite, pyrite, galena and marcasite, together with barite and a minor component of intergrown silica minerals, and are often internally zoned. A typical arrangement is for a chalcopyrite-rich lining of the interior conduit and a marcasite–pyrite–barite zone on the exterior wall. Other chimneys, even within the same field as Cu-rich examples, may be sphalerite rich. The sulfide minerals in the chimneys show complex mineral textures, including colloform and dendritic textures that indicate rapid mineral growth from a quenched solution (see Box 3.5).

The fields of hydrothermal venting are not only very inhomogeneous with respect to the style and temperature of venting and to the mineralogy of the precipitates, but are also very dynamic. Repeated surveys of the ocean floor show that chimneys are transient. Many chimneys grow and become extinct over a few years or decades. Mineral intergrowth textures indicate that even within the lifespan of activity of a single chimney, the precipitating mineral assemblage may evolve. After a chimney becomes inactive, the sulfide minerals partially dissolve into seawater and the chimney typically collapses to form sulfide-rich rubble. Large chimney fields that have been active for hundreds to up to tens of thousands of years have built up mounds of rubble of collapsed sulfide chimneys on the sea floor, the largest of which are up to about 50 m thick and comprise up to several million tons of sulfidic rock (e.g. TAG field at 26 °N on the Mid-Atlantic Ridge).

Volcanic and volcaniclastic rocks of the ocean floor surrounding active and extinct vent fields are hydrothermally altered. Drilling below mounds has revealed rock that has been pervasively altered to chlorite and quartz, and can have stockwork zones of centimetre-wide pyrite- and chalcopyrite-bearing veins (e.g. Bent Hill, Juan de Fuca Ridge, Zierenberg *et al.*, 1998; TAG site, Petersen *et al.*, 2000).

The mineral content of the black and white smokers and the turbidity of hydrothermal plumes indicate that minerals are not only precipitated in the chimneys but are also carried into the water column from vent sites. Metals are also carried away in solution. The minerals in suspension and the dissolved metals may accumulate up to many kilometres from the vent sites on the ocean floor in sediment layers that have higher than normal concentrations of ore metals, especially Fe, Mn and Zn.

There are variations in the ore metal content of sulfide deposits formed in the different tectonic environments. Sulfide accumulations formed on and in pillow-lava sequences of MORs are relatively poor in Pb and are essentially Cu–Zn ores. The most Pb-rich sulfide accumulations are in deposits on submarine arc volcanoes and along sedimented ridges, such as the Juan de Fuca Ridge offshore of the North American Cordillera.

VHMS deposits preserved on land

Individual massive VHMS ore bodies are generally small to moderately small in size (1–50 Mt ore). These are relatively high-grade ore bodies, with for instance up to 2–3% Cu and 10% Zn, which are exploited for most or all of the commodities Ag, Au, Cu, Pb and Zn. The ores also have elevated concentrations of many other elements with chalcophile geochemical affinities, including Sb, Bi and in some cases Sn.

The shape and nature of VHMS deposits in three dimensions are in many respects better known from the explored ancient examples on land than from the deposits that are currently forming on the ocean floor.

(a) Settings of VHMS deposits

VHMS deposits are known in different geological environments in which there is evidence of submarine volcanism. Multiple deposits occur in belts within specific terrains. The typical settings include:

- Ophiolites. Ophiolites are fragments of mafic ocean crust and underlying ultramafic upper mantle and have a characteristic stratigraphy (see Figure 2.7). Multiple massive sulfide deposits may occur within sequences of pillow basalts or at basalt–sediment interfaces, for instance in the Troodos ophiolite, Cyprus (Constantinou and Govett, 1978; Adamides, 2010), and the Oman ophiolite, among others. This is the 'Cyprus type' of deposit, which is typically a podiform mass of Pb-poor, Cu-rich massive sulfide with underlying silicified, hydrothermally altered, basalt that includes veins and disseminations of pyrite and chalcopyrite.
- Archaean and early Proterozoic greenstone belts. These are belts of deformed and metamorphosed rocks that were formed on older continental crust and that include thick dominantly submarine mafic to felsic volcanic successions with abundant sub-volcanic intrusions such as at Kidd Creek, Ontario, Canada (Hannington and Barrie, 1999), Crandon, Wisconsin, USA (DeMatties, 1994) and Panorama, Western Australia (Vearncombe *et al.*, 1995). The host sequences to VHMS deposits in these belts typically include a range of igneous rocks and both mafic and felsic sequences may host massive sulfide deposits.
- Mafic to felsic submarine volcanic sequences interbedded with turbidites and shales in deformed continental margin terrains such as the Kuroko district, Japan (Sato, 1977; Urabe and Sato, 1978) and the Mt Read Volcanics, Tasmania, Australia (Large, 1992). Many of these volcanic belts formed during convergent-margin tectonics. The settings include submarine volcanic arcs, as is the case of the Miocene Kuroko (= black ore)

deposits of Japan, which are lenticular Zn–Pb–Cu bodies of massive sulfide associated with intermediate to felsic calc-alkaline lavas and pyroclastic rocks, or back-arc basins in the case of the host sequences of deposits in the Cretaceous magmatic belts of the cordilleras of Peru and Mexico.

- Marine sedimentary successions including turbidites and black shales with intercalated volcanic rocks such as the Iberian pyrite belt, 'Rio Tinto', Spain and Portugal (e.g. Tornos, 2006) and Rammelsberg, Germany (Large and Walcher, 1999). These are typically Cu–Zn-rich deposits and many are stratiform rather than podiform or lenticular. They are sometimes classified as Besshi-type deposits.

In almost all cases, the host sequences of VHMS deposits are strongly deformed and metamorphosed. This is a consequence of the need to tectonically emplace sea-floor rocks onto continental crust in order to preserve them. The deposits are thus also deformed and the ore and alteration minerals are metamorphically recrystallised such that primary mineral textures and structures that formed during crystallisation at the sea floor are generally poorly preserved, if at all. Until the discovery of sea-floor hydrothermal activity, this deposit type was often argued to have formed during deformation and metamorphism. However, sulfide breccias that are likely fragmented collapsed chimneys, and primary mineral textures, including dendritic sulfides and concentric mineral banding of growth around chimneys, or framboidal pyrite that grew in the water column, are preserved in some deposits.

(b) Distribution, shape and size of ore bodies

Where VHMS deposits occur there are typically multiple ore bodies. The deposits may be spaced out at a single stratigraphic level (Figure 3.48) or more rarely stacked at multiple levels in the host volcanic sequence. Where deposits are at a single stratigraphic level, this level either marks a hiatus of volcanism or a marked change in style or composition of volcanism. Deposits at the single stratigraphic horizon are characteristically fairly regularly spaced with a spacing of between 5 and 20 km being typical (Figure 3.48). Multiple closely clustered lenses occur in some cases.

Many deposits are adjacent to faults, including caldera rim faults, where these are present, and these faults can be implied from breccias and variations in unit thickness to have been active during volcanism. Some deposits are adjacent to localised volcanic features such as rhyolite domes (Figure 3.49). Sub-volcanic intrusions are a characteristic component of the host succession, and have typically intruded about 2–4 km below the stratigraphic level of the deposits.

The ore bodies vary in shape from podiform to lenticular bodies of massive sulfide up to a few tens of metres in thickness (Figure 3.49) to laterally more extensive stratiform sheets. The podiform bodies have plan-view areas of up to about 100 by 100 metres, hence similar to the larger known vent deposits on the ocean floor.

(c) Internal structure of massive sulfide ore bodies

Massive sulfide ore bodies are inhomogeneous, whatever metals are present and whatever the ore-body shape and style. Stratiform bodies are characterised by centimetre- to decimetre-scale layers with different dominant sulfide minerals, for instance sphalerite–galena, or chalcopyrite–pyrite. Layering of ore types is most markedly developed in

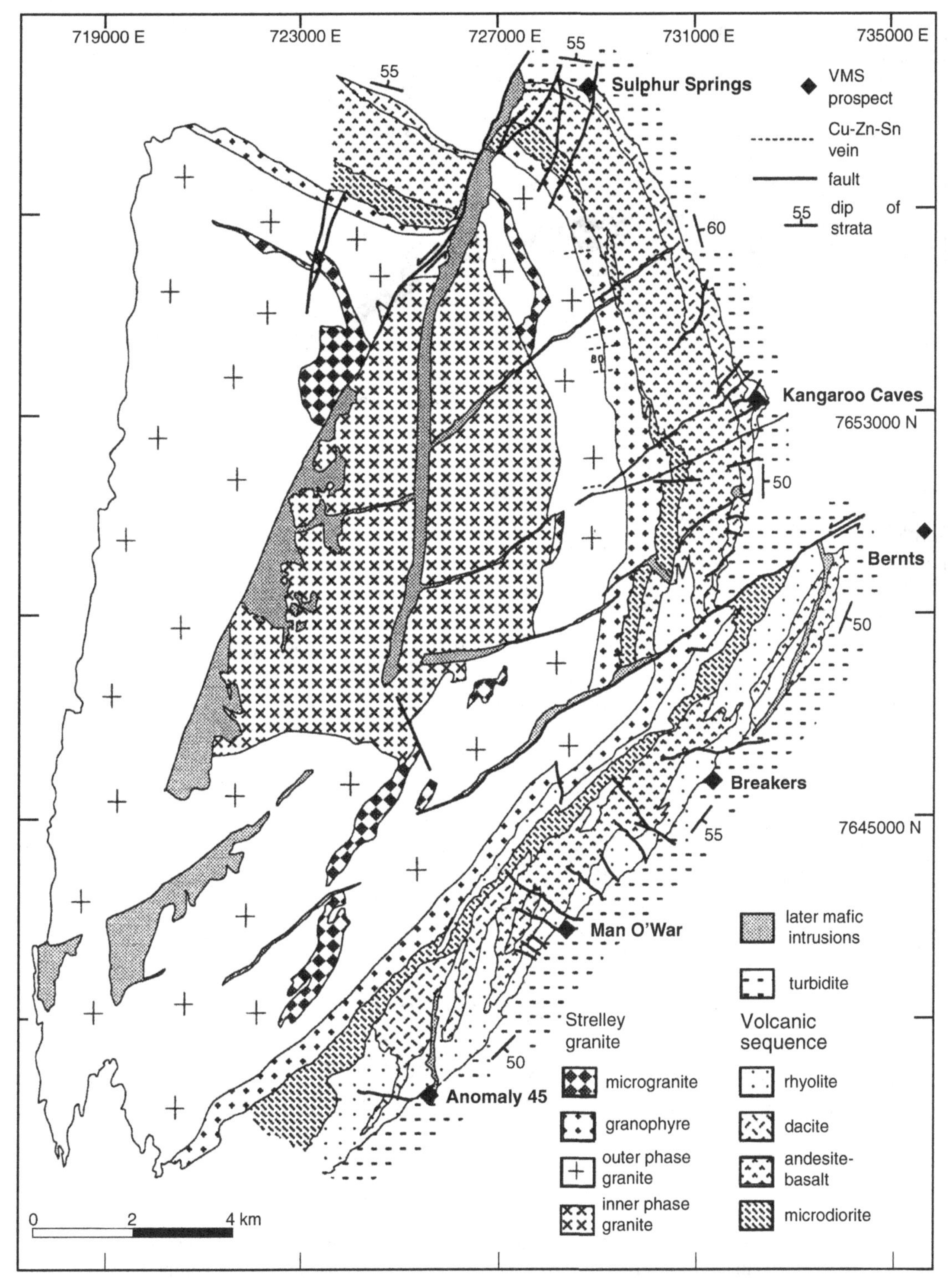

Figure 3.48 Distribution of volcanic-hosted massive sulfide deposits in the Panorama area of the Pilbara Craton, Western Australia. The Archaean greenstone belt succession is folded into a dome cored by the $\approx$ 3200 Ma Strelley granite, dips steeply to between northeast and southeast, and is composed of about 3 km of mixed mafic to felsic volcanic rocks that are at least in part co-magmatic with the Strelley granite (shaded) overlain by turbiditic metasediments. The VHMS deposits are at or close to the stratigraphic top of a volcanic sequence and spaced at intervals between 5 and 8 kilometres along strike (Brauhart *et al.*, 1998).

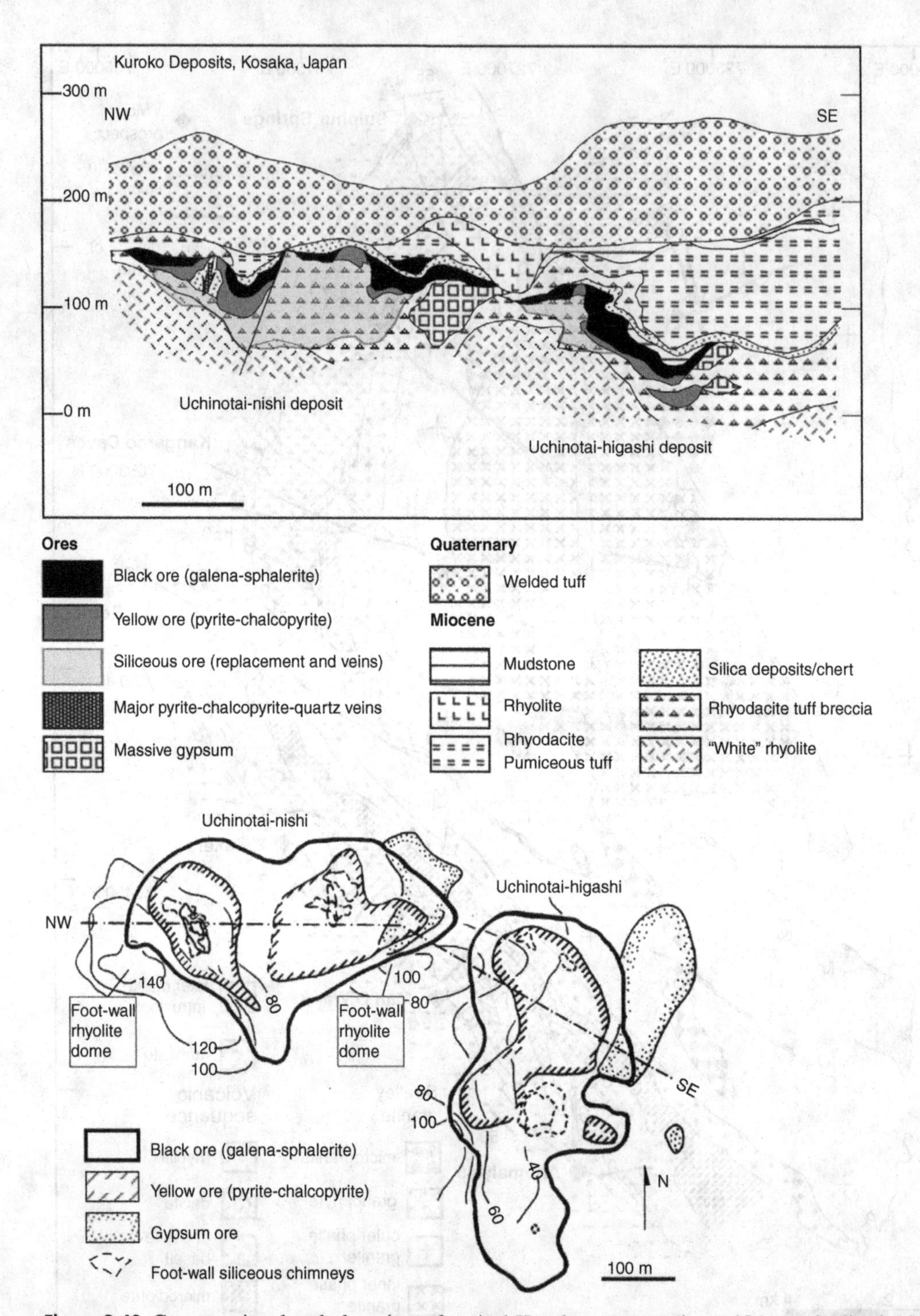

Figure 3.49 Cross-sectional and plan view of typical Kuroko-type massive sulfide ore bodies in the Kosaka deposit at the type locality Kuroko Belt in Japan. The ore bodies are hosted within a complex sequence of intermediate to felsic volcanic and volcaniclastic rocks, which include lava domes and various tuff units. Characteristic settings of ore bodies are the margins of slightly older 'white' rhyolite domes – the positions of which are shown by the contours on the map. The deposits are irregular lenses internally zoned with upper black (sphalerite–galena) ore, less extensive lower yellow (pyrite–chalcopyrite ore), and each lens is underlain by a number of discordant silicified and ore 'chimneys' (after Urabe and Sato, 1978).

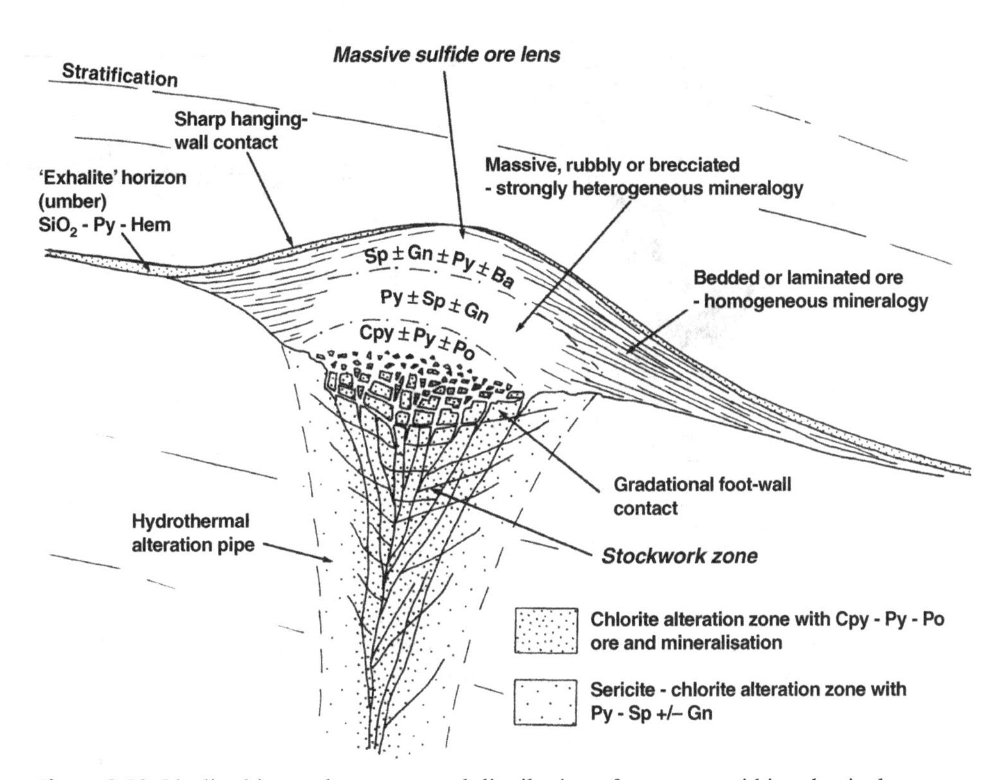

Figure 3.50 Idealised internal structure and distribution of ore types within a lenticular volcanic-hosted massive sulfide deposit. After Lydon (1988). Mineral abbreviations: Ba – barite; Cpy – chalcopyrite; Gn – galena; Hem – haematite; Po – pyrrhotite; Py – pyrite; Sp – sphalerite.

podiform and lenticular ore bodies (Figures 3.49 and 3.50). In these, there is vertical zonation of ore styles with massive sulfide underlain by a stockwork or stringer zone in which sulfide- and quartz-bearing veins cut the host-rock. Small sulfide lenses may be present a few metres below the main ore body. The massive sulfides are in many cases overlain by a layer of Fe–Mn oxide-rich chemical sediment such as ochre or umber. Thin layers of Mn-rich cherty sediment, in some cases with high concentrations of barite (barium sulfate), may extend up to several kilometres laterally away from ore bodies at the stratigraphic horizon of the ores.

There is also zonation of the metal content of ore within massive sulfide bodies. Most typically Cu is at highest concentrations in the stockwork zone and in the lower half and centre of a massive sulfide lens, whereas Zn and Pb are most abundant at the top of the massive sulfide body and at its peripheries.

(d) Ore and gangue mineralogy

Massive intergrowths of multiple sulfide minerals are characteristic of this deposit type. Pyrite is generally the dominant sulfide mineral throughout an ore body. Chalcopyrite is the main Cu mineral, although less common ore minerals include other copper sulfides

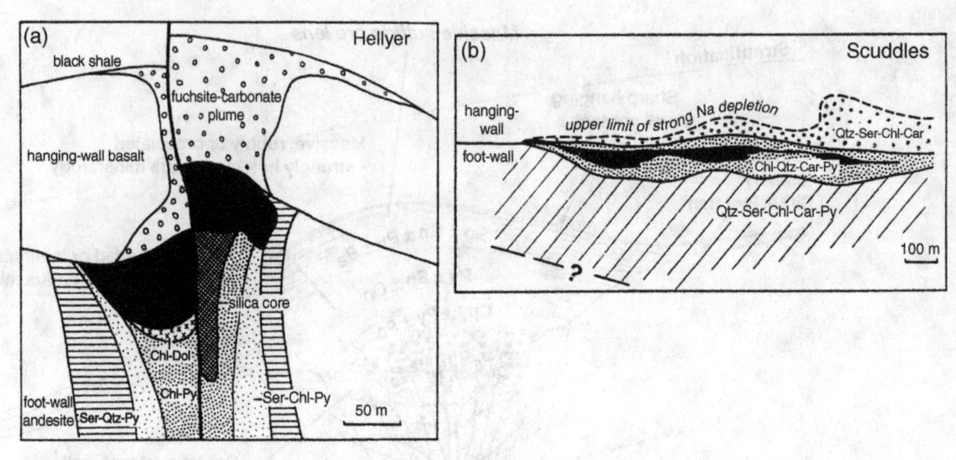

Figure 3.51 Typical patterns of alteration around VHMS deposits of lens shape with a distinct alteration (feeder) pipe in the foot-wall (a, Hellyer), or a stratiform ore body with stratabound foot-wall alteration (b, Scuddles). The white-mica–carbonate alteration in the hanging-wall is generally relatively weak and is considered to be the result of continued hydrothermal fluid flow after burial of the deposit by the hanging-wall volcanic rocks (Large, 1992). Mineral abbreviations: Car – carbonate (mostly ferroan dolomite and ankerite); Chl – chlorite; Dol – dolomite; Py – pyrite; Qtz – quartz; Ser – sericite.

such as cubanite ($CuFe_2S_3$) and tennantite ($(Cu,Fe)_{12}As_4S_{13}$). Sphalerite and galena are the main ore minerals of Zn and Pb. Gold is present as small grains of native gold and may be variably at highest concentrations either with the highest concentrations of chalcopyrite or with sphalerite and galena.

The main gangue mineral is quartz, which is ubiquitously disseminated through the massive sulfide bodies and also occurs as masses in the underlying stockwork ore. Barite and other sulfate minerals are common especially in the upper levels of the body. Magnetite and pyrrhotite are minor components of many ores.

(e) Hydrothermal alteration around the deposits

Hydrothermal alteration is recognised in and around the deposits (near-field), and also regionally (far-field) extending up to about 20 km laterally around some deposits, and to stratigraphic levels of 2–4 km below the deposits.

The near-field alteration is almost entirely in the stratigraphic foot-wall of the deposits, although weaker alteration is recognised in the hanging-wall (Figure 3.51). The alteration in the foot-wall is either in a pipe that extends below the centre of the massive sulfide lens and surrounds the stockwork vein zone (Figure 3.51a), or in a stratabound zone that extends up to a few hundreds of metres below the sulfide lens (Figure 3.51b). The latter is more common where the ore body has a stratiform rather than podiform shape.

Magnesian-chlorite–quartz–pyrite is the characteristic assemblage in the most strongly altered rocks of the stockwork zone and immediate foot-wall of massive sulfide ore. Ankeritic carbonate occurs as 'spots' in this assemblage in the foot-wall of some deposits. Sericite gradually replaces chlorite as the dominant alteration mineral with increasing distance, either away from the centre of the alteration pipe, or below the sulfide lens.

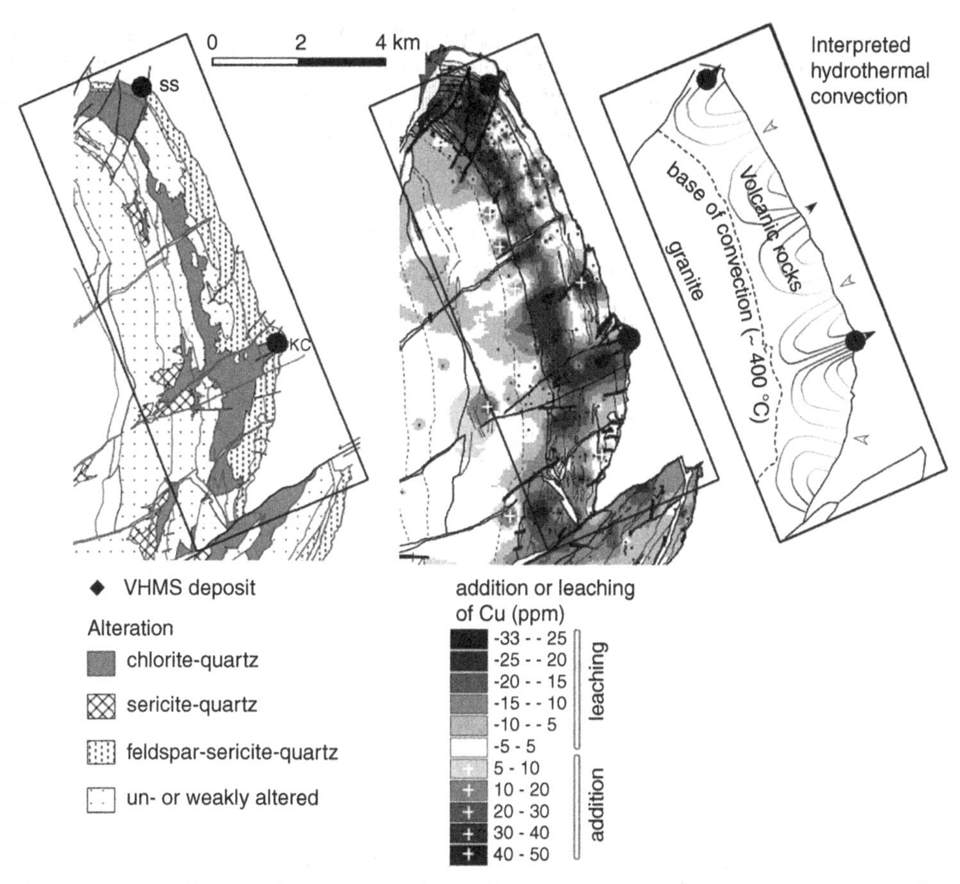

Figure 3.52 Distribution of alteration facies within the Strelley succession, Western Australia. (For a wider geological context see Figure 3.48.) Because of domal folding, the map is essentially a cross section through the sea floor at the time of mineralisation. Fluid flow is interpreted to have been in convection cells with down-flow from the sea floor through zones of feldspar–sericite–quartz alteration, reaching to about 2 km depth (chlorite–quartz alteration) and up-flow back to the sea floor at the deposits (Brauhart *et al.*, 2001).

The far-field alteration is less intense than the near-field. Characteristic minerals are sericite, chlorite and epidote. The alteration is developed as stratabound zones, often most strongly a couple of kilometres stratigraphically beneath the ore bodies, and in pipes of alteration a few hundred metres wide extending from this stratabound zone up to the stratigraphic level of the ores (Figure 3.52).

Patterns of fluid flow in VHMS hydrothermal systems

Fluid-inclusion data indicate that VHMS ores precipitated from fluids of slightly higher to slightly lower salinities than seawater (~ 3.5 wt % salts). These data in combination with alteration mineral assemblages in and around the ores indicate that the ores formed at temperatures of most commonly 250–350 °C. The temperatures and fluid salinities are thus similar to those recorded at active sea-floor vents. The hydrothermal fluid is inferred to be dominantly seawater that was heated as it convected through ocean-floor rocks.

The distribution of hydrothermal alteration stratigraphically below the ores, in combination with numerical simulations of likely patterns of fluid convection in the crust (see Box 4.1), allow interpretation of the patterns and causes of hydrothermal fluid flow that gave rise to the deposits (Figure 3.52). Fluid convection cells with lateral dimensions of up to about 10 km and vertical dimensions to 2–4 km develop in the rocks where there is a steep geothermal gradient above a sub-volcanic intrusion. The heat source driving convection is the axial magma chamber in the case of a MOR environment, but can be a mafic sill or an intermediate to felsic intrusion in other environments. Downwelling and refluxing of cold ocean water appears to be relatively diffuse and causes broad zones of sericite alteration in what would have been the upper 1–2 km of the crust. This water is heated to about 350–400 °C in the sub-horizontal basal limbs of the convection cells, where it causes widespread chlorite–quartz–epidote alteration. Fluid upwelling to the sea floor with little loss of heat is through relatively narrow pipes below the massive sulfide ore bodies and also causes strong chlorite–quartz alteration.

Composition of ore fluids and sources of dissolved metals

The vented hydrothermal fluid is seawater that has been compositionally modified by dissolution of minerals and precipitation of hydrothermal alteration minerals in the subsurface as it migrates through the rocks of the ocean floor. Compared to seawater, the hydrothermal water has lost Mg, largely as a result of precipitation of magnesian-chlorite below and laterally away from the deposits. Epidote is a major alteration mineral, especially in basalts laterally away from the deposits, and its formation involves oxidation of Fe(II) to Fe(III). This oxidation is balanced by reduction of the fluid and conversion of dissolved sulfate to sulfide.

A number of metals, including Fe, Mn, Zn and Cu, are at orders of magnitude higher concentrations in solution in the hydrothermal fluid at the vent than in seawater. Systematic sampling and chemical analysis of rocks within the hydrothermal convection cells shows that a fraction of the ore metals are on average leached, such that average whole rock Cu concentration for instance is reduced in altered rock from about 50 to 30 ppm (Figure 3.52). The mass of altered rock that the circulating water interacts with in the convection cells is large and mass-balance calculations show that sufficient metal is leached from altered rock to account for that deposited in the sulfide ore bodies. These elements are thus leached as a result of mineral–rock reactions during the migration of the fluids. Iron and manganese are in particular leached from ferromagnesium minerals such as pyroxenes. Different amounts of the metals are available for leaching from different rock types. Copper is most abundantly leached from mafic rocks such as basalts and is hence in highest concentrations in Cyprus-type deposits in ophiolites, whereas Pb is more abundantly leached from felsic volcanic sequences or clastic, terrestrially derived sediments.

Although mass-balance calculations confirm that the ore metals are likely leached during fluid convection through the crust, it should be noted that the results do not rule out a role of magmatic fluids released from the underlying intrusion and mixed in with the convecting ocean water. A magmatic-hydrothermal fluid is expected to be released from the underlying magma chamber that is driving the convection, and the presence of such a fluid is indicated at active vents by some aspects of fluid chemistry. Minor Cu–Zn–Sn veins that cut the upper levels of the sub-volcanic granitic intrusion and lower levels of the overlying volcanic sequence at, for instance, Panorama (Figure 3.48) may be marking release and upward escape of magmatic-hydrothermal fluids into the convection system. A contribution to the ore is, however, difficult to fingerprint. The ratios of ore metals in VHMS ores are for instance characteristic of the volcanic environment in which the ores form, such that ores in mafic rocks are Cu-rich whereas those in felsic rocks are Pb-rich. These relations can be explained as a result of composition of the rocks of the ocean crust irrespective of whether the metals are sourced by leaching or by a magmatic fluid. In general we would expect that greater masses of magmatic fluid would be released from calc-alkaline magmas in submarine island arcs than from tholeiitic magmas at MORs and that a magmatic component of ores hence contributes more significantly to ore formation in these settings, and may be a reason for Au- and Ag-rich ores in these environments.

Interpretation of the development of VHMS deposits

Volcanic-hosted massive sulfide deposits are formed in sub-sea-floor hydrothermal systems in which the heat source for fluid convection is a shallow intrusion that reaches to within about 2–4 km of the sea floor. The hydrothermal fluid is dominantly seawater, possibly mixed with minor magmatic water. The waters are characteristically heated up to about 350–400 °C. The temperatures of the vented waters may be limited to about 350 °C by a number of factors, including enhanced compressibility of water near its critical point (Figure 3.1 and Box 4.1) and the temperature at which ocean-floor rocks are strong enough to hold fractures open so as to allow convective fluid circulation. This temperature corresponds approximately to the transition from the brittle to ductile deformation behaviour at the typical stress levels imposed on the crust.

Where convecting hydrothermal fluids are heated to sufficiently high temperatures (200–350 °C), leaching of the metals from primary minerals and quantitative reduction of dissolved sulfate to sulfide with oxidation of Fe(II) to Fe(III) in the altered rocks is promoted, and there is thus potential to form a sulfide ore body at a vent. Transport of the metals in solution is as Cl^- complexes. Fluids which are convected through the ocean crust but are not heated to these temperatures may precipitate iron oxide lenses on discharge.

Precipitation of sulfide minerals is prompted by cooling and mixing of the hydrothermal fluid with seawater. Mixing causes first an increase in fluid pH, hence reducing the solubility of sulfide minerals, and then oxidation, which allows precipitation of gypsum and barite.

The precipitation of sulfide minerals takes place at various sites: below the vents; at the hydrothermal vents; and after discharge of the hydrothermal fluid into ocean water. Podiform and lenticular massive sulfide lenses are formed by processes similar to those observed on the ocean floor. These bodies represent mounds built up at or just below the sea floor by precipitation at the vents, including in chimneys which become inactive and collapse. The zonation of ore minerals in a lens (Figure 3.50) is a product of progressive fluid cooling on percolation through the porous sulfide mass, such that minerals that reach saturation at highest temperatures are precipitated first and hence deeper. The sequence of precipitation from a fluid that vents at above about 300 °C is: (1) chalcopyrite; (2) sphalerite + galena; (3) manganese oxides and barite. As the mound progressively grows, higher-temperature-precipitating minerals replace those that precipitated at lower temperature.

Layered, stratiform deposits are the result of a differing site and style of ore mineral precipitation that does not have an exact modern analogue. These ores may represent 'sedimented' sulfide grains that were entrained in plumes that rose up into the ocean water

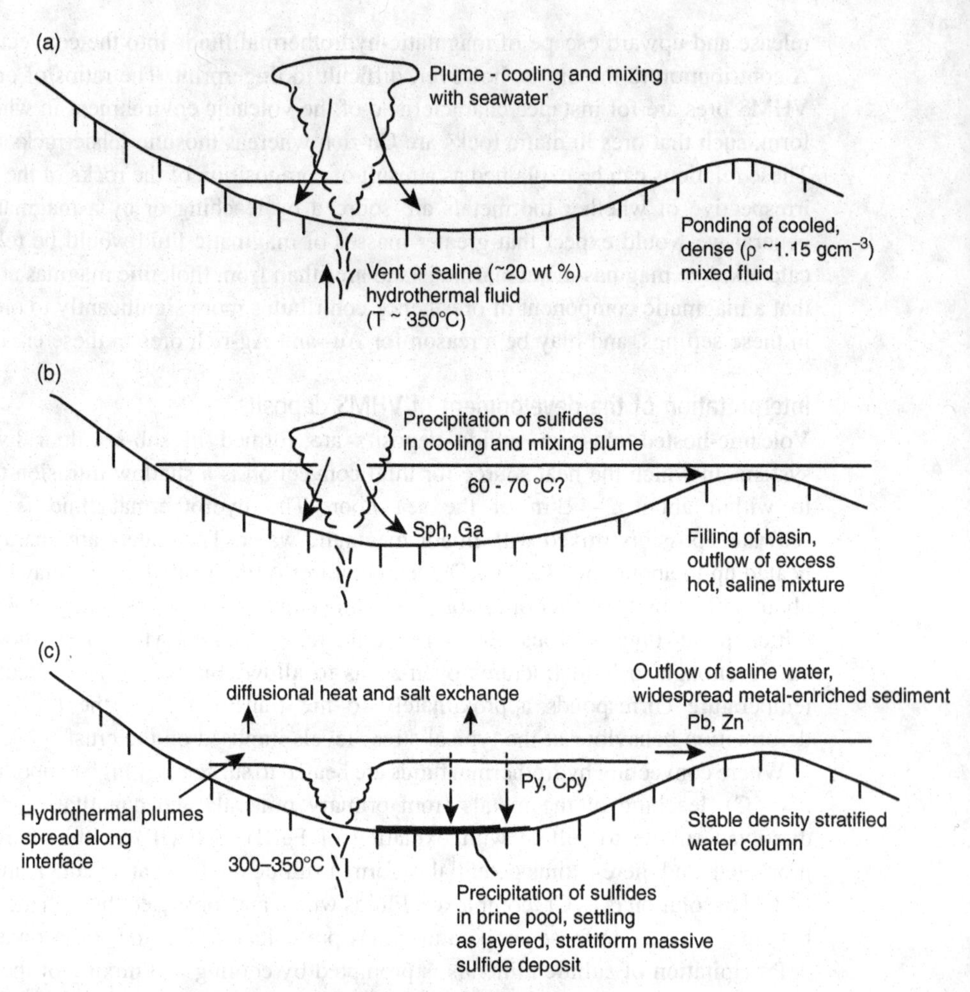

Figure 3.53 A possible time sequence leading to the formation of layered, stratiform massive sulfide deposits in brine pools on the ocean floor (after Solomon and Groves, 2000a). (a) Discharged saline hydrothermal fluids are denser after cooling than ambient seawater and pond in a closed depression on the sea floor, in which the salinity progressively increases. (b) Filling of the basin to form a stratified water column with warm saline waters below ambient seawater. Sulfide minerals begin to precipitate because of cooling and mixing along the upper interface. (c) Development of a near-steady-state brine pool. Vented high-temperature hydrothermal fluids spread out at the density interface at the top of the brine pool, lose heat and salt to the overlying ocean, and consequently precipitate sulfide minerals which sediment as layers. The repeated sphalerite–galena and chalcopyrite–pyrite layers of the sulfide deposits are likely the result of changing conditions of the fluid and of the pool. Some Pb and Zn may escape into overflow basins.

column, or were precipitated and grew largely in the water column of pooled hydrothermal fluid on the sea floor (brine pools) (cf. Figure 3.53), as is observed at the Atlantis II Deeps in the Red Sea. Transport of minerals in venting water is seen for instance in the 'smoke' of black smokers. Mineral precipitation in the water column is indicated for instance by mineral textures such as framboidal as opposed to dendritic pyrite. Analogy with the Red Sea sites suggests that pooling of hydrothermal waters in a sea-floor basin over long enough periods of time to precipitate a massive sulfide ore body would be favoured where the venting fluids are more saline than seawater and hence relatively dense.

Requirements for the formation of VHMS deposits

The distribution of active hydrothermal vents and actively accumulating sulfide deposits on the sea floor shows that high-temperature hydrothermal activity can occur around any active volcanic centres on the sea floor. The following are critical factors which are required for the formation of a VHMS deposit:

- A shallow heat source is required at relatively shallow depths (2–4 km) in order to drive strong hydrothermal convection. Numerical simulations of convection above a laterally extensive heat source such as a large sill or intrusion at a few kilometres in the crust reproduce the width of convection cells implied by the typical 5–20-km spacing of deposits at ore horizons (Figure 3.52).
- Water depths of greater than about 1 km allow the most efficient precipitation of sulfides in a massive mound or in a stratiform layer. At shallower depths, a fluid at approximately 400 °C will boil before it reaches the sea floor. The changes in solution chemistry that take place on boiling mean that much of the sulfide would precipitate below rather than at the sea floor, probably as disseminations and veins rather than massive sulfide ores. This probably occurred in some deposits, for instance at Mt Lyell, Tasmania, Australia, where lenses of disseminated Cu-rich ore in altered volcanic rocks are present stratigraphically beneath massive Pb–Zn-rich sulfide lenses. Some active hydrothermal vents are at shallower water depths, but in most cases these are not precipitating massive sulfides at the sea floor.
- Sulfides are unstable and will oxidise in contact with typical present-day seawater. Rapid burial of the deposit is thus generally required after formation to prevent oxidation and dissolution into seawater. Many ancient deposits are interpreted to have formed in resurgent calderas after a major caldera collapse but just before a new phase of magmatism during which the deposits were buried.

3.3 Syn-orogenic hydrothermal ore deposits without close spatial or temporal relations to magmatism

Three important types of hydrothermal Au and Au–Cu deposits form during regional tectonism and within the time spans of active magmatism in the host geological terrains. These are:

(i) *Orogenic Au deposits*, which are alternatively known as **lode** Au deposits, quartz-vein Au deposits, Au-only deposits or **mesothermal** Au deposits. These are vein and replacement Au deposits in metamorphic and intrusive igneous rocks, most of which have formed broadly during periods of regional metamorphism.

(ii) *Carlin-type gold deposits* are named after the type locality in Nevada, USA. Based on the host-rocks and the nature of the ore, they have been alternatively named *sediment-hosted gold deposits* and *disseminated gold deposits*. They are *replacement* ore bodies in metasomatically altered impure carbonate and calcareous siliciclastic sedimentary rocks.

(iii) *Iron oxide–copper–gold (IOCG) deposits* are hydrothermal ores most importantly of copper and gold, but which may also produce other metals as by- and co-products (U, Ag, LREEs). The distinct characteristic of these deposits is that the major Fe-bearing mineral in ore is magnetite or haematite rather than an iron sulfide mineral such as pyrite.

Ore fluids flow upward through the crust at the site of all three types of deposit. None of the deposit types are, however, clustered around igneous centres. Isotope and element ratios of ores are not closely correlated with those of nearby magmatic rocks. These observations do not rule out derivation of the ore fluid from magmas at some distance to the deposits. However, fluids of two other sources are also expected to be circulating in the crust in tectonically and magmatically active environments of the type in which the deposits form, and one or more of these other fluid types have been proposed to be the ore fluids in each deposit type, including metamorphic fluids and deeply convecting meteoric fluids (Figure 3.54).

Metamorphic fluids are those derived from devolatilisation mineral reactions that release volatiles (H_2O, CO_2, H_2S) during prograde regional or contact metamorphism in

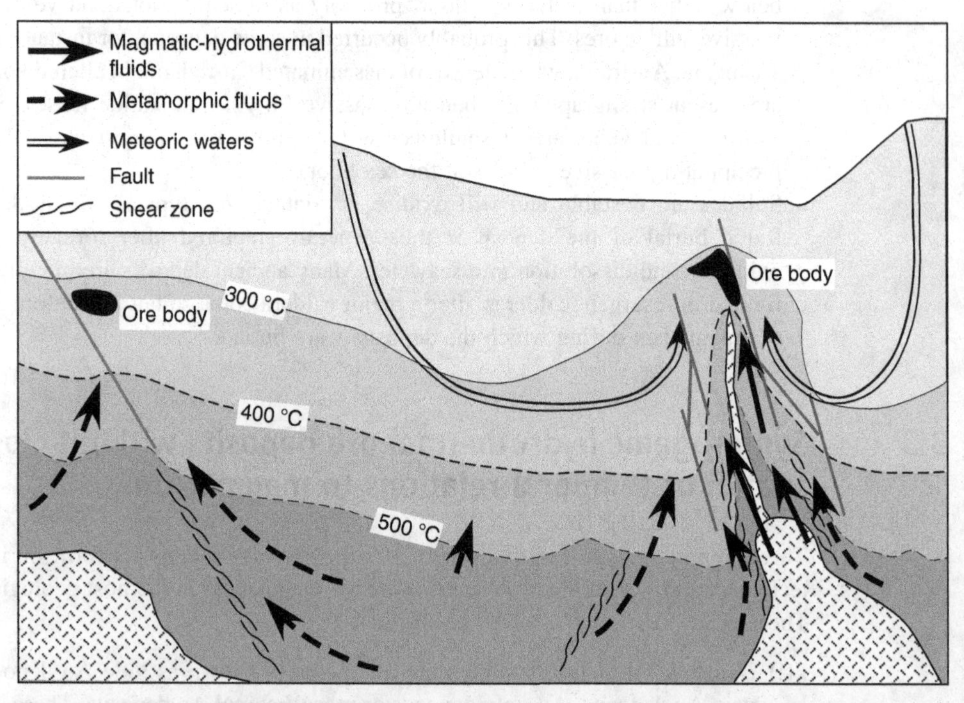

Figure 3.54 Schematic of possible sources of hydrothermal fluids and likely fluid pathways in a magmatically, metamorphically and tectonically active continental terrain with high topographic relief. Likely patterns of long-distance circulation of metamorphic, magmatic and surface-derived meteoric fluids are shown. Possible sites of ore bodies along upward-directed flow paths are also shown.

the crust or in mineral reactions in the mantle. We cannot make direct observations of the behaviour of fluids released in metamorphic reactions. It is likely, however, that immediately after a metamorphic devolatilisation reaction the fluid will be dispersed on a grain scale through the rock, and may dissolve minerals up to the limits of solubility or availability. The fluid will be at lithostatic pressure and will escape upwards through the crust. The channelways of escape are marked by many of the generations of veins that are a common component of metamorphic terrains.

Terrains of tectonically and magmatically active crust are also expected to be sites of vigorous and deep convection of surface-derived waters driven by magmatic heat, topography and seismic activity, and fluids flowing upwards after convection to depth could thus be ore fluids. It is generally assumed that the depth of convection of surface-derived waters is at about the level of the base of the seismic zone in the crust and the depths at which the highest temperatures recorded for geothermal reservoir fluids of about 300–350 °C are attained (see Section 3.2.6).

Fluids from either of these sources and from magmatic sources would thus be expected to rise through the crust at the same sites, for instance above and along the margins of intrusions. It is, however, inherently difficult to determine the origin of the ore-bearing hydrothermal fluid of deposits in these settings, and hence the source of both the fluid and of the metals. The ore fluids are not chemically distinctive and many have stable isotope ratios, pH and oxidation states that are close to those in equilibrium with average crust. Long distance of fluid flow may mean that the fluid compositions have been modified between source and deposit. In addition there are difficulties imposed from complex structural histories with multiple phases of deformation along the ore-hosting structures of many lodes such that vein and ore minerals are deformed and recrystallised and it is difficult for instance to determine exactly which fluid inclusions trap the ore fluid. Later or earlier phases of fluid migration within the lode structures may mask unique isotope or chemical signatures of the ore fluids.

3.3.1 Orogenic Au deposits

This deposit type is the source for about one-third of the Au that is mined worldwide. The deposits are mined in most cases only for Au, and only very rarely for co- or by-product As and Sb. Silver contents are notably very low in these ores and Ag/Au ratios are much lower than either average crustal abundances or the ratios in other hydrothermal ore deposit types, including VHMS and epithermal deposits. Many deposits are relatively small (less than or of order of 1 Mt ore) and these small deposits have been evaluated as resources because they are often suitable for small-scale mining. There are, however, a number of large, world-class deposits, including the world's largest gold producer at the time of writing at Murantau in Uzbekhistan.

The deposits formed in actively evolving orogenic belts and are hosted in regionally metamorphosed and intrusive igneous rocks. As a deposit class they are almost unique in that they form at relatively great pressures of between about 1.5 and 5 kbar, hence at depths in the crust of about 4 to 15 km. Although they form at moderately high temperatures, the temperatures of formation are lower than many magmatic-hydrothermal deposits and are most typically between 300 and 450°C. They are most commonly hosted in rocks metamorphosed to the greenschist facies.

Where they occur, they are often abundant and widespread throughout the host belt, although are typically clustered in ore 'camps' spaced at a few tens of kilometres distance (e.g. Figure 3.55). The large numbers of deposits in many orogenic belts formed within short periods of a few million years within the typically much longer metamorphic and deformational histories of the host orogen.

Geological and tectonic settings of orogenic Au deposits

Many orogenic belts of all geological ages host at least a few small orogenic gold deposits, but we recognise the following specific orogenic environments in which these deposits are both abundantly formed and in which larger deposits occur. All of these environments are characterised by a low-pressure facies series of metamorphism.

(a) Archaean and early-Proterozoic greenstone belts

Orogenic Au deposits are widespread to abundant in parts of most Archaean cratons. The Golden Mile at Kalgoorlie in the Yilgarn Craton of Western Australia (Clout *et al*., 1990), the Hollinger-McIntyre (Burrows *et al*., 1993) and Sigma deposit in the Superior Province of Canada (Robert and Brown, 1986), and deposits of the Barberton greenstone belt, South Africa (de Ronde *et al*., 1992) are some of the larger and best-known examples. They are also present in some early-Palaeoproterozoic terrains, including the Ashanti deposit and others in the Birimian belts of Ghana and Burkina (e.g. Oberthür *et al*., 1997).

The deposits are hosted in all rock types in the cratons, but almost all in Archaean cratons are hosted in **greenstone belts**. These are arcuate to elongate belts of deformed and regionally metamorphosed volcanic–sedimentary sequences that formed on older continental crust and that are preserved between granite–gneiss complexes and granite plutons. The larger and better preserved greenstone belts are up to a few tens of kilometres wide and typically have relatively low metamorphic grade, greenschist-facies or sub-greenschist-facies cores, and higher-grade amphibolite-facies margins against the bounding granites and granite–gneiss terrains. The majority of the Au deposits are hosted in the greenschist-facies zones, but some occur in lower- and in higher-grade metamorphic rocks. Specific rocks types within the greenstone belts are statistically preferred host-rocks, especially banded-iron-formations (BIFs) and mafic rocks (metadolerite and metabasalt). In the Superior Province in particular, deposits are concentrated along and within a few kilometres of major 'breaks', which are steeply dipping fault or shear zones traceable over hundreds of kilometres that separate blocks with differing stratigraphic histories.

The cratons have complex magmatic and deformational histories. The most common timing of ore deposit formation is late in the tectonic and magmatic history, just prior to stabilisation of the terrain as a craton and approximately coeval with the last major phase of granitic plutonism.

(b) Slate belts

Slate belts are orogenic belts in which the most widespread unit is a sequence of at least a few kilometres' stratigraphic thickness of deformed and metamorphosed turbidites that was deposited at an active continental margin. The deformation to produce belts of repeated large-scale near-upright folds was during later compressional deformation along the continental margin. Orogenic Au deposits are common in slate belts that have been largely metamorphosed to greenschist-facies conditions and have been intruded by granitoid

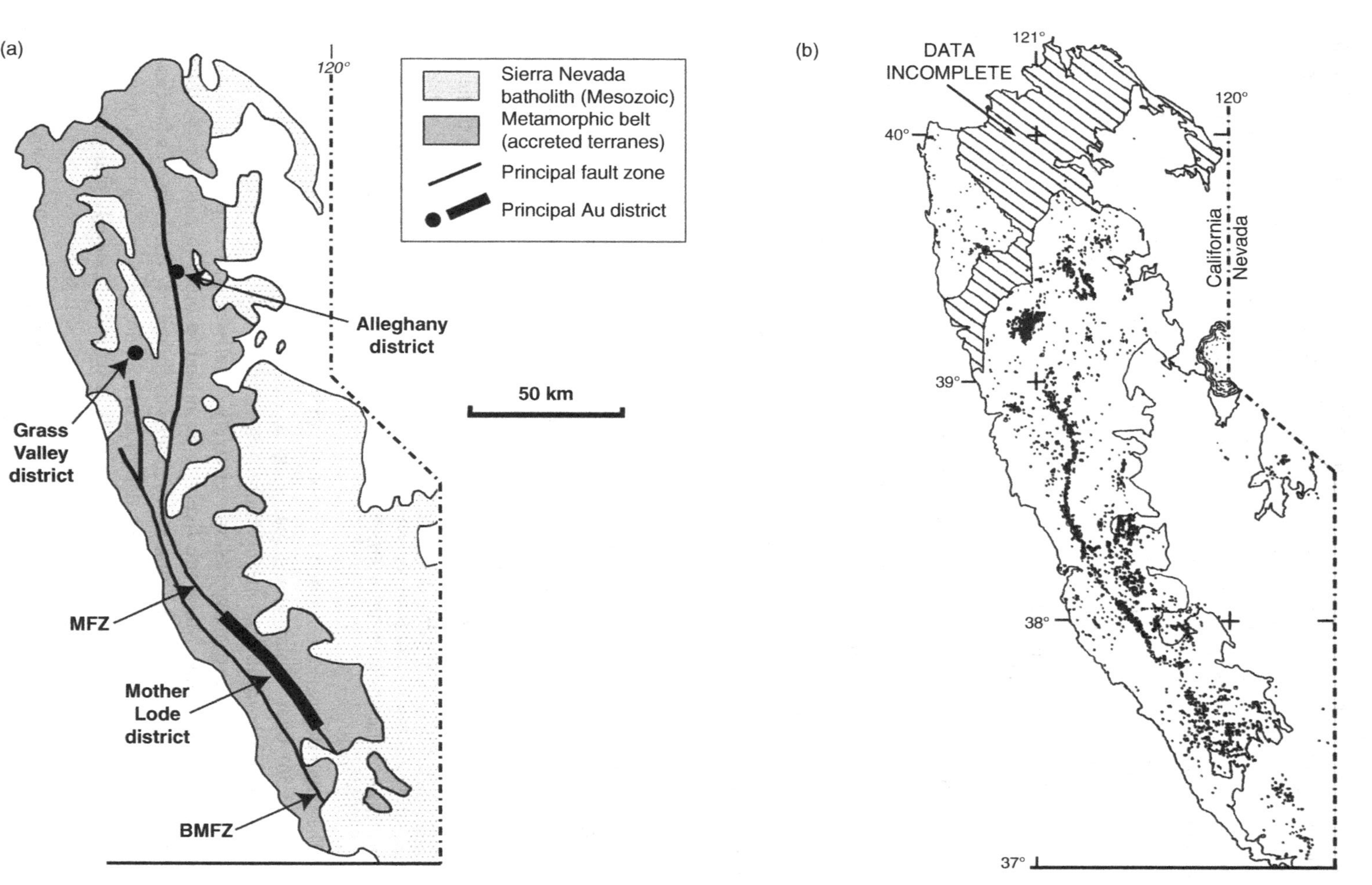

Figure 3.55 (a) Simplified geological map of the area of early-Cretaceous 'orogenic' quartz vein Au deposits in the Sierra Nevada foothills of California. The most important mining districts are labelled. MFZ = Melones Fault Zone; BMFZ = Bear Mountain Fault Zone. (b) Distribution of deposits. The deposits are widespread, but are clustered, especially along and adjacent to north-northwest–south-southeast-trending fault and shear zones. Note the concentration of deposits in the metamorphic rocks and their rarity in intrusive rocks. (After Böhlke, 1999).

plutons during or after low-grade metamorphism, including the Lachlan Fold Belt, Victoria, Australia (e.g. Phillips and Hughes, 1996), and the Meguma Terrane, Nova Scotia, Canada (Sangster and Smith, 2007). The deposits such as Macraes Flat of the Haast Schist Complex of the South Island of New Zealand (e.g. de Ronde *et al.*, 2000) are unusual in that they are in a terrain which lacks abundant syn-tectonic granitic plutons.

Within these terrains, the dominant host-rocks to the deposits are the metaturbidites, but some may also be hosted in tectonically interleaved mafic rocks or in granites near the margins of the plutons. Most of the deposits are hosted in greenschist-facies rocks.

(c) Cordilleran orogenic belts

Deposits are abundant in some segments of cordilleran style or accretionary orogenic belts, which are composed in part of deformed and deeply eroded magmatic arcs with extensive granite batholiths such as the Mother Lode, Alleghany and other major historic gold-mining districts of the western Sierra foothills, California, USA (Figure 3.55). In this environment, the deposits form at times of arc granitic magmatism but are hosted in the metamorphic rocks of the orogen rather than in and around the granites. In the case of the Californian examples they formed at the same time as the intrusion of the Sierra Nevada batholith but are most abundant along large-scale shear zones a few tens of kilometres to the west of the exposed batholith and closer to the site of the active subduction zone. The deposits form along segments of the arcs and at times in their development when and where porphyry- and epithermal-type deposits are not formed.

Characteristics of orogenic Au deposits

(a) Shape and size of deposits

Many ore bodies are lode-shaped, i.e. they are tabular or planar bodies of ore, which in some cases are single continuous quartz veins, and in others zones of closely spaced small quartz veins and adjacent hydrothermally altered rock. Typical dimensions of a lode may be 2 metres wide by hundreds of metres along strike and down dip. The ore bodies are characteristically structurally controlled rather than stratabound. Many are within or adjacent to structures and their shapes match those of the host structures and component fabrics, for instance, with an ore-body long axis parallel to the stretching lineation of a host shear zone.

The larger deposits are formed of networks of veins or mineralized shear zones (e.g. Figure 3.56) over a plan-view area of up to about 1 square kilometre. These are typically mined within a single operation. There is a spectrum of scale of these lode networks, from dense networks of relatively small-scale structures, for instance centimetre-wide veins, which are likely to be mined in bulk, to networks of thicker and more widely spaced lodes that are each selectively mined. The former is the case of many large, more recently developed, deposits. Where deposits are formed of networks of veins and lodes, the envelope of the ore body is typically a steeply plunging pipe or ribbon.

Lodes and veins are typically steeply dipping and plunging. In contrast to epithermal deposits, ore is marked by vertical continuity (Figure 3.57). Some larger orogenic Au deposits have been mined continuously from the surface to between 2 and 3 kilometres depth, with little change in ore grade with depth. Some of the deepest-mined deposits

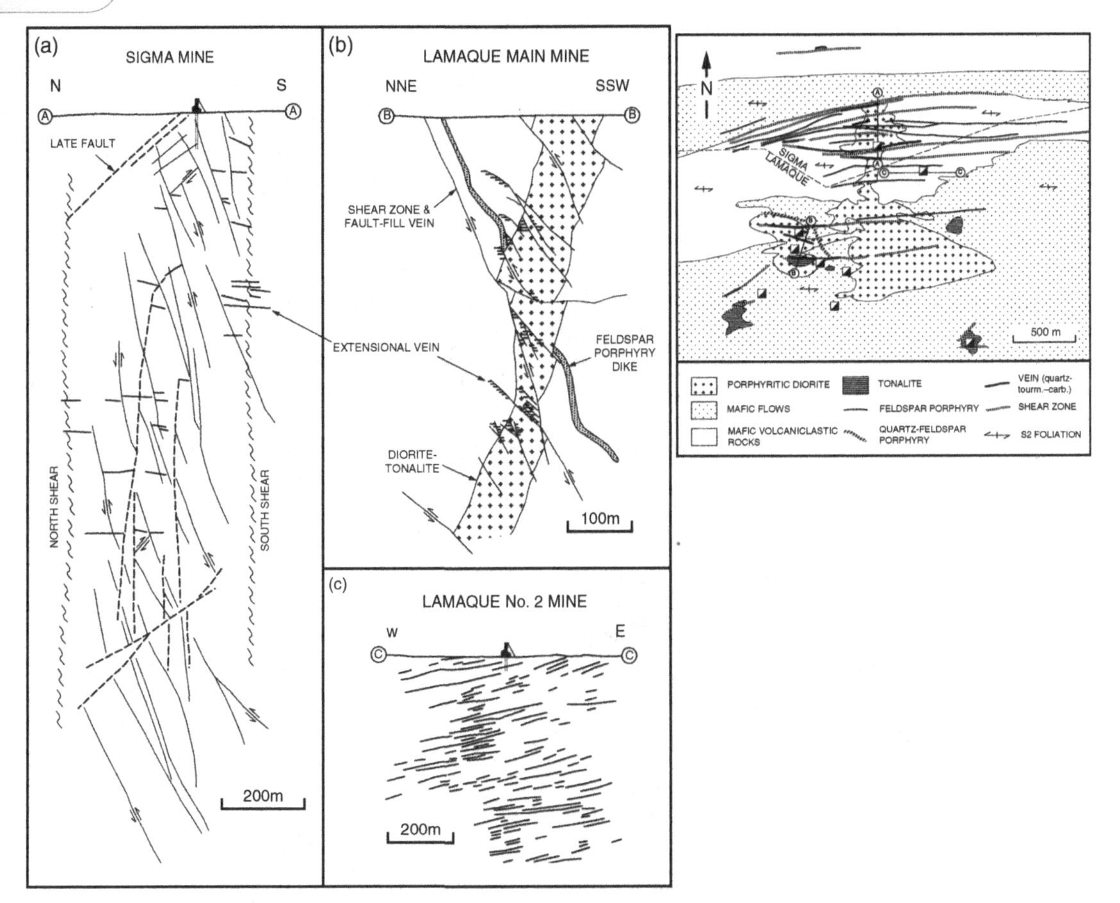

Figure 3.56 Map and sections showing the structure and setting of the network of mineralised veins at the Sigma–Lamaque deposit, Val d'Or area, Quebec, Canada. The deposits are formed of networks of individual veins that are typically 1 to 5 m thick within a volume of rock about 1 km across, 1.5 km long and at least 2 km deep. Many veins are hosted in a sub-parallel set of shear zones (steeply dipping, east–west striking 'shear zone' or 'fault-fill' veins). Others are sub-horizontal 'extensional veins' cutting massive rock (Robert, 1990).

include those of the Kola Peninsula, India, and deposits in the Superior Province of Canada, including Sigma, Quebec.

(b) Structural settings of deposits

Orogenic gold deposits are the most closely and strongly structurally controlled of all deposit types. The structural settings of the deposits are, however, very variable. Many lodes and veins are hosted along shear zones or faults, a setting in which they are typically restricted to segments of the host structure, for instance those segments along which the structure is in a specific orientation (e.g. Figure 3.58). Other structural settings of deposits include fold hinge zones (**saddle reefs**), and sheeted veins or systematically oriented vein stockworks in massive rock that is bounded by shear zones or faults. Multiple types of structure may host ore zones within a single deposit.

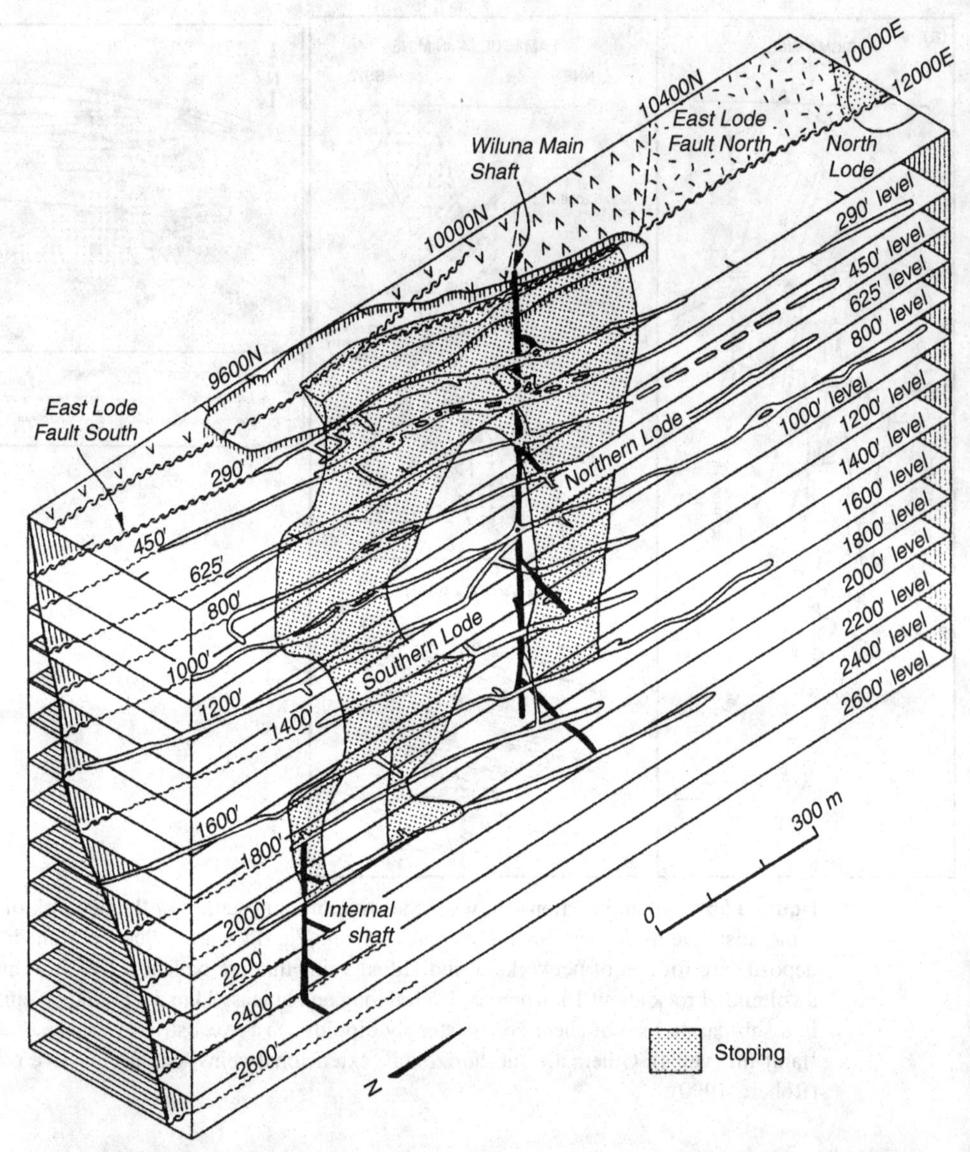

Figure 3.57 Three-dimensional image of the East Lode deposit, Wiluna. For the camp-scale setting of this deposit see Figure 3.58. The stoped rock (i.e. volumes of ore that have been mined) defines irregular, steeply dipping and steeply plunging ribbons up to a few metres thick along a fault zone (Hagemann *et al*., 1992).

(c) Mineral assemblages of ore and hydrothermal alteration zones

Veins in orogenic Au deposits are generally quartz-dominated and range from internally structureless and massive to coarsely banded or laminated. Most have a minor content of carbonate minerals, most typically calcite or ankerite, and up to a few per cent of sulfide minerals. Other vein minerals include albite, and tremolite and diopside in deposits hosted in higher-metamorphic-grade rocks. Mineral grains with euhedral terminations are rare and most quartz is interlocking-granular, although in some veins the textures indicate that

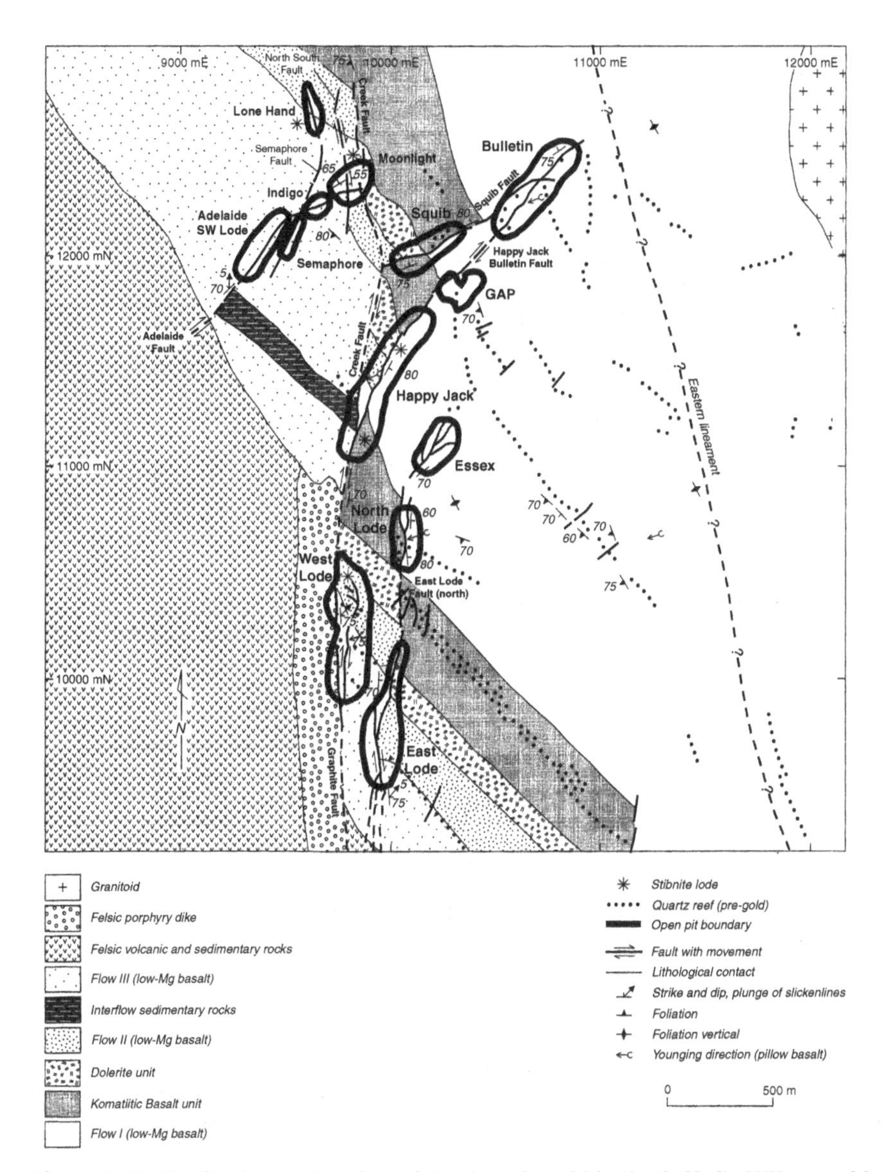

Figure 3.58 Distribution and setting of Au deposits within the 2.63 Ga Wiluna gold camp, Yilgarn Craton, Western Australia. The deposits occur within and around segments of brittle–ductile fault zones, especially at fault intersections, fault terminations and other structural irregularities (Hagemann *et al.*, 1992).

the granular texture is the result of recrystallisation of grains that may have grown as euhedrally terminated free-standing crystals.

Alteration **haloes** of widths from a few centimetres to tens of metres around lodes are characteristically developed around veins and ore zones (Figure 3.59). These alteration haloes form continuous sheaths or envelopes around the steeply dipping and plunging lodes. They are markedly zoned laterally away from the vein or ore zone with different mineral assemblages that mark increasing degrees of mineral replacement and

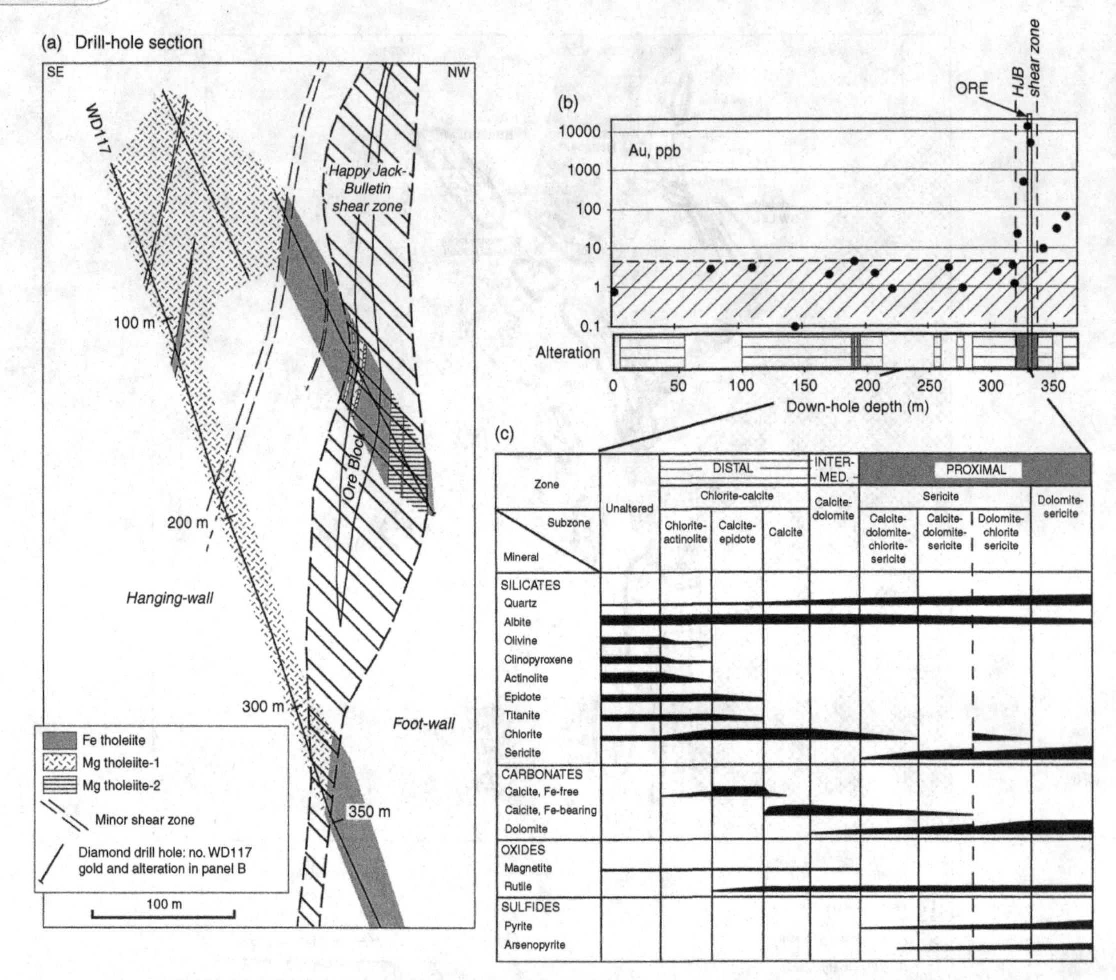

Figure 3.59 Alteration minerals and their distribution at the Bulletin gold deposit, Wiluna, Western Australia (after Eilu and Mikucki, 1998). Panel (a) shows the drill-hole section of hole WD 117. Panel (b) shows the distribution of alteration zones around the lode together with gold grade along this drill hole. The shaded band shows the typical range of concentration of gold in mafic rocks. Gold is at above-background concentrations in the shear zone and in rocks that have been altered to sericite-zone assemblages. Panel (c) shows the distribution of minerals in the different zones of the alteration halo with ore at the right-hand edge.

metasomatism with increasing proximity to the vein. Distal and proximal alteration assemblages are thus often distinguished. In contrast to the marked zonation across the alteration envelopes, the alteration assemblages vary little or only subtly along the strike length or down dip along the lode, even over depth differences of the order of a kilometre.

In most deposits the alteration zones are composed of mineral assemblages that are stable at similar pressure and temperature conditions as the metamorphic assemblages in surrounding rock. Calcite, ankerite, quartz, dolomite, sericite and sulfide minerals are the common alteration minerals in the proximal alteration zones in the common setting of deposits in greenschist-facies metamorphosed mafic or intermediate composition rocks and because Fe is present only in carbonate and sulfide minerals in these zones, the rocks often appear bleached compared to less strongly altered rock. Chlorite and calcite are dominant minerals in the distal zones. The proximal zone assemblages are similar to those of phyllic alteration of porphyry deposits, but are distinct in that they have abundant carbonate. Assemblages including biotite, albite, amphibole and diopside occur adjacent to veins in deposits hosted in slightly higher-grade metamorphic rocks of upper-greenschist and lower-amphibolite facies. Alteration in banded iron formations may be marked by massive replacement of iron oxide minerals by iron sulfides together with some carbonate minerals: that in metaturbidites may be relatively subtle and be most clearly marked by carbonate 'spotting'.

Sulfide minerals are an almost invariable component of both ore and alteration haloes and form typically around 5 vol % of ore. Pyrite is the dominant sulfide in most ores, and typically coexists with one or both of pyrrhotite and arsenopyrite. Assemblages of mixed telluride minerals are important components of some deposits. Base-metal sulfides (chalcopyrite, sphalerite, galena etc.) are rare and minor.

Gold is hosted in veins and in the adjacent most strongly altered wall-rock, commonly as native gold or electrum attached to or as inclusions in sulfide minerals. In some deposits Au is 'refractory' and may be hosted either in telluride minerals such as calaverite ($AuTe_2$) or at least part of it in small inclusions and in solid solution in the sulfide minerals pyrite or arsenopyrite, or rarely in the iron arsenide loellingite ($FeAs_2$).

(d) Ore fluid composition and conditions of ore deposition

Fluid-inclusion compositions and the vein and alteration mineralogy indicate that the hydrothermal fluid of orogenic Au deposits is a low- to moderate-salinity (2–10 wt %) aqueous fluid with significant concentrations of CO_2 in solution (mole proportion X_{CO2} = 0.05–0.25). It is further of near-neutral pH, has an oxidation state not strongly different from that of the metamorphic rocks of the host sequence, and contains sufficient concentrations of H_2S in solution to precipitate iron sulfide minerals. The zonation of the alteration assemblages shown in Figure 3.59 is the result of progressively increasing metasomatic addition of S^{2-}, CO_2 and K^+ to the rock with increasing proximity to the ore zone. Silica is also added to the ore zone in most deposits, both as the major component of the veins and into altered wall-rock.

Ore deposition occurs at metamorphic pressure–temperature conditions. Deposition within the range of 250–400 °C at about 1–3 kbar (3–10 km depth) is most common, but ore formation also occurs at temperatures up to about 600 °C and pressures of 4–6 kbar. Subtle change of alteration and ore mineralogy with depth in the deeper-mined ores indicate that temperatures were higher at deeper levels by a few tens of degrees per kilometre, as would be consistent with the steepness of a typical geothermal gradient. For instance, biotite replaces sericite as the most important Al-bearing mineral in the proximal alteration zones at the greatest depth of mining in Sigma, Quebec, and in a number of other deposits.

Interpretations of ore genesis

(a) Chemical processes of Au precipitation

Gold is expected to have very different behaviour in hydrothermal solutions than most other metals. Thermodynamic data of different aqueous complexes of gold have been derived through measurement of the experimental solubility of gold in aqueous solutions at high pressures and temperatures equivalent to those of ore formation. These data allow us to determine that Au is present predominantly as Au(I) in solution and is carried at

250–400 °C dominantly as bisulfide complexes (e.g. $Au(HS)_2^-$) in near-neutral pH H_2S-bearing fluids, although the exact stoichiometry of the Au complex is uncertain.

The association of gold ore with metasomatic addition of sulfide to the ore and alteration haloes, and the tendency of the highest-grade ore zones to be in Fe-rich host-rocks such as BIF and mafic rocks, can be explained by Au precipitation as a result of chemical reactions between an H_2S-bearing aqueous fluid and wall-rock. Iron oxides and silicates are absent in the vein and the proximal alteration zones in the wall-rock, and the iron that was contained in them has been used to form Fe-bearing sulfides, for instance pyrite and arsenopyrite. Some of the H_2S content of the fluid was thus lost as a result of interaction with the wall-rock to form the alteration assemblages. This loss of H_2S would reduce the solubility of gold bisulfide complexes in solution.

A schematic reaction between sulfide in the vein fluid and the minerals of the wall-rock is:

$$FeO(oxides/silicates) + 2H_2S \rightarrow FeS_2(pyrite) + H_2O + H_2. \qquad (3.15)$$

A reaction describing precipitation of Au carried as bisulfide complexes in solution and promoted by loss of sulfide from the fluid is:

$$Au(HS)_2{}^-{}_{(aq)} + 0.5H_2 + H^+ \rightarrow Au + 2H_2S. \qquad (3.16)$$

There are multiple possible causes of Au precipitation through Equation (3.16). In addition to loss of sulfide from the fluid, Au precipitation may also be promoted by interaction between the fluid and a chemical reductant, for instance graphite in a carbonaceous metasedimentary rock, or by acidification. Alternatively, fluid-phase separation into water-rich and gas (CO_2–H_2S)-rich phases may induce changes to fluid pH and hydrogen fugacity that induce precipitation. Phase separation at the pressures and temperatures of ore formation may occur as a result of reduction of fluid temperature and pressure as the fluid moves towards the Earth's surface or because of creation of volume on opening of fractures during seismic events. The multiple possible chemical causes of Au precipitation may explain why these deposits occur in rocks of essentially all compositions.

(b) Patterns of hydrothermal fluid flow

The pressures of ore formation, the vertical continuity of ore zones over at least a couple of kilometres, and the overall shape of larger ore bodies as continuous, steeply plunging pipes allow us to infer relatively long vertical distances of fluid transport through crystalline rocks of the upper and middle crust in these hydrothermal systems.

Quartz solubility decreases with decreasing pressure and temperature everywhere in the crust below a few kilometres depth. The ubiquity of quartz in the veins thus shows that fluid was cooling and was moving upwards rather than downwards through the crust. Fluid flow is along structures including fractures, faults, shear zones, and bedding-plane surfaces. These structures are mechanical discontinuities in the low-porosity crystalline rock masses that host these deposit types. They can act as mechanical weaknesses that open under fluid pressure and hence become zones of permeability in the rock mass.

The structural relations between veins and structures and textures show that many host structures were active during mineralisation (see Figure 3.60). The laminations of many veins are the result of repeated fracture opening during the period of hydrothermal activity. Deformation of vein minerals shows that there was movement along the host structures at least at times during vein infill or after filling. Because open space

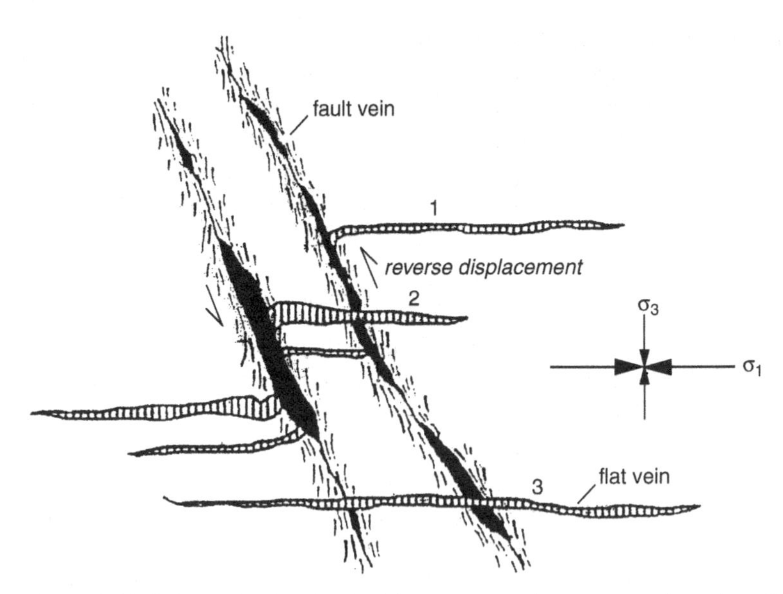

Figure 3.60 Schematic sketch of evidence from vein structures for mineralisation during a phase of deformation from the Sigma deposit, Quebec, Canada (see Figure 3.56) (after Robert and Brown, 1986). Gold is hosted in veins in steeply dipping shear zones (fault veins) and in adjacent flat veins that splay off these fault zones into undeformed country rock. The shear zones are interpreted to have first formed earlier in the structural history and to have been re-activated at the time of hydrothermal fluid flow. In contrast, the flat veins are interpreted to have formed as tension fractures at the time of mineralisation. Some of the flat veins have been deformed in the shear zones (1), others cut across the shear zones without deformation (3). These relations indicate that quartz and Au precipitation occurred over a time period while the shear zone was active and under the stress regime as indicated.

becomes relatively rapidly destroyed by rock compaction at high temperatures and high confining pressures, movement along structures may be a critical process at the depths of formation of orogenic deposits in the crust to generate permeability for hydrothermal fluid flow.

In most cases, the host-rocks of these deposits have had complex structural histories such that there are multiple orientations and generations of different types of structures. Not all structures will however be ore-bearing. Ore is hosted in those specific structures and structures in specific orientations that were permeable at the time of mineralisation. Many of the structures that host lodes are 'inherited', that is, they were present before the onset of hydrothermal fluid flow and were exploited as zones or planes of permeability. New structures, including extensional veins, formed during ore formation and in many cases channelled fluid flow. Because the new structures form in a specific orientation we can infer that the host-rocks were in a deviatoric stress field over the period of mineralisation. We can explain the distribution of lodes and their geometry at many deposits through consideration of the likely structural response of the host mass of rock to imposed deformation through analysis of which orientations and positions of structure were most favourable for failure and opening in the imposed tectonic stress regime (e.g. Figure 3.60).

(c) Origins of the hydrothermal fluid

At the site of the deposits, the ore fluid flowed upward over long vertical distances. The ore fluids have been contrastingly suggested to be magmatic-hydrothermal fluids, fluids derived from metamorphic mineral reactions, mantle-derived fluids of either magmatic or metamorphic origin, or deeply convecting surface-derived meteoric water. Two fluid sources are most consistent with the data on fluid composition and the times of mineralisation within the histories of the host terrains:

(1) Fluids from metamorphic devolatilisation reactions.

Mineralisation took place during regional metamorphism and most typically when the host-rocks were at or near peak metamorphic temperature. The host terrains would have been metamorphically active, and levels in the crust would have been heating and undergoing prograde metamorphic reactions. It is thus hypothesised that the chemically distinct ore fluids of this deposit type are fluids released in prograde metamorphic mineral reactions. Gold is thus proposed to be leached from minerals into a metamorphic fluid from a large volume of metamorphic rocks in which it was present at concentrations similar to those of the crustal average (2–5 ppb). The fluid subsequently focussed into channelways along which it escaped upwards through the crust.

One set of metamorphic reactions that would be appropriate to release a mixed H_2O–CO_2 fluid which could leach and collect Au are the reactions at the greenschist- to amphibolite-facies transition, particularly devolatilisation reactions in carbonated mafic rocks or metamarls. The fluids released in these reactions are calculated from mineral thermodynamic data to be mixed H_2O–CO_2 fluids, hence similar to those trapped as fluid inclusions in the ore veins. In view of the importance of sulfide complexing for Au solubility, the amounts of Au leached from the rocks and the Au concentration of the fluid would likely be controlled by the sulfide content of the fluid. The fluids would contain dissolved sulfide if the source rocks contained pyrite. Metamorphic reactions in a subduction zone below the crust are proposed as an alternative source of Au-bearing metamorphic fluids in these deposits.

(2) Magmatic-hydrothermal fluids.

Orogenic Au deposits are not clustered within a few kilometres of magmatic centres. However, a low-pressure facies series of metamorphism rarely develops without concurrent magmatism, and almost all metamorphic belts with orogenic Au deposits had widespread coeval magmatism. In the case of the Sierra foothills (Figure 3.55), ore formed at the same time as the large granitic Sierra Nevada batholith a few tens of kilometres inboard; in the Archaean Yilgarn craton the period of Au mineralisation overlaps with the last major period of widespread granite magmatism in the craton.

Given the long distances of upward-directed flow inferred for the ore fluids, they may thus be exsolved from granitic magmas that crystallised below the depths of deposit formation, hence at depths of greater than about 10 km. If the fluids of these deposits are magmatic-hydrothermal, we need to explain why there are differences of compositions compared to those of the fluids of porphyry and related deposit types. The high CO_2 content of the fluid is consistent for instance with the higher solubility of CO_2 in silicate magmas at higher pressures, and mixed H_2O–CO_2 fluids are commonly recorded as fluid inclusions in granites in the middle crust. Granitic magmatism is typically a result of heat advection from the mantle into continental crust by mafic magmas, and the fluid may thus have been alternatively exsolved from mantle-derived magmas that drove the granitic magmatism. In this respect, fluid exsolution from lamprophyre magmas has been proposed as lamprophyres have a close spatial and temporal association with many, but not all, orogenic Au deposits.

Box 3.7 Structural control on hydrothermal ore deposits

The concept of 'structural control' has many aspects in ore deposit geology. It is critical for exploration and for evaluation of hydrothermal ore deposits. For reviews of the role of structures in the formation of different hydrothermal ore deposit types, see papers in Richards and Tosdal (2001).

In the broadest sense, the designation of a structural control on an ore body is the rationalisation of empirical observations that the site and shape of an ore deposit are geometrically and spatially related to structures. However, the term may also be used in an interpretative sense to express the notion that the formation of some deposits is controlled by active deformation along the host structure, for instance, by seismic slip on faults or by movement along shear zones.

Empirical relations between ores and structures

The empirical sense of structural control describes geometrical and spatial relationships between an ore body and one or more structure. It is implicitly assumed that there may also be relations between the shapes and forms of an ore body and the orientations of the principal stresses at the time of mineralisation. Structural control explains for instance the orientations of stockwork veins in porphyry deposits where these are, for instance, results of combined regional stresses and stresses imposed from the magma body (Figure 3.12).

There are many and contrasting types and styles of structure that ore bodies may be related to. For instance:

- Many veins are hosted in fractures, faults or shear zones, most typically along segments of a host structure (Figure 3.58). The host structures are not necessarily major structures, and there are many cases in which ore is hosted in relatively minor faults and shear zones with little displacement (e.g. see Figure 3.63). **Ore shoots** (Figure 3.57) are often elongate parallel to the movement vector of the host structure or are along an intersection line of two intersecting structures.
- Deposits may be hosted in structures other than faults or shear zones. One common structural control is in one or multiple-stacked stratabound layers in a fold-hinge zone (see for instance figures in Sangster and Smith, 2007).
- Deposits may be stratabound but centred on a fault or shear zone that cuts the host strata (Figure 3.62), or stratabound and bound on one side by a fault or shear zone (see Figures 3.68 and 4.12). In some cases deposits are near a structure but do not overlap it and die out in close proximity (e.g. Ridley and Mengler, 2000).

Empirical relations are also apparent between regional-scale structures, such as boundaries between crustal blocks and terrains, and the sites of magmatic-hydrothermal ore deposits. The large Bingham Cu–Au–Mo porphyry deposit in Utah, USA, for instance is at the eastern limit of the area of crust that extended in the Basin and Range Province and formed at the time of onset of extension (see Figure 3.65). It is thus likely that the magmatic complex that hosts this major porphyry deposit was 'controlled' by a structure in either the lower crust or lithosphere, and that this structure is a boundary between blocks of different rheological behaviour. Relations have been investigated between structures on an even larger scale in the mantle, as mapped by tomography, for instance on the siting of diamond-bearing kimberlites (e.g. O'Neill *et al.*, 2005; Jelsma *et al.*, 2009).

Passive and active roles of structures

We can differentiate situations in which a structure plays a passive role on the site of an ore deposit, in that it provides an architecture that controls hydrothermal fluid flow or magma flow, from those in which a structure plays an active role, and in which physical processes in the development of the structure are critical controls on fluid flow.

A passive control on fluid flow is the result of the presence of the structure in the rock mass. A fracture or fault can be a plane of high porosity and hence of high permeability, especially in crystalline rocks. Many major faults are foci of high and continuous flux of gases, for instance of CO_2 and ^3He, that are presumably sourced from the lower crust or mantle. (Note that faults may, in contrast, be aquitards in porous sedimentary or volcanic rocks, most commonly as a result of clay or cataclasite smears along fault planes.) A structure does not have to be tectonically active during hydrothermal fluid flow to be a permeable channelway. A fault, especially one in crystalline rock, can further be a zone of mechanical weakness, which may be preferentially opened as a result of hydrofracture as fluid pressure, for instance, is progressively raised (e.g. Sanderson and Zhang, 1999).

Alternatively, fluid flow may be near or adjacent to a structure, rather than along the structure. Structures can control the locus of fluid flow through their control on stress distribution where rheologically contrasting rock types are juxtaposed in a regional tectonic stress field (e.g. Ridley, 1993). A common situation of hydrothermal deposits that form in marine sedimentary basins is that ores are adjacent to growth faults which are in many cases also boundaries between horsts and grabens and of facies contrasts in the host strata (Nelson, 1997, and see Section 4.2.2). In these situations faults may control fluid flow through one or more of:

- permeability distribution on juxtaposition of contrasting rock types;
- differing water depth across fault scarps in submarine basins, and hence horizontal gradients in fluid pressure;
- a temperature gradient across the fault (e.g. Matthai *et al.*, 2004) and hence a control on the geometry of free or forced fluid convection of fluids in the crust.

Active deformation as a control on fluid flow

Active structures can be preferential fluid channelways and control hydrothermal fluid flow.

- Active structures may be permeable as a result of dilatancy or the increase in the volume of a rock mass during deformation. Dilatancy occurs for instance through the opening of space on movement along irregular faults or shear zones (jogs), or between beds as a result of flexure during the tightening of folds, in the formation of saddle reefs (e.g. Cox, 2005). A rock mass can also dilate during penetrative deformation even at slow strain rates as a result of opening of small fractures. Mineral grains in a deforming mass of rock dilate on an atomic scale as atoms are displaced in a deforming crystal lattice.
- The seismic cycle has been described as a poro-elastic dislocation and involves build up of stress in rock adjacent to a fault and release of this stress. There are thus transient variations in rock stress, which may drive fluid flow around seismically active faults (Sibson *et al.*, 1988).
- Fractures may be formed during seismic failure. The fractures are permeable zones through the rock mass; they may seal after growth of minerals, but equally may be re-opened and permeability renewed during each seismic event. Permeability along a fault may thus be formed and renewed through seismic cycles (Sibson, 1987).

Although an active role of a host structure to ore is suggested where textures and fabrics show deformation during the time period of ore mineral precipitation, evidence that a structure was active during ore precipitation should not, however, be taken as proof that the deformation was the cause of fluid flow. Fluid flow will in general be along faults that are in orientations and in stress fields in which they are likely to fail (e.g. Barton *et al.*, 1995). However, fluid flow can, conversely, cause deformation. Such a feedback was first clearly demonstrated at Rocky Mountain Flats in Colorado, USA, where pumping of water into an aquifer was found to induce swarms of small-magnitude earthquakes (Hsieh and Bredehoeft, 1981). Hydrothermal fluid flow may thus induce fault or shear-zone movement, essentially as a result of weakening of the rock mass where high fluid pressures reduce internal friction in the rock mass, for instance along a fault plane. The division between essentially active roles (in which tectonic activity controls fluid flow) and passive roles of structures on fluid flow may thus not be clear based on textural and structural evidence alone.

We further need to consider the possibility that ore deposits may control the siting of structures. Analogue experiments of rock deformation have shown that shear zones nucleate at the margins of contrasting rocks units (inhomogeneities) in a rock mass. Shear zones can nucleate either where the inhomogeneity is stronger than the surrounding rock (e.g. Ildefonse and Mancktelow, 1993) or where it is weaker (e.g. Mandal *et al.*, 2004). In the case of ore bodies, hydrothermal alteration, for instance phyllosilicate-rich alteration assemblages in the foot-wall of a VHMS deposit, may provide a weakness that will guide the nucleation of a shear zone. We should not necessarily infer that a deposit is where it is because of a host or adjacent structure.

The notion that an ore deposit only forms because of movement along a structure is most specific to orogenic Au deposits, and in particular when it is assumed that the fluid source is from metamorphic devolatilisation reactions. It is argued in this case that stresses developed as a result of active regional deformation drive hydrothermal fluid flow, and at a minimum are essential for focussing large enough volumes of fluid through to form a deposit, and may also promote interaction between a fluid and wall-rock that would not otherwise take place (Sibson *et al.*, 1988). These arguments are based in part on observations in wells of changes to groundwater tables before, during and after earthquakes – which are themselves explained by the cycle of stress build-up and release through an earthquake cycle (e.g. Rojstaczer and Wolf, 1994; King *et al.*, 2006). It is suggested that there would be multiple repetitions of the stress cycle on a seismically active fault plane over the time period of formation of a hydrothermal ore deposit, and hence advection of sufficient volumes of fluid flow to form the deposit. However, the observations of groundwater flow through a seismic cycle are not only on seismically active faults, but equally on faults within the volume of aftershocks of an earthquake (Cox and Ruming, 2004), and even on faults outside of the volume of aftershocks and up to about 1000 km distance from the focus of a major earthquake (King *et al.*, 2006). Pinpointing a role of earthquake activity in the formation of an ore deposit may therefore be difficult.

3.3.2 Carlin-type gold deposits

This is a distinct type of Au deposit that was first recognised and described at the Carlin deposit in northeastern Nevada, USA (Radtke, 1985). Carlin-type Au deposits are relatively low-temperature hydrothermal **epigenetic** deposits that form at ~ 150–250 °C at estimated depths of 1 to 6 km at times of widespread regional plutonism and minor volcanism. They are formed of disseminated ore with high concentrations of Au together with As, Hg, Sb and Tl in altered and replaced carbonate-bearing unmetamorphosed sedimentary rocks that are part of a continental shelf and margin sequence of mixed siliciclastic and carbonate rock. The contained Au is largely 'invisible', and is present either as sub-microscopic grains included in pyrite and arsenopyrite or is substituted in the crystal lattices of these minerals.

A large number of significant deposits that fit into this class (Goldstrike, Jerritt Canyon among others) have subsequently been discovered within an area of about 100 000 km^2 in northern Nevada and have been determined to have formed during a short time period in the middle Cenozoic (43–35 Ma).

Examples of deposits of this class have also been discovered or recognised in a number of geological terrains elsewhere in the world. A similar geographical cluster of Carlin-type Au deposits is now recognised in Guizhou and Yunnan provinces of southwest China (Peters *et al.*, 2007; Zaw *et al.*, 2007). In this area the ores are of Cretaceous age and are hosted in an older sequence of continental shelf and margin siliciclastic and carbonate sedimentary rocks, which contain similar rock types as the sequences that host the deposits around Carlin. The Alshar deposit in Macedonia was historically mined for As and Sb and is a Miocene-aged example (Percival and Radtke, 1994; Volkov *et al.*, 2006). There are possible examples also in Sulawesi, central China and Thailand, among other areas.

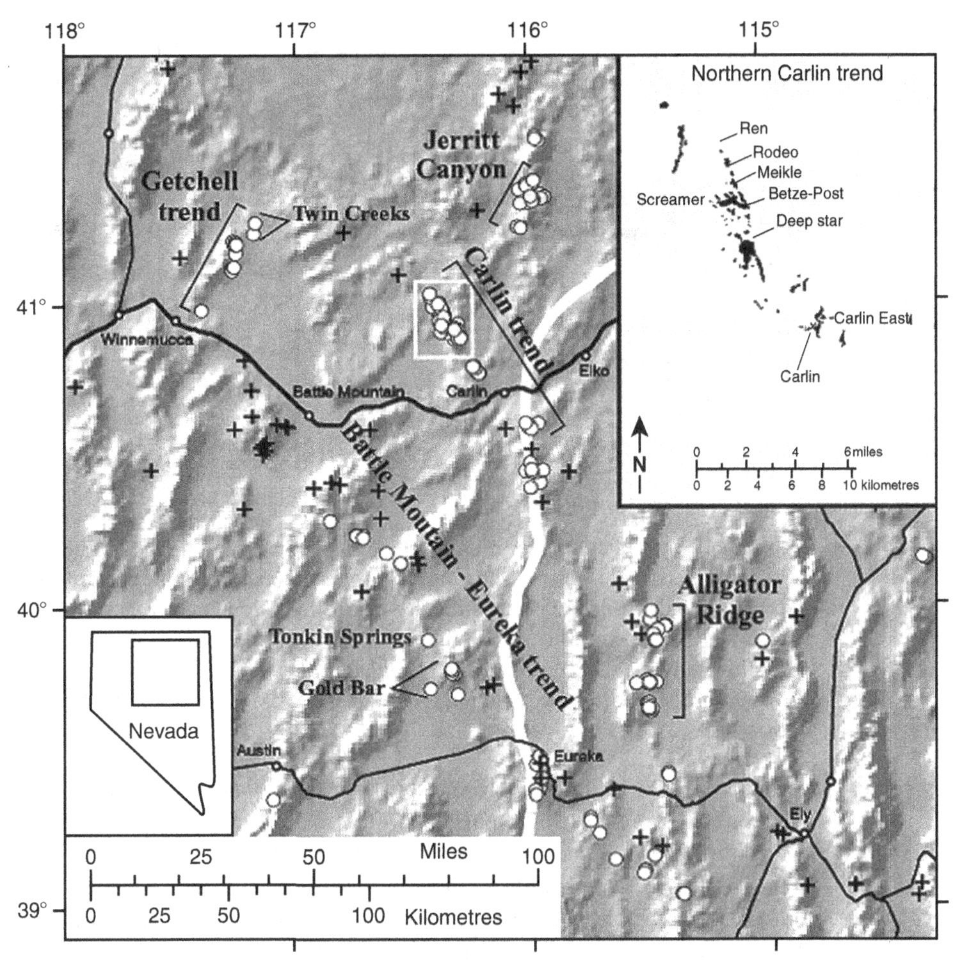

Figure 3.61 Distribution of Carlin-type Au deposits (open circles) and Au deposits of other (mainly epithermal) types (crosses) in northern Nevada (Hofstra *et al.*, 2003). The inset is a blow-up of the map of the northern Carlin trend and shows the plan-view extent of the ore bodies in this area. The figure background is a grey-scale image of the Basin and Range topography. The white solid line is the trace of the Roberts Mountain thrust fault and marks the eastern limit of a major phase of late-Palaeozoic thrust-style deformation. Note that almost all Carlin-type deposits in this area are concentrated along narrow linear zones = 'trends', that are not spatially related to either the major Palaeozoic or Cenozoic structures. Note also that there have been recent discoveries well off the trends.

Settings, distribution and shapes of Carlin-type gold deposits

The regional settings of the deposits in all areas where they have been recognised are all shelf and shelf-margin sedimentary sequences that have been folded and fault-sliced but not metamorphosed. The deposits in Nevada are in part of the Basin and Range Province that underwent large strain extensional deformation of the crust from the middle Palaeogene. Most deposits are hosted in Palaeozoic rocks and are located along linear 'trends' or zones a few kilometres wide that run oblique to Basin and Range faults and the topographically prominent associated horsts and grabens (Carlin trend, Battle Mountain – Eureka trend, Figure 3.61).

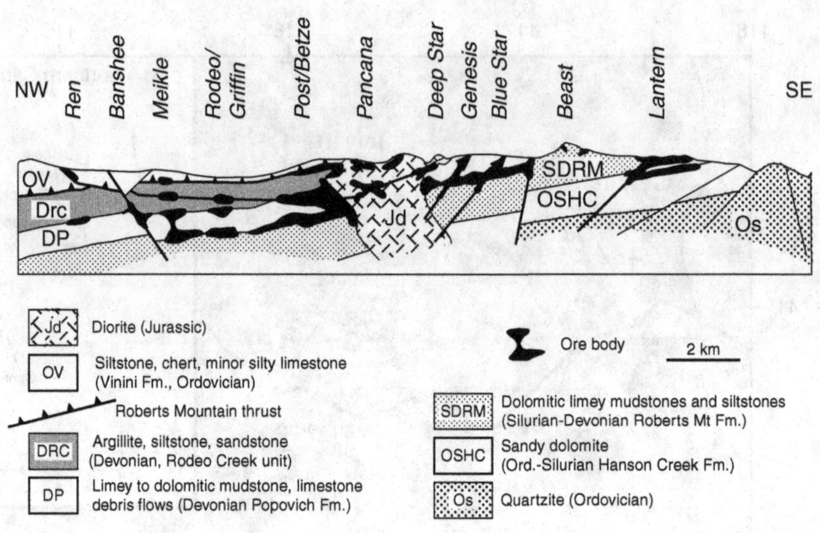

Figure 3.62 Simplified long section (northwest–southeast) along the northern Carlin trend (see Figure 3.61 for location) showing the projected position and extent of the Carlin-type deposits. The deposits are partly structure-controlled (along and adjacent to faults) and partly stratabound within specific 'favourable' horizons, e.g. the lower Popovich Formation (Hofstra and Cline, 2000). It is difficult to determine whether structures controlled fluid flow because 'feeder zones' below the ore zones have not yet been recognised.

These trends have no marked surface expression, for instance fault zones, but mark boundaries between blocks of rock of different composition in the deeper crust. The cluster of Cretaceous-age Carlin-type Au deposits in southwest China is hosted in a similarly older sequence of continental-shelf and -margin siliciclastic and carbonate sedimentary rocks that contains similar rock types as the sequences that host the deposits around Carlin. Linear trends of deposits have, however, not been recognised in this area.

Carlin-type Au deposits are to variable degrees both stratabound and structurally controlled. A proportion of the deposits are strongly stratabound and stratiform, with alteration and mineralisation defining laterally extensive (> 1 km), flat-lying bodies at specific 'favourable' levels in the stratigraphic and structural succession (Post-Betze, Blue Star, Figure 3.62): in some there are multiple, stacked ore bodies in the relatively shallow-dipping but deformed host sequences. The ore bodies may however be bounded by or may spread out laterally away from faults along favourable stratigraphic units (e.g. Banshee and Lantern in Figure 3.62) or may be centred on fold-axes of kilometre-scale open folds or structural domes. Some deposits are, in contrast, irregular zones that cross-cut the stratigraphic sequence (e.g. Meikle, Figure 3.63). Some of these discordant deposits are centred on mappable steeply dipping faults, but the largest part of the discordant Meikle ore body follows faults with almost unmappable small offsets rather than the larger faults of a set. Higher-grade Au within stratabound ore bodies may similarly be along swarms or networks of fractures and minor faults, many of which are nearly invisible in the intensely altered rocks.

The most common host-rocks to Carlin-type ore are impure shaly limestones and limey shales, and more rarely impure dolomitic limestones. At the Alshar deposit, the major host is a tuffaceous dolomite. Karst breccias in these carbonate rocks are a common

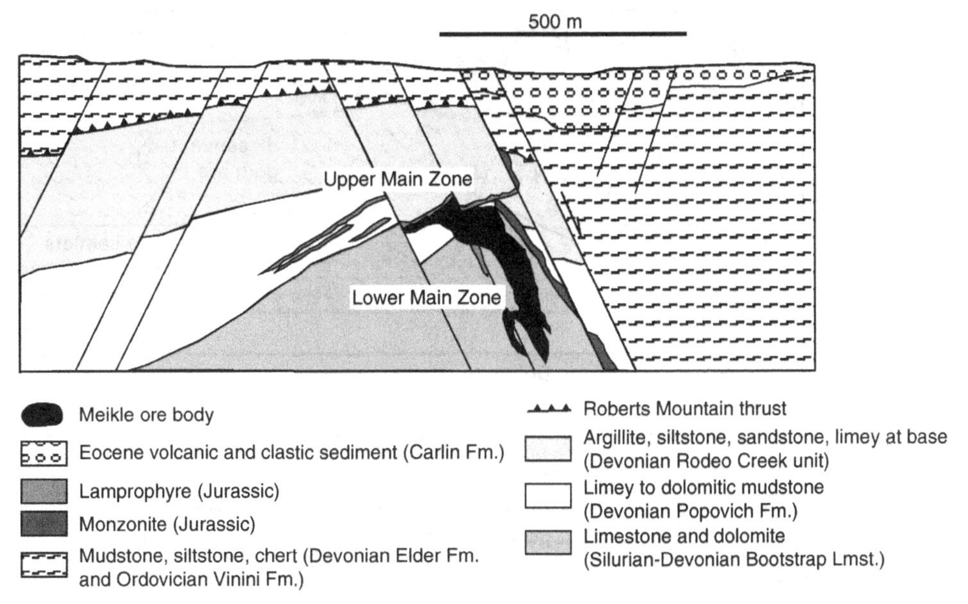

Figure 3.63 Cross section through the Meikle ore body of the northern Carlin trend (see Figure 3.64) showing the irregular shape of the ore body, its elongation parallel to Basin and Range normal faults, but position between rather than along major faults (Emsbo *et al.*, 2003).

component of the host-rocks to ore and host ore at some deposits, including at Meikle (Figure 3.63). Ore also occurs in lesser abundance in other rock types, particularly carbonaceous shales, sandstones and conglomerates, and in igneous intrusions, including granitoid stocks and mafic dykes.

Some of the deposits have close spatial and temporal relations to magmatic activity. Many of the Nevadan deposits are associated with clusters of small, intermediate to felsic intrusions of a similar age as the deposits. The Carlin trend of deposits overlies a line of plutons that are imaged geophysically, the tops of which are estimated to be at a few kilometres depth and which are the presumed feeder intrusions to the minor intrusions in the ore bodies. No similar small or underlying larger intrusions have been recognised, however, at the deposits in the Guizhou Province of China, and the deposits are distant from known areas of magmatism of the same age. In contrast, the Alshar deposit in Macedonia is centred above a shallow-level porphyritic stock and the ore is zoned over a distance of about 2 kilometres with respect to relative abundances of Au, Sb, As and Tl.

Nature of ore and of hydrothermal alteration at Carlin-type Au deposits

Carlin-type ores are 'Au-only' and have concentrations of Ag often lower than those of Au, and concentrations of Cu and other base metals are either little enriched or are unenriched above average crustal concentrations. The ores are, however, strongly enriched in As, Sb, Hg and Tl, and these elements have been extracted as co-products from some. Some ores also have marked enrichment in Ba in barite, but there is not a clear paragenetic association of barite and Au, pyrite and arsenopyrite, and it may have been precipitated either at an earlier or later stage in the history of the rocks than the Au. There is no systematic zonation of ore metals within or around most ore bodies.

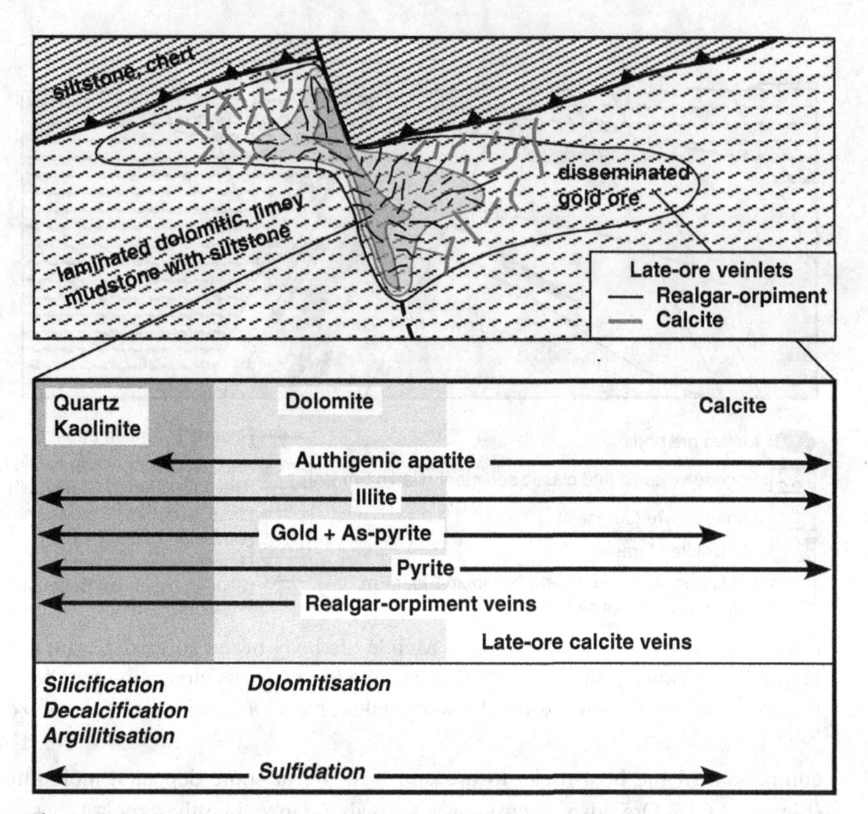

Figure 3.64 Schematic cross section of a Carlin-type ore body showing the position and extent of disseminated ore and zones of alteration and 'late-ore' veinlets, modelled on the Jerritt Canyon deposit (after Hofstra and Cline, 2000). The lower panel shows the mineral distribution from the centre of the ore zone to the edge of altered rock.

The characteristic alteration types and styles in and around ore zones can be described as decalcification, dolomitisation, silicification and argillic alteration or kaolinisation. Many of these types of alteration involve replacement of the host-rock minerals such that earlier sedimentary fabrics are preserved. Calculations of the mass balance during alteration show significant overall mass loss and strongly altered rocks are in many cases strongly porous. Decalcification affects impure carbonate rocks and to a lesser degree dolomite, and involves dissolution of carbonate grains and replacement by newly grown quartz and clay minerals. Silicification includes the formation of compact masses of often vuggy jasperoidal silica. Argillic alteration is most important where siliciclastic and igneous rocks are included in the ore bodies, and together with jasperoidal quartz appears to mark the most intense alteration in these hydrothermal systems (Figure 3.64), although there is generally not a systematic relationship between high Au grade and a specific facies of alteration.

Fine-grained (1–50 μm) pyrite and arsenopyrite are characteristic of Carlin-type ore and form a few per cent of ore as disseminated grains. There is generally a strong correlation between ore grade and the abundances of these sulfide minerals. The pyrite may have multiple petrogenetic stages with early, syn-sedimentary or diagenetic pyrite overgrown by As-rich ore-stage pyrite. The ore-stage growths may themselves have complex growth zones on a micrometre scale that are marked by large variations in the contents of As and Sb, and some zones in the pyrite are arsenical pyrite with abnormally high contents of up to about 10% As. The mineral growth zones can be correlated between samples across an ore body and may thus mark the passage of fluids of different composition through the deposits. Other common sulfide minerals include marcasite, stibnite, realgar and orpiment.

Gold is in some ores petrographically visible as sub-micrometre-sized grains in pyrite but is more commonly almost entirely petrographically 'invisible' in unweathered ore and is present in these cases in concentrations of up to several thousand ppm in ore-stage arsenian pyrite and in arsenopyrite. These high concentrations of invisible gold are either sub-microscopic, nanometre-sized grains hosted in the sulfide grains or are ions in the crystal lattices of the host minerals. There is a strong correlation between Au and As content of the pyrite such that most Au is in specific thin growth zones of the zoned sulfide grains.

There are generally no veins in the ore bodies that formed at the same time as disseminated ore minerals. Stibnite–realgar-bearing veins are sporadically developed and are a marker of the presence of ore. These veins are likely formed as a result of dissolution of fine-grained ore minerals during a later phase of groundwater flow, transport of As and Sb within the ore body, and reprecipitation.

Conditions of Carlin-type ore formation and nature of the ore fluid

Temperatures of ore formation of between 150 and 250 °C at 1–6 km depth are inferred from the few fluid inclusions that are trapped in fine-grained replacive ore minerals, alteration mineralogy, the mineral reactions of alteration, and estimates of the depth of erosion from reconstructions of the topography at the time of ore formation. The ore fluid is inferred to be weakly to moderately acidic in view of the widespread and pervasive carbonate dissolution in the ores and the presence of kaolinite as opposed to illite in alteration assemblages. The available fluid inclusions indicate fluid temperatures consistent with these estimates and have trapped low-salinity aqueous fluids with low but detectable CO_2 ($X_{CO2} < 0.04$).

There is evidence from markers of maximum temperature such as vitrinite reflectance that ore zones reached temperatures a few tens of degrees higher than surrounding rocks. The ore hydrothermal fluids can thus be interpreted to have been flowing upwards through the rock pile. Continuous fluid pathways marking this upward flow and extending from below to above ore bodies, for instance 'feeder zones' into the ore bodies, have, however, not yet been clearly identified.

Constraints on ore genesis

Critical points in a genetic model for Carlin-type Au deposits have been difficult to determine. This is in part because of complex multi-phase orogenic histories of the host sequences and the many hydrothermal and metasomatic imprints on the rocks during sedimentation, diagenesis and at different stages in the regional orogenic histories.

(a) The regional and tectonic setting of mineralisation

The determination of the timing of mineralisation was a significant step in constraining the tectonic setting of Carlin-type Au deposits in Nevada. The age of mineralisation along the Carlin trend of about 40 Ma is just before or at the onset of Basin and Range extension. The age shows that there is spatial and temporal correlation with a widespread and distinct episode of calc-alkaline, intermediate to felsic magmatism that occurred in the crust that

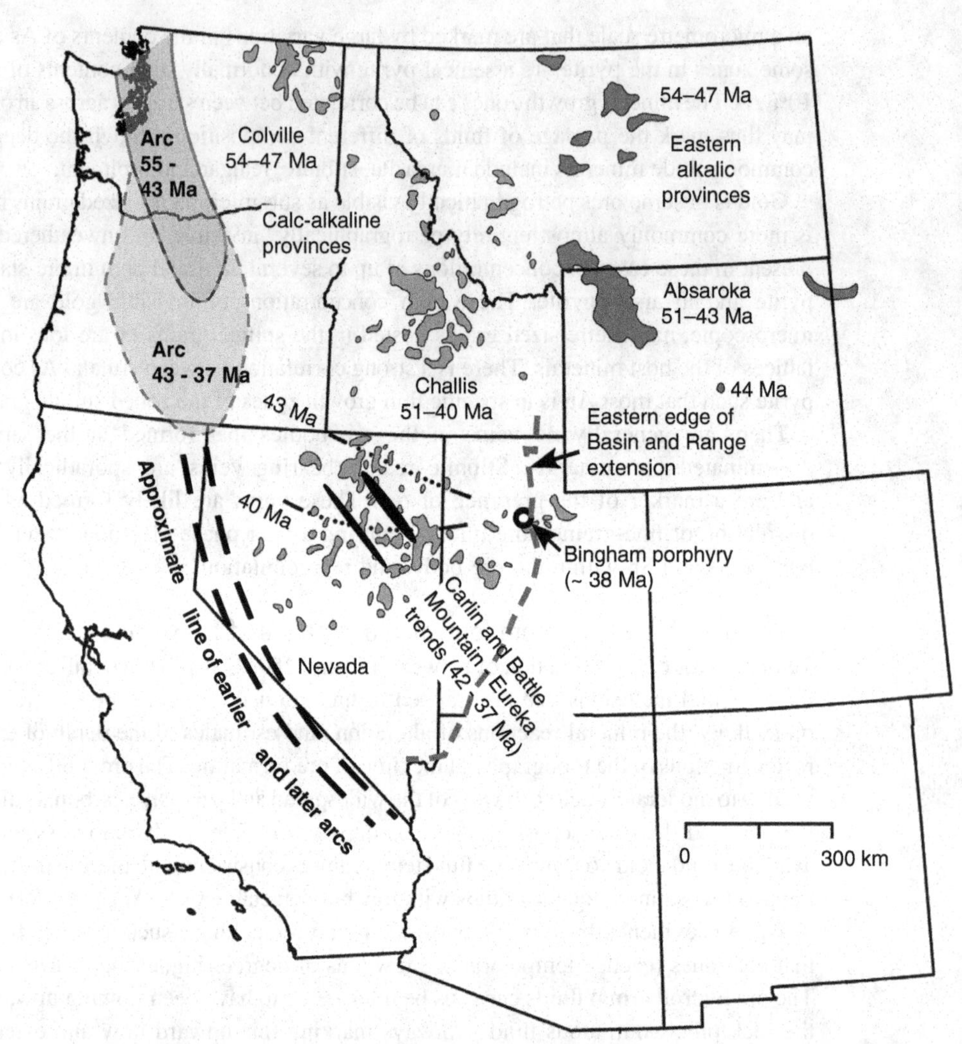

Figure 3.65 Position of the Carlin-type deposits (shown as ellipses marking the two main trends) relative to the distribution of 55–37 Ma, intermediate to felsic, Palaeogene intrusive and volcanic rocks shown in the grey shades. Magmatism migrated progressively southwards over this period in the hinterland to the line of the magmatic arc along the continental margin of the western USA, and was synchronous with ore formation in northeast Nevada when it reached this area (Ressel and Henry, 2006). The magmatism in and around the Bingham porphyry deposit is also a part of this event. The period of the magmatism in Nevada was a time of subduction along the west coast of the continent, but a hiatus in arc magmatism along this segment of the continental margin (e.g. Dickinson, 2004).

had been deformed in the Laramide orogen in the hinterland to the general line of the longer-lived magmatic arc along the western margin of North America (Figure 3.65). This magmatism occurred when the arc itself was not magmatically active. The timing does not, however, allow us to distinguish whether the ore fluids were sourced from the magmas or whether the magmas provided heat to drive convection of external groundwaters.

(b) The processes of mineralisation

Apatite fission track data and indicators of organic-matter maturation show that the ore deposits are hosted in volumes of rock up to a couple of kilometres across that reached higher temperature than the surroundings. The deposits are thus hosted where sufficient hydrothermal ore fluid flowed upwards through this level in the crust to significantly affect rock temperature. The feeder pathways of upward flow are, however, not clearly recognised at the deposits. Some ore bodies follow faults which can be thus inferred to be the pathways of the upward flow. In others, however, ore occurs as apparently floating, stratabound masses, in some cases as bodies stacked above each other within a few hundred metres of the structural pile, and zones of hydrothermal alteration or ore connecting these stacked bodies have not been recognised. Although we cannot define the flow path, the stratabound ore bodies indicate that the ore fluids reacted preferentially with specific stratigraphic units. Rock composition and rock texture are likely controls on the sites of the stratabound ore bodies. The preference for impure carbonate host-rocks is most likely a result of the chemical 'reactivity' of these rocks in the ore fluid, and in particular the rapidity at which carbonate minerals are dissolved. One effect of carbonate dissolution would be a 'reaction–infiltration feedback' in which the pervasive secondary porosity allows infiltration of the ore fluid into rock further from the fluid channelway, hence allowing further reaction.

The chemical causes of Au precipitation are at least in part similar to those in orogenic Au deposits. Iron, which is a minor component of carbonate minerals and also clay minerals in the sedimentary rock, reacts with dissolved H_2S and As species in the fluid to form pyrite and arsenopyrite. The fixing of H_2S in sulfide minerals causes a reduction in the solubility of gold bisulfide complexes. The precipitation of Au in sulfide crystal lattices may be a result of precipitation at disequilibrium conditions, or may be because the nucleation of native Au is difficult at the relatively low temperatures of mineralisation.

(c) Origins of the ore fluid

The concentration of Au ores in the Carlin area of Nevada is exceptional. Large volumes of fluid or fluid of abnormally high Au content migrated through this part of the upper crust over a relatively short period of time. There is a clear temporal and spatial relationship between magmatism and Carlin-type Au deposits in Nevada, although a similar relationship with magmatism is not apparent at the deposits in China.

Stable isotope (H, C, O, S) compositions of ore minerals such a carbonates, quartz and kaolinite, and also compositions of fluids in fluid inclusions, are widely varying, even within a single deposit. In general, however, the scatter of isotope compositions of minerals indicates that both surface-derived (meteoric) and 'deep' fluids migrated through the deposits and both can hence be inferred to have contributed to the development of the ores. The different fluids may have migrated either at different times in the tectonic history of the rocks, or as mixtures. Because multiple fluid types flowed through the rocks, it is more difficult to convincingly demonstrate which fluid carried and precipitated the Au.

The timing of fluid migration and the compositions of the ore fluid constrains it to be one of:

(i) Meteoric waters driven by convection to depth as a result of high-heat flow during the magmatic event at the onset of the Basin and Range extension. Carlin-type deposits form at between about 1 and 6 km depth, and hence no deeper than the depths that surface-derived waters are known to be convected to in active geothermal belts (see Section 3.2.6). If this is the ore-fluid origin, the abnormally high concentrations of S, As and Au in the ore fluid were likely leached into the convecting waters from sulfidic marine shales of the continental-margin sequence, and the presence of such shales in the host sequence may be a requirement for formation of Carlin-type ores.

(ii) Fluids released either during contact metamorphism at the time of intrusion of the granitic magmas or magmatic fluids derived from the granitoid magmas where these crystallised at relatively great depth in the crust. The ores in Nevada are associated with calc-alkaline magmatism that did not form in a typical arc setting (Figure 3.65). The magmas may have had different sources or evolutions than those that give rise to the more common porphyry deposits along convergent margins. If either magmatic or metamorphic sources are the fluid origin, Carlin-type deposits could thus be shallow-level expressions of hydrothermal systems similar to those of orogenic Au deposits. In this respect it is noted that there are close similarities of the chemistry of the ores and of the molecular make-up and the chemistry of the fluids in fluid inclusions with those of orogenic Au deposits, including low Fe and Cu contents and relatively high concentrations of volatile elements such as As and Sb.

3.3.3 Iron oxide–copper–gold (IOCG) deposits

IOCG deposits are hydrothermal ore deposits exploited primarily for Cu and Au in which the major hydrothermal Fe mineral is one or both of the low-Ti, iron oxide minerals magnetite or haematite rather than one of the iron sulfide minerals pyrite or pyrrhotite. The iron oxide minerals make up more than 10% by mode of ore, and in many cases occur together with Fe-rich silicate minerals, most commonly grunerite or Fe-actinolite. Copper is most commonly present as chalcopyrite, or more rarely as bornite or chalcocite. Other characteristics of the deposits are:

- Hydrothermal alteration is quartz-poor and quartz veins are rare.
- One or more of the LREEs, Ag, As, Co, Mo, Ni and U are at concentrations significantly higher than average crust in ore and are in some cases extracted as by-products.

As is the case with orogenic Au deposits and Carlin-type Au deposits, IOCG deposits are temporally associated with magmatism, but are not closely spatially associated with specific intrusions or intrusive centres. The settings of some of the deposits classified in this class are similar to one of porphyry, skarn or epithermal deposits, but other deposits of the class are not so closely related to intrusive or volcanic centres.

The recognition and characterisation of IOCG deposits as a deposit group is relatively recent and stemmed from the discovery in 1975, and subsequent evaluation, of the world-class Olympic Dam deposit in South Australia, which includes resources of Cu, Au, Ag, U and LREEs (Reeve *et al.*, 1990; Reynolds, 2001). Since the definition of this deposit class, some deposits that were known earlier have been reclassified and include vein-hosted deposits in coastal central Chile. Other IOCG deposits have been discovered through targeted exploration. As will be discussed further below, our understanding of these deposits is not yet to a level at which many aspects of the genesis of the deposits can be explained, or even to the level at which we can define which deposits belong to the type (Box 3.8) or common geological and tectonic environments of deposit formation. The processes that lead to ore formation, control deposit siting, generate the ore fluid and control its composition, and the chemical reactions and pressure–temperature conditions of ore mineral precipitation, are also incompletely defined.

Olympic Dam as the type example of an IOCG deposit

(a) Setting of the Olympic Dam deposit

The Olympic Dam deposit in the Gawler Craton of South Australia is an exceptionally large and complex ore body (> 2.3 Gt resources) of early Mesoproterozoic age. It is hosted in one member of a regionally extensive suite of batholiths of intracratonic syenogranites of A-type (anorogenic) affinity that intruded to shallow levels in the crust beneath units of volcanic and volcaniclastic rocks, at most a few million years before the hydrothermal event that formed the ore body at Olympic Dam and nearby sub-economic iron oxide ores. The Olympic Dam ore body is hosted in a large composite unit of closely spaced and overlapping haematite-rich breccias known as the Olympic Dam Breccia Complex (Figure 3.66). Within the complex, there are regularly shaped and typically elongate and steeply dipping to sub-vertical individual breccia bodies of variable size. The breccias can taper or thicken with depth and can pinch and swell. The breccia complex covers an irregular area of about 4 by 3 km in plan view, is surrounded by a halo of less intensely brecciated granite, and extends over several hundred metres depth. The upper levels of the breccias were eroded and the remaining levels were covered by later sedimentary rocks.

Different types of breccia are mapped within the complex, based mainly on clast mineralogy and the intensity of brecciation. Haematite is, however, the dominant matrix mineral throughout the breccia. There is a central funnel-shaped haematite–quartz breccia pipe that is barren of sulfides and of sericite and which is marked by distinct high barite (2–5%) content. The surrounding breccias are broadly zoned from those with dominantly haematite-altered clasts or heterolithic breccias with mixtures of intensely hydrothermally haematite-altered and granite clasts in a haematite matrix in the centre, to less-intensely disrupted rock with jigsaw-style granite clasts in a haematite matrix towards the periphery.

Clasts and blocks of volcaniclastic tuffs that have presumably been derived from an eroded volcanic cover are widespread in the breccia and have similar styles of hydrothermal alteration as the granite clasts. Their presence suggests that the breccia formed at most a few hundred metres below the surface. Dykes and small stocks of lamprophyre, basaltic and felsic rock intrude the breccia and are in part also incorporated into the breccia. These thus give evidence of a magmatic input during brecciation.

(b) Ore at the Olympic Dam deposit

Copper and gold ore with 1–6% Cu and about 0.5 ppm Au is developed within a portion of the volume of the breccia complex, although much of the rest of the complex has elevated concentrations of the ore elements. Most ore occurs in an irregular, undulatory, but overall flat-lying ore zone a few tens of metres thick with concentrations of copper

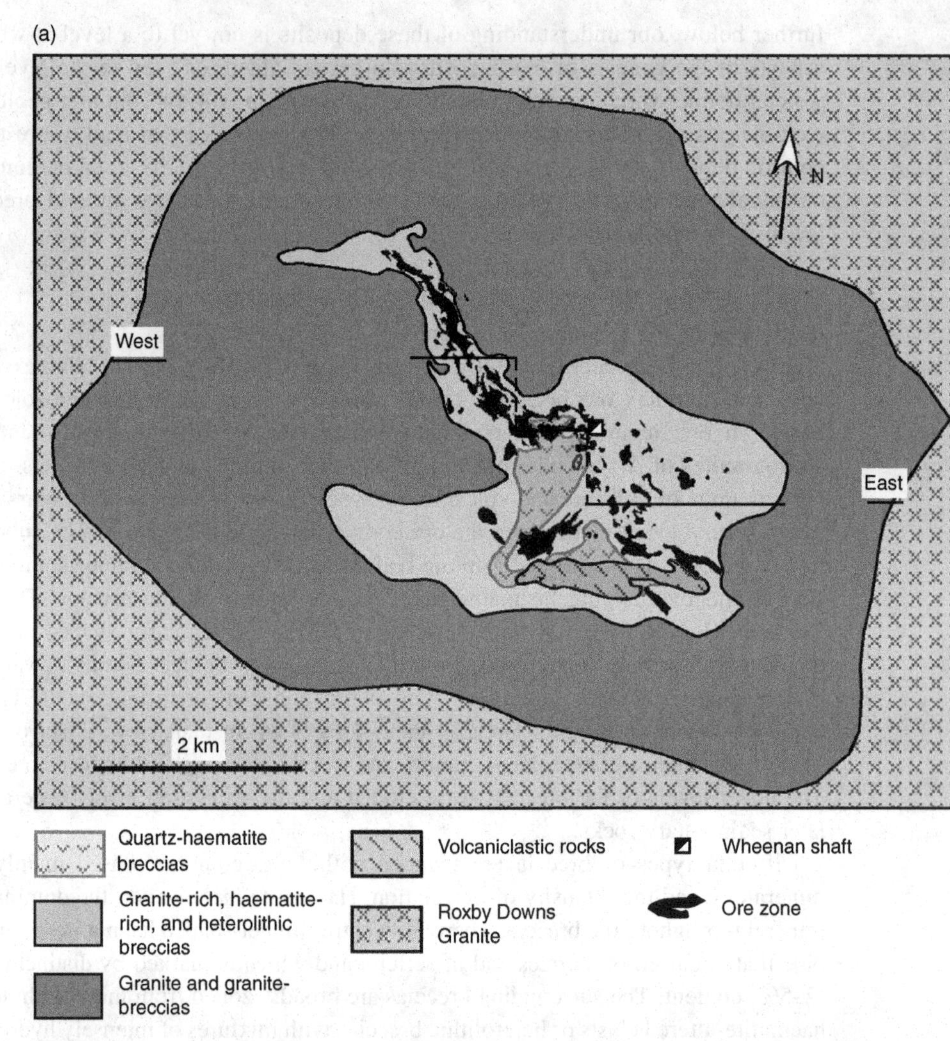

Figure 3.66 Plan (a) and cross section (b) through the Olympic Dam Breccia Complex, which is the host to the Olympic Dam IOCG deposit (after Reeve *et al.*, 1990 and Reynolds, 2001). In general the intensity of brecciation decreases outwards from the centre. Brecciated rock extends over 1 km beyond the limits of ore. The majority of ore is in sub-horizontal lenticular ore bodies within the interior of the breccia complex that straddle the bornite–chalcopyrite interface. The central and latest haematite–quartz breccias are barren of Cu and Au. The irregular bodies of volcaniclastic rocks are 'slumped' down during repeated brecciation and are interpreted to have formed a now eroded volcanic sequence that overlay the Roxby Downs Granite. Reprinted with the permission of the Australasian Institute of Mining and Metallurgy.

sulfide minerals (Figure 3.67). The Cu ore minerals are mainly disseminated and intergrown with haematite in the breccia matrix. Repeated brecciation means however that the ore and gangue minerals also occur in breccia clasts. Gold occurs as fine grains with the copper sulfide minerals. Other minerals in the ore bodies and in surrounding intensely altered rock include pyrite, sericite, fluorite, barite and siderite.

(b)

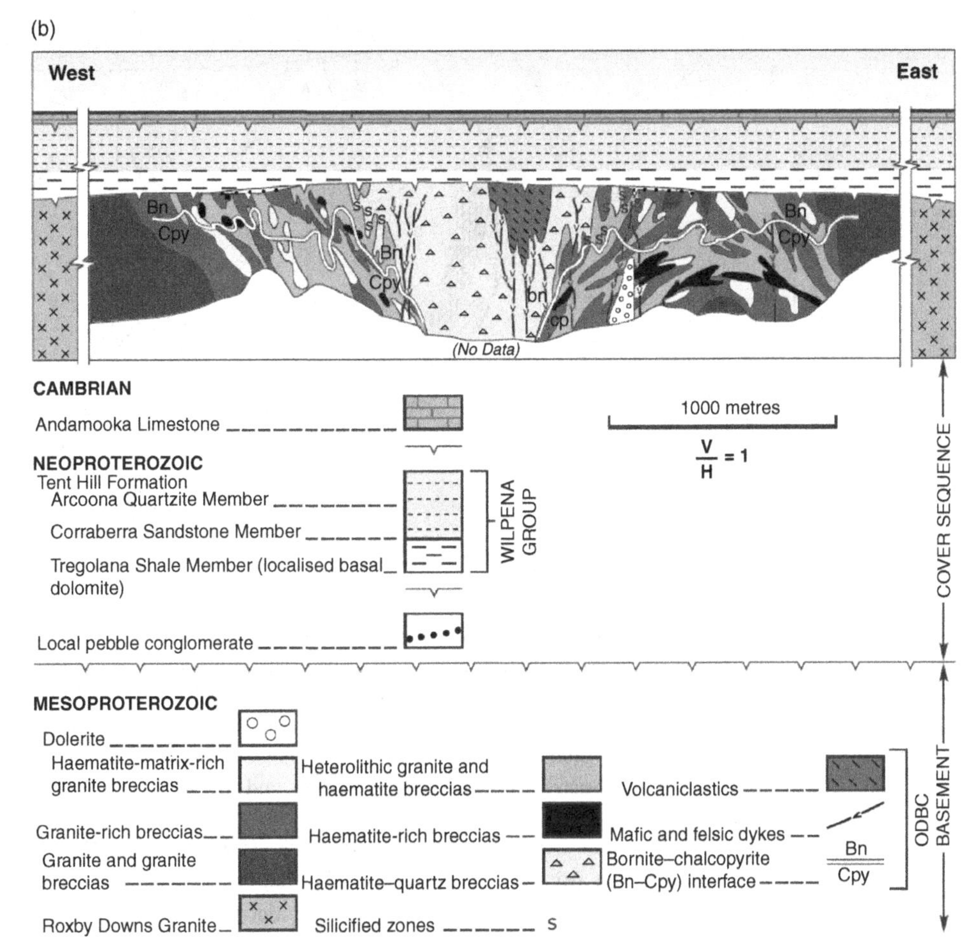

Figure 3.66 (*cont.*)

There is vertical zonation of a number of ore and gangue minerals through the breccia. Uranium and REE resources are a few tens of metres above the Cu ore zones. The Cu ore is vertically zoned with a generally sharp overall sub-horizontal but undulating interface between chalcopyrite-bearing ore below and bornite–chalcocite-bearing ore above. Magnetite is preserved in the deepest explored levels below the more peripheral areas of the breccia, although often is rimmed by haematite.

(c) Processes of formation of the Olympic Dam deposit

The host breccias of the Olympic Dam deposit are interpreted to be hydrothermal and possibly magmatic-hydrothermal diatreme breccias that formed through repeated hydrothermal activity over a prolonged period of time, with many hydrothermal eruptions breaching the surface, probably to form a maar.

The temperatures of ore formation are estimated as 300–400 °C. The irregular sub-horizontal ore bodies and bornite–chalcopyrite interface have shapes which have some resemblance to the shapes of ore bodies of high-sulfidation epithermal deposits (e.g.

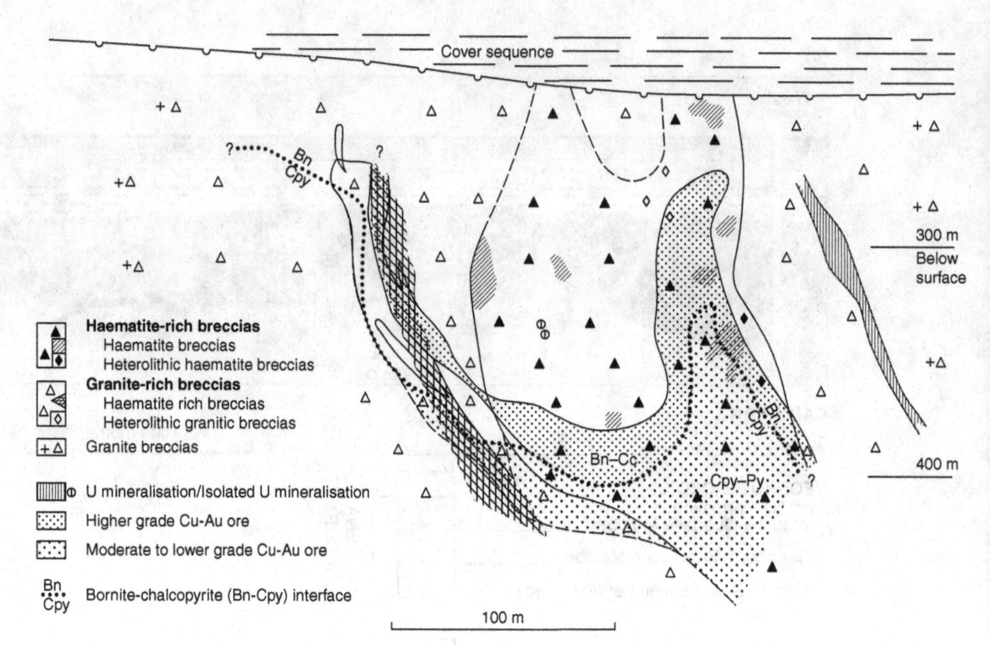

Figure 3.67 Cross section of an ore body at Olympic Dam (after Reeve *et al.*, 1990). As is typical, Cu–Au ore transects multiple breccia units and straddles the undulatory bornite–chalcopyrite interface with the highest grades in the bornite zone above the interface. Uranium mineralisation is distinct and occurs as lenses above the Cu–Au ore zone. Mineral abbreviations: Bn – bornite; Cc – chalcocite; Cpy – chalcopyrite; Py – pyrite. Reprinted with the permission of the Australasian Institute of Mining and Metallurgy.

Figure 3.29), in which fluid mixing at an unstable sub-horizontal fluid interface is a critical process in ore mineral precipitation. The nature of the interface in the case of Olympic Dam, with chalcocite and bornite above chalcopyrite, suggests higher-oxidation-state near-surface water and more-reduced deeper fluid.

Ore formation post-dates the host Roxby Downs Granite by a few million years, but isotope data, in particular of Nd isotopes (ε_{Nd}) indicates that the hydrothermal fluid was derived neither directly from the host granite nor indirectly from a melt that formed by differentiation of the host granite. The isotope data suggest rather that the fluid was derived from or equilibrated with mafic rocks. Intrusion of lamprophyre and basalt dykes during brecciation indicates that mafic magma was present at depth, as is commonly the case below a large granitic pluton, and geophysical data (gravity and magnetic data) indicate that a body of several kilometres lateral extent of dense Fe-rich, presumably mafic rock is present at about 4 km below the ore body.

The fluid inclusion record indicates multiple fluids in the ore, including high-temperature hypersaline brine with more than 40 wt % salt, low-density aqueous vapour and paragenetically late, lower-salinity, lower-temperature aqueous fluids. There is thus also similarity of the fluids with those of high-sulfidation epithermal deposits. We can infer phase separation of an intermediate-salinity fluid to brine and vapour as it rises. Late fluids of similar composition may have cooled and risen without intersecting the two-phase field. Phase separation and the formation of a low-density aqueous vapour may have been a contributory factor to repeated hydrothermal eruptions.

The variable characteristics and commonalities of other IOCG deposits

In respect to the size, overall form of the host breccia complex, and association with large granite batholiths of A-type affinity, Olympic Dam is unique among known IOCG deposits. Other deposits have different and very variable forms and very variable geological settings, and do not share some of the mineralogical characteristics of the Olympic Dam deposit.

(a) Styles and settings of other IOCG deposits

Two large Mesoproterozoic IOCG deposits in Australia (Ernest Henry, Queensland and Prominent Hill, South Australia) are essentially composed of steeply plunging, slightly irregular but overall elliptical breccia pipes about 200 m in diameter that cut through low metamorphic grade units that include both volcanic and sedimentary rocks. The host-rocks at the Prominent Hill deposit include dolomite and dolomitic shale. Magmatism was synchronous with metamorphism in both areas.

There is an extensive belt of Jurassic to early-Cretaceous IOCG deposits in the coastal cordillera of Chile and southern Peru with close spatial and time relations to magmatism. The belt includes deposits of a wide range of forms. There are swarms of steeply dipping fault-hosted vein-like ore bodies (2–30 m wide, 1–5 km strike length and continuous down dip to over 500 m) in the area around Tocopilla, which cut a diorite intrusion of similar age (Figure 3.68a). Other deposits in this belt include composite stratabound mantos (blankets), veins, breccias and stockworks in sequences of hydrothermally altered andesitic and dacitic volcanic and volcano-sedimentary rocks (e.g. Candelaria, Figure 3.68b).

A number of major deposits of late-Archaean age are recognised in Brazil in rock sequences that have undergone complex histories of deformation. The Salobo deposit is formed of sub-parallel, stratabound lenses hosted in biotite–magnetite schist with layers rich in fayalite, grunerite and garnet. Tourmaline is pervasive through the ore body. The ore may be in part a deformed breccia. The gangue assemblages indicate that the deposit either formed at relatively high temperatures (> 500 °C) or was metamorphosed after formation. The presence of significantly less metamorphosed rocks within about 1 km of the deposits may indicate that the deposit formed in a 'hydrothermal metamorphic' event in which hydrothermal fluid heated a large column of host-rock. The Sossego–Sequerinho deposit is in breccia with biotite-altered clasts, possibly derived from granite, in a carbonate–chalcopyrite–chlorite matrix.

(b) Mineralogy and distribution of hydrothermal alteration

Hydrothermal alteration in and around IOCG deposits has been characterised based on the occurrence of specific minerals that are prominent in hand specimens. Many IOCG deposits occur within regionally extensive (kilometres to tens of kilometres) volumes of rock that have been affected by sodic or sodic-calcic hydrothermal alteration (e.g. Figure 3.68b) and in which original rock textures appear 'washed out'. The difference between these two alteration facies may be related to the host-rocks: sodic alteration is marked by pervasive replacement of original minerals by albite or scapolite, especially in rocks of felsic composition; sodic-calcic alteration is marked by albite–actinolite, diopside or rarely garnet and epidote, with in some cases scapolite, and is most typically developed in mafic rocks. Where these alteration types are associated with IOCG

(a)

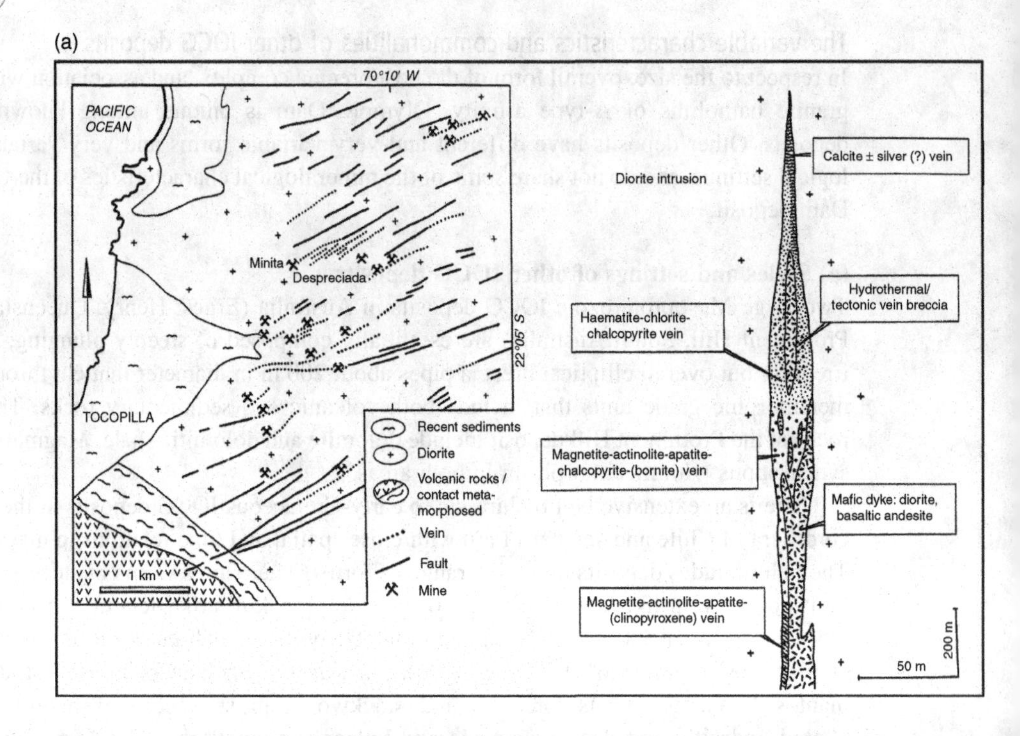

(b)

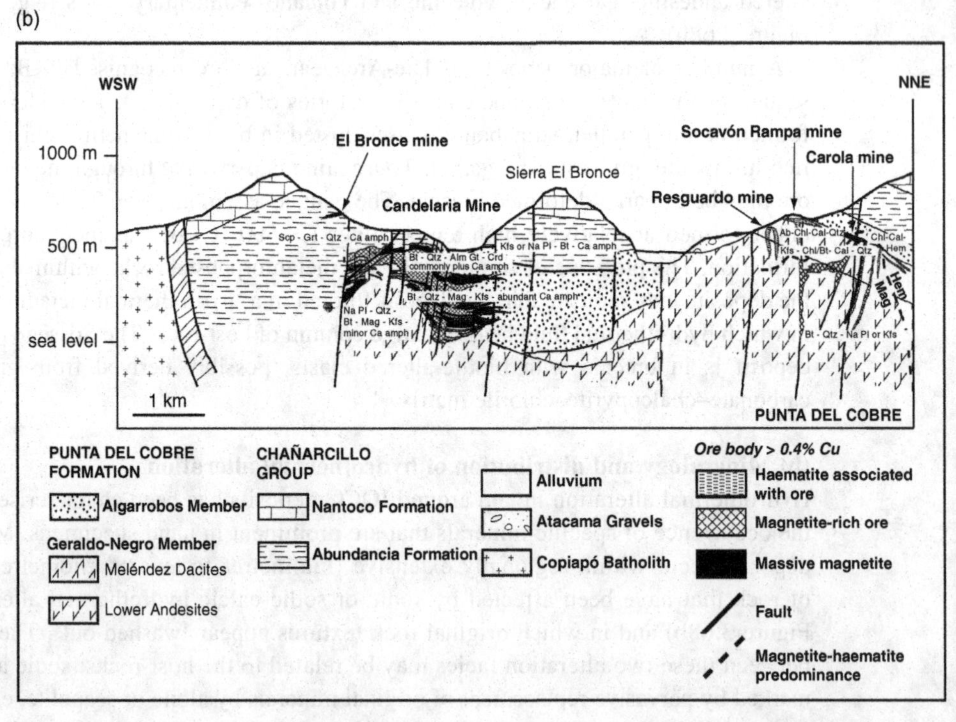

Figure 3.68 Styles, settings and distribution of alteration mineral assemblages of IOCG deposits in the Mesozoic belt of the coastal cordillera of Chile. (a) Vein swarm at Tocopilla

deposits, the altered rocks include up to 10% iron oxide minerals, either magnetite, or disseminations of haematite, to give 'red rock'. Another alteration type that has been recorded and described around IOCG deposits is calcic – with actinolite, diopside, hornblende, grunerite and garnet and generally a few per cent magnetite or haematite.

The sodic and sodic-calcic alteration that has occurred in the rocks surrounding IOCG deposits may, however, be coincidental and not directly related to ore genesis. The alteration assemblages in the ore bodies themselves are generally termed 'potassic' with one or more of the minerals K-feldspar (in felsic rocks), biotite (in mafic rocks) or sericite being major minerals, although the ore bodies are in many cases not enveloped by a single alteration facies. This type of alteration is not identical to potassic alteration in porphyry deposits and a typical gangue mineral assemblage may be K-feldspar–sericite–magnetite–quartz±biotite–actinolite–chlorite. Vertical zonation of alteration minerals over a few hundred vertical metres through and around an ore body is characteristic, for instance with actinolite at depth and chlorite at shallow levels, or zonation through biotite, sericite and argillic alteration zones. Assemblages of chlorite–sericite–calcite or dolomite–quartz and haematite at shallow levels are termed 'hydrolytic' alteration facies. Advanced-argillic alteration has however not been recognised at or above any of these deposits. The iron oxide mineral in the ore is variably magnetite or haematite, and vertical zonation, with haematite at shallow depth making way to magnetite at greater depth, is recorded at a number of deposits (Figure 3.68).

(c) Ore-fluid conditions and chemistry

The mineral assemblages indicate that IOCG deposits formed or were recrystallised at moderate to high hydrothermal temperatures (300–500 °C). Multiple hydrothermal fluid types are recorded from fluid inclusions, as at Olympic Dam. Most typically there are high-temperature brines with up to about 40 wt % salts, gas-rich fluids, and lower-temperature low- to moderate-salinity fluids. In contrast to the record at Olympic Dam, the vapour-rich fluid is CO_2-rich rather than aqueous in many deposits.

Interpretations of processes of IOCG ore genesis

The variability of the settings, associated rocks, ore-body forms and nature of the ore of these deposits, beyond the common mineralogical components of ore and general characteristics of hydrothermal alteration, means that models of the genesis of these deposits

---

Caption for Figure 3.68 (*cont.*) within a diorite pluton (Sillitoe, 2003). The schematic section through a vein shows typical lateral and vertical zonation of vein infill and the presence of a mafic dyke at depth in the vein fracture. (b) Deposits in the Candelaria–Punta del Cobre region (after Marschik and Fontboté, 2001). The Candelaria deposit is formed of manto-like stratiform breccia and replacement ores with multiple sets of discontinuous veins and veinlets. The deposits at Punta del Cobre include breccia and veins. Note the similar vertical zonation of iron oxide as at Tocopilla. The Algarrobos Formation is volcaniclastic: the Chanarcillo Group is dominantly calcareous with some interbedded volcaniclastic rocks. Mineral abbreviations: Ab – albite; Alm Gr – almandine garnet; Bt – biotite; Cal – calcite; Chl – chlorite; Crd – cordierite; Grt – garnet; Hem – haematite; Kfs – K-feldspar; Mag – magnetite; Qtz – quartz; Scp – scapolite.

are incomplete. The common petrological characteristics allow some inferences to be made about the processes and environments of ore formation.

At temperatures of 300–500 °C the ore mineral assemblage magnetite (or haematite)–chalcopyrite is stable over a range of chemical conditions of relatively low f_{H2S} (or f_{S2}) compared to the typical pyrite–chalcopyrite assemblage of porphyry deposits. A relatively low sulfide concentration in the ore fluid is thus implied. The low sulfide fugacity may be reflecting a low sulfur concentration or, if the oxidation state is high, sulfur may be present but as sulfate.

Upward-directed fluid flow in the ores is implied from phase separation in the hydrothermal fluid. The marked vertical zonation of ore and alteration assemblages within and around the ore bodies indicate vertical temperature and redox gradients. These may either be results of fluid cooling or mixing, but in either case can be most easily explained if ore formation occurred within a few kilometres of the surface. Cooling, reduced pressure, or mixing could cause ore precipitation from the fluid.

Most deposits are hosted at least in part in breccias. The causes of many of the breccias are, however, unclear. Hydrothermal eruption diatremes that vent to the surface are expected to extend only to about 500 m depth if the fluid is dominantly water; however, fluid composition may allow deeper-sourced eruption if the composition is such that the fluid expands more strongly than water as pressure decreases. Carbon dioxide and fluoride contents may both be important in this respect. Brecciation may however have occurred without venting of steam to the surface. Some breccias may be collapse breccias, for instance formed as a result of space created by alteration and mineral dissolution reactions in strongly reactive fluids.

Interpretations of settings of IOCG ores

The Mesozoic IOCG deposits of Chile are within a tholeiitic to calc-alkaline magmatic arc that was dominated by relatively mafic (diorite) magmatism. The ores, however, formed within a restricted time period of subduction magmatism in this arc. There is an antithetic relation in the Chile arcs between IOCG and porphyry deposits.

Older IOCG deposits are spatially and temporally associated with intracratonic magmatic belts, or with regionally extensive magmatic events in cratons. The associated magmatism is in all cases relatively alkaline, and although the igneous rock compositions are consistent with arc magmatism, it is not clear that the magmatic belts were arcs. Minor mafic and ultramafic dykes intruded into the Olympic Dam ore body at the time of formation and geophysical data imply a larger mafic body at about 4 km depth. Mafic magmatism may be a required component of the setting of IOCG ores.

Ore-fluid sources

There is as much uncertainty over the sources of the hydrothermal fluid or fluids, and hence also of dissolved metals, in genetic models proposed for IOCG deposits as for Carlin-type and orogenic Au deposits. Isotope ratios of ore minerals may not be direct tracers of fluid source where the composition of the fluid is partially or completely reset as a result of chemical interaction with rock along flow paths.

There are a number of constraints on ore-fluid source:

(i) IOCG deposits are spatially and temporally linked to magmatic activity.
(ii) Fluid temperatures reached temperatures higher than 400 °C, especially early in ore paragenesis.

(iii) Fluids were saline and in most cases include CO_2 in solution. The salinity of the brines in IOCG deposits (up to 40 wt %) are higher than almost all surface-derived evaporite waters, which have salinities of up to about 30 wt % except in the unusual case of extreme evaporation beyond halite saturation. Fluids have relatively low sulfide contents compared to those at other hydrothermal ore deposit types.
(iv) Regionally extensive sodic and sodic-calcic alteration indicates widespread lateral migration of high-temperature fluids.
(v) The mineralogy of hydrothermal alteration in the ores overlaps that of IOCG and porphyry ores. Both involve addition of potassium. Magnetite is a common and important mineral of many potassic alteration zones of porphyry deposits, especially in barren cores of deposits. These cores may contain up to about 20% modal magnetite, e.g. at Bajo de l'Alumbrera, Argentina.

In the case of IOCG deposits there are two proposed ore-fluid sources. The first source comprises magmatic-hydrothermal fluids, most probably derived from mafic or intermediate magmas at depths of up to several kilometres below the deposits. A magmatic source would explain the relatively high fluid temperatures. Maximum measured temperatures of present day geothermal waters are about 375 °C, and hydrothermal fluid temperatures of greater than this are thus easiest to explain as being cooled magmatic fluids than heated surface-derived waters. The high salinities of brines are easiest to explain as a product of phase separation in a hot, salt-bearing fluid rising from depth. Magmatic fluids derived from a deep magma chamber would likely have significant concentrations of CO_2 in solution.

There is, however, little information about the possible nature of a magmatic source for the fluid. If the ore fluids are magmatic-hydrothermal, a full ore genetic model would include explanation of why ore-fluid compositions are distinctly different at IOCG than at other magmatic-hydrothermal classes such as porphyry and epithermal deposits. The contrast in isotope ratios of ore minerals and host granite at Olympic Dam for instance shows that there is not a direct chemical relationship between the ore and the granite, but does not rule out a source from a deeper magmatic body, including the large, dense, presumably mafic body that is inferred from geophysical data to be present below the ore. Although lamprophyre and mafic dykes intruded at the same time as ore formation at Olympic Dam, they are strongly altered and cannot be used to infer the geochemical nature and petrogenetic affinity of the deep mafic body. Porphyry stocks or similar intrusions that may be derived from a major magma body at depth have not been mapped at other IOCG deposits.

The second proposed ore-fluid source comprises saline brines convected to depths of a few kilometres and heated to temperatures of greater than 300 °C in the aureole of shallow intrusions. If the ore fluids are surface-derived brines, by analogy with VHMS deposits, ore fluids could convect through rock above and around magma chambers, and the ore metals would be leached from country rock through which they advect. A high-salinity fluid formed through evaporation of meteoric or ocean water is proposed in particular as the ore fluid, in view of the widespread sodic-calcic alteration that is almost ubiquitously associated with IOCG deposits, and is used as a guide to exploration. The Mesozoic deposits in South America formed at desert latitudes and are hosted in rock sequences that include evaporites. However, the ores are hosted in an immature arc that may not have risen as a continuous land mass above sea level and it is not known whether there was a widespread potential source of brines at the time of ore formation.

Evidence in support of convecting brines being the ore fluids comes through observations on alteration and hydrothermal processes in geothermal fields in which evaporitic brines are a major geothermal fluid. The best studied example is the Salton Sea of southern California, USA, which is a stratified hydrological system in which the deeper levels of the geothermal field have hot brines of evaporitic origin with salinities of 25–30 wt % and temperatures of up to 360 °C at depths of about 3 km. This analogy is particularly appropriate to Olympic Dam in view of its redox interface between oxidised and reduced fluids and hence possibly between a saline deep fluid and a less-saline shallow fluid. There is widespread sodic-calcic alteration in the geothermal field with a similar scale of footprint as is recorded around IOCG deposits. There are relatively high concentrations of Cu and Zn in solution and there is active or recent precipitation of sulfide minerals in veins and in altered rock at temperatures of about 300 °C, although dominantly of Pb–Zn sulfide minerals rather than Cu and Au. The geothermal brines have, however, relatively low Fe contents.

The significance of sodic-calcic alteration to the formation of IOCG deposits is, however, equivocal. Although this style of alteration is widespread and strong as often flat-lying zones around many IOCG deposits and camps, the ore bodies are not clearly centred within this alteration, and the ores themselves are associated with pipes of potassic alteration with biotite or sericite (e.g. Ernest Henry). We thus infer that there was likely a second fluid of contrasting composition which formed the potassic alteration and by inference the ore. If convecting saline brines are not the ore fluid, we should expect some IOCG deposit to have no associated district-scale sodic-calcic alteration, and this has yet to be recognised.

Box 3.8 What deposits should be filed in the IOCG class?

Different deposits which have been classified into the IOCG class have contrasting styles of mineralisation. There is not, however, universal consensus as to what the characteristics or combinations of characteristics are of an IOCG deposit, and our knowledge is still actively developing through new discoveries and studies. There is thus also uncertainty as to which deposits should be included in this class. Compilations and discussions of the deposit class have parsed differently (Hitzman, *et al.*, 1992; Porter, 2000; Williams *et al.*, 2005; Groves *et al.*, 2010). Deposits have been included as IOCG which in other studies have been variably classified as iron-skarns (compare for instance Williams *et al.*, 2005 with Meinert *et al.*, 2005), iron oxide-rich copper sulfide ores in carbonatites (Groves and Vielreicher, 2001), and massive iron oxide apatite ores in felsic volcanic rocks (e.g. Nyström and Henriquez, 1994). Each of these deposit types share some but not all of the characteristics of the type-example IOCG deposits such as Olympic Dam and clear boundaries between classes based on specific characteristics are difficult to hold. The Oak Dam East deposit in the Gawler

Craton near Olympic Dam is, for instance, a stratiform, massive iron oxide breccia without copper sulfide minerals or Au, but one level in the breccia contains ore-grade Cu concentrations (Davidson *et al.*, 2007).

One significant contrast between the different classification schemes is over whether massive iron oxide apatite ores in volcanic terrains (Pea Ridge, Missouri, USA; Kiruna, Sweden; Chilean iron belt) should be included within the IOCG class. These magnetite or haematite ores include, in most cases, concentrations of apatite which holds REEs, but lack significant Au or Cu content. The mechanisms of the formation of these ores are themselves debated, and they have been proposed to be hydrothermal exhalative, hydrothermal replacement, and orthomagmatic, in the last model formed from an immiscible iron oxide-rich melt that separated from a silicate melt (cf. Section 2.2.3). They are associated with regionally extensive Na–Ca alteration. On a regional scale, these deposits occur in many of the same terrains and formed over broadly the same time periods as IOCG deposits. The Chilean iron belt includes, for instance, ore bodies formed within the same magmatic arc as deposits that are classified as IOCG (e.g. Sillitoe, 2003), and small IOCG deposits have been recognised around Kiruna in northern Sweden, although these latter appear to have formed a few tens of million years later.

Ores at the Plio-Pleistocene El Laco volcano in the Chilean Andes are informative in this respect. Massive iron oxide ore bodies are exposed at the present surface, which are variably interpreted as iron oxide-rich lava flows (e.g. Nystrom and Henriquez, 1994) and hydrothermal replacements of silicate volcanic rocks (e.g. Sillitoe and Burrows, 2002). Alteration is pervasive in rock down to a few hundred metres below the iron ores and resembles the Na–Ca alteration that is present around IOCG deposits with scapolite and albite as important minerals. The altered rocks also include magnetite-rich veins. However, no concentration of Cu and Au in either the altered wall-rock or the veins has yet been recognised (Naranjo *et al.*, 2010). A similar relationship is apparent at Kiruna in which IOCG-style alteration and iron oxide veins occur in the stratigraphic foot-wall of the now steeply dipping iron oxide ore body, but there are no known Cu–Au ores in this volume of altered rock (e.g. Hitzman *et al.*, 1992).

The observations in Sweden and in Chile suggest, therefore, that the two deposit types are not directly linked. The uncertainty in assignment of the ores may thus be in part because different processes may give rise to iron oxide-rich ores and cause potassic and sodic-calcic alteration. Mineral assemblages are controlled by pressure, temperature and composition of the environment, and are not products of process. IOCG deposits and the iron oxide alteration associated with iron oxide apatite ores may be formed from fluids of similar composition at similar ranges of temperature, by chance, or because the fluids in both cases equilibrated with similar Fe-rich S-poor rocks. The term deposit 'clan' has also been proposed to express the concept that specific geological environments may be suitable for formation of multiple different deposit types, and hence that a number of deposit types may have similar assemblages and form in a single terrain over similar time periods, but be only indirectly linked by ore genetic processes (Groves *et al.*, 2010). Geological environments may be prospective for both deposit types.

However, different relations between Cu–Au ores and massive magnetite ores are observed at the Pea Ridge iron deposit in Missouri, USA (Sidder *et al.*, 1993; Seeger *et al.*, 2001). This is in many respects a typical massive iron oxide ore body that transects a sequence of mainly rhyolite lava flows. However, the iron oxide ore hosts smaller breccia pipes of hydrothermal origin which have close-to-economic grades of Au and REEs. Rocks surrounding the iron oxide ores have assemblages of actinolite–quartz–pyrite–chalcopyrite–magnetite–apatite, hence similar to alteration assemblages as surround many of IOCG deposits. The two ore types are potentially here genetically related.

Questions and exercises

Review questions and problems

3.1 Low-sulfidation epithermal Au deposits and orogenic (lode) Au deposits are similar in that quartz-rich veins are an important host of Au in both deposit types. Imagine that you found a loose boulder-sized block that included a quartz–sulfide vein with Au. Describe and discuss the various criteria that you would use in the field to determine whether the vein was derived from a low-sulfidation epithermal or an orogenic Au deposit.

3.2 (i) Define the difference between 'low-sulfidation' and 'high-sulfidation' as these terms are applied to epithermal deposits.

(ii) Provide a sketch or a description of where low-sulfidation and high-sulfidation deposits are typically found relative to the components of a volcanic centre. The sketch section should extend horizontally and vertically a few kilometres.

(iii) Provide sketches or descriptions of the typical shapes and forms of low-sulfidation and high-sulfidation deposits on an ore deposit scale.

3.3 Describe the chemical and physical processes that are involved in:

(i) the generation of a hydrothermal ore fluid through second boiling in a crystallising felsic pluton,

(ii) the precipitation of ore minerals from such a hydrothermal fluid in porphyry deposits.

3.4 Massive sulfide bodies which are ores for multiple chalcophile elements can form in ocean-floor (VHMS) environments or in mafic intrusions. Describe and explain the observations you would make (in the field and in the laboratory) to determine which of the two ore deposit types a sample of massive sulfide ore comes from.

3.5 Write schematic chemical or mineral reactions and short explanations for each of the following chemical processes in hydrothermal ore genesis. Give labelled sketches showing the geological setting in which each of the processes takes place:

(i) the formation of moderately acid hydrothermal fluids (pH 4–5) in phyllic alteration zones of porphyry deposits,

(ii) the formation of extremely acid waters ($pH < 2$) in high-sulfidation epithermal deposits,

(iii) the formation of higher metal grades in skarns than in porphyry ores.

3.6 State the types of ore, and give a sketch that shows the pathways of hydrothermal fluid flow that explains the formation of hydrothermal ore that can be observed actively taking place:

(i) at back-arc ridges on the ocean floor,

(ii) in hot-spring pools in some geothermal fields,

(iii) beneath fumaroles and solfataras in volcanic craters.

3.7 Carlin-type and orogenic (lode) Au deposits have very similar ore chemistry – co-enrichment of Au with As and Sb, low concentrations of Ag and base metals. Provide labelled sketches or paragraphs which summarise and contrast the following features and geological settings of the two deposit types:

(i) the tectonic environments of formation of the deposits, including relations to magmatic activity;

(ii) the metamorphic environments of ore deposit formation;

(iii) the typical host-rocks to ore;

(iv) the shapes of the ore bodies.

3.8 If the proposal that orogenic Au deposits are precipitated from metamorphic fluids is correct then one factor controlling their formation is the release of sulfide from minerals at the same time as water is released in metamorphic dehydration reactions in order to carry Au in solution as bisulfide aqueous complexes.

Write balanced chemical reactions for metamorphic devolatilisation breakdown of pyrite (FeS_2) to:

(a) pyrrhotite (assumed formula FeS),

(b) magnetite (Fe_3O_4).

(In both cases assume that water is present and that the released sulfur may be either sulfide or sulfur dioxide in solution.)

What chemical factors would promote or buffer the reactions you have written?

3.9 Is Au carried in solution in the modified seawater that forms VHMS deposits dominantly as bisulfide or as chloride aqueous complexes?

An equation for the dissolution of Au as a bisulfide complex is given in Section 3.3.1. Write a similar reaction for dissolution as the Au(III) chloride complex $AuCl_4^-$.

Draw a schematic diagram of pH vs. f_{O2} showing the direction of increasing solubility of the gold bisulfide and the gold chloride complexes.

Given that seawater has a pH of about 8 and is strongly oxidised and that the hydrothermal fluid of most black smokers has a pH of between 3 and 4 and has f_{O2} low enough that sulfide species will dominate over sulfate species, discuss whether the chloride complex may be a significant carrier of Au in these hydrothermal systems.

3.10 An unusual rock type that occurs in some sequences of sedimentary and intercalated igneous rocks that have been metamorphosed to medium-grade (amphibolite facies) is cordierite–anthophyllite schist. These schists have approximately equal modal abundance of the two minerals and only minor content of other minerals. They are the result of isochemical metamorphism of hydrothermally altered rock. Given the chemical formulae of the minerals (cordierite = $(Mg,Fe)_2Al_3Si_5AlO_{18}$; anthophyllite = $Mg_7Si_8O_{22}(OH)_2$), determine the most likely mineralogy of the pre-metamorphic rock type and what ore deposit type you might expect to find in the vicinity of these schists.

Discussion questions

3.11 Copper in particular and also metals such as Sn and Ag are most likely carried as chloride complexes in magmatic-hydrothermal fluids. The chloride content of the aqueous fluid exsolved from the magma may thus be a control on metal grade in ores. What are the various possible controls on the chloride content of an exsolved magmatic-hydrothermal fluid in an arc or similar convergent-margin magmatic setting?

3.12 Would source plutons to major magmatic-hydrothermal ore deposits be depleted or enriched in the ore metals?

3.13 Do we expect porphyry and high-sulfidation epithermal deposits to form truly synchronously during magmatic degassing? Give reasons for and against this proposition with a brief explanation of your reasoning for each point.

Further readings

The following articles give background to hydrothermal systems, the geological processes behind the formation of the ore deposit types described and discussed in this chapter and discussions of ore fluid source where this is debated.

(1) The development and chemical behaviour of hydrothermal fluids:

Brimhall, G. H. and Crerar, D. A. (1987). Ore fluids: magmatic to supergene. *Reviews in Mineralogy* **17**, 235–321.

(2) Discussions on the origins of breccias in magmatohydrothermal ore systems:

Norton, D. L. and Cathles, L. M. (1973). Breccia-pipes – products of exsolved vapour from magmas. *Economic Geology* **68**, 540–546.

Perry, V. D. (1961). The significance of mineralized breccia pipes. *Mining Engineering* **13**, 367–376.

Sillitoe, R. H. (1985). Ore-related breccias in volcanoplutonic arcs. *Economic Geology* **80**, 1467–1514.

(3) Porphyry deposits – aspects of their development: in particular what combination of factors control the formation of world-class deposits?

Cooke, D. R., Hollings, P. and Walshe, J. L. (2005). Giant porphyry deposits: characteristics, distribution, and tectonic controls. *Economic Geology* **100**, 801–818.

Richards, J.P. (2011). Magmatic to hydrothermal metal fluxes in convergent and collided margins. *Ore Geology Reviews* **40**, 1–26.
Sillitoe, R.H. (2010). Porphyry copper systems. *Economic Geology* **105**, 3–41.

(4) Greisens – analysis and discussion of the causes of Sn precipitation in greisens:

Halter, W., William-Jones, A.E. and Kontak, D.J. (1996). The role of greisenization in cassiterite precipitation at the East Kemptville tin deposit, Nova Scotia. *Economic Geology* **91**, 368–385.
Heinrich, C.A. (1990). The chemistry of hydrothermal tin(-tungsten) ore deposition. *Economic Geology* **85**, 457–481.

(5) Polymetallic veins and vein fields – what are the causes of the metal zonation?

Audétat, A., Günther, D. and Heinrich, C.A. (2000). Causes for large-scale zonation around mineralized plutons: fluid inclusion LA-ICP-MS evidence from the Mole Granite, Australia. *Economic Geology* **95**, 1563–1582.

(6) High-sulfidation epithermal deposits – differing studies of the evolution of fluids after they are released from high-level magma chambers:

Berger, B.R. and Henley, R.W. (2011). Magmatic-vapor expansion and the formation of high-sulfidation gold deposits: structural control on hydrothermal alteration and ore mineralization. *Ore Geology Reviews* **39**, 75–90.
Heinrich, C. (2005). The physical and chemical evolution of low-salinity magmatic-fluids at the porphyry to epithermal transition: a thermodynamic study. *Mineralium Deposita* **39**, 864–899.
Mavrogenes, J., Henley, R.W., Reyes, A.G. and Berger, B. (2010). Sulfosalt melts: evidence of high-temperature vapor transport of metals in the formation of high-sulfidation lode-gold deposits. *Economic Geology* **105**, 257–262.

(7) Low-sulfidation epithermal deposits – relations between the deposits and the behaviour of geothermal systems:

Simmons, S.F. and Browne, P.R.L. (2000). Hydrothermal minerals and precious metals in the Broadlands–Ohaaki geothermal system: implications for understanding low-sulfidation epithermal environments. *Economic Geology* **95**, 971–999.

(8) VHMS deposits – analysis of the variability of the environment of ore formation in modern and ancient examples:

de Ronde, C. E. J., Massoth, G. J., Butterfield, D. A., *et al.* (2011). Submarine hydrothermal activity and gold-rich mineralization at Brothers Volcano, Kermadec Arc, New Zealand. *Mineralium Deposita* **46**, 541–584.
Large, R.R. (1992). Australian volcanic-hosted massive sulfide deposits: features, styles, and genetic models. *Economic Geology* **87**, 471–510.
Tornos, F. (2006). Environment of formation and styles of volcanogenic massive sulphides: the Iberian Pyrite Belt. *Ore Geology Reviews* **28**, 259–306.

(9) Orogenic gold deposits – what can we infer about the origin of the hydrothermal fluids?

Groves, D. I., Ridley, J. R., Bloem, E. J. M., *et al.* (1995). Lode-gold deposits of the Yilgarn Block: products of late-Archaean crustal-scale overpressured hydrothermal systems. *Geological Society of London Special Publication* **95**, 155–172.

Phillips, G. N. and Powell, R. (2010). Formation of gold deposits: a metamorphic devolatilization model. *Journal of Metamorphic Geology* **28**, 689–718.

Ridley, J. R. and Diamond, L. W. (2000). Fluid chemistry of orogenic lode gold deposits and implications for genetic models. *Reviews in Economic Geology* **13**, 141–162.

(10) Carlin-type gold deposits – contrasting proposals for the sources of the hydrothermal ore fluid:

Emsbo, P., Hofstra, A. H., Lauha, E. A., Griffin, G. L. and Hutchinson, R. W. (2003). Origin of high-grade gold ore, source of ore fluid components, and genesis of the Meikle and neighboring Carlin-type deposits, northern Carlin trend, Nevada. *Economic Geology* **98**, 1069–1106.

Muntean, J. L., Cline, J. S., Simon, A. C. and Longo, A. A. (2011). Magmatic-hydrothermal origin of Nevada's Carlin-type gold deposits. *Nature Geoscience* **4**, 122–127.

(11) Iron oxide–copper–gold (IOCG) deposits – efforts to constrain the ore fluid source:

Baker, T., Mustard, R., Fu, B., *et al.* (2008). Mixed messages in iron-oxide-copper-gold systems of the Cloncurry district, Australia: insights from PIXE analysis of halogens and copper in fluid inclusions. *Mineralium Deposita* **43**, 599–608.

Barton, M. D. and Johnson, D. A. (1996). Evaporitic source model for igneous related Fe-oxide–(REE–Cu–Au–Ag) mineralization. *Geology* **24**, 259–262.

Pollard, P. J. (2006). An intrusion-related origin for Cu–Au mineralization in iron oxide–copper–gold (IOCG) provinces. *Mineralium Deposita* **41**, 179–187.

4 Hydrothermal ore deposits II: sedimentary environments

Hydrothermal ore deposits in sedimentary basins are the dominant source worldwide of Pb, Zn, Co and U, and are significant sources of Ag and Cu. The major types of deposit are base-metal sulfide deposits and uranium deposits. Some of the types of uranium ore are not 'hydrothermal' based on the criterion that is commonly used to classify ore deposits in sedimentary basins – that the mineralising fluid was hotter than ambient temperatures. These ores are rather the result of groundwater flow at near-ambient conditions. They are, however, considered together with hydrothermal uranium ores because of the commonalities among the different deposit types.

The deposits in sedimentary basins are either **syngenetic**, and formed at the same time of sedimentation of the host-rocks, or epigenetic and formed up to hundreds of million years after sedimentation. Most of the host basins were not sites of magmatism over the times of mineralisation. Neither magmatic-hydrothermal fluids nor magmatic heat-driving fluid flow are thus considered to be factors in the formation of these ores. As will be discussed in this chapter, the metal-carrying solutions migrated distances of up to a few to hundreds of kilometres through the sedimentary rocks of the basin and in some cases also through underlying basement.

The relatively low temperatures of fluids in sedimentary basins, with maximum temperatures of between about 100 and 150 °C in a basin of a few kilometres thickness, and more rarely up to about 250 °C, mean that in general metal solubilities are low in groundwaters in sedimentary basins. Specific geochemical environments are required for efficient transport and precipitation of ore minerals from migrating low-temperature waters.

4.1 Hydrothermal fluids in sedimentary basins

Non-magmatic hydrothermal fluids that are indigenous to sedimentary basins are here called basinal waters and include waters of ocean and meteoric origin that have infiltrated from above into the basin, and connate and diagenetic waters that are derived from within the basin. Unconsolidated sediments have pore volumes of between 10 and 70% and below the water table these pore spaces are generally saturated with fluid. Large volumes of basinal waters are thus present in pore space in incompletely cemented sedimentary rocks. Permeability can be high, especially where primary porosity is preserved, and fluid fluxes can be high where there is both high permeability and a strong driving force for flow. Metal contents of meteoric waters and seawater are low. The metals are thus inferred to have been dissolved into the basinal hydrothermal fluids either during fluid migration or into fluids released in diagenetic reactions. In either case they are thus derived by leaching of the strata in the basin or from basement rocks.

4.1.1 Chemical characteristics of basinal waters

Fluid salinity

The salinities of present-day basinal waters range from those such as rainwater with very low salt content (tens to hundreds of ppm TDS – total dissolved solids) to brines with several times the salt concentration of seawater (up to 40 wt % TDS). The salts are dominantly chloride and sulfate salts with Na^+ and Ca^{2+} in most cases the dominant cations. High-salinity waters can be formed as a result of strong evaporation of any basinal waters, most commonly of either meteoric waters or seawater, or through dissolution of sedimentary evaporites where these are encountered by migrating waters.

Analyses of pore waters sampled from depths of up to a few kilometres in modern sedimentary basins, in combination with theoretical considerations of metal solubility, show that salinity is an important control on metal transport, particularly of the base metals Pb, Zn and Cu, which can be present at concentrations of up to hundreds of ppm in those basinal brines which have salinities many times that of seawater (Figure 4.1). These three metals all form strong chloride complexes in low-temperature aqueous solutions. The control on metal solubility in saline water can be expressed through dissolution reactions such as:

$$ZnS_{(sph)} + 2NaCl_{(aq)} + 2H^+{}_{(aq)} \leftrightarrow ZnCl_{2(aq)} + H_2S_{(aq)} + 2Na^+{}_{(aq)}. \quad (4.1)$$

In this reaction sphalerite is dissolved to form a two-fold coordinated zinc chloride complex in solution. The mass action relation of this reaction is such that sphalerite solubility will be proportional to the square of the concentration of NaCl in solution with all other factors held constant.

Redox state

Sedimentary environments and hence also basinal fluids have a wide range of redox states, from highly oxidised in equilibrium with atmospheric free oxygen to strongly reduced in equilibrium with organic carbon. The stability of many minerals and the solubility of redox-sensitive elements are both controlled by redox state. The settings and distribution of organic matter in sedimentary rocks is therefore an important control on chemical processes in sedimentary basins, including ore mineral precipitation.

Organic matter can occur in sedimentary rocks as:

- Indigenous solid remains of organisms that were sedimented and buried with the host sediments. The remains are modified by diagenetic reactions (organic-matter maturation) on burial and heating, and can be in the form of kerogen or graphite, in for instance black shales, or in some cases coal.
- Live organisms – bacteria and other micro-organisms are abundant in most soils and also in many unconsolidated sediments. Chemotrophic bacteria are known to occur in pore spaces of sediments to temperatures of a little over 100 °C, hence to about 4-km depth.

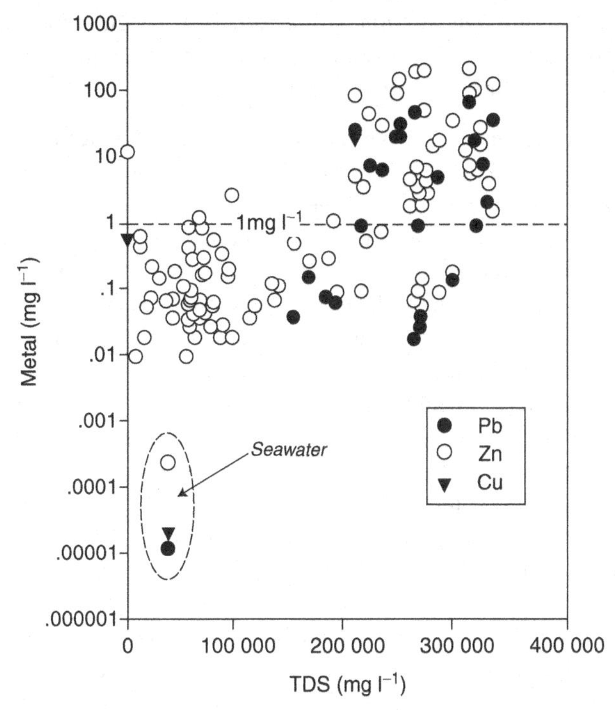

Figure 4.1 Base-metal contents of saline groundwaters in modern sedimentary basins and comparison with seawater. TDS = total dissolved solids and is a measure of fluid salinity – seawater TDS ≈ 35 000 mg l^{-1}. Note that basinal waters have orders of magnitude higher concentrations of the metals than seawater, and there is a threshold of about 15 % dissolved salts above which many basinal waters carry relatively high concentrations (> 10 ppm) of one or more base metal (after Hanor, 1994).

- Humic acids and similar soluble organic compounds that are formed during decay of organisms in soils and in near-surface sediments.
- Liquid and gaseous hydrocarbons, including biogenically produced methane, and petroleum and hydrocarbon gas produced by maturation of organic matter in the oil and gas windows. Solid bitumen can be formed from pore-space petroleum by loss of volatile compounds.

The redox state has a particularly marked control on the precipitation of base-metal sulfides, as these minerals are only stable under reducing conditions. Sulfur is most abundant in solution as sulfate in most near-surface waters, including in ocean water, in which it is the most abundant anion after chloride. Few groundwaters have high sulfide contents, and high sulfide concentrations would reduce the capacity of a fluid to carry base metals (as is demonstrated by the reverse of Equation (4.1)).

The reduced sulfur in sulfide minerals in sediments and sedimentary rocks may either be:

(i) Sulfide that was released directly from organic matter into solution, for instance during the generation of oil.

(ii) Derived from sulfate through reaction with a reductant, which in most cases will be organic matter of some form. The sulfate may be either (a) sulfate in solution or (b) a component of evaporite minerals such as gypsum.

Whichever of the processes in the above list is the source of reduced sulfur in sulfide minerals, organic matter is likely involved in its production. Sulfide from sulfate is the result of chemical reaction between an aqueous or mineral oxidant (sulfate) and a reductant (organic matter). Redox reactions are kinetically slow, even where all components are in aqueous solutions, and because of low temperatures in sedimentary basins, slow reaction kinetics can influence whether aqueous sulfide is formed by reduction of sulfate even where the ingredients are available and thermodynamic conditions are favourable. At low temperatures (up to about 100 °C), however, a number of genera of bacteria, including thermophilic genera, such as *Thermodesulfobacterium*, reduce sulfate to sulfide in or adjacent to their cells and make use of the chemical energy (ΔG_r) released as an energy source for metabolism. They thus promote the formation of sulfide in bacterial sulfate reduction. Note, however, that although bacteria and other organisms can catalyse a thermodynamically favoured reaction, they cannot act against the thermodynamic driving force, and hence cannot form sulfide in an environment in which it would not be stable. Bacteria can survive only up to temperatures a little higher than 100 °C, and bacterial sulfate reduction is thus not possible at higher temperatures. At temperatures greater than about 140 °C, however, the sulfate to sulfide reaction is kinetically faster even without catalytic bacterial mediation and takes place inorganically through thermochemical sulfate reduction (Figure 4.2).

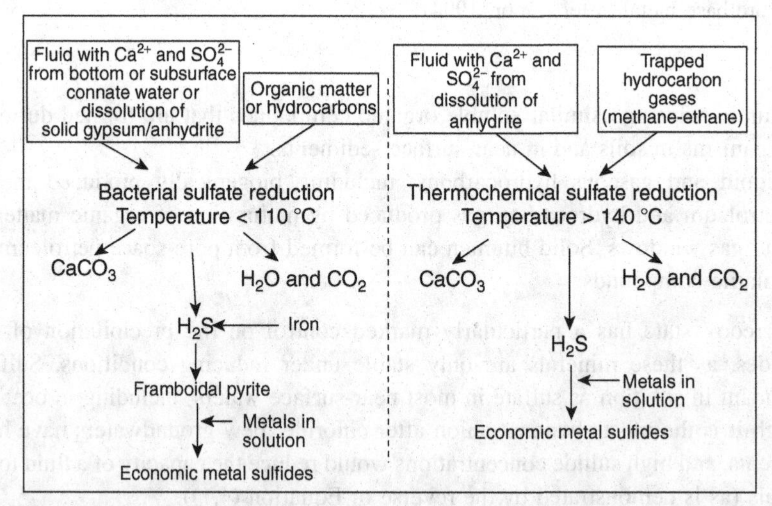

Figure 4.2 Steps in bacterial and thermochemical sulfate reduction of dissolved sulfate to the end products of base-metal sulfide minerals such as galena or sphalerite. Each panel shows a likely sequence of steps through intermediary states (Warren, 2000) and distinguishes the typical pathways at low temperatures in which base-metal sulfides replace intermediary pyrite and at high temperatures in which they may be precipitated directly from solution.

Ore metals in basinal waters

It can be inferred that the metals carried in basinal hydrothermal fluid are predominantly leached from the strata in the basin or from underlying basement rocks during fluid migration. Ocean waters, for instance, have less than nanomolar concentrations of both Zn and Pb (Figure 4.1). Most sedimentary rocks are not abnormally enriched in the metals of economic interest. The metals are present at ppm concentrations, either in solid solution in the common rock-forming minerals or in accessory minerals. The different ore metals may be held in different sedimentary minerals. The concentrations of metals in a migrating basinal fluid will be controlled not only by interaction with minerals that contain the metals, but also by 'metal availability', for instance, whether diagenetic mineral reactions release the metals into the pore fluid. The mineralogy of aquifers in particular can thus be an important control on the concentrations of metals in groundwaters and hydrothermal waters in basins.

As is discussed further in the following sections, there are empirical relationships between sedimentary-basin hydrothermal ore deposit types and the type of sedimentary rocks that form the major aquifers in the sedimentary basin – Zn-rich deposits occur in carbonate-dominated sequences, Pb-rich in sandstones and Cu-rich deposits in sequences with either or both of continental red-bed sandstones and basaltic lavas (Figure 4.3). These relations suggest that the metal content of the fluid is controlled by metal availability in specific rocks of the basin.

The minerals that host the metals in the sedimentary rocks and the mineral reactions that release them into a migrating fluid are incompletely known, but we can speculate on likely hosts. Likely hosts of Zn are carbonate minerals, and Zn would be potentially released on replacement for instance of calcite by dolomite. Lead can be in relatively high concentrations in feldspars in clastic sediments and these are in many basins replaced by clay minerals during diagenesis, hence potentially releasing Pb. Copper can be in relatively high concentrations adsorbed onto haematite or iron hydroxide coating of quartz grains in red-bed sandstones and these coatings may be dissolved during syn-diagenetic fluid flow. Copper is sourced from ocean-floor basalts to form volcanic-hosted massive

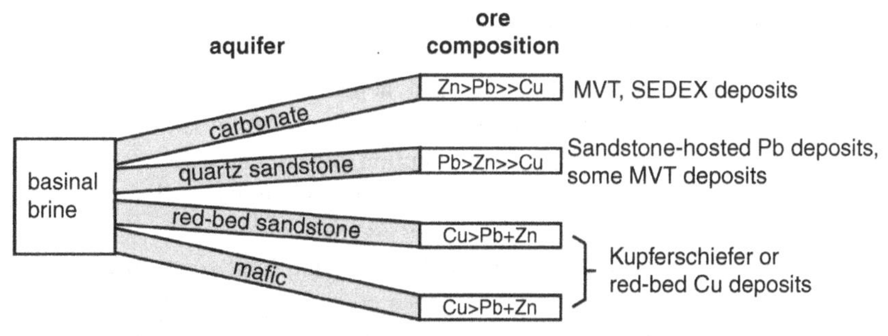

Figure 4.3 Empirical relations between aquifer rock type and metal content of sediment-hosted hydrothermal ore deposits and examples of each deposit type. See text for discussion of diagenetic mineral reactions that may contribute to these relations (modified from Sverjensky, 1989). The ore composition indicates the relative economic value of the metals rather than the concentrations: many deposits mined primarily for Cu (Section 4.2.3) have larger tonnages of Zn than Cu, but are primarily Cu ores.

sulfide (VHMS) deposits, and may thus also be sourced from mafic rocks interbedded with sedimentary rocks in a basin, most likely when primary mafic silicate minerals are replaced by lower-temperature hydrous minerals such as chlorite and epidote.

4.1.2 Large-scale fluid flow in sedimentary basins

It was discussed in Section 3.1 that movement of metal-bearing fluids is required to form large ore deposits. Large-scale movement of connate and meteoric waters through sedimentary rocks of the host basin is occurring in the present day, for instance in the Great Artesian Basin of Australia, in which meteoric waters flow from surface recharge to discharge over distances of over 1000 km and over times of up to one million years, reaching depths of about 4 km and temperatures of about 100 °C before rising back up to the surface at springs.

Subsurface fluid flow follows Darcy's law that flux is proportional to the product of the hydraulic conductivity (an expression of permeability) and the gradient of hydraulic head, where the hydraulic head (or 'head') is the height to which a free column of water will rise. In an isotropic medium, Darcy's law is such that the direction of fluid flow is down the gradient of hydraulic head. The pathways and patterns of migration of fluid in basins can thus be understood through consideration of permeability and gradients in hydraulic head, which provide the driving forces for flow.

Permeability structure of sedimentary basins

Our understanding of the patterns and controls of fluid flow in sedimentary basins builds on observations that have been made and principles that have been developed in particular by the hydrocarbon and groundwater extraction industries. There are many analogies between flow paths of ore-forming hydrothermal fluids and those of hydrocarbons. In uncompacted or poorly compacted sedimentary rocks, the major pathways for fluid flow are permeable rock units. The permeabilities of rock units vary by order of magnitude depending on the rock type, texture and petrography. We typically differentiate aquifers, which have high permeability and are permissive of fluid flow, and aquitards or aquicludes, which have low permeability and transmit orders of magnitude lower fluid fluxes. Fluid pathways are hence often sub-parallel to stratigraphic layering.

Permeability can be strongly influenced by textural and mineralogical modifications during cementation, diagenesis and compaction. Primary porosity is typically occluded with compaction and cementation and hence permeability generally decreases with increasing depth of burial. It is possible, however, for porosity to reform during diagenesis, for instance as a result of dissolution of cement minerals in basinal waters, or replacement of calcite by dolomite in carbonate units, and this is differentiated as secondary porosity.

Fracture permeability, including fault planes and fault zones, becomes a more important component of permeability and hence control on fluid flow at greater depths where rock permeability is generally lower. In contrast to the assumptions that are made for flow of magmatic-hydrothermal and other deeply derived hydrothermal fluids, fault zones in sedimentary basins are not necessarily zones of high permeability. Faults often form aquicludes and can seal oil reservoirs, for instance, because of a lining of clay smear or smear of compacted cataclasite. Low-permeability faults can compartmentalise sedimentary aquifers such that they no longer act as permeable units that allow long-distance flow.

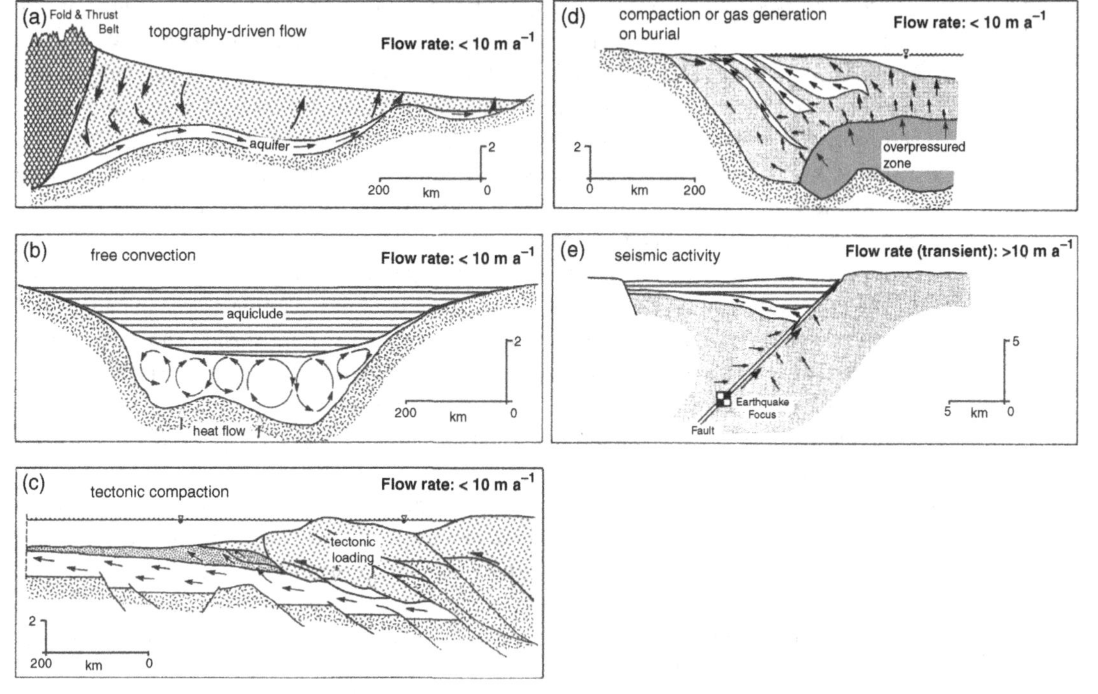

Figure 4.4 Some scenarios for large-scale groundwater flow in a sedimentary basin based around different driving forces (after Garven, 1995). See text for further discussion.

Driving forces for large-scale fluid flow in sedimentary basins

In contrast to the case of solutions derived from magmatic and metamorphic sources, which are at pressures approximately corresponding to rock pressure (lithostatically pressured) and rise through the crust, surface-derived waters in and just below sedimentary basins are generally hydrostatically pressured (or hydro-pressured) such that the fluid pressure is approximately equivalent to the weight of the overlying water column up to the water table. This is, however, not invariably the case for connate and diagenetic waters where sediment compaction is too fast to allow water in pore spaces or gas produced by organic matter maturation, for instance, to escape. Fluid pressures in fine-grained sedimentary rocks that are being actively buried and heated can thus rise significantly above hydrostatic pressures to give 'overpressure'.

We can distinguish different causes of gradients in the hydraulic head and hence different driving forces for basin-scale fluid flow (Figure 4.4). These include:

- Topography (Figure 4.4a): the water table is generally higher under areas of high topography. The hydraulic head is thus higher at any altitude under high topography and the horizontal component of groundwater flow thus generally follows regional topographic gradients. Flow is away from a mountain or upland belt. A common pattern of basin-wide topography-driven groundwater flow involves recharge and downward flow in an upland area, near-horizontal flow following the topographic slope and upward flow and discharge at topographic lows.

- Thermal convection: thermal expansion of water means that the fluid pressures at any depth are lower where temperatures in the column of water are higher. With high geothermal gradients and hence sufficiently strong vertical gradients in fluid density, free convection can occur. Convection may either be localised, for instance above an igneous intrusion below the basin, or in multiple convection cells as illustrated in Figure 4.4b.
- Salinity: highly saline fluids are dense and will tend to flow downwards and displace less saline waters upwards in a similar fashion as for thermal convection.
- Sediment compaction: this may occur as a result of progressive basin fill and loading by sediments (Figure 4.4d) or may be driven by tectonic loading during emplacement of nappe units along thrust faults, for instance (Figure 4.4c). Compaction can cause fluid overpressure at depth in a basin or within specific strata in the basin. The net effect of overpressure will be movement of waters upwards and outwards from basins.
- Diagenetic reactions (the release of water from minerals) and hydrocarbon maturation (with the production of hydrocarbon gases) raise the internal fluid pressure of pores and hence drive fluid flow.
- Seismic activity (Figure 4.4e): deformation during an earthquake and during build-up and release of stress through the earthquake cycle ('poro-elastic dislocation') causes changes in rock permeability and porosity through compaction and opening of pore space and opening new fractures or re-opening fractures, and consequently causes changes in fluid pressure. Seismic activity can thus be a localised and transient driver of fluid flow (see also Box 3.7).

4.2 Base-metal deposits in sedimentary basins

Ores of dominantly base-metal sulfide minerals in sedimentary basins may be syngenetic, syn-diagenetic or epigenetic. The ores have the following common characteristics:

- They are hosted in sedimentary basins which have no, or only a minor component of, volcanic and intrusive rocks. Basalt lava flows and intrusions are a common component of the rift-phase fill of rift basins, but these are generally much lower in the basin fill than the base-metal ores.
- Syngenetic ores are generally stratiform and stratabound. Syn-diagenetic and epigenetic ores are generally stratabound, although may transgress the host units at a low angle, or at a high angle where they formed on either side of a fault plane.
- The ores formed at relatively low temperatures of between about 60 and 250 °C.
- In many cases, the ore fluids have salinity higher than seawater.
- Many ores have an intimate relationship with either abundant sedimentary organic matter or with migrated hydrocarbons (oil or gas).

Classes of sedimentary-basin base-metal deposits are distinguished on the basis of metal content, host-rock type and the timing of mineralisation relative to sedimentation. The three main classes are:

(i) Mississippi Valley-type Pb–Zn deposits (MVT)
(ii) SEDEX Pb–Zn(–Ag) deposits
(iii) Kupferschiefer or red-bed Cu deposits.

The oldest examples of all of these deposits types are early Proterozoic, and all thus formed after the Great Oxidation Event and the consequent changes in ocean chemistry (see Box 5.1).

4.2.1 Mississippi Valley-type (MVT) Pb–Zn deposits

Mississippi Valley-type deposits are epigenetic Pb–Zn sulfide deposits in indurated carbonate rocks which contain a few per cent to up to about 10 wt % of each of Pb and Zn, with approximately equal grades of the two or dominated by Zn. By-products from some ores include one or more of Ag, Ge, Cd, Cu, barite ($BaSO_4$) and fluorite (CaF_2).

Almost all deposits are in Phanerozoic sedimentary basins. The type locality for these deposits is the composite Midcontinent Basin of Palaeozoic age in central USA (= Mississippi Valley) in which many camps of large deposits occur (Figure 4.5)

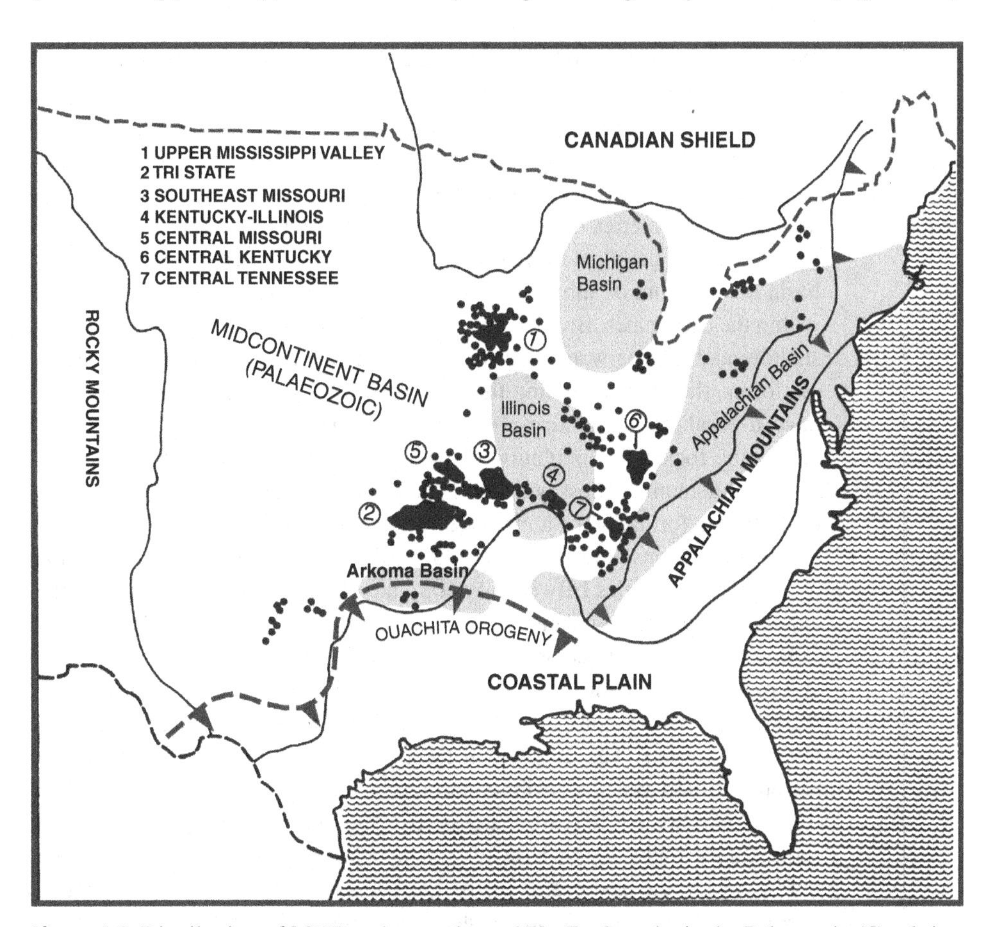

Figure 4.5 Distribution of MVT carbonate-hosted Pb–Zn deposits in the Palaeozoic (Cambrian to Permian) Midcontinent Basin of USA, with names of the major ore fields (modified from Sverjensky, 1986). The grey areas show the depo-centres of the basins of this period and mark the areas of most hydrocarbon resources. MVT ores are dominantly in intervening thinner sequences on ridges and over basement highs. The basins were deformed along their southern and southeastern margins during the late Palaeozoic Appalachian and Ouachita orogenies.

(Ohle, 1996). The deposit type is however widespread around the world and occurs in many Phanerozoic basins. Other important examples include those in the late-Palaeozoic Midlands Basin of Ireland including Navan, Tynagh among other (e.g. Hitzman and Beaty, 1996), ores in Palaeozoic basins of the North West Territories of Canada (Pine Point), in southern Poland (e.g. Leach *et al.*, 1996), in the Mesozoic Tethyan orogenic belts of Mediterranean Europe, Iran and North Africa (e.g. Sardinia, Boni 1985) and in the Canning Basin of Western Australia (McCracken *et al.*, 1996; Tompkins *et al.*, 1994; 1997). The deposits in the Midlands Basin of Ireland are sometimes considered as a separate sub-class in view of a more closely stratiform ore-body shape and higher inferred fluid temperatures during ore mineral precipitation at some deposits. Economically less-important Pb-rich Pb–Zn sulfide deposits occur in sandstones ('sandstone-hosted Pb deposits') rather than in carbonates but share many other characteristics of deposit setting with MVT deposits.

Typical characteristics of MVT deposits

(a) Geological settings of deposits

MVT deposits occur as clusters of multiple deposits and occurrences in ore districts in large intracratonic basins (as in Figure 4.5) or broad rift basins formed either in the interior of a continent or at a passive continental margin. In all cases they are hosted in basins that have up to a few kilometres fill of dominantly shallow marine sedimentary rocks which includes platform-facies carbonates, including reefal carbonate rocks, as an important component of basin fill. Gypsum or anhydrite evaporites are often present, for instance as layers or lenses within the carbonates. Broad rift basins have horst- and graben-style fault blocks marked by shallow and deeper water facies of sedimentation, the former being reefs and carbonate platforms, the latter including thick sequences of shales in some cases. Many host basins have a basal unit of coarse clastic sedimentary rocks that unconformably overlie older crystalline rocks. The evolution of many of the basins has been influenced by tectonism after sedimentation. Most typically the basins have been affected by inversion, and in some cases by the formation of a mountain belt along a basin margin.

Reconstructions of the palaeolatitude of the basins at the times of ore formation show that most were at latitudes between about 5 and 35 degrees north or south of the equator, as is expected from the abundance of reef carbonates and related sedimentary facies.

Radiometric and palaeomagnetic ages show that most MVT ores formed between ten and more than a hundred million years later than deposition of the host sediments. In the Mississippi Valley, ore is Devonian to Permian but much is hosted in Cambrian to Ordovician rocks. The deposits in Ireland exceptionally formed within a time period of about 5 million years of sedimentation.

Ores are hosted in carbonate units. The lowest carbonate unit above the clastic sedimentary rocks at the base of the succession is typically an important host for ore, but in most of the host basins ore bodies occur in carbonates units at different levels in the succession. Many deposits are hosted in specific and characteristic stratigraphic and facies settings in the carbonate units, although the characteristic settings vary from district to district. Reef-facies carbonates are common and repeated hosts for ores. Most of the major ore bodies in southeast Missouri, for instance, are hosted in a single stratigraphic unit, and within this unit in a single algal reef facies that forms a ribbon-like unit between fore-reef and reef talus to one side and a lagoonal facies to the other side (Figure 4.6).

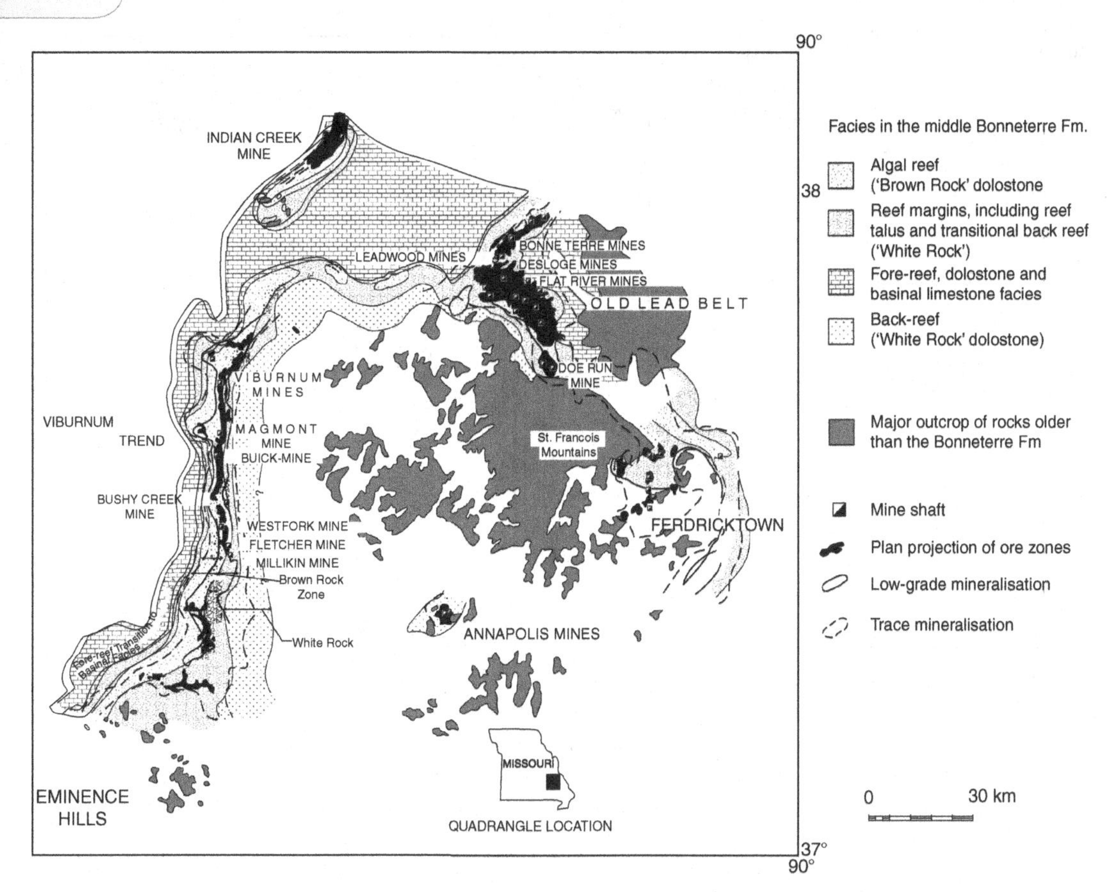

Figure 4.6 Stratigraphic setting of MVT Pb–Zn mineralisation in the Cambrian Bonneterre Formation in southeast Missouri. Most deposits ring a basement high and occur largely in the algal reef facies of the host formation, which is interpreted to have been deposited in reefs that fringed islands in the Cambrian shallow sea of the area (Ohle, 1996). The Bonneterre Formation is a limestone over most of the basin, but is generally dolomite in the vicinity of MVT ore deposits.

Many ores are either near the margins of the basin or are above basement highs in the interiors of the basin (Figures 4.6 and 4.7). Deposits in some of the host basins are characteristically adjacent to extensional faults that were active during basin formation, and may be either on the upthrown or the downthrown block. They may also be hosted in subtle structures such as low-amplitude anticlines, for instance draping an underlying basement high (Figure 4.7b). Karst structures are important hosts for ores (Figure 4.7a). The karst may be stratabound along an exposure surface, for instance at the eroded top of a carbonate unit, or be collapse karst within the host carbonate units. Some ores are hosted in karst that formed along an exposure surface and hence after an episode of sub-aerial exposure of the host marine strata.

(b) Nature of MVT ores

Most MVT ore bodies are small- to medium-sized irregular massive lenses and worm-shaped bodies of a few million tonnes of ore that are stratabound in massive carbonate

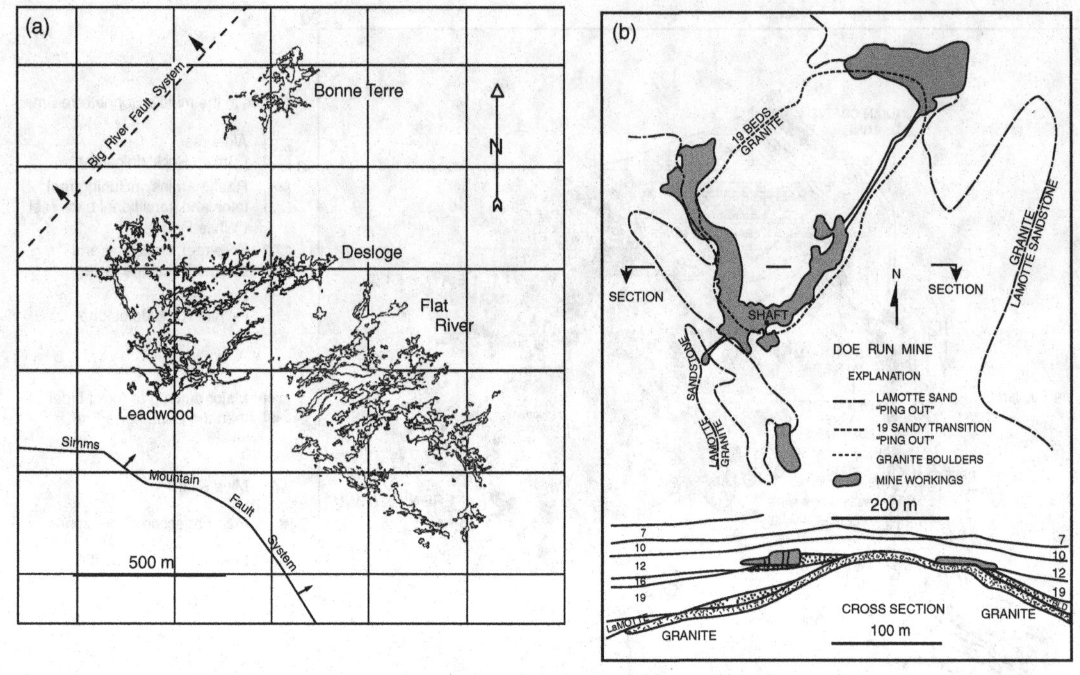

Figure 4.7 Variability in the geometry and setting of MVT ore bodies (after Ohle, 1996). (a) Plan views of the Bonne Terre and neighbouring deposits in the Old Lead Belt of southeast Missouri Fletcher Mine (see Figure 4.6), showing worm-shaped karst-fill style of ore bodies. (b) Doe Run deposit in the Viburnum Trend (see Figure 4.6) showing the position of the stratabound ore zones above stratal pinch outs around an emergent basement high. The numbers in the cross section refer to sedimentary units within the host carbonate Bonneterre Formation. The Lamotte below the Bonneterre Formation is composed dominantly of sandstone.

units (Figure 4.7). Although most ore bodies have irregular shapes and are not stratiform, they are stratabound and hence overall typically flat-lying in the largely undeformed host sedimentary units. Some ore bodies or parts of larger ore bodies follow and spread out into host carbonate strata adjacent to faults that cut the host sequence, and in these cases the bodies tend to have more podiform rather than stratiform shapes.

The host-rocks are invariably massive but strongly porous carbonates. Most host-rocks are limestones that have been dolomitised, often in irregular-shaped bodies that cut across the host strata. Zebra dolomite with parallel, slightly irregular, thin bands of white saddle dolomite and dark replacive dolomite, which may have formed through selective dissolution and replacement of parts of the primary fabric of the limestone, can be a distinct local form of dolostone. Dolostone is not, however, an invariable host to ore, and even where it is the host to ore, pre-, syn- and post-ore dolomite is recognised, and the ores are thus not necessarily spatially and temporally related to dolomitisation The most common alteration in and around ores is dissolution of the host carbonate, often with associated collapse brecciation. There are columnar, stratiform and complex cave-like breccia bodies, in some cases with complex reticulate shapes.

MVT ores have characteristically simple ore and gangue mineralogy. The major ore minerals are sphalerite, galena, marcasite and pyrite, with minor chalcopyrite and other

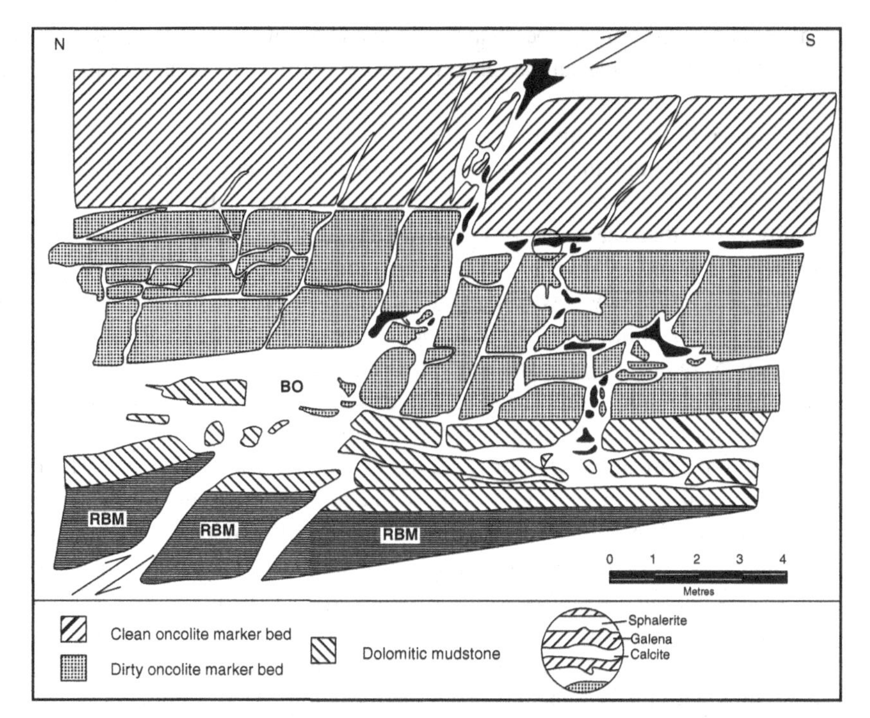

Figure 4.8 View of ore structures within part of the Cadjebut MVT deposits, Western Australia (Tompkins *et al.*, 1997). The unshaded spaces are breccia ore (BO) and dilation ore veins in which galena and sphalerite rim vein walls and breccia clasts (as shown in inset). The centres of some breccia spaces are filled with post-ore calcite (black). Breccia ore overprints rhythmically banded marcasite (RBM), which contains repeated centimetre-scale bands of mainly marcasite and calcite formed by pervasive dissolution and replacement of a carbonate unit and is a product of an earlier phase of ore mineral precipitation.

sulfide minerals in some cases. In addition to dolomite, barite and fluorite are distinctive gangue minerals around ores in many deposits. Ore textures are very varied and include ore-matrix breccias in which the ore minerals have grown between breccia clasts of the host-rock (e.g. Figure 4.8), rock-matrix breccias in which smaller host-rock fragments between breccia clasts have been replaced by sulfide minerals, and ore mineral growths as replacements of carbonate and evaporite minerals. Banded sulfide ore and replacement of a distinct fold form or 'teepee structures' that results from the volume change on reaction of anhydrite by gypsum give evidence of replacement of sulfate evaporite beds in some ores.

The sulfide mineral textures can be complex and varied with coarse euhedral crystals, banded sulfides, fine-grained colloform growths and dendritic galena, for instance. Many of the textures indicate mineral growth into the open space of secondary porosity that formed before precipitation of the ore minerals. There are complex, systematic, paragenetic histories of progressive filling of the open space, as recorded by growth of one ore or gangue mineral over earlier minerals and for instance by sphalerite with different coloured growth zones. Dolomite is the latest growth in vughs between the ore sulfides in many ores. Mineral growth can also be interspersed with episodes of brecciation, such that early phases of matrix-infill ore minerals are themselves brecciated (Figure 4.8). In some cases there is evidence that ore minerals deposited onto rather than precipitated onto breccia clasts, for instance in 'snow-on-roof' textures in which sulfide minerals preferentially coat the upper surface of clasts.

(c) Nature of the ore fluid

The ore fluids and the fluids that precipitated the other minerals in the paragenetic sequences are sodium–calcium–chloride brines. Fluid salinities are in all cases significantly higher than seawater (10–30 wt % salts) and are most frequently slightly lower than halite saturation (~ 25 wt % salts). Temperatures of the fluid are between 60 and 160 °C, with a peak frequency between about 110 and 130 °C, but some of the deposits in central Ireland appear to have formed at higher temperatures of up to about 250 °C.

Many of the host basins and sequences also contain significant hydrocarbon resources, although the major hydrocarbon resources are concentrated elsewhere in the host basins than the sulfide ores. In many deposits there is evidence that oil and possibly gas was present in the ore at the time of formation. Fluid inclusions of hydrocarbon fluids may be present together with inclusions of the saline brine in ore minerals. In some deposits there is oil preserved in pore space, and in the Admiral Bay deposit of Western Australia, the ore immediately underlies an oil charge in an anticline. These relations indicate either that oil or gas migrated together with the saline aqueous hydrothermal fluid, or that oil was present in the rock (e.g. in an oil reservoir) when the saline brine infiltrated. The common temperatures of ore formation correspond to those of the 'oil window' and hence are consistent with the presence of oil in the rocks during ore formation.

What can be inferred about ore genesis at MVT deposits?

MVT ore bodies formed during specific periods within the history of the host basins. The widespread development of the ores and petrographic features of the host carbonate rocks such as regional resetting of palaeomagnetic poles and regionally uniform cement stratigraphy show that the ores formed at times of migration of brines on a large scale through the sedimentary basins. The pathways of ore-fluid brine migration were controlled by the same range and different styles of permeability as typically control oil migration, including primary porosity of aquifers, faults, secondary porosity where dolomite has replaced limestone, and karst cavities in limestones. The porosity in secondary dolomite may have formed as a result of dissolution during ore fluid flow or have formed earlier and have been an important control on permeability.

The depths of ore formation are inferred to be less than about 2 km, from reconstructions of basin subsidence, filling and uplift histories. Fluid inclusions show that the ore fluids were at temperatures of between about 60 and 150 °C and hence in most cases significantly hotter than ambient temperatures at the depths of mineralisation. Similar temperatures are indicated by measures of diagenetic maturation of organic matter, such as the colour of conodonts (the 'conodont alteration index'). Comparisons of these indices in ores and in surrounding rock show that maximum temperatures were higher in the ores. The relatively high temperature gradients away from ore bodies imply relatively rapid upward fluid movement from deeper in the basin, and also that fluids were cooling at the site of the deposits.

(a) The sources and evolution of the ore fluids and contained solutes

The saline ore fluids are very similar in composition to saline pore waters at a few kilometres depth in many present-day sedimentary basins, especially those that include evaporite units.

They are thus also similar to the so-called 'oil-field brines' or saline basin waters that are characteristically found with oil in many basins, for instance below oil charges. The geochemistry of the ore fluids, and particular differences and similarities between the salt content of the ore fluids and of seawater, show that the salinity of the fluids was generally generated by evaporation of seawater or other surface waters, or less commonly by dissolution of sedimentary halite into groundwater. Whether the fluid was originally meteoric, connate or ocean water cannot in general be differentiated. Whichever is the fluid source, a palaeoclimate suitable to promote strong evaporation of surface waters and form high-salinity basinal fluids is inferred to be required for the genesis of these deposits.

The composition of the waters was influenced by interaction with the rocks, it migrated through, such that, for instance, the Na to Ca ratio is lower than seawater as a result of dissolution of calcite. The fluid salinities are high enough to carry tens to hundreds of ppm Pb and Zn in solution. Dissolution of calcite may have provided a major component of dissolved Zn. The Pb isotope compositions of the fluid in some ore fields indicates that at least some of the dissolved Pb was sourced from the basement underlying the basins or from clastic sedimentary rocks that included material eroded from the basement. The ore metals can thus be inferred to have been sourced from rocks along fluid flow paths, but exact sources are undefined. The complex parageneses of the ores suggest that the metal content of the fluid entering the ore bodies may have varied strongly over the time period of fluid flow.

(b) Causes of ore mineral precipitation and sources of reduced sulfur

MVT ores precipitate along short segments of long fluid flow paths. We can further infer from regional resetting of palaeomagnetic poles and regionally uniform cement stratigraphy that fluid flow was much more widespread through the host succession than the ores. Although base-metal sulfide may precipitate from a metal and sulfide-bearing ore fluid as a result of pH increase on dissolution of carbonate (through the same reactions that promote ore precipitation in skarn deposits, see Section 3.2.3) or by fluid cooling, neither process can explain the localisation of MVT ores in stratabound pockets of generally restricted vertical extent in the host carbonate units, and neither is thus likely to be the major cause of ore precipitation.

Mixing of aqueous fluids of different compositions could have been a cause of ore mineral precipitation, at least in some deposits. Mixing of brine which contained high concentrations of the metals with lower-salinity water could reduce the solubility of metal chloride complexes if the metals are carried as polyvalent complexes. Mixing of reduced, sulfide-bearing water and an oxidised, metal-bearing brine could also cause sulfide precipitation.

The sites and localisation of the ore bodies are, however, better explained if ore mineral precipitation was prompted by a localised input of sulfide into a migrating ore fluid that contained, at least at some times, relatively high concentrations of Pb and Zn. Ore precipitated along a flow path specifically where sulfide was generated or was available.

The different generic reaction pathways that will produce sulfide in sedimentary environments were discussed in Section 4.1.1 (Figure 4.2). Sulfide could have been generated by reduction of sulfate in solution, for instance on interaction with liquid or gaseous hydrocarbons, by dissolution and reduction of sedimentary sulfate (from e.g. gypsum), or directly from maturing organic matter. There is evidence for each of these processes in at least some MVT deposits. Pseudomorphs of sulfides after sulfate evaporite minerals have been recognised at a number of deposits including Cadjebut (Western

Australia), and reservoired pore-space oil occurs in and around ores at, for instance, Admiral Bay (Western Australia) and Pine Point (Yukon Territory, Canada). Hydrocarbon inclusions are common in ore minerals in many deposits. Sulfur isotope ratios of the ore minerals also indicate that there are variable and multiple possible sources of sulfide in these ores. The ratios vary widely between different ore districts, and in some cases there are large variations in the ratio even within individual grains within a sample. The different isotope ratios confirm that sulfide sources or reaction pathways to produce sulfide were different in the different deposits, and may also have changed during the evolution of a deposit. A number of processes thus appear to be able to contribute to the precipitation of the ore sulfides, with different combinations of processes in the different MVT ore fields around the world.

Driving forces of large-scale basinal ore fluid flow

The tectonic and hydrogeological cause of hydrothermal fluid flow through a part or all of a sedimentary basin is an important part of an ore genetic model for sedimentary basin hosted deposits. In order to make interpretations of the driving forces of regional and widespread ore fluid flow in a basin, we need to know the timing of mineralisation and to be able to reconstruct the palaeogeography and tectonic setting at the time of fluid movement. Where we can make these reconstructions it is seen that different deposits of this ore class likely formed as a result of different hydrological processes. A specific hydrological and hence tectonic setting is not critical to ore formation, so long as long-distance flow of saline, metal-bearing fluids occurred through the basin.

(a) The Mississippi Valley deposits

For the deposits in the Mississippi Valley region of the central USA, mineralisation coincided in time with the Alleghanian–Ouachita orogeny along the southern and southeastern margins of the basin (Figure 4.5) and it is inferred that there was high topography along these margins of the basin at the time of ore formation. The most likely major driving forces for fluid flow in the basin are thus either tectonic loading or topography-driven flow associated with the orogeny, with distances of flow paths of perhaps a few hundred kilometres (Figure 4.9). The feasibility of this proposal has been demonstrated by numerical simulation of fluid and heat flow in the host basins at the time of ore precipitation (Box 4.1).

The permeability architecture of the basin would have controlled the patterns of fluid flow within an **ore district**. For instance, fluid flow may have been forced upwards through carbonate strata where these overlie a stratigraphic pinch out of the permeable basal sandstone of the Midcontinent Basin, and hence form ore bodies above a basement high, as illustrated in Figure 4.7b. Clear feeder zones, for instance of ore vein networks marking fluid flow paths below the ore bodies, are not generally recognised, but dolomitisation of the host carbonate is often in irregular discordant zones that envelope the ore zones. It is interpreted that these discordant dolomites may mark the pathways along which ore fluids were forced to migrate upwards through the stratigraphic sequence. The same pathways would have been used by any fluids that migrated before the metal-rich ore fluids, and also by hydrocarbons.

(b) Other MVT ore districts

Although a topographic driving force for basin-wide fluid flow is inferred for the Mississippi Valley deposits, this force is probably not applicable for many other MVT

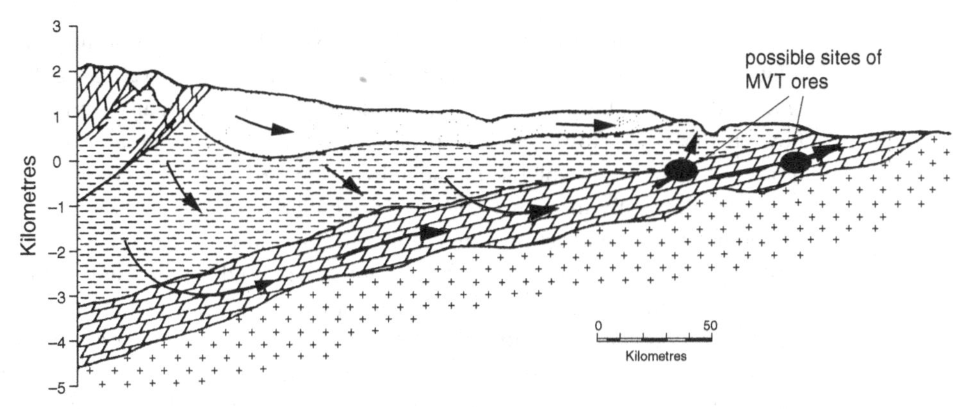

Figure 4.9 Suggested basin-scale fluid flow that gave rise to MVT deposits in the Permian in the Mississippi valley area driven by the topography of the Alleghanian–Ouachita orogeny along the southeast margin of the basin. The carbonates near the base of the sequence dipped towards the mountain chain as a result of tectonic loading and are interpreted in this reconstruction to have been the main aquifer and to have channelled groundwater flow. Note that the arrows indicate fluid movement and do not necessarily imply that fluids flowed the whole length of the paths implied, but may have been displaced from pore space within the basinal strata under the driving force of the topographic head. The deposits formed where flowing fluid is forced upwards, for instance over horst blocks and at the edge of the basin (Garven, 1995). The relatively thin basal sandstones of the basin are not shown in this reconstruction and may also have been aquifers and major fluid pathways.

districts. For the deposits of the Canning Basin, Western Australia, there is no evidence of a topographically high terrain along a basin margin that could have driven cross-basin fluid flow. The deposits are constrained from radiometric dating and reconstructions of the burial–thermal history of the basin to have formed at the end of a period of basin extension and burial. The timing and the deposit setting at the edge of a 6–8-km-thick shale-filled basin suggest that fluid flow may have been related to expulsion of fluids from the centre of the deeper basin, possibly as a result of increased pore fluid pressure on compaction or on oil and gas maturation on heating during burial. The formation of the deposits in the Midlands Basin of Ireland at higher temperatures (up to about 250 °C) within a few million years of sedimentation and adjacent to active normal faults suggests that large-scale fluid flow in the basin may have been the result of thermal convection under abnormally high geothermal gradients during extensional tectonics.

Box 4.1 Applications of computational simulation of crustal fluid flow in ore deposit geology

Hydrothermal ore deposits form one part of a hydrothermal system that is much larger than the deposit (Figure 3.2). In the case of deposits hosted in sedimentary basins, the hydrothermal systems may encompass a whole sedimentary basin. We can in general track out only a small fraction of the fluid pathways from observations on rocks and fluids and must thus make interpretations about the nature of the whole hydrothermal system.

Fluid flow in the crust is governed by physical laws and constitutive (governing) relations – see for instance summaries of Cathles and Adams (2005) and Ingebritsen *et al.* (2006). It is thus in principle possible to predict flow in any environment. Fluid flow in the crust, however, is a result of multiple and complex coupled physical processes and in view of the mathematical complexity of the coupling, predictions are most commonly made with numerical simulations.

Computational numerical simulation of fluid flow in the crust involves finite-element or related techniques and the approaches used are in many respects similar to those in simulations of the atmosphere applied in forecasting climate change or the weather. Numerical simulations involve both spatial and temporal discretisation, hence division of the mass of rock into elements of finite size and approximately regular shapes, which are assumed to have uniform properties of, for instance, porosity, permeability and heat capacity, and division of time into finite steps (Figure 4.10a and b). The finite spatial discretisation means, for instance, that thin rock units and structures cannot be accurately included. Simulations also require setting of boundary conditions around the model volume and the setting of starting conditions, and in this respect the models are less well constrained than simulations of weather because we must run models from imprecisely known starting situations at times in the past rather than from the present. Element sizes and the time length of steps depend on the situation of the simulation, but sizes are typically of about a kilometre for simulations of the whole crust, although with increasing computing power we can make more detailed simulations. Three-dimensional models are much more costly of computer power than two-dimensional models, and many models are in two-dimensions and thus assume, like many interpretations of geological cross sections, that there is no variability in the third dimension.

The multiple and, in many cases, coupled equations expressing the physical laws of flow and related processes are solved for each time step. Many of the equations that govern fluid flow in the crust, especially of a single fluid phase through porous rock (porous media), are well established, for instance Darcy's law relating flux to permeability and gradient in a hydraulic head. The form of other equations and numerical values of parameters in them, however, are less well established, for instance equations governing flow of two coexisting fluids, such as gas and a liquid that commonly coexist in high-temperature magmatic-related hydrothermal systems (as discussed for instance by Ingebritsen *et al.*, 2010). We cannot make experimental or empirical measurements of many of the parameters involved in the constitutive relations, especially those involving large spatial and temporal scales at temperatures and pressures appropriate to the interior of the crust, and although equations are improving and parameters being better constrained there will always be uncertainty in model results. It is perhaps most difficult to develop laws and relations that express how fluid flow may be influenced by other physical processes, including rock deformation and mineral dissolution and precipitation. Deformation, for instance, induces changes in pore volume and fracture aperture, hence induces changes in both permeability and the fluid pressure that may be driving fluid flow; conversely, fluid pressure affects deformation through the role of effective pressure on rock and fracture strength. Fluid flow and deformation are thus coupled and for further discussion see for instance Chapter 2 of Ingebritsen *et al.* (2006). It has been proposed additionally that seismic failure can

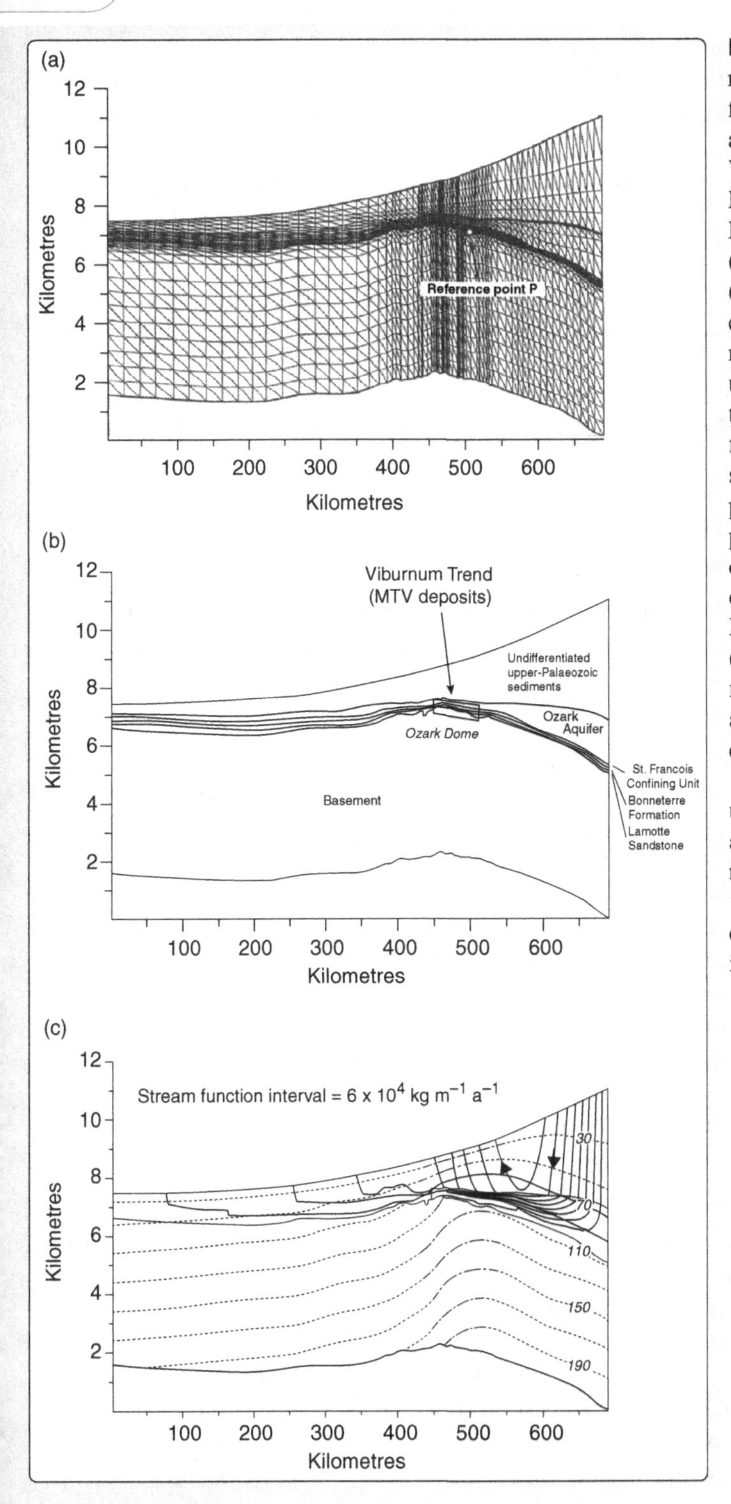

Figure 4.10 Modelling scheme and results of simulation of possible fluid flow shown in Figure 4.9 associated with MVT of the Viburnum Trend in the southeast Missouri district of the Midcontinent Basin of the USA (after Appold and Garven, 1999). (a) Discretisation into finite elements of the cross-sectional reconstruction of the basin, underlying basement and topography at the time of ore formation as shown in (b). Each stratal unit is assigned appropriate physical properties such as permeability, porosity and heat capacity. The position of the major ore deposits around the Ozark Dome or basement high are shown. (c) Example model results showing fluid flow lines (arrowed curves) and temperatures (labelled contours) after approximately 1 million years of flow. Note the updoming of the isotherms in the area of up-flowing waters, hence modelling temperatures of around 100–110 °C for the area of the MVT deposits. Stream function interval is a measure of fluid flux.

cause opening of fractures and hence development of connectivity for fluid flow (e.g. Sibson, 1996). Realistic modelling of seismic failure would be extremely difficult computationally because of condensed time periods over which the process would likely have an effect.

Despite their inherent imperfections, numerical simulations of hydrothermal fluid flow have been useful in interpretations of ore deposits and can be viewed as serving one of two purposes:

- To constrain and demonstrate the viability or otherwise of genetic models, or to compare genetic models of an ore deposit type or a specific deposit. We can investigate questions such as whether a proposed scenario of fluid flow at a deposit type is possible or likely. The effect of changing multiple variables can be examined to determine conditions that would be most favourable for ore deposit formation.
- Exploration targeting in a terrain in which the geometry and nature of rocks at depth is reasonably established. These simulations investigate questions of where ore deposition is most likely to have taken place given assumptions of fluid source, driving force for flow and chemical processes of ore mineral precipitation.

Early applications of modelling of hydrothermal systems from fluid source to sink built on established techniques of modelling of groundwater flow with the additional complexity for modelling hydrothermal systems of coupling fluid flow and heat transport and flow in a geothermal gradient. Applications included:

(i) Modelling of fluid flow in the Midcontinent basin of the USA to investigate patterns of basin-wide flow that may form MVT deposits (Section 4.2.1) under the driving force of a topographic head from the Appalachian–Ouachita orogen (Figure 4.10) (Garven *et al.*, 1993; Appold and Garven, 1999). A feasible model was one that simulated both focussed flow through the sites in the basin of the ores and the fluid temperatures of ore formation.

(ii) Modelling of unconformity-related uranium deposits (Section 4.3.1), in particular whether it was possible to simulate the temperatures estimated for ore formation at the known approximate timing of ore formation within the burial history of a basin (Raffensperger and Garven, 1995). For instance it was determined that a minimum basin thickness of about 3 km is likely required even in an environment of high geothermal gradients to promote the style of convection shown in Figure 4.20 and that the permeability and anisotropy of the basal sandstone in the host basin probably have the strongest control on ore formation.

Although the results are informative and useful to our understanding, especially when viewed with respect to specific questions about the formation of the ore deposit type, they should not be taken as proof that the model fluid flow was what took place in nature. To what degree are the simplifications of geometry and of physical relations valid? Could fluid flow patterns be significantly influenced by processes neglected in the modelling? Although we can closely reconstruct the likely large-scale geometry of the basin, there may be complexities at depth which are not resolved by seismic surveys, for instance, and we need to consider whether these would be influential. In each of two cases above, there has been questioning of the applicability of the results. In the case of the simulation of regional fluid flow at MVT deposits for instance:

- Do the regional flow models really simulate the temperatures of ore formation (Deming and Nunn, 1991)?
- Is the assumption of instantaneous build-up of a topographic head critical to the results, when in reality the orogeny would have built up progressively and hence progressively imposed an increasing hydraulic gradient?

A uniform fluid composition was assumed in the early models, and a layered distribution with increasing salinity with depth in later models. There may, however, have been spatial variations in fluid salinity.

Could there have been internal generation of fluid pressure in the basin, such as through compaction or hydrocarbon generation (Cathles and Adams, 2005)?

All of these additional complexities could in principle be incorporated into the models and their roles thus investigated. However, with increasing complexity we have greater difficulty discerning which parameters are most significant to controlling ore formation and siting.

In view of our understanding of which processes may control the formation of different ore deposit types, three topics of particular interest for more complex models are the coupling of deformation and fluid flow, the flow of fluids that boil or condense, and the dissolution, transport and precipitation of chemical components (reactive flow).

Simulation of the flow of fluids that boil or condense has been improved with more accurate representations of the physical properties of water and saline aqueous solutions. One insight from subsequent modelling of fluid flow above cooling magmas in situations appropriate to VHMS deposits on the sea floor (Section 3.2.7) is the recognition of self-organised behaviour, and that convection patterns are controlled as much by the properties of water as by the exact geometry of the system and the distribution of heat. The temperatures in the up-flow zones of convection are buffered to be about 400 °C, whatever the temperature of the underlying magma chamber (Jupp and Schulz, 2000; Conmou *et al.*, 2008). This result can be intuitively understood: the temperature of up-flow is the temperature range over which water has the highest compressibility at the relevant pressure, and because of the high compressibility a small area of uprise of fluid at these temperatures is more likely to magnify and develop into a convection cell.

4.2.2 SEDEX Pb–Zn–Ag deposits

SEDEX deposits are in some respects similar to MVT ores. They are stratiform or stratabound Pb–Zn sulfide ores hosted in sedimentary rock of early-Proterozoic to Mesozoic age, with Ag as a significant co-product in many ores and in some cases minor Cu. Many ores are also associated with major nearby accumulations of barite. There are, however, clear distinctions in the setting and timing of ore formation between the two deposit types. SEDEX ores formed syngenetically or during early diagenesis in fine-grained carbonaceous clastic sedimentary rocks that are hosted within mixed sequences of dominantly fine- to medium-grained clastic and carbonate rocks. In contrast to MVT deposits, many SEDEX deposits are large and are isolated or widely scattered deposits in their host basins. Multiple ore bodies formed at different times during basin filling in some basins.

The Carboniferous Red Dog and nearby deposits in the Brooks Range of Alaska are presently the world's largest producer of zinc (e.g. Lange *et al.*, 1985; Moore *et al.*, 1986). Proterozoic Pb–Zn–Ag deposits are an important and distinct sub-class of SEDEX stratiform deposits that occur in many middle-Proterozoic sequences of carbonate-bearing fine-grained sedimentary rocks, including the McArthur Basin with the Century and HYC/McArthur River deposits and the Mt Isa Basin with the Mt Isa deposit in northern Australia (e.g. Broadbent *et al.*, 1998; Large *et al.*, 2004), and Sullivan in the Belt Group in British Columbia, Canada (Deb and Goodfellow, 2004). Many SEDEX deposits have been metamorphosed, and Broken Hill-type deposits, which are named after the type locality of the Broken Hill deposit, New South Wales, Australia (e.g. Haydon and McConachy, 1987) have sometimes been classified separately on account of their setting in high-grade metamorphic rocks, their spatial association with metamorphosed volcanic rocks, and also because ore is at the same stratigraphic horizon as unusual strata-parallel tourmaline-rich and Mn-rich units that are interpreted to be 'exhalite' horizons or chemical sediments formed from vented hydrothermal fluids. The settings and geochemistry of these metamorphosed deposits are, however, similar in most respects to other Proterozoic deposits, even though they are distinctly pyrite-poor and F-rich, and they are thus generally considered to be Proterozoic Pb–Zn–Ag deposits that have been overprinted by amphibolite- or granulite-facies metamorphism.

The term 'SEDEX' stands for 'sedimentary exhalative' and implies ore mineral precipitation from hydrothermal fluids that vented into seawater in a similar fashion as is interpreted for ocean-floor VHMS deposits (Section 3.2.7), although at generally lower temperatures. In many cases, however, there is evidence that sulfides were precipitated diagenetically or through replacement of minerals in unconsolidated sediment, for instance textures that show base-metal sulfides grew later than diagenetic pyrite. The term SEDEX is used with knowledge that its genetic implications may not always apply.

The distinction between SEDEX deposits and sediment-hosted massive sulfide deposits (SHMS, Section 3.2.7) is not clearly defined. The term SHMS implies that a localised magmatic source of heat drove hydrothermal fluid convection, whereas convection as a result of magmatic heat would not primarily be the case at a SEDEX deposit.

Some typical characteristics of SEDEX deposits

(a) Geological settings

SEDEX deposits are hosted in long-lived sedimentary basins hundreds of kilometres broad that are either continental rift basins or basins formed along passive continental margins. The basins may contain up to several kilometres' thickness of sedimentary rocks in stacked unconformity-bounded packages. Marine-facies sedimentary rocks are generally dominant in the immediate foot-wall and hanging-wall of ores. Where the host sequences have not been strongly deformed, the basins are seen to have been segmented into multiple, fault-bounded blocks with shallows and troughs, and the ores can be seen to be hosted in all cases in narrow troughs of between about 10 km and 100 km wide, and are typically on the immediate downthrown side of trough-bounding syn-sedimentary faults,

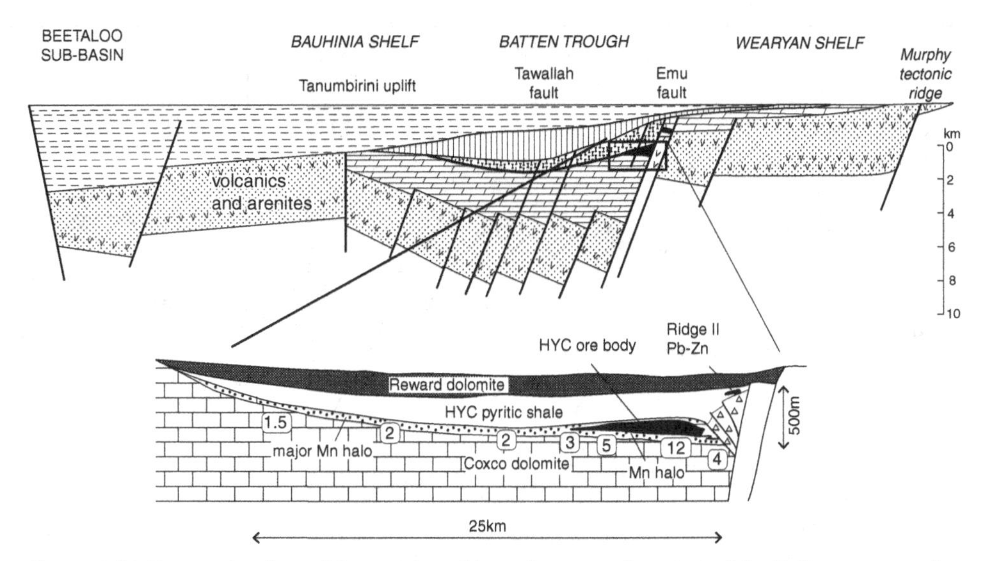

Figure 4.11 Reconstruction of the stratigraphic and structural setting of the Palaeoproterozoic HYC Pb–Zn deposit in Northern Territories, Australia (modified from Plumb, 1987 in Solomon and Groves, 2000b and Large *et al.*, 1998). The upper panel is a cross section of about 500 km. The dark shading shows the Barney Creek Formation of shales and fine-grained carbonate sediments, which hosts the deposit. The lower panel shows the position of the ore body adjacent to the half-graben bounding Emu fault. The numbers in the lower panel are wt % MnO in dolomite in the ore sequence and mark the extensive geochemical halo around the ore body.

near what was probably the deepest part of the trough basin (Figures 4.11 and 4.12). Although it is clear that the ores are hosted in deeper water facies that are time equivalents of adjacent shallower marine facies such as carbonate platforms (Figure 4.12), the water depths of sedimentation during ore deposition are poorly constrained between about 100 m and over 2 km.

The host-rocks and rocks immediately below and above the ores are in all cases dominantly grey- to black-coloured organic-rich carbonaceous shales and siltstones with up to a few per cent TOC (total organic carbon). The host-rock sequence can also include cherts, carbonate-bearing shales and siltstones, sandstones, sulfate-evaporites, thinly bedded distal carbonate turbidites and talus breccias (Figure 4.12). Fine-grained tuffite horizons are interbedded with some ores and indicate distant volcanism in the region at the time of sedimentation and ore formation. Foot-wall and hanging-wall units within the same basin-fill succession include a variety of marine sedimentary rocks including limestone, dolomite, shales, sandstones and fault-talus breccias. Evaporites are present in the host sequences above and below some of the Proterozoic deposits of northern Australia. Some of the host basins include mafic intrusions and volcanic rocks. At the Sullivan deposit there is thick mafic sill that intruded soon after sedimentation in the immediate foot-wall of the deposit and its intrusion may have influenced that nature of the ore at this site. Other host basins have mafic volcanic rocks much deeper in the succession that presumably extruded during an early rift phase of basin development.

Almost all Phanerozoic SEDEX deposits are spatially and temporally associated with major accumulations of bedded to massive barite. The barite bodies associated with the Red Dog and nearby deposits in the Brooks Range of Alaska are some of the world's

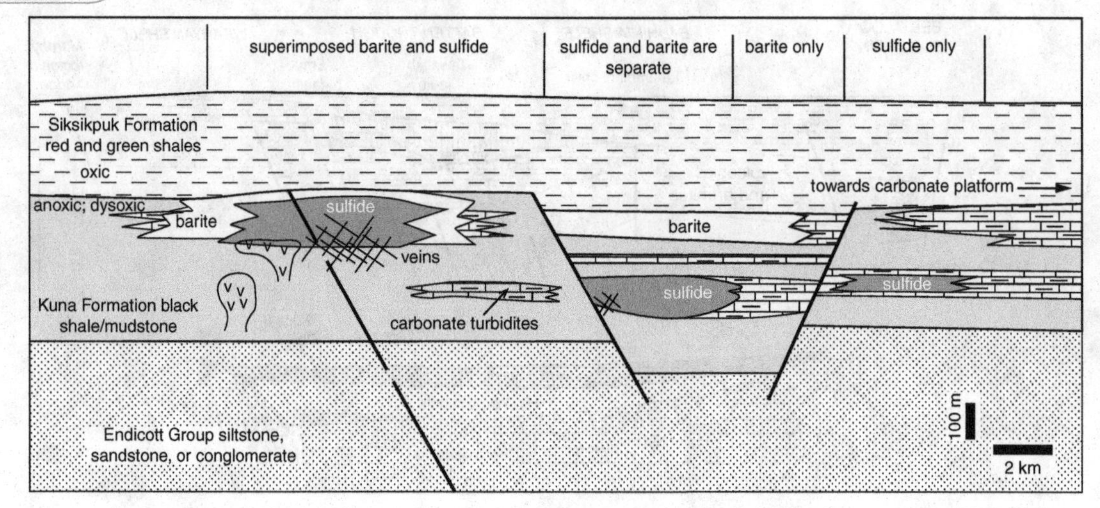

Figure 4.12 Local setting of the Carboniferous Red Dog Pb–Zn deposits, Brooks Range, Alaska. The deposits are lenses of massive sulfide, adjacent to syn-sedimentary faults and in many cases below or adjacent to massive barite bodies in a sequence of shales and mudstones and distal carbonate turbidites in what is interpreted to have been the deepest part of an intracratonic basin which was bounded by carbonate platforms (Kelley and Jennings, 2004).

largest accumulations of barite and in different cases either surround and overlie the Pb–Zn sulfide ore bodies or occur stratigraphically above the ore bodies in the same structural position adjacent to and on the downthrown side of a trough-bounding fault (Figure 4.12).

(b) Nature of the ores

The ores are stratiform and stratabound, but in many cases a broad and diffuse sub-vertical hydrothermal feeder zone extending up to a few hundred metres beneath the main ore zones is recognised, for instance by the presence of networks of mineralised veins (Figure 4.12). The sulfide ores may be massive or bedded. The Red Dog deposit (Figure 4.12) is composed of ore lenses of up to about 100 m thick that are internally unbedded and within which the ore ranges from sulfidic chert to massive sulfide with over 70% by mode sulfide minerals. In contrast, many other SEDEX deposits are bedded (Figure 4.13). These bedded deposits contain multiple, 'stacked', stratiform layers of ore spaced over a few tens of metres of stratigraphy, each up to a few metres thick and extensive over up to a few kilometres laterally. Each bedded ore body is itself characteristically layered on a millimetre to centimetre scale with laterally continuous interlayering of massive sulfide-rich ore and fine-grained silicate–carbonate dominated layers, which lack or have only a minor component of base-metal sulfide minerals but may contain a few per cent of pyrite. These thin sulfide laminae may show evidence of soft-sediment deformation such as load-casting and slump folds. The Sullivan deposit has both massive and bedded facies of ore, with a gradation between the two; the former is closer to the implied hydrothermal fluid vent and the latter at a few hundred metres distance.

Grades within ore lenses can be above 10% Pb and 10% Zn, generally with Zn at higher grade than Pb. A gradual but marked metal zonation is recorded along the length of stratiform ore bodies, most notably in the Zn:Pb ratio, such that Pb becomes relatively less

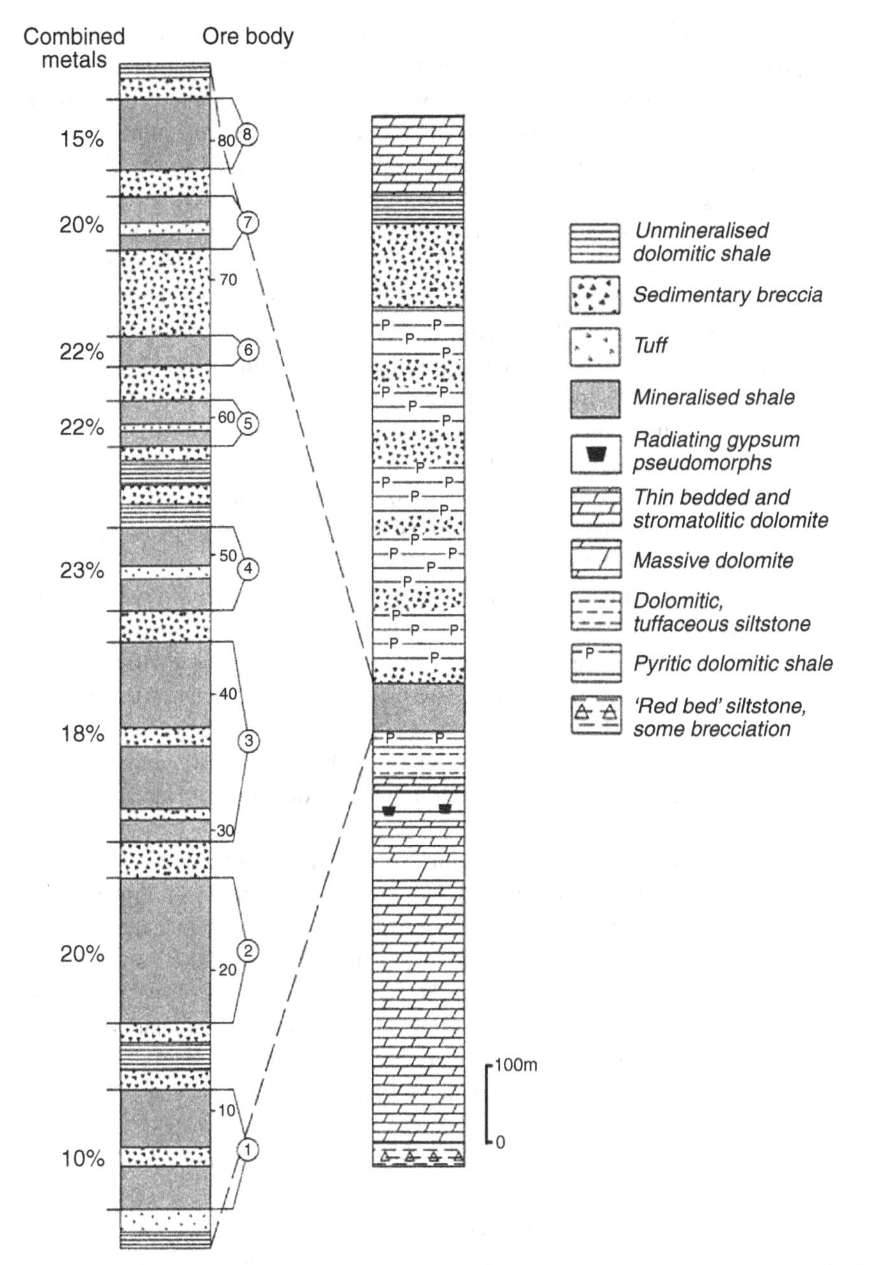

Figure 4.13 Stratigraphic setting and nature of ore at a typical banded Proterozoic SEDEX deposit, the case of HYC, Northern Territory, Australia (cf. Figure 4.10) (after Logan *et al.*, 1990 and Large *et al.*, 1998). Mineralised shale units are themselves banded on a millimetre to several-centimetre scale with lamellae of massive galena–sphalerite interbedded most typically with similar thickness quartz–carbonate silt and quartz–carbonate, fine turbidite beds. The thicker interbeds of sedimentary breccia are interpreted to be derived from the adjacent margin of the fault-bounded horst shown in Figure 4.10. Metres above the base of the lowest ore body (lens) of mineralised shale are to the right of the detailed section. Reprinted with the permission of the Australasian Institute of Mining and Metallurgy.

abundant with distance from the inferred hydrothermal 'feeder zone'. Ore minerals are sphalerite and galena with variable amounts of pyrite and in some deposits pyrrhotite. Some ores are major producers of silver, containing up to a few hundreds of ppm, mainly present in solid solution in galena or as trace, silver-bearing sulfosalts.

In unmetamorphosed bedded ores such as at HYC, sphalerite and galena are mostly very fine-grained (tens of μm), although the sulfide minerals are significantly coarser in ores which have been overprinted by even low grades of metamorphism. Ore mineral textures include base-metal sulfides rimming framboidal fine-grained diagenetic pyrite, and infilling pore space. The ores at Red Dog are in many areas breccias, for instance with clasts of sulfide in a sulfide matrix or clasts of host-rock cherty-carbonaceous shale in a sulfide matrix. The minerals in the breccia matrices and in veins in the Red Dog ores show evidence of growth into open space such as euhedral growth zones. Barite, chert and sphalerite pseudomorphs of worm tubes occur in the Red Dog ores. Indigenous organic matter as fine specks and filaments of kerogen is a component of the interleaved calcareous and shaly sedimentary rocks. Bitumen, which may be solidified migrated oil, is present in some deposits. The Century deposit has been interpreted as a now-drained oil reservoir that was charged at the time of mineralisation.

The minerals of the host sedimentary rocks are little altered in and around ore. Sideritic (Fe-rich) carbonate has been recorded as a widespread mineral beneath many stratiform ore lenses and may be an indicator of upward seepage of hydrothermal fluid. The ore horizons are however geochemically distinct over distances of several kilometres along strike from the ore bodies (Figure 4.11). They are marked by anomalously high contents of Zn, Ba (as barite), Mn (as manganese carbonate) and Tl, and also by distinct C and O isotope compositions of carbonate minerals compared to elsewhere in the host unit.

(c) Nature of the ore fluid and conditions of mineralisation

The nature of the ore fluids and the exact temperatures of mineralisation have been difficult to define because the general lack of veins and the fine grain size of the ore minerals in unmetamorphosed ores means that fluid inclusions are rare. Ore-fluid exhalation temperatures are estimated to be less than 250 °C, most probably between 100 and 200 °C. There is no evidence of boiling of the ore fluid on exhalation. The few analyses of ore fluids show the presence of relatively saline waters with 5–25 wt % NaCl, and also aqueous solutions with salinity similar to or less than seawater. The high-salinity fluids are mostly likely strongly evaporated seawaters. Methane-rich inclusions are present together with aqueous inclusions in minerals in the ore zones at Red Dog.

Genetic models for mineralisation

SEDEX ores form where hydrothermal waters either exhale as a plume from a thick sequence of sediments into the bottom waters of marine troughs or half-grabens, or infiltrate through unconsolidated and porous sediments below the sea floor of these grabens. Although the ores formed in fault-bounded troughs or half-grabens, reconstruction of basin architecture indicates that many of these troughs may have been at most a few hundred metres deep. The focus of the ore-fluid up-flow or feeder zone may be at the base of a fault scarp, but contours of Pb/Zn ratios of ore suggest that the focus was in many cases closer to the centre of the fault-bounded narrow host basin.

The thicknesses of the ore bodies, most notably at Red Dog, suggest long-lived ore precipitation. Thinly laminated ore presumably formed where ore mineral precipitation

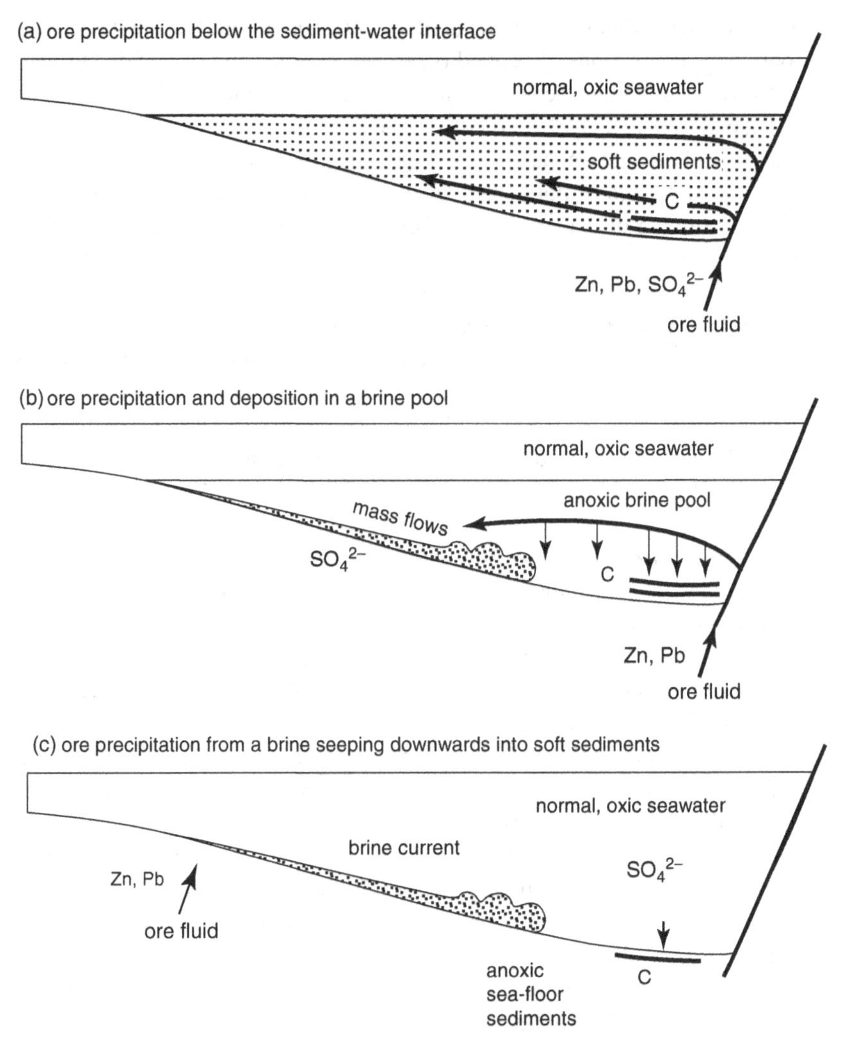

Figure 4.14 Schematic diagrams showing possible modes and sedimentary sites of formation of a SEDEX Pb–Zn deposits based on the setting and characteristics of the HYC deposit, Northern Territory, Australia (modified from Large *et al.*, 1998). The possible sources of metals, sulfate and organic reductant (= C) are indicated. Sub-sea-floor sulfide deposition (a) has been proposed to explain textural evidence for precipitation of base-metal sulfides after diagenetic pyrite. Precipitation in an exhalative brine pool (b) explains the wide geochemical anomaly and stratiform nature of the ores. The model in (c) is a compromise that in principle allows for both the textures and the extensive geochemical anomaly (model proposed by Sangster, 2002).

was punctuated by input of clastic sediments into the basin, for instance from turbidite flows sourced from the basin margins.

(a) Ore mineral precipitation

The textures of the ore sulfide minerals give conflicting evidence as to whether the ore minerals precipitate in the water column and settle on the sea floor (exhalative formation) or precipitate within soft sediments (diagenetic formation) (Figure 4.14). The fine, laterally continuous laminations of bedded ores, with interbedding of sulfide laminae with silicate- and carbonate-bearing beds in many ore bodies, sedimentary textures such as load clasts and the lateral extent of metal enrichment of the ore horizon beyond the limits of ore (Figure 4.11), suggest exhalation and spreading of the ore fluid over the sea floor. However, ore mineral growth in pore space of unconsolidated sediments is implied by ore textures such as rimming of diagenetic pyrite by base-metal sulfides and by geochemical data such as sulfur isotope ratios. The large variability of sulfur isotope ratios ($\delta^{34}S$) across growth zones of single ore sulfide grains implies grain growth in a chemically closed system in which there was limited chemical exchange with a large reservoir of sulfide, hence most likely in pore space. A possible reconciliation of the conflicting interpretations from the two groups of data is the proposal that the mineralising brine was exhaled, but after exhalation seeped downwards at least locally into soft, unconsolidated sediment (Figure 4.13c). Alternatively, fine-grained ore minerals that precipitated in the water column may have been dissolved and reprecipitated during early diagenesis.

(b) Temperatures and chemical conditions of ore formation

If the fluid was exhaled into the bottom of an ocean-floor trough it may have been ponded in pools of hot, reducing, saline brine on the sea floor similar to those presently active in the mid-ocean ridge environment in the Red Sea (see Section 3.2.7). The estimated ore-fluid temperature of around 200 °C is derived from fluid inclusions in veins and is thus the temperature at or below the point of exhalation and, if there was a brine pool, much of it would have been significantly cooler due to heat loss to the overlying ocean column; by analogy with the Red Sea brine pools, SEDEX pool temperatures may have been at about 70 °C. Much of the sulfide may thus have precipitated at these lower temperatures.

Whatever the setting of ore mineral growth was, a euxinic or anoxic environment at or immediately below the sediment–water interface is indicated by the high organic content of the host sediment unit and this environment may be critical to promote precipitation and deposition of sulfide minerals. Reduced environments are not uncommon in sediments and overlying bottom water in troughs on the ocean floor with restricted water circulation. They form where accumulation of organic matter overwhelms the ability of oxygen to diffuse or advect from higher in the water column. Many routes to sulfide production are possible in reduced sedimentary environments. Sulfide may be produced by reduction of sulfate in solution in the ore fluid by reaction with sedimented organic matter or biogenic gas (Figure 4.14a), be produced as a result of reduction of sedimentary sulfate minerals in a 'reduced pool' of seawater (Figure 4.14b), or be produced from sedimentary sulfate in reduced pore water in unconsolidated sediments (Figure 4.14c). The sulfur isotope data on sulfide minerals suggest that at least a component of the sulfide was produced through reduction of sulfate in solution in pore-space waters.

A possible present day analogue – cold methane seeps

The abundance of barite associated with many ores, the presence of methane inclusions in ore minerals at Red Dog and the formation of the ores at the foot of submarine fault scarps suggest an analogy of the SEDEX ore-forming environment with modern 'cold methane seeps' that are active in a number of continental-margin basins. The seeps typically occur at water depths down to about 1500 m at the base of fault scarps or other steep scarps on the ocean floor of continental margins, with relatively thick sequences of sedimentary cover. They can be sites of high organic productivity, even where below the photic zone. The methane in the fluids at these seeps indicates input of organically derived gases or interaction between the fluids and sedimentary organic matter, but it is unclear whether the seeping fluids are derived through compaction and organic matter maturation or are refluxed from a shelf. The Ba in the fluids is interpreted to be sourced from diagenetic mineral reactions in sediments of the basins. Although there are similarities between the chemistry and settings of these seeps with the exhaling fluids and environments proposed for the formation of SEDEX deposits, neither seeps with elevated temperatures similar to those inferred for exhalation at SEDEX deposits nor seeps of saline waters have yet been recognised.

Fluid flow patterns generating SEDEX deposits

All of the basins that host known SEDEX deposits have been to an extent truncated by later deformation, and reconstruction of the crustal environment over the full extent of the large hydrothermal systems that are inferred is thus difficult. We use indirect evidence to infer the driving forces of ore fluid flow.

The ore fluid likely had a significantly higher salinity than seawater. Although there are evaporites in the host succession below some of the deposits, this is not the case for all, and it is thus more likely that the high salinity developed in a marginal marine environment. The Red Dog deposits formed when Alaska was at low latitudes and evaporitic environments along the margins of the host basin at the time of ore formation are thus expected. If this is the ultimate source of the ore fluids, recharge was thus potentially in a marginal marine environment tens to hundreds of kilometres distant from the narrow, deeper marine troughs that host the ore bodies. The fluid may have leached Pb and Zn distant from the ore body along long fluid flow paths.

If the temperatures of the ore fluids before exhalation were around 200 °C, it is implied that they descended to depths of greater than about 5 km, and hence deeper than the known depth extents of some of the host basins. It is, however, possible that fluid flow occurred during a time of abnormally high geothermal gradients. This is consistent with the presence of mafic intrusive and volcanic rocks in the succession in some of the Phanerozoic host basins (Selwyn Basin, Yukon Territory).

Long distances of fluid flow and high fluid temperatures suggest free convection through the rocks of the sedimentary basin driven largely by high geothermal gradients. However, aquifers along which the fluid flowed have not been identified. In the Proterozoic basins of northern Australia, the aquifers may have been thick terrestrial sandstone units that are present a few kilometres below the ore-hosting strata (e.g. Figure 4.11). Cross-stratal flow from depth to the sea floor at the site of the ore deposits may have been either along trough-bounding faults or adjacent to these faults. Hydrogeological compartmentalisation of the aquifers by steeply dipping normal faults may have focussed up-flow.

4.2.3 Kupferschiefer or red-bed copper deposits

These are generally large, laterally extensive, stratabound Cu deposits, with Co as an important by-product in some deposits, and provide about 30% of the world's supply of Cu. Minor amounts of Ni, Ag, Zn, Pb and U are characteristic, with some ores having greater tonnages of Zn than Cu, although it is uneconomic to mine and extract. Localised, potentially economic, concentrations of Au and platinum-group elements (PGEs) have been discovered in and around some Cu ores. The deposits are hosted in continental rift basins close to stratigraphic boundaries between terrestrial, largely red-bed sandstones, and overlying organic-rich shales of shallow-marine or lacustrine origin. The host sedimentary sequences are thus distinct from those of MVT and SEDEX deposits.

Major ore fields of this type include the Kupferschiefer (= copper shale) in Permo-Triassic rocks in the intracontinental rift Zechstein Basin in southern Poland and central Germany (e.g. Oszczepalski, 1999; Pieczonka *et al.*, 2008), the Zambian Copperbelt deposits of Zambia and adjacent Congo in a late-Proterozoic intracontinental rift basin that has been overprinted by deformation and low- to medium-grade metamorphism (e.g. Selley *et al.*, 2005) and the White Pine deposit (Michigan, USA) in the Mesoproterozoic Midcontinent Rift of the North American shield (White, 1968; Brown, 1971). The deposits in Michigan are hosted in the same basin and stratigraphically just above historically important 'native Cu' ores of metallic copper in basalts, largely in flow-top breccias of lava flows and along faults that cut the lavas.

Characteristics of Kupferschiefer–red-bed-type copper deposits

(a) Geological settings

The major hydrothermal Cu deposits in sedimentary rocks are sited within a few kilometres of the margins of intracontinental basins in which there is both a rift and a sag phase of fill. The rift phases of basin fill are up to a few kilometres thick and are composed of basalt lava flows and a mixture of sedimentary rocks which are at least in part terrestrial sandstones (Figure 4.15). The host-rocks of the ore bodies are at or above the top of this rift phase and include terrestrial sandstones, organic-rich shales and massive dolostones. The make-up of the host sequences indicate that they were deposited at low latitudes in a desert climate, and this inference is confirmed by estimates of palaeolatitude from palaeomagnetism.

The stratigraphic setting of the ore bodies within the host basin is characteristically at or near a flooding surface above red-bed continental sedimentary rocks, which include, in different Cu ore districts, fluviatile conglomerates and sandstones (e.g. White Pine) or desert aeolian sandstones (e.g. Kupferschiefer, Figure 4.16). The flooding surface is marked by underlying dune sandstones and overlying evaporites and dolostones in the case of the Kupferschiefer. The black shales marking flooding are laterally extensive over hundreds of kilometres and may have formed in either a lacustrine or a lagoonal marine-margin setting.

In the case of both the Kupferschiefer deposits and those of the Zambian Copperbelt, individual deposits extend up to 20 km along strike, and multiple deposits are distributed at the same horizon over hundreds of kilometres of strike distance (Figure 4.15). Extensive exploration in the Zechstein Basin that hosts the Kupferschiefer deposits has shown that the ore bodies do not extend as far down dip and the deposits can thus be inferred to have formed along a trend parallel to the palaeoshoreline.

In contrast to SEDEX and some MVT deposits, red-bed Cu deposits are not closely associated with growth faults or lateral facies variations of the host strata, although both the Kupferschiefer deposits and Zambian deposits are in strata that were deposited within about 10–20 km of the basin margin at the time of sedimentation.

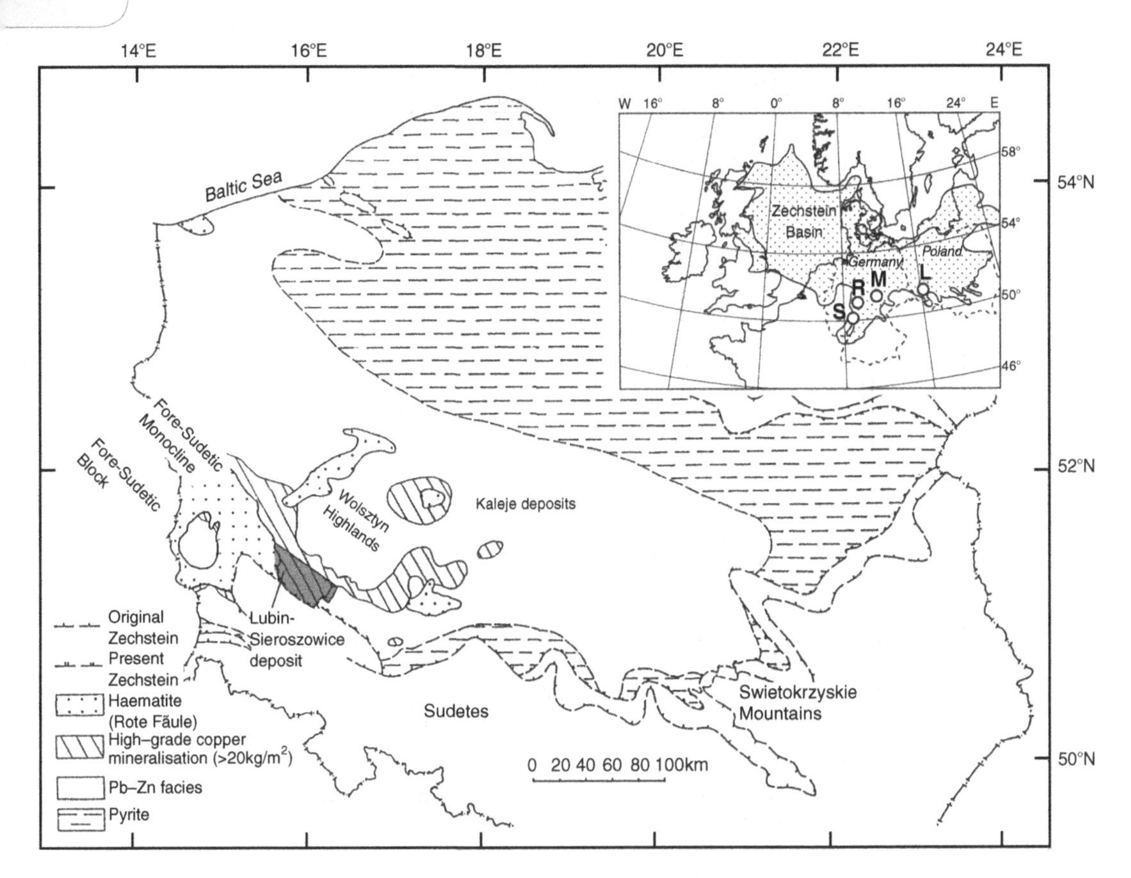

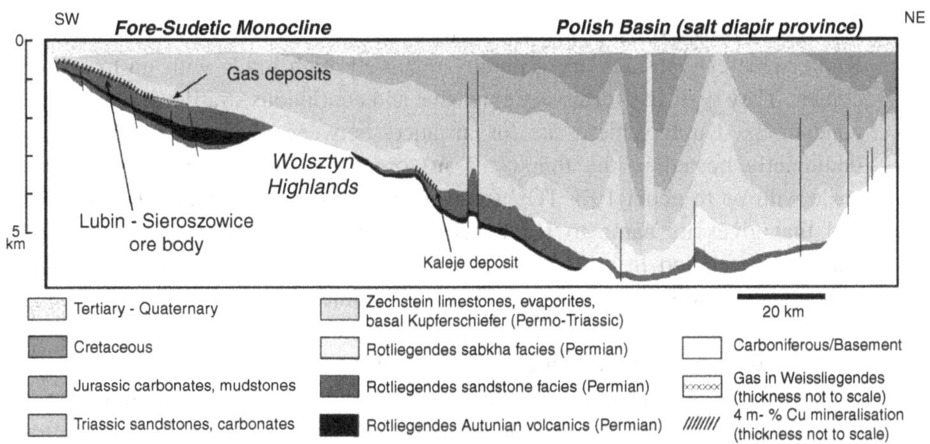

Figure 4.15 Map and cross section of the Permo-Triassic Zechstein Basin of northern Europe and settings of copper ores in the basin. The inset map shows the main ore fields along the southern margin of the basin (L = Lubin, M = Mansfeld, R = Richelsdorf, S = Sangerhausen). The Kupferschiefer shale is continuous as a thin layer across the basin and is the main Cu ore, but is mineralised only near the southern margin of the basin, and the various alteration facies at different positions in the basin and their relations to ore are shown in the main map (Oszczepalski, 1999). The cross section from the mineralised southwest margin of the basin into the centre (Hitzman *et al.*, 2005), shows the position of the Cu-ores at the top of red-bed sandstones, and the local accumulation of gas at the same horizon.

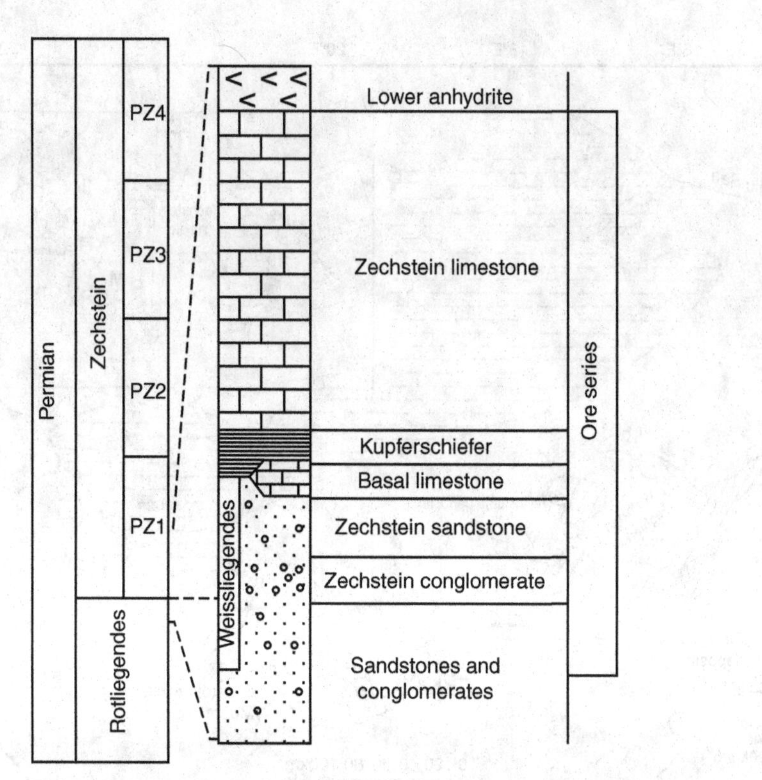

Figure 4.16 Detailed stratigraphic section across the Kupferschiefer ore zones of the Zechstein Basin (Oszczepalski, 1999).

(b) Nature of the ore

Kupferschiefer-type ores are generally relatively high grade with up to a few per cent copper. They typically form very extensive and continuous stratiform thin sheets. The ore zone in the Kupferschiefer is, for instance, between about 1 and 5 m thick and is dominantly hosted in the thin (< 1 m) organic-rich (bituminous) carbonate-bearing shale with up to about 10% TOC (total organic carbon) that marks the flooding surface and that gives its name to Kupferschiefer. The nature of the organic matter in the Kupferschiefer and in other ore-hosting shales and sandstones is varied and includes disseminated indigenous kerogen, flecks of bitumen in pore space which can be interpreted as degraded oil, bitumen and oil as inclusions in veins, and more rarely also accumulations of hydrocarbon gas and live oil.

Although ore bodies in the Kupferschiefer districts are apparently stratiform within the shale, in actuality they cross-cut units at a low angle over distances of kilometres or more (Figures 4.17 and 4.18), and 'move out' of the Kupferschiefer shale to a few metres above or below into overlying dolostone and underlying sandstone. Where the ore bodies are higher or lower in the sequence they are typically relatively wide and of low grade (Figure 4.17). At White Pine the ore is in the lowest few metres of a thicker grey to black shale-dominated unit but similarly extends into the upper metre of the underlying fluviatile sandstones. In the Zambian Copperbelt deposits are in part in organic-rich sandstone and are unusually up to 30 m thick and likewise transgress the stratigraphy at

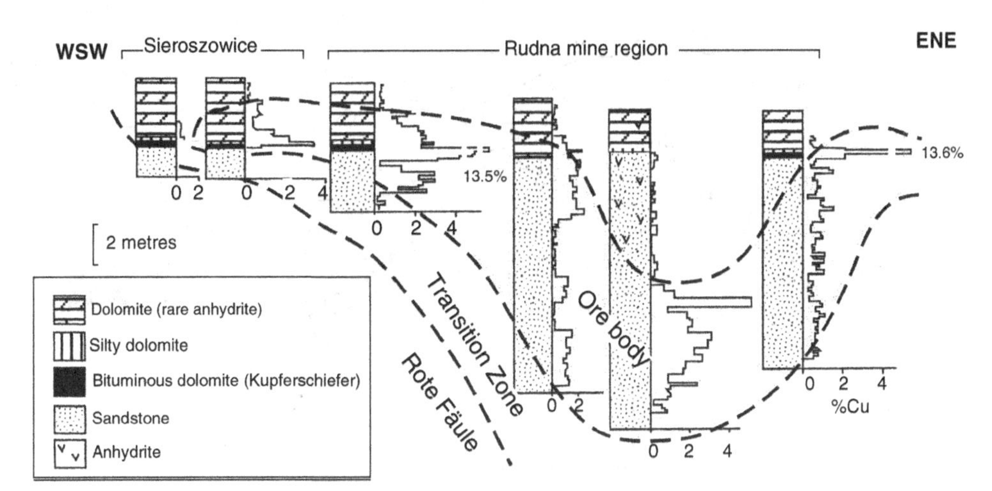

Figure 4.17 Position and grade of copper ore in profiles through the Kupferschiefer in a section through the deposits at Lubin (see Figure 4.15). Note the high grades of the ore where it is hosted in the organic-rich shale (bituminous dolomite), and generally lower grades and thicker ore zones in sandstone (Wodzicki and Piestrzynski, 1994).

low angles. The sites of some ore bodies resemble oil traps, for instance the upper few metres of sandstone units in open anticlines.

There are typically multiple Cu ore minerals in the ore bodies, including in different deposits, native copper, chalcocite, bornite and chalcopyrite in addition to pyrite and minor sphalerite and galena. Ore minerals are disseminated as fine grains in the host shale and as fill to small-scale (centimetres) fractures and veins, including bedding-parallel fractures in finely laminated shales. Ore is also disseminated in the coarser-grained host units and the ore mineral textures show that the ore minerals precipitated in primary pore space between clasts. A significant proportion of the copper minerals are hosted in small-scale fractures and faults. These may have precipitated as a result of remobilisation of earlier disseminated ore or during hydrothermal events at times of basin inversion and the formation of steep reverse faults in and adjacent to the ore bodies.

There is characteristically zonation of the major copper ore minerals laterally and vertically through the thin, sheet-like ore bodies. In the Kupferschiefer ores the zonation from base to top is chalcocite (Cu_2S) → bornite (Cu_5FeS_4) → chalcopyrite ($CuFeS_2$) → sphalerite and galena → pyrite (FeS_2). At White Pine, zonation is vertically from native copper at the base of the ore zone through chalcocite to minor bornite and then pyrite. The horizontal direction of ore mineral zonation is generally such that chalcocite is nearest to the margin of the basin and to the palaeoshoreline.

There is also zonation of alteration of the host sedimentary rocks. In the Kupferschiefer deposits, the host shales are black, organic-rich in the ore and on the pyrite side of the ore bodies, but are red (Rote Fäule) because of the presence of fine-grained disseminated haematite over areas of many kilometres of extent on the chalcocite side of the ore zone (Figure 4.18). Within the ore bodies the kerogenous organic matter of the black shale has been chemically altered and is oxidised, and has a higher O:C ratio and a lower H:C ratio than typical organic matter. Organic matter has been completely oxidised and dissolved from the red shales below the ore bodies.

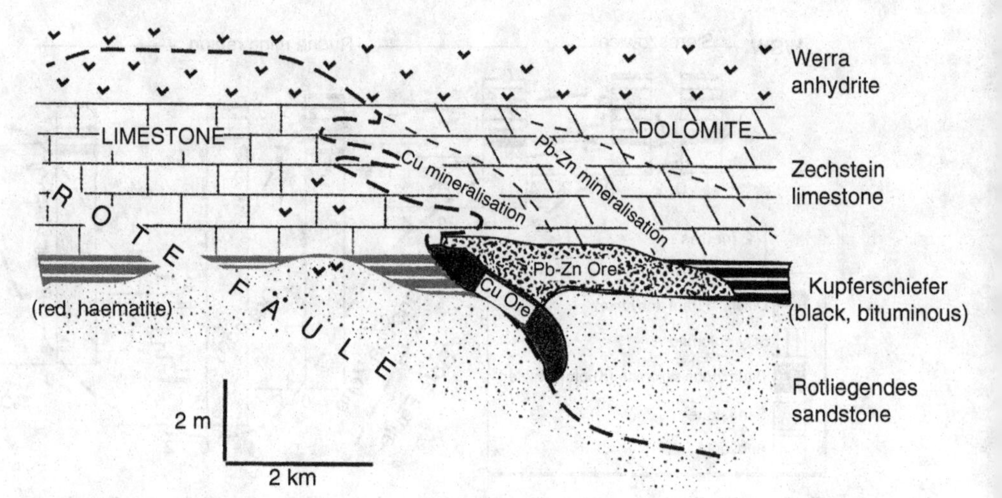

Figure 4.18 Schematic section through the Kupferschiefer with extreme vertical exaggeration showing vertical and lateral ore and alteration zonation, including Cu and Pb–Zn ores, and low-grade mineralisation. Rote Fäule is a red-coloured facies of shale and sandstone. Note the low angle discordance of ore to sedimentary units. The margin of the sedimentary basin is to the left of the figure (Rentzsch, 1974).

Interpretation of environments and causes of Cu precipitation at Kupferschiefer-type deposits

The low-angle, slightly transgressive form of the ore bodies demonstrates an epigenetic timing of mineralisation. The textural setting of ore minerals in the interstitial spaces of coarser-grained host-rocks shows that the major phase of mineralisation was syn-diagenetic. Radiometric age dating has indicated an age of mineralisation tens of millions of years after sedimentation in the case of the Kupferschiefer deposits. Depths of mineralisation between about 1 and 3 km are thus inferred based on this timing and similar relative timings are inferred in other fields. Temperatures of mineralisation are inferred to be 100–150 °C based on fluid inclusions, species of clay minerals and the degree of maturation of organic matter in the host shales. A saline ore fluid is preserved in fluid inclusions, but the composition of the fluid is poorly constrained in some deposits.

The critical common feature at the sites of ore deposition is the redox boundary between oxidised and reduced sedimentary rocks. The lateral and vertical zonal distribution of ore minerals, gangue minerals and the chemistry and distribution of organic matter in the host sediments shows that ore straddles this redox boundary. The ore and alteration minerals indicate increasing degrees of reduction from the haematite facies, which lacks organic matter, through the zones of the ore of relatively oxidised minerals (e.g. chalcocite) to relatively reduced minerals (chalcopyrite) to pyrite + bitumen-bearing shale. The ore fluid entered the site of the deposit as an oxidised fluid and became reduced on interaction with organic matter, which was itself largely oxidised to CO_2. Ore precipitation at all Kupferschiefer-type deposits is thus a result of flow of oxidised saline ore fluids with high concentrations of copper that have been leached from basin rocks into contact with reduced sedimentary rocks.

Different sources of sulfide contributed to ore mineral precipitation at the different deposits. A local source of sulfate is present, for instance, in the immediately overlying

Werra anhydrite at the Kupferschiefer deposits, and reduction of sulfate to sulfide in solution may thus be a contributing factor in the precipitation of the ore minerals. In contrast, there is no major accumulation of sulfate in the host sequence near the White Pine deposits. The ores at White Pine are sulfur-poor with native copper as a major ore mineral, and the sulfide incorporated in chalcocite and bornite was most likely derived by dissolution of pyrite from the host shale by the oxidised ore fluids.

Interpretations of fluid flow at Kupferschiefer deposits

In both the Zechstein Basin and at White Pine, ore fluid flow was syn-diagenetic at temperatures of order of 100 °C at up to about 40 million years after sedimentation. The timing implies mineralisation after burial of the host-rocks to greater than about 1 km depth by sediments of the sag phase of fill of the basin. The great lateral extent of the mineralisation at one stratigraphic horizon implies a regional event of fluid flow. The position of deposits and their distribution strung out parallel to the basin margin and around basement horsts (Figure 4.15) suggests that the geometry of the strata below the ore horizon controlled ore fluid pathways.

One possible case of ore fluid flow is fluid expulsion from deeper levels in the basin centre during basin compaction (Figure 4.19). Expelled fluids would have been oxidised because they were derived largely from red-bed sandstone sequences and contained dissolved sulfate, and on compaction would have migrated upward and towards the basin margin where they interacted with the ore shale horizon. In the scheme shown in Figure 4.19, the ore fluids are redirected down dip along strata immediately beneath the poorly permeable shale so as to form the extensive sheets of near-stratabound ore. The presence of ore along fractures and veins in the shales shows that the shales may have been overpressured at the time of mineralisation. Overpressure, if it was present, would have favoured strata-parallel flow of fluid below the unit. There are gas accumulations at this horizon up and down dip of the Kupferschiefer deposits, which, in view of the stratigraphic setting, were most likely derived by maturation of organic matter in the shales and downwards gas migration.

Basin compaction provides limited volumes of fluid, and mass-balance calculations show that although there would have been sufficient Cu in the dewatering sedimentary rocks to provide the mass of Cu in the ores, fluids derived only from the underlying basin rocks would need to have Cu concentrations higher than has been recorded in basinal brines. The chemical requirements for ore formation would be easier to meet if there was a larger volume of water. This larger volume could be recharged basinal waters that flowed across the host basin through the sandstone units underlying the shales, in which case fluid flow would have been most likely driven by topographic gradients.

4.3 Uranium deposits in sedimentary basins

The majority of uranium resources are ores formed by syn-diagenetic migration of low-temperature waters through sedimentary basins, in particular through high-permeability sandstones in intracontinental basins. Many of these basins are largely or totally terrestrial. Different types of deposit are distinguished on the basis of sedimentological setting of the host-rocks, ore-body form and inferred relative timing, depth and temperature of

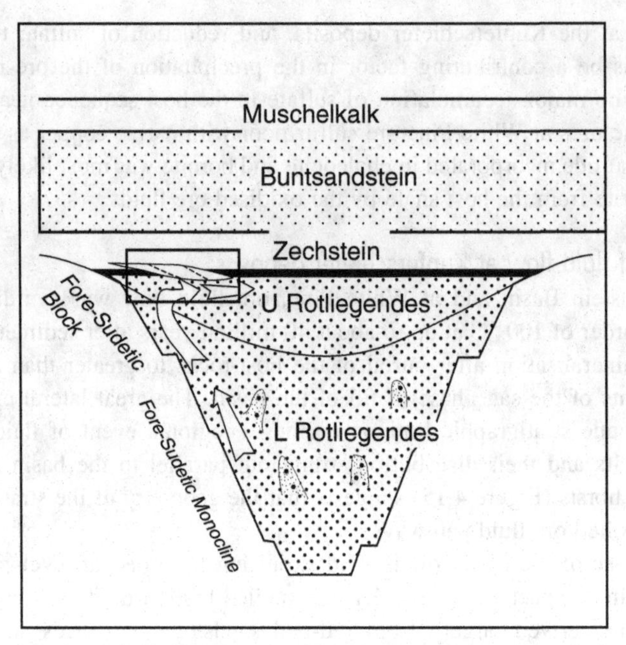

Figure 4.19 Vertically exaggerated cross section through the southwest part of the Zechstein Basin (Fore-Sudetic Monocline of Figure 4.15) at the time of ore formation, with interpretation of the source and fluid flow patterns in the Kupferschiefer. The box shows the area shown in Figure 4.18. In this interpretation the ore fluids are interpreted to be connate waters expelled from the underlying red-bed sandstones during compaction. The ore mineral zonation indicates fluid movement towards the basin centre and stratigraphically up through the sequence. At the scale of a deposit, the fluid flow would have been near parallel to the sedimentary units because the Kupferschiefer shales would have been an aquitard (Jowett, 1992). However, mass-balance calculations based on the volume of rock and tonnages of ore in the Lubin Basin of Poland show that the ore fluid would have needed an unreasonable minimum of about 120 ppm Cu in solution to deposit the Cu in the Kupferschiefer (Cathles, 1993). An alternative proposal for ore fluid flow invokes similar fluid flow patterns in the vicinity of the deposits but topographically driven recharge of larger volumes of fluid from an opposing margin of the basin (e.g. Brown, 2006).

mineralisation. Commonalities among the deposit types include that ores are all epigenetic, fine-grained precipitates of uranium(IV) oxide and silicate minerals in pore space or as replacements of earlier minerals, and are all related to leaching of uranium from rocks in the host basin, transport by basinal waters and precipitation in the ore bodies. In addition, all ores are associated with distinct diagenetic imprints in surrounding rock. Fluid reduction on interaction with organic matter is a common critical control on ore mineral precipitation.

Geochemistry of uranium

Uranium is lithophile. Its average crustal abundance is about 1.5 ppm with highest concentrations in felsic rock types. In normal rock types uranium is dominantly present as a trace element in a number of accessory minerals, including the phosphate minerals apatite and xenotime. Common ore minerals include: carnotite, $K_2(UO_2)_2(VO_4)_2{\cdot}3H_2O$;

uraninite (or pitchblende), UO_2 or poorly crystalline mixed uranous-uranyl oxide (~ U_3O_8); and coffinite, $U(SiO_4)_{1-x}(OH)_{4x}$.

Uranium has two dominant natural oxidation states, uranous – U(IV) (relatively reduced) – and uranyl – U(VI) (oxidised). A number of uranyl complexes are strongly soluble in oxic water, hence in waters that have equilibrated at surface conditions, for instance uranyl carbonate (e.g. UO_2CO_3) and uranyl chloride (e.g. $UO_2(Cl_2)$). The solubility of U decreases strongly with reduction in conditions in which uranous species are dominant, and is many orders of magnitude lower under conditions at which sulfides and organic carbon are stable.

Uranium is strongly leached from minerals during weathering because of its high solubility in oxidised waters. Shallow groundwaters in granites, for instance, can have U concentrations of up to 20 ppb. Grades of uranium ores are typically 0.1–0.5% U_3O_8, which is thus an enrichment of greater than 100 times over typical granite concentrations of 5–10 ppm. A few ore bodies have pockets or zones of significantly higher grades of up to about 20%.

Types of sedimentary-rock-hosted uranium deposits

The different types of uranium deposits in sedimentary rocks and sediments are:

(a) *Unconformity-related uranium deposits.* These are irregular bodies of ore straddling or just below an unconformity at the base of an intracontinental sedimentary basin and are hosted either in basal sandstones of the basin or in basement rock. Ore precipitation is inferred to be epigenetic at relatively high basinal temperatures of around 120–200 °C.

(b) *Tabular uranium deposits.* These are extensive and relatively thin, flat-lying ore bodies within fluvial arkosic sandstones in interior continental basins. They formed at relatively low temperatures (less than 80 °C, and possibly as low as about 40 °C) relatively soon after sedimentation and during early diagenesis.

(c) *Roll-front uranium deposits.* These are similarly low-temperature ores hosted in fluvial arkosic sandstones in interior continental basins, most characteristically in sandstones interbedded with finer-grained sedimentary rocks. The ore bodies have the distinct shape of a stratabound, relatively narrow ribbon with a cross-sectional shape of a crescent or a ‘u’ on its side.

Many tabular and roll-front deposits are also ores for vanadium, and some were mined for vanadium from carnotite, for instance, and others for radium, before uranium first became an economic commodity during the 1940s. Some of the lower-grade ores of these two types are of economic interest because they are amenable to *in-situ* leach mining in which solvents (e.g. sulfuric acid or sodium carbonate) are pumped through the ore zones to extract uranium into solution without physical mining of the ore.

4.3.1 Unconformity-related uranium deposits

These are irregular coffinite- and uraninite-bearing ore bodies that are generally sited just below or straddling a major unconformity at the base of an intracratonic sedimentary basin (Figure 3.63). Important examples are all in Palaeoproterozoic rocks with the major fields being in the approximately 1800–1550 Ma Athabasca Basin of Saskatchewan,

Canada (Jefferson *et al.*, 2007) and in a sequence of stacked, slightly older basins in the Pine Creek inlier of Northern Territory, Australia (Coronation Hill, Jabiluka, etc.; Solomon and Groves, 2000c). The deposit class includes some very high-grade ores with up to about 20% U_3O_8 (e.g. Cigar Lake, Saskatchewan) and these high-grade deposits are the largest accumulations of uranium in sedimentary-rock-hosted deposits. The Oklo deposits of Gabon, which were the sites of natural nuclear reactors after high-grade uranium accumulation, are in sandstones in a similar intracratonic basin, but the ores are stratabound well above the basal unconformity (Gauthier-Lafaye and Weber, 1989). Some examples in both Canada and Australia are polymetallic with variable amounts of arsenide and sulfide minerals with anomalous concentrations of base metals Ni, Co, Cu, Pb, Zn and Mo and precious metals Au, Ag and PGEs. Some deposits have been exploited in particular for Au as a co-product.

Nature and setting of ores

These deposits are irregular elongate pods up to tens to hundreds of metres across of disseminated to locally massive uranium minerals variably along the unconformity or as approximately stratabound replacement ore bodies in moderately to steeply dipping units in the underlying basement. The basal strata of the basins form the hanging-wall to the ore bodies and are in all cases comprised of a few-hundred-metre-thick succession of terrestrial, poorly cemented and porous, largely fluvial quartz sandstones with thin conglomerate horizons. Felsic and mafic volcanic rocks are present higher in the host sequences in the Pine Creek inlier, but not in the Athabasca Basin. Foot-wall basement rocks can be metamorphic, volcanic or sedimentary. In all cases, however, carbonaceous rocks are present in the basement in close association with ore. Ore bodies are often elongate along or adjacent to faults that cut the unconformity.

The deposits are characterised by distinctive patterns of alteration in ore and in surrounding rock (Figure 4.20). The host-rocks to ore are desilicified and are now clay-rich or chlorite–sericite-rich. The immediately overlying rocks are silica-rich, forming a silicified cap. A pipe-like 'friable zone' in which cement has been dissolved from quartz sandstones occurs above many ore bodies.

The ores are difficult to date, but radiometric dates on uraninite indicate that mineralisation took place tens to hundreds of million years after sedimentation, and hence in all cases dominantly in the Palaeo- or Mesoproterozoic. Based on fluid-inclusion data, temperatures of mineralisation are estimated to be about 120–200 °C. Depths of mineralisation are poorly constrained but from reconstructions of the sedimentary-basin sequence are most likely of order 2–4 km.

Genesis of ores

The aquifers for hydrothermal fluid flow at these deposits were the basal sandstones of the basins, but flow may have extended into underlying basement, particularly through weathered rock within a few tens of metres below the unconformity and along fractures. The relatively high temperatures of mineralisation and the timing of up to a few hundred million years after sedimentation suggest that flow of heated fluids took place after significant burial and that high heat flow in the host basins may be a critical factor forcing fluid flow by driving convection of groundwater in the host sandstones of the basin (Figure 4.20). The strong alteration in a pipe above the ore bodies indicates upward flow

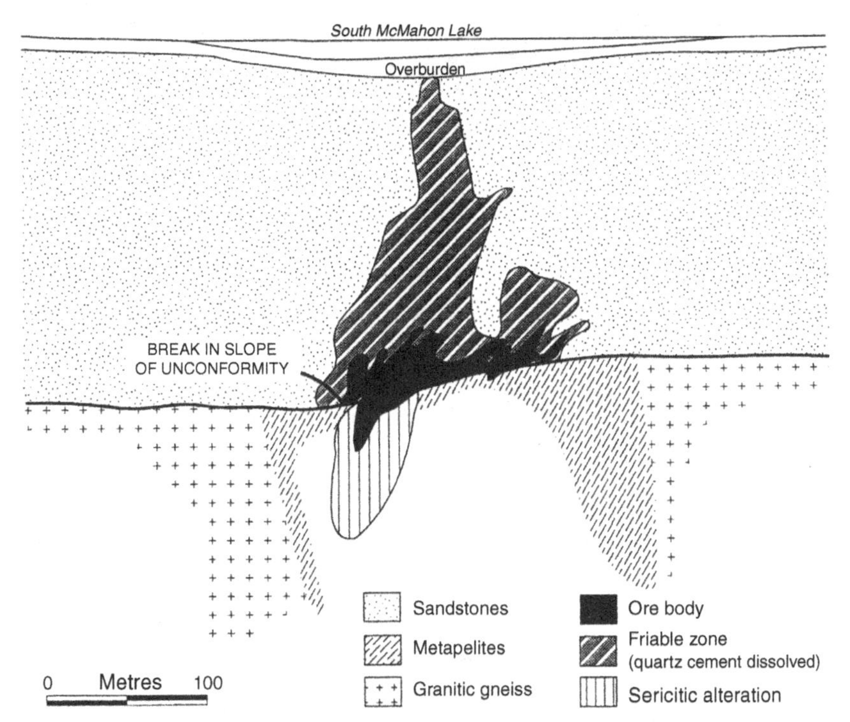

Figure 4.20 Typical setting of a Proterozoic unconformity-related U deposit, specifically the Midwest deposit, Athabasca Basin, Saskatchewan (Ayres *et al.*, 1983).

of heated convecting water above the deposits. If fluid flow is as shown in Figure 4.21, these pipes thus indicate that the deposits formed at sites of flow convergence within convection cells. Whether ores formed straddling or underlying the unconformity may reflect a vertical component of fluid flux at the site of mineralisation.

Oxidised groundwater will leach U from minerals in aquifer rocks. There is petrographic evidence for this from dissolution of apatite, xenotime and replacement of heavy-mineral grains in sandstones in the host basin, for instance those in heavy-mineral laminae along bedding surfaces in sandstone. The U will precipitate if the fluid interacts with organic-rich reduced rocks. In the flow scheme of Figure 4.21, reduced rocks are graphitic schists that are truncated by the unconformity. Chemical interaction with reductants may be either on fluid flow through the graphitic shales and schists or on interaction with a flux of CH_4 from below the unconformity and derived by maturation of the organic matter in shales.

Although similar geochemical and hydrogeological settings developed later in Earth's history, unconformity-hosted uranium deposits are restricted to rocks of early-Proterozoic age (2.4 to 1.6 Ga). The reasons for their development at this stage in Earth's history are uncertain although are likely related to the changes in atmosphere and surface water chemistry at the Great Oxidation Event (GOE) (see Box 5.1). The time of ore formation is from and up to several hundred million years after the GOE, and the first step of oxidation of the atmosphere. The early Proterozoic may have been a time in which terrestrial sandstones contained high concentrations of readily leachable detrital U-bearing minerals that had been eroded from rocks such as granites and felsic volcanic rocks, including

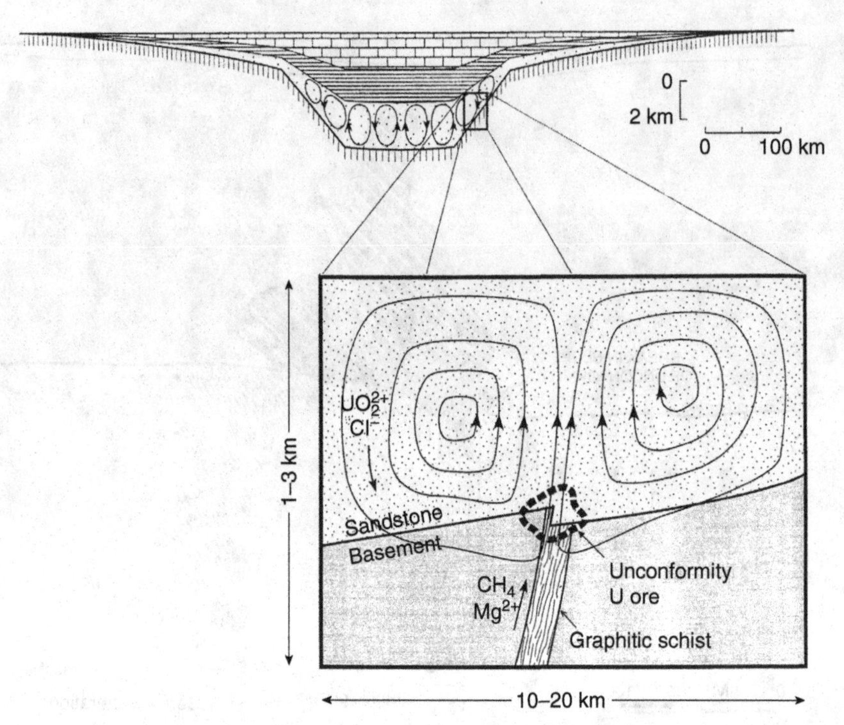

Figure 4.21 Conceptual hydrogeological model for the formation of an unconformity-related U deposit involving free convection in the thick permeable sandstone units above the unconformity, driven by regional high heat flow with position of the convection cells controlled by irregularities such as fault steps on the unconformity surface (Raffensperger and Garven, 1995).

uraninite, whereas in more modern environments these minerals do not survive in weathering and terrestrial sedimentary environments.

4.3.2 Tabular uranium deposits

These are tabular and flat-lying, generally relatively low-grade ores in which fine-grained coffinite, uraninite and uranium complexed in amorphous organic matter have precipitated in pore spaces of fluvial sandstones. Carnotite and vanadium chlorite and oxide minerals are intergrown with the uranium oxide minerals and are the main hosts of vanadium in these ores. Ore bodies occur scattered over a few tens of kilometres in clusters and trends in the host basins. The important ores in the Colorado Plateau region of southwestern USA (Grants-Laguna region, New Mexico; Uravan region, Colorado etc., e.g. Northrop and Goldhaber, 1990) are of Triassic and Upper Jurassic age. Other important deposits in the world which are probably of this type include those in the Arlit area of Niger in Carboniferous to Jurassic rocks, and those in late-Cenozoic rocks in the Frome Embayment of the Great Artesian Basin in South Australia.

The deposits in the Triassic Chinle Formation and the Upper Jurassic Morrison Formation in the San Juan and Paradox basins of the Colorado Plateau have been described and investigated in detail and the description here is based on the characteristics of these ores.

Setting and nature of ores

The host basins developed in the interiors of continents, although there are marine sedimentary units deeper in the sequences. The basins developed in arid climatic settings in which water was input as both river water and groundwater from adjacent highlands. One characteristic of the host sequence is a significant presence of vitric felsic tuff, input into the basin as, for instance, ash fall. At the stratigraphic level of the deposits the rocks are fluvial sandstones and mudstones, with lacustrine deposits in the basin depo-centre (Figure 4.22). That the lake in the basin centre was an alkaline and saline playa lake is indicated by diagenetic minerals characteristic of alkaline conditions in the sandstones, with K-feldspar, analcime, clinoptilolite and smectite in zones outwards from the site of the lake. The clustering of deposits is along trends parallel to and approximately following the line of the lake shore at one stage in its development and also in the position of the change in palaeoslope and boundary from alluvial fan to playa-lake deposits. Later burial has caused at least localised heating and overprinting of the diagenetic features that formed during the lifetime of the playa lake and circulation of associated saline and alkaline groundwaters.

The ore bodies are essentially flat lying and tabular bodies of disseminated uranium minerals with dimensions of up to about 1 km in plan view and up to a few metres thick. They generally appear parallel to the bedding of the host, but can cross-cut units at a low angle. Many of the deposits, and also the mineralised horizons that are traceable along strike of the ore bodies, cut up section towards the basin centre (Figure 4.22). The deposits occur within thick arkosic sandstone units but there is no clear stratigraphic control on their exact siting except for a general relationship with coarser-grained, channel-fill facies. Some ores occur as vertically stacked bodies a few tens of metres apart in the same sandstone unit.

The ore minerals precipitated early in the sequence of diagenetic infilling of pore space. Coffinite, for instance, coats detrital quartz grains and is intergrown with V-chlorite. The coffinite is in many deposits green to black in colour as a result of included amorphous organic matter. There are distinct styles and mineralogies of diagenesis in the rocks around the deposits, both above and below and also along strike and down dip in the ore horizon. Diagenetic dolomite is important above and below ore, but carbonates are rare in the ore zone. Authigenic kaolinite and smectite are also characteristic of rocks surrounding ore. The ore trends are marked by low magnetic susceptibility as a result of partial dissolution of iron from detrital ilmenite, particularly above the ore zones.

Although plant debris is a widespread component of the host sandstones, there is not a close relationship between ore and this material. Rather, many but not all of the ores are closely associated with concentrations of up to about 3% by mode of a distinct form of organic matter. This is amorphous solid organic matter that lines pore spaces and that lacks a microscopic cellular structure that would be indicative of indigenous kerogen. The chemistry of this organic matter, including its high oxygen and low hydrogen contents, indicate that it is formed of humic substances, or 'humates', that precipitated from humic acids that were in solution in groundwaters. Humic acids are derived from decaying plant matter in soils and in sediments at temperatures below those of the oil window ($< 80\ ^\circ C$) and break down at higher temperatures.

Genesis of tabular uranium ores

The timing of ore precipitation during early diagenesis, the tabular form of the deposits and their tendency to cross stratigraphic units at a low angle imply ore formation either

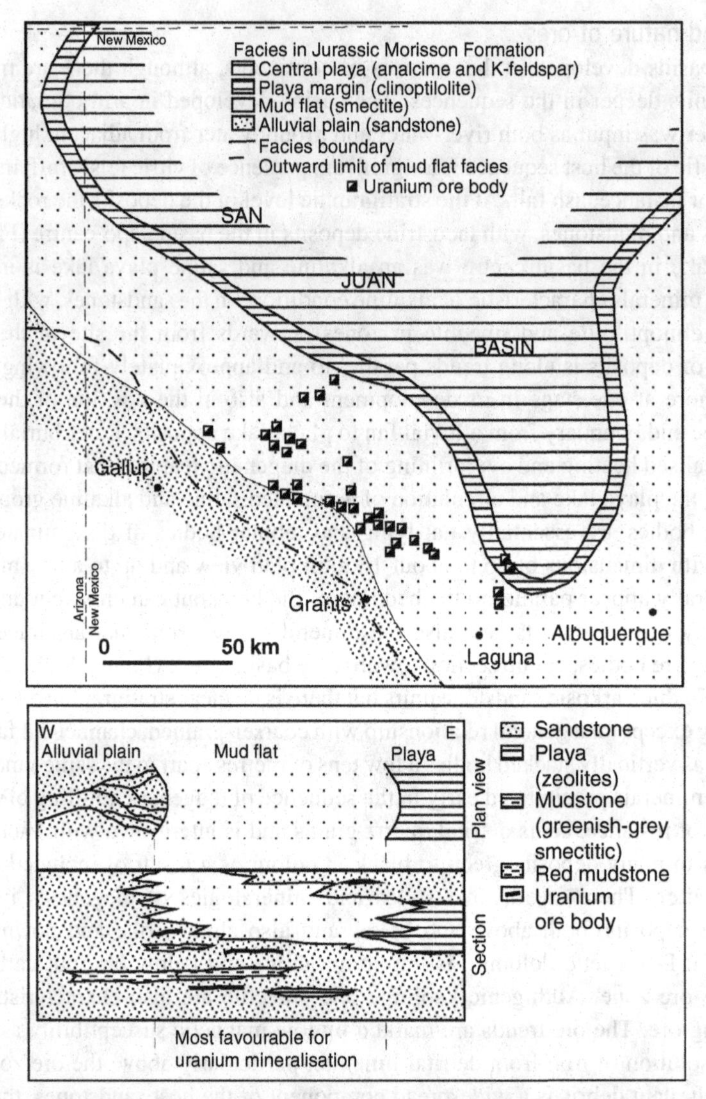

Figure 4.22 (a) Map of the distribution of facies and uranium deposits in the Jurassic Morrison Formation in the Grants–Laguna area of the southwest part of the intracontinental San Juan Basin, New Mexico, USA. Note the concentration of deposits in the area of the mud-flat facies along a broad trend parallel to the palaeoshoreline of the playa lake. (b) Schematic palaeogeography of facies in the Morrison Formation and section showing the position of tabular uranium deposits in sandstones beneath or about the mud-flat facies (after Turner-Peterson, 1985). Reprinted by permission of the American Association of Petroleum Geologists.

at the water table or at a gravitationally stable interface between groundwaters of contrasting composition, for instance between a dilute water above and a saline water below. Many aspects of the geochemistry and diagenesis of the host sediments support the latter interpretation.

Calculations of groundwater flow paths in the basins at the time of deposition of the playa lake sediments and comparison with similar modern intracontinental basins with

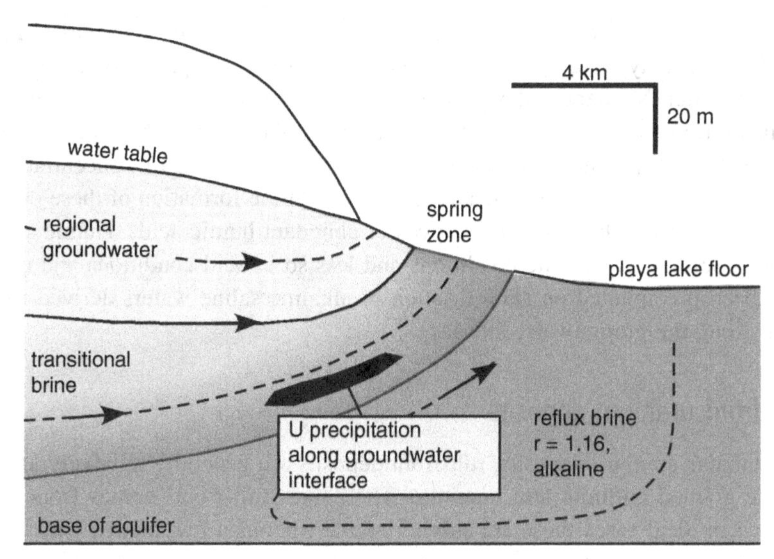

Figure 4.23 Possible groundwater flow paths during the genesis of a tabular U deposit. The groundwater flow is based on observations at the saline Lake Tyrrell, Victoria, Australia (Macumber, 1992) where evaporative concentration of inflowing surface and spring water ensures the production of dense saline brines that reflux downwards and outwards from below the lake basin. Uranium precipitation is assumed to be along the interface between saline reflux brines and low-salinity shallow regional groundwater. Lake Tyrrell is marginally acidic rather than alkaline as inferred for the playa lake of the San Juan Basin (see Figure 4.22), but the groundwater flow patterns and the position of the interface will be independent of this difference and the patterns at Lake Tyrrell comparable to those inferred in the San Juan Basin (Sanford, 1994).

enclosed evaporite lakes such as in southern Australia allow interpretation of the fluid flow paths that may have given rise to ore precipitation (Figure 4.23). The sub-horizontal interface along which ore precipitation took place can be inferred to have been between deep saline waters that were derived either from the basin-centre playa lake or through dewatering of underlying marine sedimentary rocks, and overlying near-surface dilute waters that were flowing down the topographic gradient towards the basin centre. Discharge in these hydrological settings is where the interface intersects the land surface and is typically at the shore of the lake. These inferences, in combination with the stratigraphic positions and shapes of the ore bodies, suggest that they formed at depths of probably less than about 200 m below lake-shore mud flats or adjacent sediment fans and a few kilometres up the topographic slope from the lake shore.

Uranium may be leached from weathering of clasts of vitric felsic tuff that are a significant component of the sediments of the basins, but adjacent basement includes U-rich rocks and these may thus be an additional uranium source. Uranium is expected to be carried in an oxidising groundwater, and hence in waters above a groundwater interface, and to be precipitated on interaction with a reductant. In these deposits the reductant was most likely the concentration of humic substances that had previously or simultaneously precipitated at the groundwater interface. High concentrations of organic matter are not present in all ore bodies, but temperatures during later burial of greater than about 80 °C may have at least in part destroyed a pore-space accumulation of humic organic matter. Methane and other hydrocarbon gases released from the humic substances on maturation may have induced the diagenetic replacement of Fe–Ti oxide minerals that is commonly apparent in the overlying strata. Precipitation of concentrations of pore-space humic substances may be a critical factor in the formation of these ores. Mud flats and shallow saline lakes can be sources of abundant humic acids. Humic substances are strongly soluble in alkaline conditions and less so in acid conditions and may therefore have been precipitated on neutralisation of alkaline saline waters derived from the lake basin along the groundwater interface.

4.3.3 Roll-front uranium deposits

Like tabular uranium deposits, roll-front deposits are generally relatively low-grade ores of fine-grained coffinite and uraninite. The name 'roll-front' comes from the ore-body shape – in ideal cases these are sub-horizontal ribbons a few metres wide which have a crescent shape in cross section. The roll-front is hosted in sandstone beds between less-permeable sedimentary units (Figure 4.24) in fluvial or littoral (strandline and deltaic) sedimentary sequences. In reality the ore-body shapes are not all horizontal 'u's, and many ores are more irregular stratabound lenses. Stacked lenses may be present in interbedded sequences of lenses of fine and coarse-grained fluviatile sediments as are typical of braided river depositional systems.

Roll-front uranium ores are mostly of Mesozoic and Cenozoic age. They have been an important uranium source in the USA, and include some of the ores in Triassic and Jurassic sedimentary rocks of the Colorado Plateau and also the ores in Cenozoic continental sedimentary rocks in Wyoming and inland of the Gulf of Mexico coast of southern Texas.

Nature and setting of the deposits

The ore minerals are dominantly fine-grained precipitates in pore spaces of sandstones and as coatings and replacements of fragments of sedimentary plant matter. The deposits are marked by a change in redox state of the host sedimentary unit across the ore bodies, with 'grey', reducing facies with organic matter and some disseminated pyrite ahead of the crescent and 'yellow', oxidising facies with disseminated goethite and other ferric oxy-hydroxide minerals behind the ore zones (Figure 4.24). Calcite cement can be more abundant in the sandstones adjacent to the ore bodies, but there are no other marked changes in mineralogy associated with the ore zones.

The host sedimentary basins in Wyoming and Texas have neither been buried nor significantly tilted since deposition and the host strata remain close to their original orientation. Ore formation can thus be determined to be generally within a few hundred metres of the palaeosurface at temperatures of at most about 50 °C, and a few kilometres down stratal dip of the original depositional edge of the host units.

Genesis of the deposits

The sites and forms of roll-front ore bodies within and extending across aquifer units indicates that mineralisation was the result of pervasive flow of groundwater in pore space

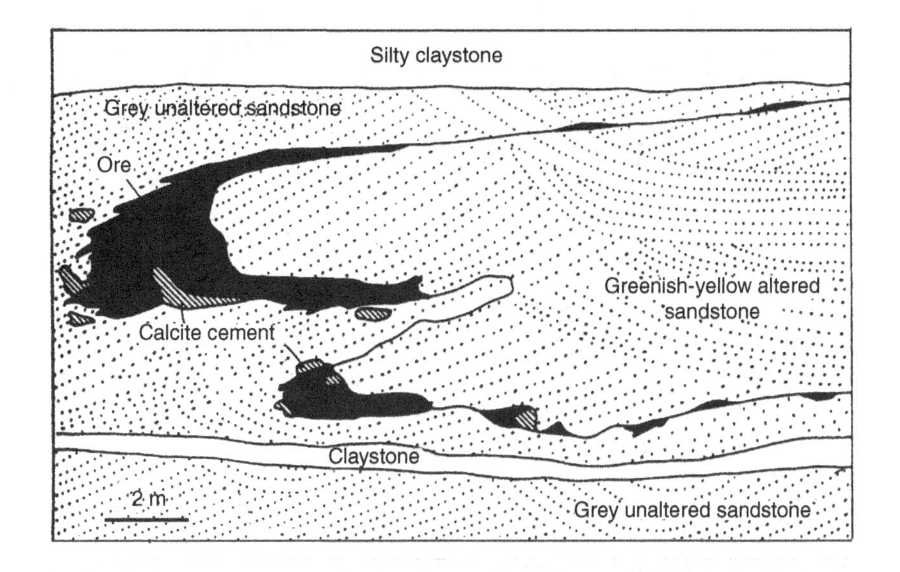

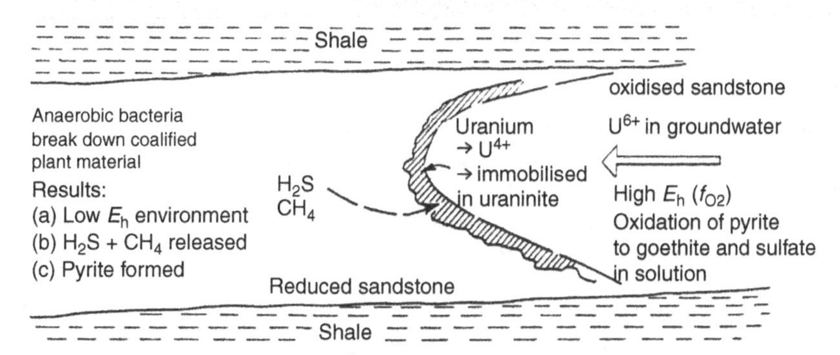

Figure 4.24 Typical setting, shape and petrographic characteristics of a roll-front U deposit (Harshman, 1972), as based on examples from the Shirley Basin, Wyoming. The lower panel summarises the chemical processes of formation and migration of a roll-front deposit (after Rackley, 1976).

through the host units. The exact timing of mineralisation is in most cases difficult to determine, and may potentially be several tens of million years after deposition of the host sedimentary rocks. Groundwater flow paths through the aquifers would have been parallel to the strata and approximately horizontal, with recharge a few kilometres up dip, possibly near the depositional edge, and discharge is assumed to be a few kilometres down dip (Figure 4.25). As is typical of fluid flow in sedimentary basins, flow paths may be controlled by the permeability of faults or permeability contrasts across faults.

The alteration around the deposits implies that ore fluid was oxidised. In this respect it is noted that the deposits in western USA are in terrains that were in a semi-desert climate both during sedimentation and the likely times of fluid flow. Uranium is leached up-flow of the ore body and is transported as U(VI) complexes. Although a clear depletion of U up-flow of the ore zones in the host sandstone is not apparent, some leaching of uranium from detrital U-rich minerals such as apatite may have taken place. Alternatively

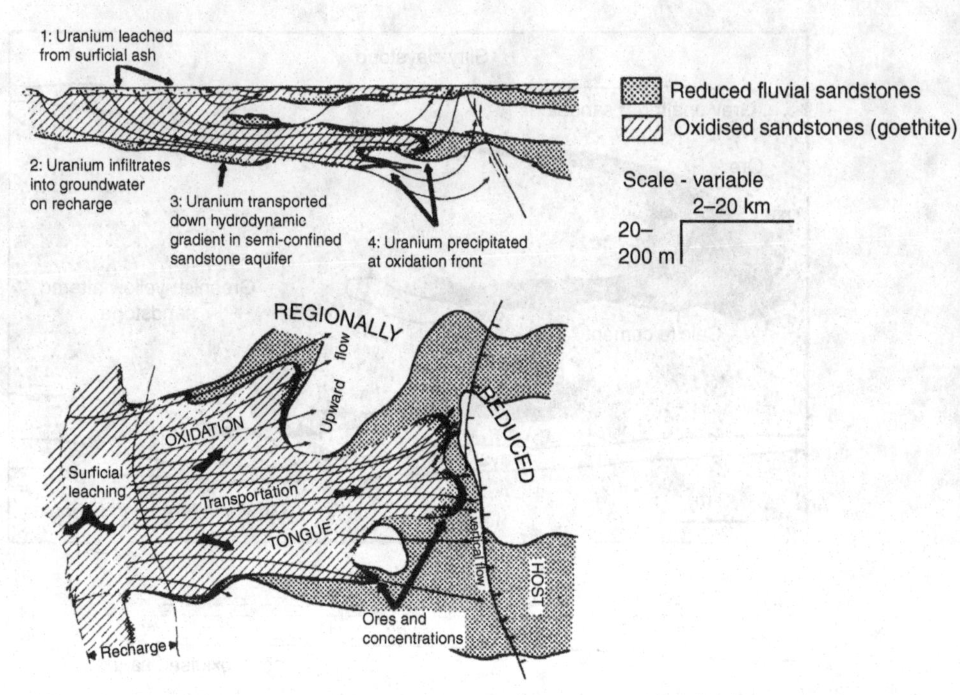

Figure 4.25 Section and map view of groundwater flow in the formation of roll-front U deposits in continental clastic sedimentary rocks. Surface derived waters leach U, e.g. from volcanic rocks in or above the basin. U is precipitated as shown in Figure 4.24 at the interface between primary, reducing and altered, oxidised sediments. This front migrates with time through the aquifers (Nash *et al.*, 1981).

U may have been leached from surficial sediments that could have been leached on infiltration at the time of groundwater flow, including veneers of vitric felsic tuffs, or from weathered U-rich basement rocks such as granites that are exposed at the edge of the host basins in Wyoming, for instance (Figure 4.25). The uranium in the roll-front deposits on the Colorado Plateau was likely derived from nearby earlier tabular ores.

Uranium is precipitated from solution as U(IV) minerals where the fluids are reduced by reaction with and oxidation of sedimentary organic matter and diagenetic sulfide minerals (Figure 4.24). The reaction takes place over a zone of finite width along the flow path. The ore-forming processes are, however, analogous to chemical processes in chromatographic columns, and continual groundwater flow will mean that the uranium minerals are redissolved as oxidised fluid impinges from behind the front. The redissolved uranium will be transported down-flow and reprecipitated in the reduced environment that prevails on the down-flow side. With time, therefore, the ore body migrates down-flow, although at a rate much slower than the flow rate of the groundwater. Over time, it also develops a lobate ore front in cross section, the exact ore-body shape being a function of the distribution of permeability and hence fluid flux in the host strata. On a district scale, multiple tongues are developed in plan view within favourable strata because of larger-scale variations in permeability and many of the ore bodies are elongate approximately parallel to the direction of flow rather than perpendicular to it (Figure 4.25).

Questions and exercises

Revision questions and problems

4.1 The solubility of Zn as $Zn^{2+}_{(aq)}$ at neutral pH is 0.02 μg l^{-1} at 25 °C, 0.38 μg l^{-1} at 100 °C and 3.2 μg l^{-1} at 200 °C.

What volume of pure water would be required to precipitate 10 Mt of Zn through simple cooling of water saturated in sphalerite at 100 to 25 °C?

What volume is required to precipitate the same tonnage of Zn through cooling from 200 to 100 °C?

This is a realistic question because some Zn deposits are formed in sedimentary rocks as a result of precipitation from waters at these temperatures.

Review questions and problems

4.2 Ore minerals in both Mississippi Valley-type (MVT) and Kupferschiefer (red-bed) Cu deposits are considered to precipitate from groundwaters that have migrated long distances in sedimentary basins. Outline, with reference to example ore fields, the possible causes of and controls on this groundwater flow in the ore-hosting sedimentary basins.

4.3 Write schematic chemical or mineral reactions and a sentence of explanation for the following:

(i) Why high salinity of migrating basinal water is an important factor for the formation of base-metal sulfide ores in sedimentary basins;

(ii) Why organic matter is an important factor in the precipitation of base-metal (Cu, Pb, Zn) sulfide ores in sedimentary basins.

4.4 Mass balance in the formation of roll-front uranium deposits:

(i) If 50% of a concentration of 10 ppm U in sandstones is leached by migrating groundwater, what will be the average grade of a 5-m-wide roll-front ore deposit 5 km down the flow path if the efficiency of precipitation of U is effectively 100%?

(ii) What will be the grade of a 5-m-wide deposit 15 km down the flow path?

(iii) What will be the grade of a 5-m-wide deposit that is oriented in map view at 15° to the direction of the flow path (see Figure 4.25)?

Discussion questions

4.5 Speculate on the control that steps in the evolution of the biosphere and atmosphere and ocean oxidation (Box 5.1) played on the distribution through Earth's history of:

(i) SEDEX deposits (Section 4.2.2) – the earliest are late Palaeoproterozoic (~ 1650 Ma) and the youngest Jurassic.

(ii) MVT deposits (Section 4.2.1) – these have formed abundantly over a number of periods in the Phanerozoic, but are rare in earlier periods of Earth history with possible examples preserved from the Palaeoproterozoic.

(iii) IOCG deposits (Section 3.3.3) – the earliest are late Archaean, they are most abundant in the late Palaeoproterozoic, and formed locally and at specific periods up to the present.

(iv) The oldest known major phosphorite deposits (Section 5.1.3) are early Cambrian in age.

Further readings

The following articles give background to hydrogeological, geological and geochemical processes in the formation of hydrothermal ores in sedimentary basins.

(1) Fluids in sedimentary basins – their origins, flow paths and controls on chemistry:

Garven, G. (1995). Continental scale groundwater flow and geologic processes. *Annual Review of Earth and Planetary Sciences* **23**, 89–117.

Hanor, J. S. (1994). Origin of saline fluids in sedimentary basins. *Geological Society Special Publication* **78**, 151–174.

(2) MVT Pb–Zn deposits – some discussions of the chemical and geological controls on ore precipitation:

Anderson, G. M. (1991). Organic maturation and ore precipitation in southeast Missouri. *Economic Geology* **86**, 909–926.

Anderson, G. M. (2008). The mixing hypothesis and the origin of Mississippi Valley-type ore deposits. *Economic Geology* **103**, 1683–1690.

Basuki, N. J. and Spooner, E. T. C. (2002). A review of fluid-inclusion temperatures and salinities in Mississippi Valley-type Zn–Pb deposits: identifying thresholds for metal transport. *Exploration and Mining Geology* **11**, 1–17.

Sverjensky, D. A. (1986). Genesis of Mississippi Valley-type lead–zinc deposits. *Annual Reviews of Earth and Planetary Sciences* **14**, 177–199.

Wenz, Z. J., Appold, M. S., Shelton, K. L. and Tesfaye, S. (2012). Geochemistry of Mississippi valley-type mineralizing fluids of the Ozark plateau: a regional synthesis. *American Journal of Science* **312**, 22–80.

Wilkinson, J. J. (2010). A review of fluid inclusion constraints on mineralization of the Irish ore field and implication for the genesis of sediment-hosted Zn–Pb deposits. *Economic Geology* **105**, 417–431.

(3) SEDEX deposits – a possible analogy with cold methane seeps on the ocean floor:

Torres, M. E., Bohrmann, G., Dubé, T. E. and Poole, F. G. (2003). Formation of modern and Palaeozoic stratiform barite at cold methane seeps on continental margins. *Geology* **31**, 897–900.

(4) Kupferschiefer deposits – analysis of the role of organic matter in ore mineral precipitation:

Bechtel, A., Gratzer, R., Püttmann, W. and Oszczepalski, S. (2002). Geochemical characteristics across the oxic/anoxic interface (Rote Fäule front) within the Kupferschiefer of the Lubin–Sieroszowice mining district (SW Poland). *Chemical Geology* **185**, 9–31.

Jowett, E. C. (1992). Role of organics and methane in sulfide ore formation, exemplified by Kupferschiefer Cu–Ag deposits, Poland. *Chemical Geology* **99**, 51–63.

Selley, D., Broughton, D., Scott, R., *et al.* (2005). A new look at the geology of the Zambian Copperbelt. *Economic Geology 100th Anniversary Volume*, Hedenquist, J. W., Thompson, J. F. H., Goldfarb, R. J. and Richards, J. P. (eds.), Colorado, Society of Economic Geologists, pp. 965–1000.

(5) Uranium deposits in sedimentary basins – analyses of the chemistry and hydrogeology of their formation:

Hansley, P. L. and Spirakis, C. S. (1992). Organic matter diagenesis as a key to a unifying theory for the genesis of tabular uranium–vanadium deposits in the Morrison Formation, Colorado Plateau. *Economic Geology* **87**, 352–365.

Hobday, D. K. and Galloway, W. E. (1999). Groundwater processes and sedimentary uranium deposits. *Hydrogeology Journal* **7**, 127–138.

Komninov, A. and Sverjensky, D. A. (1996). Geochemical modelling of the formation of an unconformity-type uranium deposit. *Economic Geology* **91**, 591–606.

5 Ore deposits formed in sedimentary environments

Ore deposits form in sedimentary environments as a result of one of two generalised geological processes: either as a result of mineral precipitation from solution in surface waters, most commonly from sea water or lake waters; or as a result of physical accumulation of ore minerals during processes of sediment entrainment, transport and deposition. The theoretical backgrounds to ore formation through each of these two processes are discussed separately followed by detailed descriptions of some important deposit types formed through each process.

5.1 Chemical precipitation from surface waters (hydrogene deposits)

Compositions of surface waters and ore mineral precipitation

A chemical sedimentary ore is one in which the concentration of ore minerals accumulated as a result of precipitation from solutions at the Earth's surface. All natural surface waters are multi-element solutions. Average seawater contains about 3.5 wt % salts, comprising of order weight per cent levels of Na and Cl, and tens to hundreds of ppm of a number of elements of economic interest (Figure 5.1).

As discussed further below, the concentrations of some elements in ocean waters have varied significantly through geological history. Element concentrations also vary with depth through a seawater column, particularly concentrations of those elements that are incorporated into organic matter, for instance as nutrients (Figure 5.2). Nutrient elements are released from organisms as they decay on descent through the water column and hence are typically at highest solute concentrations in deeper water. There is a zone of relatively low oxygen concentrations (oxygen minimum zone) at a few hundred metres in most oceans, which forms at the depth of greatest rates of organic matter decay and hence consumption of oxygen.

Seawater is close to saturation with respect to some minerals, including calcite and quartz. Changes of either water chemistry (e.g. chemical reduction, increased salinity due to evaporation) or conditions of the environment (temperature) may therefore induce authigenic mineral precipitation, either in the water column or in water-saturated bottom sediments. Mineral precipitation may be biologically mediated, especially where it occurs in bottom sediments.

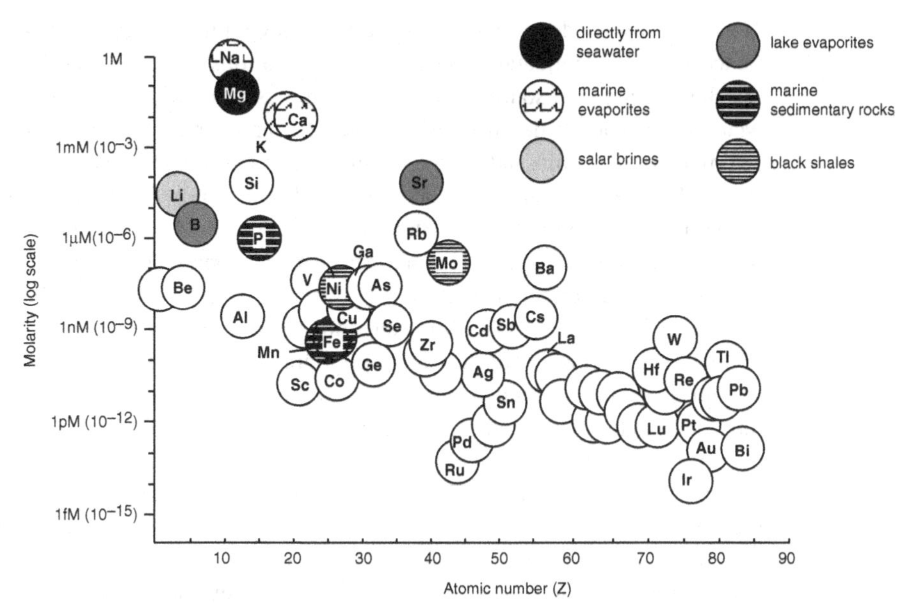

Figure 5.1 Concentration of the elements in average seawater (compiled from the Geochemical Earth Reference Model (GERM) database; see earthref.org/GERM). The shaded circles indicate elements or elements of mineral salts that are extracted either from natural surface waters or from sedimentary chemical precipitates of seawater and other surface waters, and are differentiated by the dominant source of each of these elements in sediments and sedimentary rocks. Note that the concentrations are averages in present-day oceans, vary somewhat in the present-day oceans and likely varied through geological time.

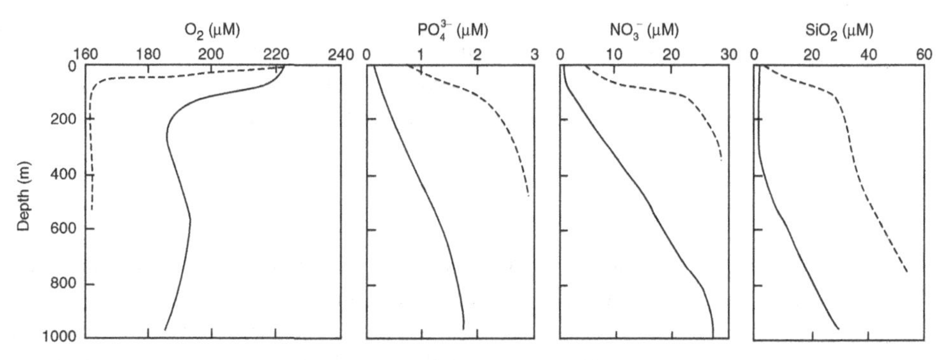

Figure 5.2 Concentration of some components in an ocean water column on the upwelling Peru–Chile continental margin (dashed lines), which has abnormally high organic productivity in surface waters, and a typical near-shore ocean column represented by the east coast of Australia (solid lines). (Data from CSIRO, Division of Oceanography.)

Major types of chemical sedimentary ores

There are a number of different ore deposit types for different commodities that are formed through chemical processes in sedimentary environments. These are listed here; the first three deposit types are described in greater detail in following subsections.

(i) Ironstones are by definition sedimentary rocks with more than 15 wt % FeO and are hence Fe-rich sedimentary rocks (= iron formations). Ores formed from ironstones are the major world source of Fe. These are described in detail in Section 5.1.1.

(ii) Manganese deposits, described in Section 5.1.2, are manganese oxide or carbonate-rich sedimentary rock units and are the world's major source of Mn. Manganese nodules on ocean floors at bathyal depths are composed dominantly of fine-grained manganese oxide minerals with high concentrations of metals adsorbed onto grain surfaces, and are a possible future source of a number of metals including Mn, Ni and Cu.

(iii) Phosphorites, described in Section 5.1.3, are sedimentary rocks with between 40 and 90% modal apatite (specifically, francolite $\approx Ca_{10}(PO_4)_5(CO_3)F_2$).

(iv) Evaporites. Marine evaporites are important ores of minerals of many elements that are in relatively high concentration in seawater (common salt, potash, magnesium carbonate, see Figure 5.1). Evaporites formed in lakes in some enclosed continental basins in desert climates (salars) in tectonically and volcanically active regions such as eastern China, Chile and adjacent countries, Nevada, USA, and are the dominant source of boron minerals, strontium minerals and sodium carbonate minerals such as trona and soda (e.g. Risacher and Fritz, 2009). Lithium is dominantly extracted from brines that are interstitial to evaporite minerals in continental evaporite basins.

Nitrate minerals (salitre) and iodine are extracted from minerals at or near the Earth's surface that precipitate from fog droplets that condense on rock and regolith surfaces in the hyper-arid Atacama desert of northern Chile and southern Peru (e.g. Ericksen, 1981). These condensates are ultimately derived from sea spray and evaporation from the nutrient and biologically rich upwelling cold surface waters of the Humboldt Current of the adjacent Pacific Ocean.

(v) Black shales. Some organic-rich black shales have ore-grade concentrations (~ 1%) of a number of metals, most importantly Mo and Ni, for instance in early Cambrian formations in southern China (e.g. Lott *et al.*, 1999). The metals are considered to have been sequestered from seawater onto grain surfaces (adsorbed) or into amorphous organic matter that was on or just below the sea floor.

5.1.1 Iron ores in ironstones

Although ironstone is a sedimentary rock with an unusually high concentration of iron, it is itself rarely ore and ore bodies are the result of epigenetic upgrading and a further stage of concentration of iron after sedimentation. The formation of Fe-rich sedimentary rock is, however, a critical first stage in the genesis of sedimentary-rock-hosted iron ores, and is therefore discussed in detail here.

Worldwide there are two contrasting widespread types of ironstone:

(i) Oolitic (ooidal) ironstones – with oolites predominantly of chamosite (Fe-rich chlorite of approximate composition $Fe_5AlSi_3AlO_{10}(OH)_8$ with about 40 wt % FeO), goethite, or limonite (FeOOH) in a matrix of Fe-rich minerals, including iron carbonate (siderite) (e.g. Van Houten and Bhattacharyya, 1982). Jurassic ironstones interbedded with carbonate units were historically important ores in France (Lorraine) and England (Northamptonshire) but little ore of this type is currently exploited.

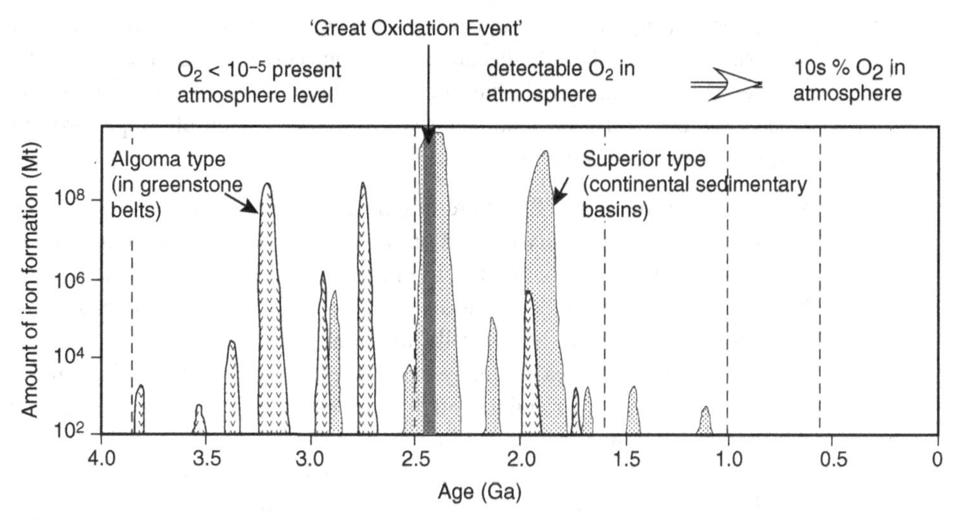

Figure 5.3 Abundance of cherty iron formations of Algoma and Superior types through geological time (after Bekker *et al*., 2010), and major events in the evolution of oxygen in the atmosphere. Note the log scale and the predominance of iron formations within about 200 million years after the GOE. The events in the evolution of the atmosphere are described and discussed in more detail in Box 5.1.

(ii) Cherty iron formations – with iron oxides (haematite, magnetite) and more rarely iron carbonate (siderite) and iron silicate minerals (e.g. greenalite, stilpnomelane etc.) together with fine-grained quartz. Cherty ironstones include thin units of fine-grained Fe-rich sediments that are being deposited at present in marine settings, in particular on volcanic seamounts (e.g. Hein *et al*., 1994). The economically important ores, however, are hosted in banded iron formations (= BIF, itabirite, jaspillite), units which can be of order 100 m thick and which are almost all of Archaean and Palaeoproterozoic age and are present in most of the major shield regions of the world, including Brazil, India, South Africa, Australia and the Lake Superior region, USA.

At present, almost all the world's iron is sourced from ores in early Precambrian cherty iron formations.

Cherty iron formations

Major cherty iron formations were almost entirely formed in late-Archaean and Palaeoproterozoic times (Figure 5.3) with a few units formed through the Mesoproterozoic and Neoproterozoic eras. They are in particular present in most Palaeoproterozoic terrains in the world, and are abundant in some. Two types are distinguished based on geological setting:

- *Algoma-type* iron formations are sedimentary units in sequences of submarine volcanic rocks in greenstone belts, most commonly of Archaean age. They are generally of limited strike extent (up to tens of kilometres) and thickness (a few metres), but can be up to about 100 m thick.
- *Superior-type* iron formations are laterally extensive (hundreds of kilometres of strike length and original surface extent) in units up to hundreds of metres thick that typically

include interbedded, thin units of Fe-rich shale. These occur in sedimentary basins of marine strata of late-Archaean or Palaeoproterozoic age that were deposited over igneous and metamorphic basement. The basin successions typically include carbonates, shales, sandstones and some volcanic rocks. Multiple Superior-type iron formations are present in many of these basins.

Banded iron formations (BIFs) are the most widespread facies of cherty iron formation. These are characteristically banded on multiple scales ranging from metres (macrobanding) to centimetres (mesobanding) to millimetres (microbanding). The meso- and microbanding is composed of alternations of Fe-rich minerals and quartz or cherty quartz. Haematite or magnetite is generally the dominant iron mineral. Macrobanding can be correlated over distances of many kilometres in an iron formation and is marked by Fe-rich shale bands. Granular iron formations (GIFs) constitute a sedimentary facies which lacks the finer-scale banding and in which textures indicate deposition from currents. This facies is most commonly formed of transported and reworked sand- and conglomerate-sized clasts of iron-formation material.

Chemically, a typical cherty iron formation is composed predominantly of Fe_2O_3 and SiO_2, in most cases with sub-equal amounts of each, and hence up to about 30% Fe. Other major elements are present in concentrations much lower than crustal average, most importantly CaO, MgO and Al_2O_3.

Iron-formation facies can be classified also on the basis of the dominant Fe mineral:

- oxide facies – magnetite or haematite;
- carbonate facies – siderite;
- silicate facies – greenalite (Fe-serpentine), minnesotaite (Fe-talc), stilpnomelane etc.

It is not clear, however, that any of the Fe minerals that are presently in the rocks were the minerals of primary sedimentation. Diagenetic modification and recrystallisation during low-grade metamorphism has overprinted all iron formations and has caused recrystallisation and new mineral growth. Haematite is in many cases the paragenetically earliest Fe mineral. However, recent Fe-rich sediments on volcanic seamounts suggest that the primary Fe minerals may have included cryptocrystalline iron hydroxide minerals (ferrihydrite) and Fe-rich clay minerals (nontronite) and hence that haematite may have been produced by recrystallisation of these minerals in a diagenetic environment. Textures of the silica minerals indicate that they were primarily precipitated as cryptocrystalline quartz (chert) and have recrystallised to quartz in most cases.

Depositional environment of cherty iron formations

Because there are no close modern analogues of cherty iron formation sedimentation similar to that of Archaean and Palaeoproterozoic basins, it is difficult to define the environment of deposition and the controls on sedimentation. We reconstruct environments based on a combination of geological setting, sedimentary structures and rock chemistry.

Although sedimentary structures indicative of clast transport and resedimentation are present in some iron formations, the textures, mineralogy, chemistry, lateral extent and presence of planar, finely laminated beds of most Superior-type BIFs indicate deposition by settling of Fe and silica mineral grains that formed as chemical and/or biologically mediated chemical precipitates in the water column, although the minerals may have been recrystallised in diagenetic reactions after sedimentation.

The sedimentological settings of the formations, with lateral continuity of strata of up to several hundred kilometres, interbedding with shales and carbonate units, together with facies reconstructions of underlying and overlying clastic and carbonate units, imply that the iron formations were deposited in marine sedimentary basins on stable shelves or on slopes seaward of stable shelves. One requirement for the environment of deposition is that there was little input of volcanic or terrigenous material, either of which would have provided an Al-bearing component to the sediment. The macro-interbeds of Fe-rich shale do have a significant volcaniclastic component and presumably mark episodes of volcanogenic input onto the shelf or slope.

The role of ocean chemistry in the genesis of banded iron formations

Iron is present at very low concentration in modern-day oceans ($\approx$ 1 nM, Figure 5.1), dominantly as Fe(III). The low concentrations are the result of the low solubility of Fe(III) ions and complexes relative to oxide and hydroxide minerals (cf. Figure 3.3). Because of the low concentration of iron in normal seawater, iron minerals would be very unlikely to precipitate as a major component of sediments over large areas. Most seawater in the modern oceans is oxic with a few ppm of molecular O_2 in solution. However, two known extensive modern oceans and ocean basins (the Black Sea, and the Cariaco Basin offshore of Venezuela) include anoxic (with no molecular oxygen in solution) and euxinic (with sulfide in solution) water at depth. The Black Sea is layered with 20–200 m of oxic surface waters over euxinic deeper waters. Iron concentration in these basins is significantly higher than in normal ocean water. In the Black Sea it has a marked peak in concentration at the top of the anoxic layer, where it is present in solution as Fe(II) species at several orders higher concentration than typical seawater (up to 10 μM, Figure 5.4). It is also present as concentrations of very-fine-grained particulate matter near the redoxcline. These modern-day anoxic basins give an analogue for the chemical behaviour of ancient oceans and hence for the deposition of iron formations.

As the atmosphere before the Great Oxidation Event (GOE) at about 2.42 Ga lacked free oxygen the oceans of the early Earth would have been largely anoxic (see Box 5.1). Concentrations of Fe are thus predicted to have been present in seawater in the Archaean as in modern-day anoxic oceans. In the Proterozoic, the oceans were most likely stratified with free oxygen and hence low Fe concentrations in near-surface waters, but anoxic Fe-rich waters at depth (Figure 5.4). The oceans are interpreted to have been sulfide-poor at these stages in the Earth's evolution, and Fe concentrations would not have been as strongly buffered by pyrite precipitation. They could thus have been as high as the peak concentrations in the Black Sea throughout the water column below a near-surface oxygenated zone.

If parts or all of the oceans had high dissolved Fe(II), precipitation of iron(III) hydroxide minerals, e.g. goethite ($Fe^{III}OOH$), would occur wherever and whenever there was oxidation of Fe(II) to Fe(III). There are three mechanisms by which dissolved Fe may have been oxidised in the water column over a stable shelf or continental slope. (i) If the ocean was chemically stratified with respect to oxygen content, with free oxygen derived through photosynthesis in an upper zone but absent below, Fe would precipitate on transport of water upwards across this chemocline. (ii) Iron(II) was biogenically oxidised without free oxygen in the photic zone, for instance by Fe-oxidising photosynthetic

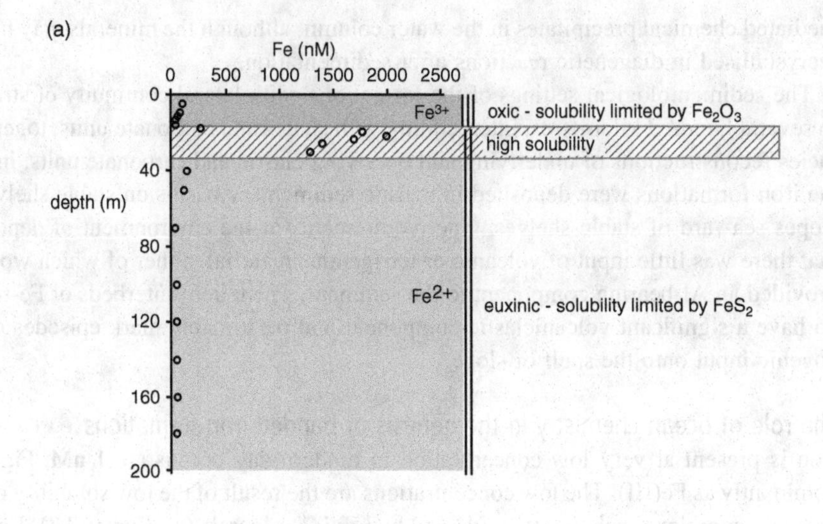

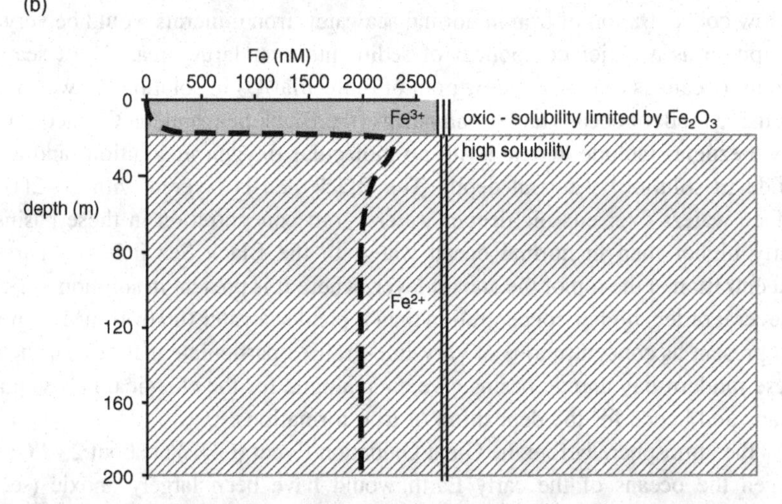

Figure 5.4 (a) Distribution of Fe with depth in a profile in the Black Sea and controls on its solubility. The oxic–anoxic boundary in this section is at about 20 m depth, immediately below which iron is soluble as Fe(II) species. Iron solubility reduces at greater depth because of increased H_2S content of the water, and precipitation of Fe with sulfide in pyrite (after Millero and Söhn, 1992). (b) Possible distribution of Fe with depth in Palaeoproterozoic oceans. It is assumed that there is limited H_2S in solution in oceans at this time, and hence that Fe concentrations below the oxycline are only weakly buffered by precipitation of pyrite.

anoxygenic bacteria. (iii) Oxidation of Fe(II) to Fe(III) was photochemically driven by ultraviolet (UV) radiation in the photic zone of ocean water (Figure 5.5). In any of these cases, movement of deep water onto shelves is required to maintain a flux of precipitation.

The first mechanism may be applicable after the GOE, and the peak of Fe formation at this time or soon afterwards may have been a unique stage in ocean evolution. The decrease in the number and size of iron formations with time from about 2.45 Ga to about 1.8 Ga in the Palaeoproterozoic (Figure 5.3) was possibly either a result of drawdown of the Fe reservoir of the ocean over time with progressive oxidation or of increasing sulfide

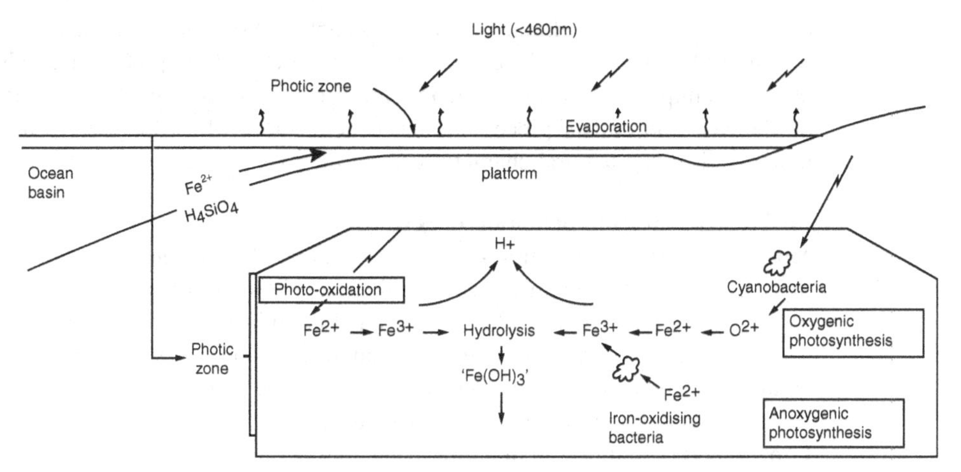

Figure 5.5 Possible mechanisms of precipitation of Fe and SiO_2 from seawater to form oxide-facies BIF (Morris, 1993). Fe^{2+} in upwelling water is oxidised to Fe^{3+} either through reaction with O_2 produced by photosynthesis in the photic zone, directly by anoxygenic photosynthesis, or by photochemical processes. Fe^{3+} is weakly soluble and once formed is precipitated as hydroxide. In this model, SiO_2 is precipitated largely as a result of evaporation on the shallow platform.

concentrations in ocean water. The other two mechanisms could have also been operative before the GOE and can thus explain chemical precipitation of iron in Algoma-type iron formations in the Archaean. Iron and SiO_2 would have been input into ocean waters in solution in hydrothermal fluids exhaling on the sea floor in association with sea-floor volcanism, in many cases remote from the site of precipitation of these elements. Weathering and transport through rivers is an alternative source of these components in ocean water.

The origins of the millimetre- to centimetre-scale iron–silica banding in BIFs are uncertain. Annual climatic oscillations (varves) are a possible cause of the fine banding, but this has not been demonstrated through determinations of the frequency of the bands. Lower-frequency periodic upwelling of deep ocean water may alternatively bring Fe-rich water from an ocean basin to the photic zone on a platform and be a process leading to fine microbanding in BIFs. The fine microbanding could alternatively be the result of early diagenetic recrystallisation in water-rich porous sediment, for instance through breakdown of Fe-rich clay minerals such as nontronite to haematite and quartz.

Iron ore in iron formations

There are various grades and types of iron ore classified by their chemistry and physical behaviour. The most valuable ores are those which are composed almost entirely of iron oxide and hydroxide minerals and which can be fed directly into an iron furnace to produce pig-iron and eventually steel. These have Fe contents of greater than about 64 wt % (equivalent to Fe_2O_3 greater than about 95 wt %) and concentrations of a number of 'contaminants' (Si, Al, Ca, P, etc.) below threshold values set by the requirements of steel making. A low-phosphorous content ($< 0.1\%$) has in particular been critical as P will combine with Fe in steel and result in a brittle product. Phosphorous can be present in the ore either as trace apatite or in solid solution in goethite. Aluminium is present in solid solution in goethite or as a component of the shale interbeds that are typical of iron formations (see Figure 5.8) and Al content is also critical in smelting and needs to be below about 2% to prevent 'freezing' of the furnace, i.e. solidification at the temperature of operation, during smelting. Ores which do not have the correct characteristics for direct feed into a furnace need to be blended with other ores, or to be processed and **beneficiated** to remove impurities. Which Fe-rich rock is ore depends on both the grade of the ore and on the ease and cost of processing and beneficiation, which themselves depend on the mineralogical and textural characteristics of the rock.

(a) Unenriched iron-formation ores (taconites)

Most banded iron formations contain 20–35% FeO. These are generally insufficiently rich in Fe to constitute economic ore. So-called taconite ores typically have iron contents at the upper end of this range (30–35% Fe) and are mined where economic conditions are suitable. The economic extraction of iron from taconites is favoured where mineral textures and other rock properties are such that the iron oxide minerals can be easily separated from gangue, for instance where the rock contains coarse-grained magnetite formed by recrystallisation of earlier haematite during greenschist- or amphibolite-facies metamorphism and which can be separated from quartz and other minerals in a magnetic separator without the need to crush to a very-fine grain size.

(b) High-grade *in situ* enriched iron ores

Most BIF-hosted iron ores are the result of *in situ* epigenetic upgrading to greater than about 56% Fe and the formation of rock dominated by iron oxide and hydroxide minerals. The content of silica in particular is thus very much less than in the primary BIF and can be less than 5%. *In situ* upgrading is demonstrated by the preservation of the macrolayering and mesolayering of the BIF, and rarely microlayering, in the ore bodies. In some of the deposits in India, 'corestones' of unenriched or weakly enriched iron formation are recognised in the interior of ore bodies. There are generally sharp irregular contacts between ore and unenriched BIF across which the layering is traceable. Ore bodies are large bodies that can be up to several kilometres in length. They are stratabound and dip parallel with the host BIF, although can have complex shapes, for instance where the host sequences are folded and cut by faults. Ore bodies are often restricted to one level within the BIF unit and may be bounded above and below by one of the prominent shale interbeds within the unit.

Different types of ore are, however, distinguished based on mineralogy and on texture, and the common end-member mineralogies of high-grade iron ores are (Figure 5.6):

- *Microplaty haematite* and *martite–microplaty haematite* (mpl-H and M–mpl-H) ores. Martite occurs in the original iron bands of the BIF and haematite as fine-grained, often felted platy crystals up to a few tens of micrometres in diameter that appear to have grown into pore space produced by dissolution of silica. Pore space is abundant (up to 30%) also in these ores.
- *Martite–goethite* (M–G) ores (martite = haematite pseudomorph after magnetite). The textures of this ore type have similarities with those of mpl-H and M–mpl-H ores. Martite forms pseudomorphs after original magnetite in the Fe-bands of the BIF and goethite is largely in highly porous, felted masses with up to about 30% pore space that have replaced the chert and mixed silica–iron silicate bands of the BIF. A high P content ($> 0.07\%$) is common in these ores and will reduce the ore's market value if it is not removed. This type of ore is most common in Western Australia, but is also recorded in the other major iron-ore provinces.
- *Coarse-grained haematite* ores (H). These may be massive ores with coarser-grained (< 100 μm) granuloblastic or micaceous textures of haematite and less pore space than

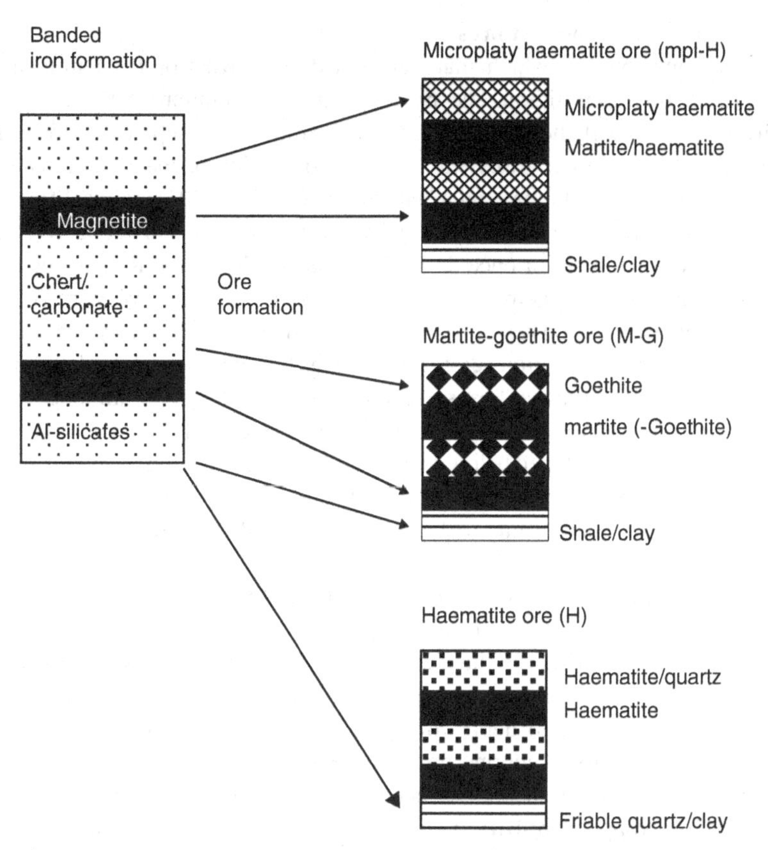

Figure 5.6 Different types of iron ore in BIF differentiated by iron mineralogy showing the preservation of the primary layering of the BIF by different combinations of new minerals and volume reduction, which is most marked in the replaced primary silica layers (after Clout and Simonson, 2005). In most cases the paragenesis of replacement has multiple stages and is hence more complex than shown here.

M–mpl-H ores. This type of ore is characteristic of many Brazilian deposits where the host sequences are more strongly deformed and metamorphosed. Magnetite is a locally important constituent of this ore type.

- '*Hydrated ore*' is goethite-rich ore that occurs at and below the current ore surface to depths of up to a few tens of metres and in which primary BIF textures are largely overprinted. This is the result of near-surface weathering of the exposed upper levels of an ore body.

Many ores show complex paragenetic histories that indicate that they developed through growth and replacement of different iron minerals and mineral assemblages at different times. A single ore body may be composed of zones with different types of ore (e.g. Figure 5.8). In addition, there are locally bodies of altered BIF with different mineral assemblages than those of the major types of ore. These parts of the BIF are typically enriched to sub-ore concentrations of Fe. Blocks with magnetite–apatite and magnetite–siderite (or ankerite)–apatite are recognised, for instance below and around some of the M–mpl-H ores of the Hamersley Province and Yilgarn Craton (Mt Tom Price, Figure 5.8b; Koolyanobbing).

(c) High-grade transported ores

There are some ores in which material eroded from BIFs or from iron ores has been concentrated on redeposition. The processes of concentration in these ores are in part mechanical and in part chemical, and involve dissolution and reprecipitation of Fe minerals on a local scale. These ores are essentially a form of **ferricrete**, that is, a cemented iron oxide or iron hydroxide regolith material. Channel iron deposits (CID) are **pisolitic** to **pelletoidal** ores of dominantly goethite with some haematite, which occur along river channels or palaeochannels that drain or once drained BIF units (Morris and Ramanaidou, 2007). This type of ore is especially common in the Hamersley Ranges of Western Australia. Their formation and preservation may have been promoted in this province by limited erosion since the Mesozoic and a long period before and up to the present of a warm, relatively dry climate. The Yandi CID, for instance, forms a sinuous ore body almost 50 km long and up to 50 m deep. Conglomeratic ores or **canga** are talus slope breccias of dominantly goethite-rich ore that are commonly developed below scarps or hills of exposed iron formation or iron ore, and also where iron formation was exposed before deposition of recent valley fill sediments or in some cases below unconformity surfaces that developed much earlier.

Geological settings of high-grade *in situ* enriched iron ores

The majority of ore known in all iron-ore provinces of the world is restricted to above the depth of weathering, which in the tectonically stable cratonic settings in which most iron formations occur may be as deep as 400 m. Martite–goethite ore bodies in particular extend from the surface to depths of up to about 200 m, although may be covered by recent sediments. Cross sections through the Hamersley Ranges of Western Australia show that ore bodies or bodies of economic or sub-economic M–G ore occur at the majority of outcroppings of BIF (Figure 5.7), although these ore bodies are not continuous along the lengths of the outcrops of the units.

Ore bodies of all types are often **structurally controlled** (see Box 3.7). They may be sited, or instance within synclines and complex trains of folds, or their foot-walls or hanging-walls may be delimited by faults or dykes (e.g. Figure 5.7).

The settings of many haematite-bearing ores (M–mpl-H and H) are most complex. These ores may extend as irregular bodies that extend deeper than known M–G ore and down to at about 400 m depth. Although most are above the 'depth of weathering', parts of some are ore bodies at depth which are separated from the surface by unenriched BIF (Figure 5.8a), and some extend into sub-ore grade replacements of BIF below the depth of weathering (Figure 5.8b). Ores of this type in a number of areas in the world are sited beneath regional Palaeoproterozoic unconformities and can be interpreted or demonstrated to have formed well before the development of the present topography, including examples in Western Australia, South Africa and in the Negaunee Iron Formation of Michigan, USA. The presence of clasts of haematitic ore in late Palaeoproterozoic conglomerates overlying such an unconformity in the Hamersley Range demonstrates early formation at least in this case.

Processes of epigenetic *in situ* enrichment of iron formation

In principle the upgrading of an iron formation to iron ore may be the result of SiO_2 leaching, Fe addition or a combination of these. Mass-balance and mineral textures together with the stratigraphic thinning of BIF units within the ore body by 30–50%

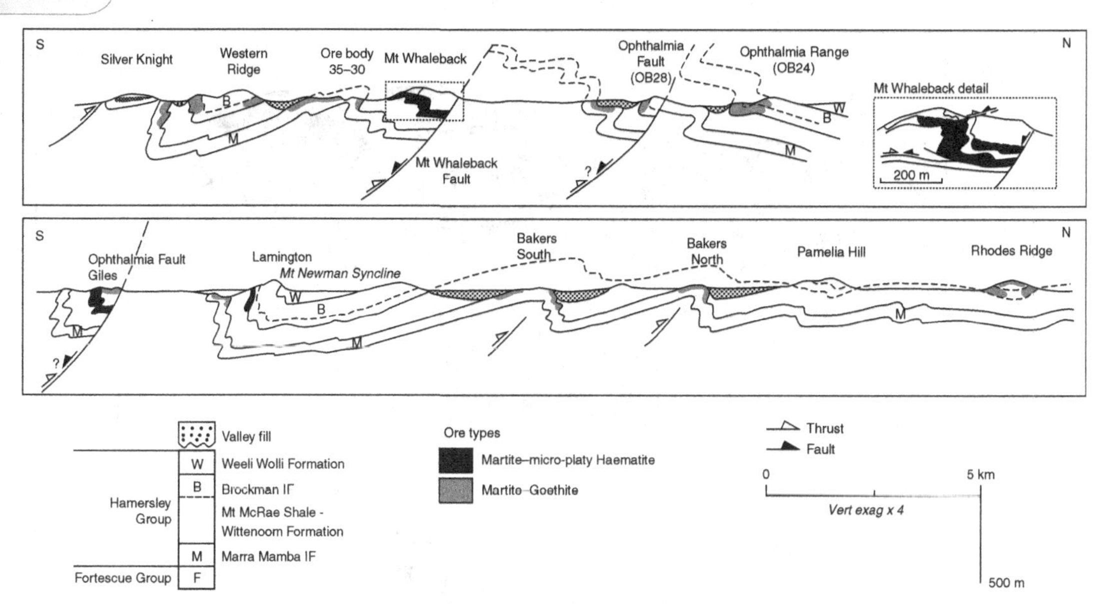

Figure 5.7 Two simplified north–south cross section across segments of the Hamersley iron-ore province of Western Australia showing the Brockman and Marra Mamba iron formations that are the hosts of the major iron ores and the positions and extents of known iron ores (after Morris and Kneeshaw, 2011). Note that some of the M–G ores shown are sub-economic.

(e.g. Dales Gorge Member at Paraburdoo, Figure 5.8a), and the high porosity of, in particular, the silica bands of M–G and M–mpl-H ores that are replaced by goethite or microplaty haematite, show that Fe enrichment predominantly involves leaching of silica and other impurities, together with oxidation of Fe. Precipitation of externally derived Fe is at most of lesser importance.

The settings of the M–G ores and the abundance of hydrated goethite in these ores show that they most probably formed at low temperatures, most likely as the result of deep **lateritic** weathering, which in most cases took place beneath a long-lived surface close to the present topographic surface. This enrichment to depth is promoted by long periods of tectonic stability: in the case of the Hamersley Range of Western Australia, deep weathering below the present land surface has been active at least episodically from the Cretaceous to the present. In addition, a warm climate and sufficient topography to promote groundwater flow through the BIF to dissolve quartz and flux it away from the BIF will promote ore formation. For further discussion and explanation of the chemistry and controls on deep lateritic weathering see Section 6.1.

The processes of formation of *in situ* M–mpl-H and H ores are less clear and have been proposed to be either a result of **supergene** or weathering processes that were active to usually great depths in the past or as a result of interaction with hydrothermal fluids which come from depth (**hypogene** enrichment). Sequential enrichment through both types of process has also been proposed (Figure 5.9).

If these haematite-rich ores were formed through supergene processes below ancient unconformities in the past, their mineralogy, textures and the positions of ore bodies imply that they may at least in part have formed originally as M–G ores through

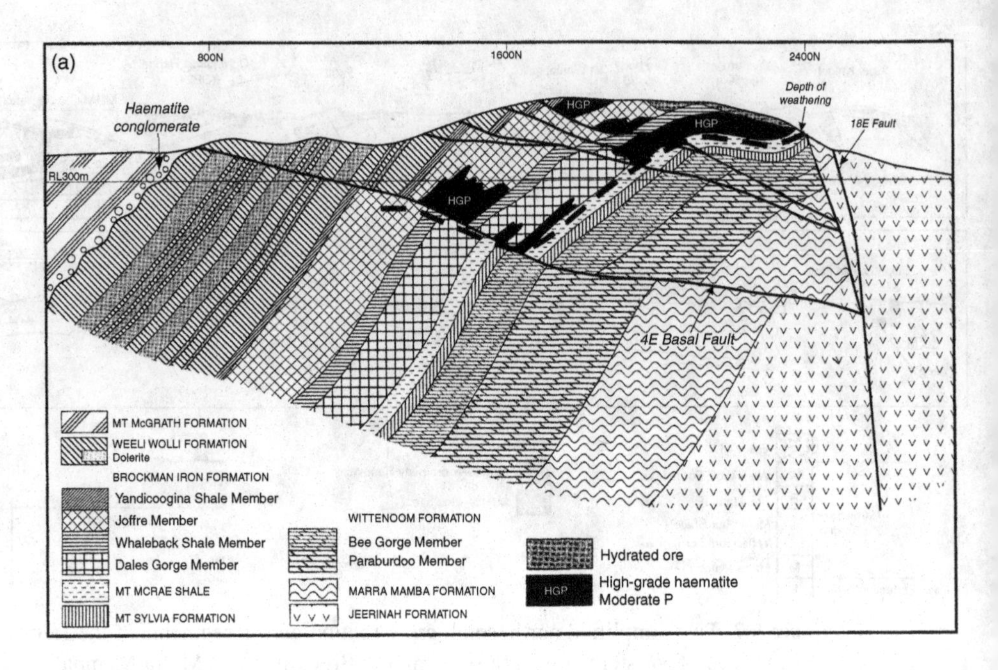

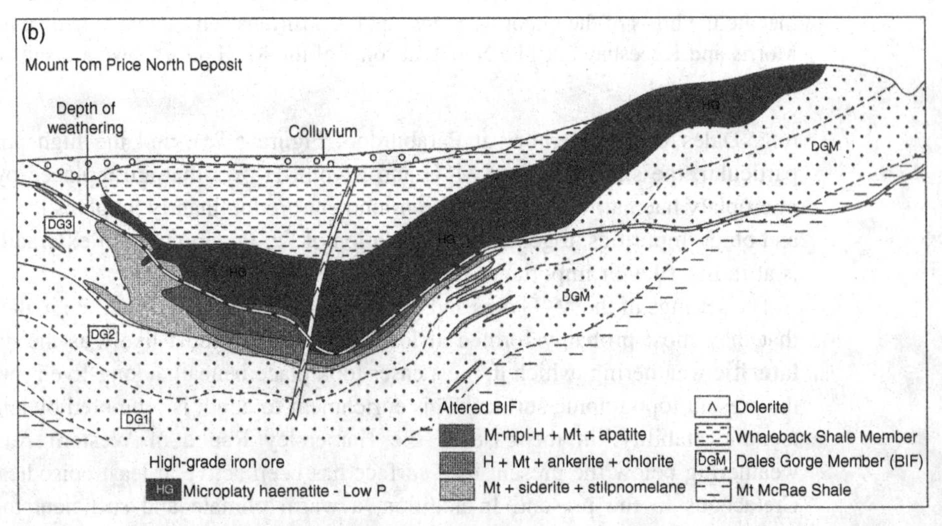

Figure 5.8 Cross sections through major high-grade haematitic iron-ore deposits in the Hamersley Province of Western Australia. (a) The Paraburdoo 4 East ore deposit (Taylor *et al.*, 2001). The ore is dominantly stratabound replacement of the Dales Gorge Member of the Brockman Iron Formation. Note the thinning of the Member where it is replaced in the upper ore body. Some ore bodies are delimited by faults: these are interpreted as rotated Palaeoproterozoic normal faults. All ore is above the depth of weathering, but deep ore bodies are not continuous up to the land surface but are separated from it by enriched BIF. (b) The Mount Tom Price North Deposit (Thorne *et al.*, 2004). Iron ore is almost entirely stratabound within the stratigraphically thinned unit DG3 of the Dales Gorge Member IF and has a foot-wall against the underlying shale macroband DG2. BIF adjacent to the ore below the depth of weathering is altered to various assemblages with magnetite, Fe-bearing carbonates and Fe-silicate minerals. Mineral abbreviations: H – haematite; M – martite, Mt – magnetite.

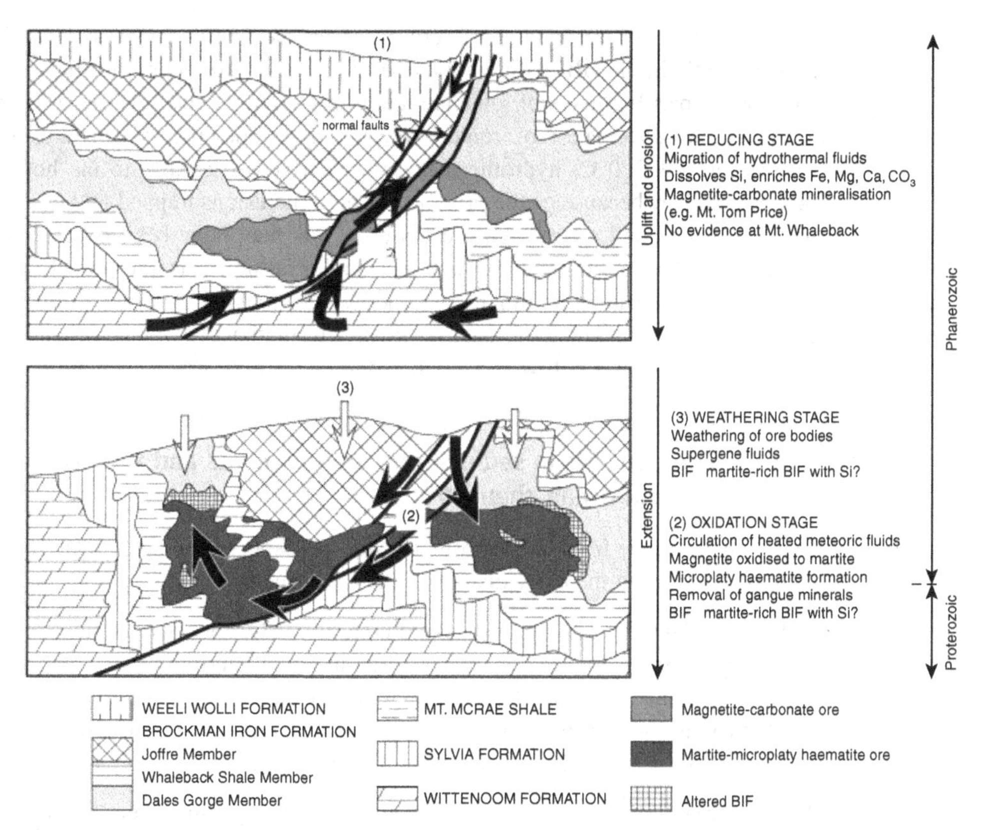

Figure 5.9 Proposal for multiple phase genesis of M–mpl-H and related iron-ore types in the BIFs of the Hamersley Basin, Western Australia. (1) An early hypogene stage of upward-flowing hydrothermal fluids at about 200 °C during regional tectonism caused alteration to relatively reduced magnetite–carbonate assemblages at depth and adjacent to faults. (2) Later, downward percolation of surface-derived waters caused oxidation and replacement of these carbonate-bearing iron formations to form haematite-rich high-grade iron ores. Recent near-surface weathering beneath the present land surface has caused hydration and reconstitution of the ores above the depth of weathering (after Taylor *et al.*, 2001).

weathering reactions below a land surface. Such weathering would have taken place where and whenever BIF was exposed to the atmosphere and the climate was humid tropical or seasonally humid tropical. The irregular shapes of the ore bodies are presumably related to patterns of groundwater flow, possibly at times of relatively large topographic relief such that water flowed from the surface through rocks at several hundred metres of depth, to leach silica from BIF at these depths. The presence of ore bodies that are within the rock pile and not continuous to the surface could be a result of structural and fracture control of water flow. Microplaty haematite would be formed largely from goethite by dehydration and recrystallisation at higher temperatures (100–140 °C) during burial or regional tectonism. Massive haematite could have been formed by recrystallisation at higher temperatures.

If ore formation at several hundred metres depth involved addition of Fe as well as leaching of SiO_2, and oxidation of magnetite to haematite, this may have been aided by the relatively high electrical conductivity of magnetite-rich rocks and development of galvanic cells in the BIF with conduction of electrons from depth to the surface charge, balanced in part by fluxes of ions, including migration downwards of Fe^{2+} ions.

The proposed formation of iron ores through hypogene processes invokes flow of high-temperature (> 200°C) hydrothermal waters derived externally to the host BIF and is proposed in part because of the presence of fluid inclusions trapped at these temperatures within veins in and around the ore bodies. Flow may have been over large distances, potentially across a sedimentary basin in a similar fashion as is interpreted for many base-metal sulfide deposits in sedimentary rocks (see Section 4.1.2). The chemical reactions of enrichment are poorly defined, and may involve formation of carbonate-rich, sub-ore-grade rock such as is known below and around some of the high-grade haematitic ore bodies.

Box 5.1 Ore deposits in sedimentary environments and the evolution of atmosphere and ocean chemistry

Important evidence from ore deposits that the atmosphere had a significantly different composition during the early history of the Earth includes:

- The abundant unweathered pyrite and uraninite (UO_2) pebbles in late-Archaean and early-Proterozoic palaeoplacers deposited in fluvial environments (Section 5.2.2).
- The abundance of banded iron formations in the Archaean and early Proterozoic which must have formed through chemical precipitation of iron, presumably as Fe(II), from solution in ocean water (Section 5.1.1).

The interpretation has therefore developed that the early atmosphere lacked molecular O_2 and that the oceans had an oxidation state low enough that Fe(II) dominated over Fe(III) (e.g. Cloud, 1973; Holland, 1984). Some time in the early Proterozoic a threshold was reached in an evolution towards a surface chemistry more similar to that of the modern day, with sufficient O_2 in the atmosphere and in solution in the oceans that Fe(III) became the dominant form of Fe in surface environments. This evolution is most likely a result of biogenic production of oxygen from CO_2 at the Earth's surface in photosynthesis.

A number of lines of evidence from the geochemical distribution of elements that have multiple oxidation states support this interpretation, including evidence from the relative abundances of the rare-earth elements (REEs) Ce and Eu in sedimentary rocks (e.g. Bekker *et al.*, 2010; Maynard, 2010). Many modern-day sediments for instance have 'Ce anomalies' and lower or higher normalised abundances of Ce than of adjacent REEs in the periodic table (La and Pr), which form because Ce has differing chemical behaviour than the other REEs, as it exists as Ce(IV) in addition to Ce(III) in oxidising conditions. Cerium anomalies are, however, absent from Archaean sedimentary rocks, presumably because Ce(IV) was not formed in the reducing surface conditions that prevailed in the Archaean.

These interpretations have been in essence proven with the discovery and evaluation of natural 'mass-independent fractionation' (MIF) in many Archaean and earliest Proterozoic sedimentary rocks, in particular of the multiple sulfur isotopes. Sulfur has four stable isotopes (^{32}S, ^{33}S, ^{34}S and ^{36}S). Differences in the ratios of these isotopes in different phases are the results of partitioning and can be described using the δ notation – see Box 2.1.

Isotope fractionation that results from either equilibrium distribution between phases or kinetic processes is always such that the degree of fractionation is proportional to the mass difference of the isotopes. The fractionation of $^{34}S/^{32}S$ by any process at the Earth's surface and in the interior should therefore be approximately double that of the fractionation of $^{33}S/^{32}S$. Significant mass independent fractionation of sulfur is however recorded for many samples of Archaean sedimentary rocks. We define $\Delta^{33}S$ as the difference between the measured $\delta^{33}S$ and the $\delta^{33}S$ expected by comparison with $\delta^{34}S$ (e.g. Farquhar *et al.*, 2000). There is a sharp temporal cut-off between the Archaean and earliest Proterozoic regime in which sulfur in rocks had variable $\Delta^{33}S$ and the modern regime with no MIF ($\Delta^{33}S = 0$) (Figure 5.10). There was thus a clear change in behaviour at about 2450 Ma.

MIF of sulfur isotopes occurs as an effect of solar UV radiation in the atmosphere, which also occurs at the present, causing photolysis (spallation or photochemical dissociation) of molecules of volcanically produced gases such as SO_2 (e.g. Farquhar *et al.*, 2010). The important requirements to produce the geological-record MIF shown in Figure 5.10 are the strength of the photolysis process and the likelihood of preservation of the signature in sedimentary minerals. There are two reasons why the

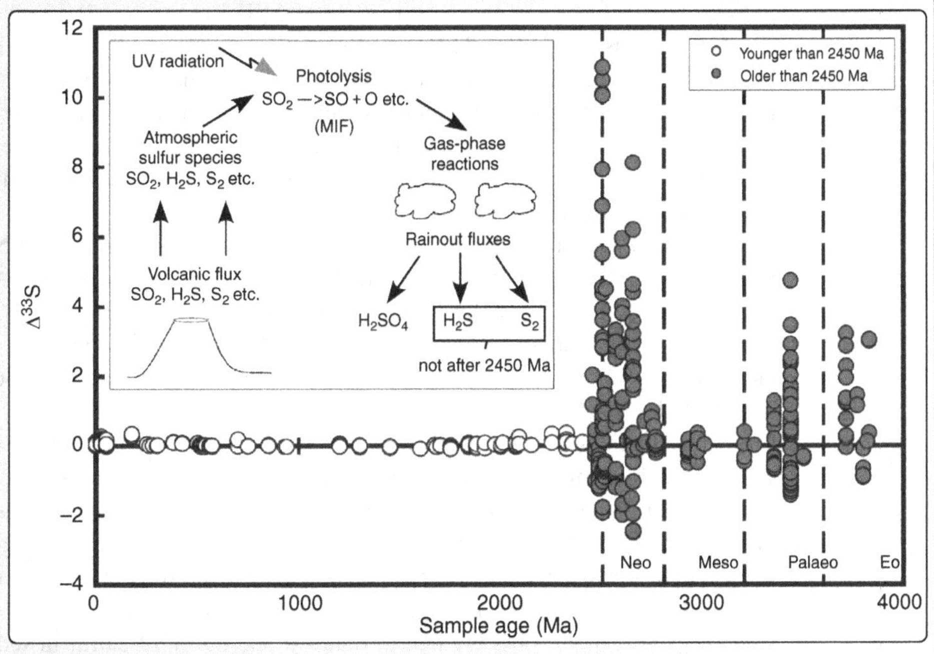

Figure 5.10 $\Delta^{33}S$ as a measure of MIF of sulfur isotopes between different reservoirs in the Earth versus time showing the clear change in behaviour just after the end of the Archaean era (Farquhar *et al.*, 2010). The inset shows a schematic of pathways of sulfur in circulation from volcanic sources through the atmosphere to incorporation in sediments, which can explain the behaviour (modified from Pavlov and Kasting, 2002). See text for discussion.

change in sulfur isotope signature at ~ 2450 Ma marks a threshold of oxygen content of the atmosphere: (1) oxygen and ozone absorb UV radiation of sufficiently high energy to cause SO_2 photolysis and thus strongly suppress photolysis in the lower atmosphere; and (2) at the present day all sulfur that precipitates into the oceans from the atmosphere is converted into sulfate (H_2SO_4) before incorporation into minerals. The different isotope signatures of the different gas species in the atmosphere are therefore homogenised in the oceans. Preservation of MIF is only possible if minerals precipitate from multiple sulfur species such as sulfide and polyatomic, zero-valent species (S_2, S_8, etc.), and is hence only possible where these species are not oxidised in the oceans (e.g. Pavlov and Kasting, 2002; Farquhar and Wing, 2003).

The constraints from the history of MIF, together with the range and average values of $\delta^{34}S$ in sedimentary rocks through time, tell us about the evolution of sulfur reservoirs in the Earth's surface environments (Farquhar *et al.*, 2010). We now have a fairly precise time line and knowledge of evolution of the sulfur and oxygen chemistry of the oceans and atmosphere (see also Figure 5.3), and of how the steps in the evolution link to the contributing factors of the evolution of life and, possibly, climate. The evolution of ocean chemistry over time from the early Archaean is considered to be:

- The atmosphere prior to 2450 Ma was dominated by CH_4 and N_2 and had molecular oxygen concentrations that are estimated to be at most about 2 ppm (less than 10^{-5} present levels) based on the preservation of MIF isotope fractionation of sulfur (Pavlov and Kasting, 2002). The oxidation state was buffered by the coexistence of CO_2 and CH_4, and was such that sulfide was the dominant form of sulfur in the oceans, although low concentrations of sulfate of less than 200 μM were present (compared to modern concentrations of about 30 mM). The early oceans were sulfur-poor because of lower rates of weathering dissolution of sulfide minerals in anoxic surface waters.
- Production of O_2 in the photic zones of the ocean by oxygenic photosynthetic bacteria (cyanobacteria) commenced at some time probably in the late Archaean. Initially, oxygen produced by photosynthesis would have been largely removed from solution by redox reactions such as the oxidation of Fe(II) to Fe(III), probably close to the sites of formation, although molecular oxygen may have built up to detectable levels at least locally and episodically in the late Archaean.
- The Great Oxidation Event (GOE) in the early Proterozoic at about 2.42 Ga, at which time biogenically produced oxygen built up levels in the atmosphere that exceeded 10^{-5} times the present concentrations, as documented by the cessation of preservation of MIF. The upper levels of the oceans equilibrated with the atmosphere and hence also become oxic at this time. However, deeper levels of the ocean remained anoxic, probably throughout the Proterozoic. The oceans also remained relatively sulfur-poor, possibly due to removal and burial of sulfur in sedimentary pyrite (e.g. Farquhar *et al.*, 2010). The average $\delta^{34}S$ in sedimentary rocks over this period is 'heavy' and well above zero. As sulfur input from magmatic sources would have had $\delta^{34}S$ ~ 0, there must thus have been a sink for 'lighter' sulfur, most probably through burial in deep ocean sediments.

– Drawdown of Fe(II) from solution in the deeper oceans, in part by precipitation of BIFs with the last major BIFs forming at about 1.85 Ga.
– Slow or episodic build-up of oxygen levels in the atmosphere and oceans during the Proterozoic era. Uraninite is present in sedimentary rocks deposited after the Great Oxidation Event (GOE), and may have been essentially stable because oxygen levels remained low. Oxygen levels in the atmosphere reached levels close to those of the present in the late Proterozoic. There may be periods into the Phanerozoic in which at least some ocean basins became anoxic, for instance because of restrictions to ocean circulation (e.g. Eastoe and Gustin, 1996).

There is evidence that other aspects of the chemistry of the oceans have also varied through time, including the make-up of its salinity, such that at different periods halite rather than gypsum was the first mineral to precipitate in evaporites (Lowenstein *et al.*, 2005). These variations may have affected the abundance of sulfate minerals such as barite in marine environments.

5.1.2 Sedimentary-rock-hosted Mn deposits

Manganese, like iron, has multiple oxidation states in natural environments. The geochemical behaviour of Mn in aqueous solutions has similarities to that of Fe, including:

- The dominant form of manganese in reducing near-surface waters is Mn(II), which is relatively soluble.
- There are two oxidised forms, Mn(III) and Mn(IV), that form in sequentially higher oxidation states. Mn(IV) is the dominant form in environments that are in equilibrium with an oxygenated atmosphere. Both forms are weakly soluble in typical waters and precipitate in surficial environments as commonly occurring manganese oxide and hydroxide minerals, e.g. hausmannite (Mn_3O_4) and pyrolusite (MnO_2).

There are, however, also differences in geochemical behaviour of Mn compared to that of Fe:

- The boundary between the conditions at which Mn(II) and Mn(IV) species are dominant is at higher oxidation levels than the equivalent boundary for Fe species, and Mn is thus soluble in a wider range of surficial environments than is Fe.
- Mn(II) does not readily bond with sulfide, but has almost exclusively lithophile behaviour. The major Mn minerals that precipitate in surficial environments are carbonates and oxides.

Manganese is added to seawater in sub-sea-floor hydrothermal systems, for instance at mid-ocean ridges, and also in solution and in particulate matter in river waters. As a consequence of the low solubility of Mn(IV), Mn is at low concentrations (≈ 0.5 nM, see Figure 5.1) in normal seawater. Manganese concentrations in solution as Mn(II) can be at least three orders of magnitude higher in anoxic seawaters (< 20 μM, e.g. the Black Sea, Figure 5.11). Additionally, high concentrations of particulate Mn as fine-grained

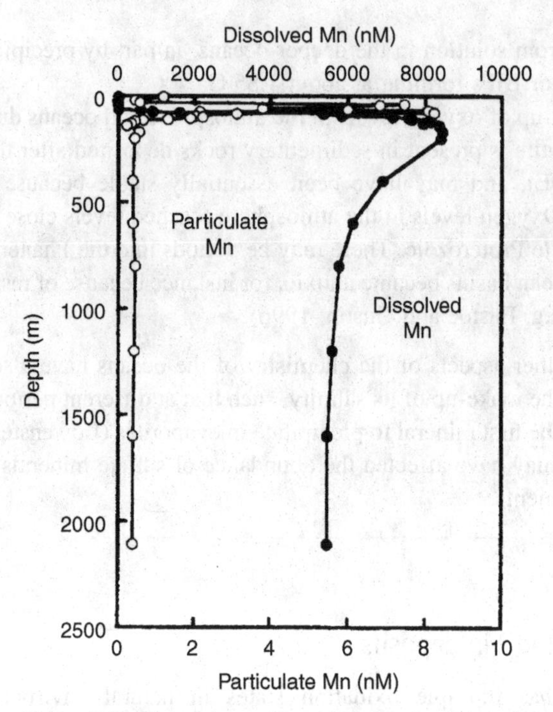

Figure 5.11 Example profiles of concentrations of particulate and of dissolved Mn in the Black Sea (Lewis and Landing, 1992). The peak in particulate Mn concentrations just above the near-surface drop-off of dissolved Mn is typically formed of fine grained manganese(IV) oxide and hydroxide minerals.

manganese oxide and hydroxide minerals can form and occur just above the oxycline in a stratified water column as a result of transport and dispersion of Mn across the redox gradient (Figure 5.11). In contrast to the behaviour of Fe, however, high concentrations of solute Mn are not restricted to a narrow depth interval below the oxycline of modern anoxic basins, but continue to depth. Because Mn does not precipitate in sulfide minerals under **euxinic** (sulfide-present) conditions its concentration in both anoxic and euxinic waters is generally controlled by the solubility of the manganese carbonate mineral rhodochrosite.

Locations of sedimentary Mn deposits

The majority of Mn supply and the majority of known Mn resources are from a small number of large, stratiform deposits of late-Archaean to Tertiary age hosted in sequences of shallow marine sedimentary rocks that formed distant from active volcanism. By analogy with iron formations, these are sometimes known as manganese formations. Many of the largest known deposits are of early-Proterozoic age, and the ca. 2.2 Ga Hotazel Formation of the giant Kalahari manganese field in South Africa includes about 50% of reserves (e.g. Beukes and Gutzmer, 1996; Gutzmer and Beukes, 1996). Other large deposits include the late-Cretaceous deposits of Groote Eylandt, Northern Territory, Australia (Bolton *et al.*, 1990) and the Neogene Nikopol deposits of southern Ukraine (Kuleshov, 2011).

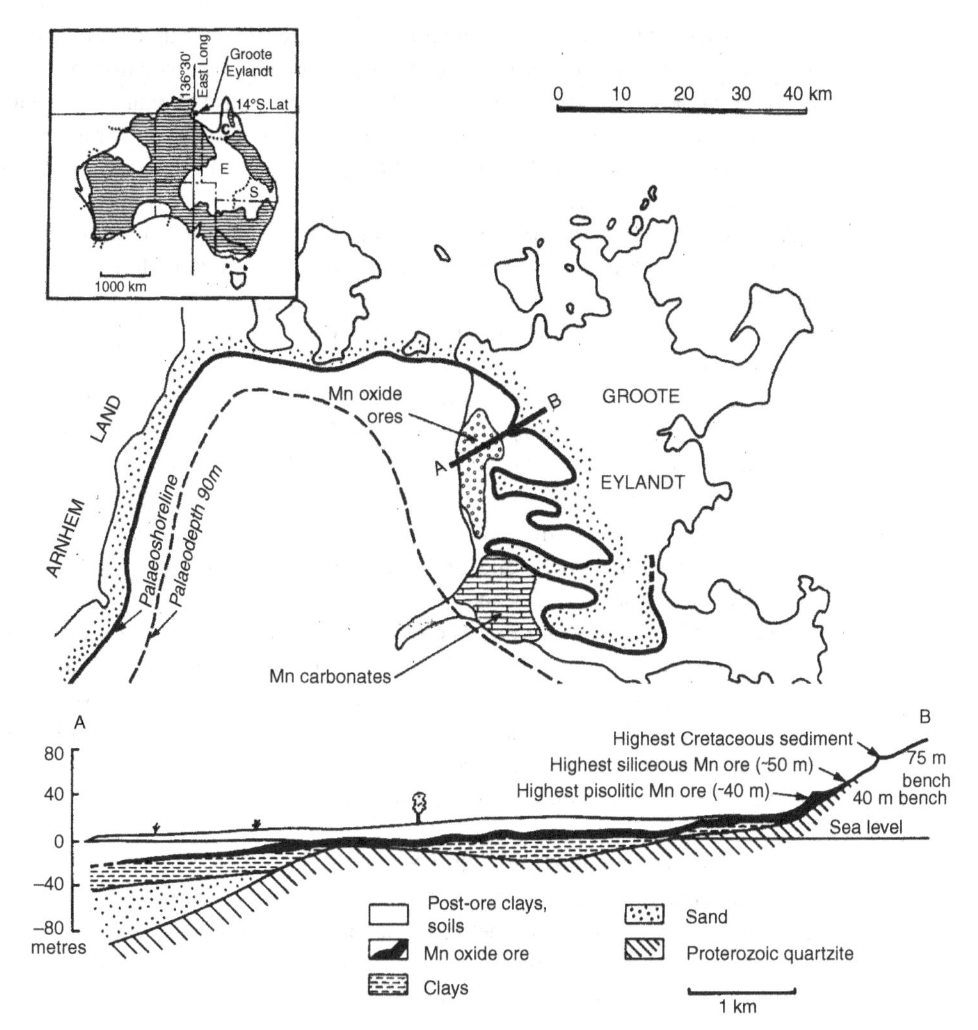

Figure 5.12 Geological map and cross section through the late-Cretaceous Mn deposit of Groote Eylandt, Australia (Bolton *et al.*, 1990). The deposits formed close to the shoreline of a relatively restricted marine basin. There is a zonation of mineralogy and texture of the Mn ores from the palaeoshoreline seaward. The richest ores are the pisolitic manganese ores, which presumably formed in an environment in which bottom sediment was continually reworked by currents. Reprinted with the permission of the Australasian Institute of Mining and Metallurgy.

Characteristics and settings of sedimentary Mn ores

Many Mn ores are essentially sedimentary rocks that have undergone little chemical modification since deposition. The stratigraphic settings of all Mn ores indicate deposition in relatively shallow water on continental-shelf environments, in some cases in semi-restricted basins. Although the sedimentary units are laterally extensive, they are much less so than iron formations, and typical dimensions are from less than 5 m to about 30 m thick with lateral continuities of over tens of kilometres, as for example at Groote Eylandt (Figure 5.12).

These sedimentary rock units contain one or more Mn-bearing mineral as the dominant component, most commonly either pyrolusite (MnO_2) or rhodochrosite ($MnCO_3$), more rarely braunite ($Mn^{2+}Mn^{3+}{}_6SiO_{12}$) or other Mn-bearing carbonate or silicate minerals. Ore grades are thus between about 25% and 40% MnO. Other major chemical components of the ores include SiO_2, Fe_2O_3 and CaO, and gangue minerals are variably quartz, haematite, apatite and clays.

A number of the major ore fields have separate carbonate and oxide ore bodies at the same horizon. In other deposits there are relatively abrupt boundaries between ores with different Mn minerals. In all cases the manganese carbonate ore formed down palaeoslope of the manganese oxide ore, and the oxide ores are interpreted to have precipitated in relatively shallow waters.

The early-Palaeoproterozoic manganese ores are interbedded with BIF, in the case of the Hotazel Formation, for instance, with three Mn-ore horizons 5–45 m thick interbedded with similar thicknesses of variably carbonate-, oxide- and silicate-facies iron formation. The Mn-ore horizons are generally massive and lack the conspicuous banding that is characteristic of iron formations. The mixed iron and manganese formation overlies intermediate volcanic rocks and underlies dolostone or is unconformably overlain by later-Palaeoproterozoic shales and quartzites.

Phanerozoic Mn ores are most commonly immediately associated with organic-rich shales, glauconitic shales, sandstones, cherts and carbonate units including marls, limestones and platform dolostones. Some deposits are immediately above a major unconformity. In some deposits, manganese oxide minerals are characterised by **pisolitic** and oolitic textures, for instance at Groote Eylandt (Figure 5.12). These geologically recent Mn ores are Fe-poor compared to the Proterozoic ores.

Mn ore deposition in marine environments

The sedimentary deposition of concentrations of Mn minerals is interpreted to be the result of precipitation on a shelf of fine particles of manganese(IV) oxide minerals that grow in the water column after oxidation of Mn(II) to Mn(IV) above an oxycline (Figure 5.11). The simplified reaction is:

$$Mn^{2+}{}_{(aq)} + O_2 \rightarrow MnO_2. \quad (5.1)$$

The manganese oxide precipitates may either accumulate on the sea floor under oxic waters, or, where they sink into deeper anoxic bottom waters, the Mn may be recycled back to Mn(II) through biologically mediated reactions at or below the sea floor. A proportion of the Mn(II) reacts with organically produced CO_2 to form manganese (II) carbonate minerals, particularly rhodochrosite. Carbon isotope ratios in rhodochrosite in these deposits show depletion of ^{13}C relative to bicarbonate in solution in seawater and thus imply that much of the included carbon was sourced through oxidation of organic matter. Pisolitic and oolitic textures in some ores indicate that manganese oxide minerals precipitated onto nuclei in agitated shallow water.

The flux of terrestrial sediment into the ocean must remain low to allow accumulation of high-grade Mn ore. The early-Palaeoproterozoic deposits formed over a similar span of Earth history as Superior-type iron formations and are interpreted to have formed in essentially the same environment as the interbedded BIFs. They thus formed on a continental shelf at a time in Earth's history after the GOE during which the oceans were stratified with respect to redox state. There was thus a large reservoir of reduced Mn in solution in relatively deep ocean water which, for instance, periodically or locally upwelled onto a shelf. The BIFs that are interbedded with manganese formations have only slightly higher Mn contents than typical Superior-type iron formations, whereas Fe is a major component of the Mn ores. The interbedded BIF and manganese formations may thus be a result of relatively subtle oscillations of water depth, with Mn only depositing when water was shallowest and concentrations of oxygen reached a threshold value.

The formation of an Mn deposit during more recent periods of Earth history requires development of anoxic or euxinic waters in an ocean basin. The presence of glauconite and a high organic content in associated shales are both indicative of chemically reducing conditions in ocean waters at or above the sediment interface. The development of anoxic water is most likely to occur in a semi-enclosed shallow marine basin but may also occur in a zone of extended oxygen minimum water, such as forms in an ocean column below areas of high organic productivity in ocean surface waters and hence high fluxes of decaying organic matter. This setting is implied by the association of Mn ores with cherts and black shales. A marine transgression is indicated by the stratigraphic positions of some ores above unconformities and may be important to form anoxic conditions in semi-enclosed basins through promoting high organic productivity and release of organic matter from buried soils and flood-plain sediments. In contrast to oceans of the Palaeoproterozoic, anoxic ocean basins of the present day have high concentrations of Mn but low concentrations of Fe(II) and very little Fe is available to be precipitated along with Mn.

Role of weathering and secondary enrichment of manganese formations

In a similar fashion as for ironstones, secondary enrichment, either as a result of hydrothermal activity or as a result of weathering, has upgraded parts of many manganese formations through leaching of Ca and SiO_2 and through conversion of manganese carbonate minerals to manganese oxides. The upgrading is, however, typically by at most a few per cent MnO, and such upgrading is not a necessary requirement for economic ore. Weathered manganese formations are characterised by a dominance of hydrous Mn(III)- and Mn(IV)-bearing phases such as manganite ($Mn^{3+}O(OH)$) and todorokite ($(Ca,Na,K)(MgMn^{2+})Mn^{4+}{}_5O_{12}xH_2O$).

5.1.3 Sedimentary-rock-hosted phosphorus deposits

The major world sources of phosphate for production of fertilisers are marine sedimentary rock units with between 40 and 90% modal apatite (specifically, francolite $\approx Ca_{10}(PO_4)_5(CO_3)F_2$). Phosphorite or phosphorus rock is a sedimentary rock with greater than 9% P_2O_5, hence about a hundred times enrichment over average rock. The Permian Phosphoria Formation of the Rocky Mountains of northern USA is a major example of phosphorite and contains about five times the amount of phosphorus than is presently in solution in all ocean waters.

Phosphorus is moderately abundant in seawater, with total concentrations of micro-moles (Figure 5.1). The largest reservoir in the oceans is, however, organic matter. Phosphorus is an essential element for life, and is in relatively high concentrations (up to about 1%) in both hard parts of macro-organisms (e.g. bones) and as an essential

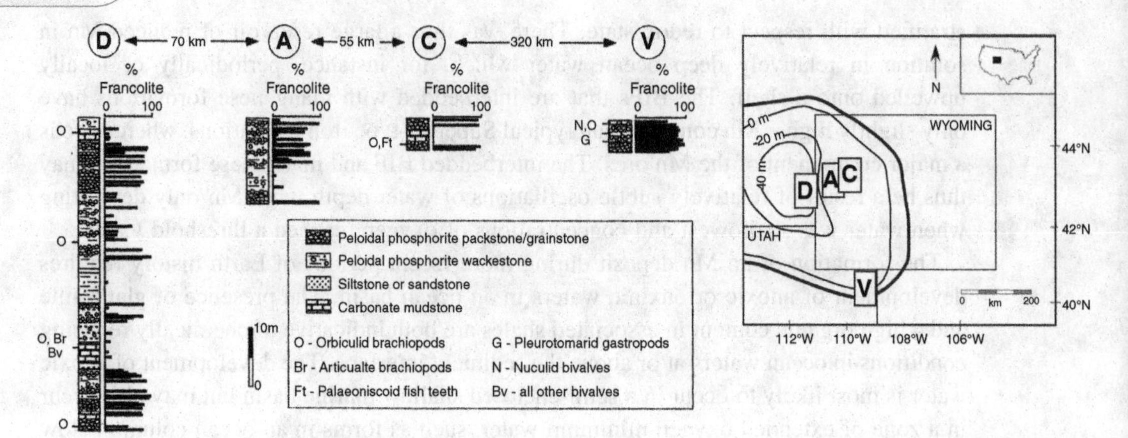

Figure 5.13 Setting and measured sections of the Permian Phosphoria Formation in different parts of the Phosphoria Basin of the western USA. Overall palaeoslope is up from west (seaward) to east (landward) (Hiatt and Budd, 2003). Macrofossils are rare throughout most of the basin, and the forms indicate dysaerobic waters. Near-shore shallow-water conditions are indicated at the eastern locality (V).

component in molecules such as adenine triphosphate (ATP) and adenosine diphosphate (ADP) in the cells of all organisms, including unicellular organisms.

Nature and settings of phosphorite units

Sedimentary phosphate ores occur in sequences of dominantly shallow marine sedimentary rocks and are stratiform units up to a few metres thick and laterally extensive over distances of up to about 100 km in shallow marine sedimentary sequences (Figure 5.13). Associated sedimentary units include one or both of reduced fine-grained sediments (glauconitic shales, black shales and carbonate mudstones) and biogenic chert (Figure 5.13). The nature of many of the host sequences shows that they are condensed sequences in which there was limited sediment accumulation over relatively extended periods of time in shallow marine environments that had little terrestrial sediment input. The 1–30-m-thick Permian Phosphoria Formation, for instance, is estimated to have been deposited over a time period of about 6 million years.

Textures of phosphorites include: nodular or peloidal aggregates of fine-grained apatite in organic-rich shale, indicating that apatite grew diagenetically in the sediment column, rather than was precipitated or was sedimented direct from seawater; apatite oolites, indicating reworking by currents; and shales with thin apatite laminae. The ores can have high concentrations of a number of trace elements, including Se and U.

Many phosphorite units formed at low palaeolatitude at less than about 40 degrees from the equator. They formed from the Cambrian to the Miocene and predominantly in restricted periods of Earth history, in particular the Cambrian, Permian and Miocene eras.

Formation of phosphorite

Authigenic phosphate minerals are actively precipitating in a range of marine sedimentary environments in the oceans, including on shelves and in deep marine basins. Information

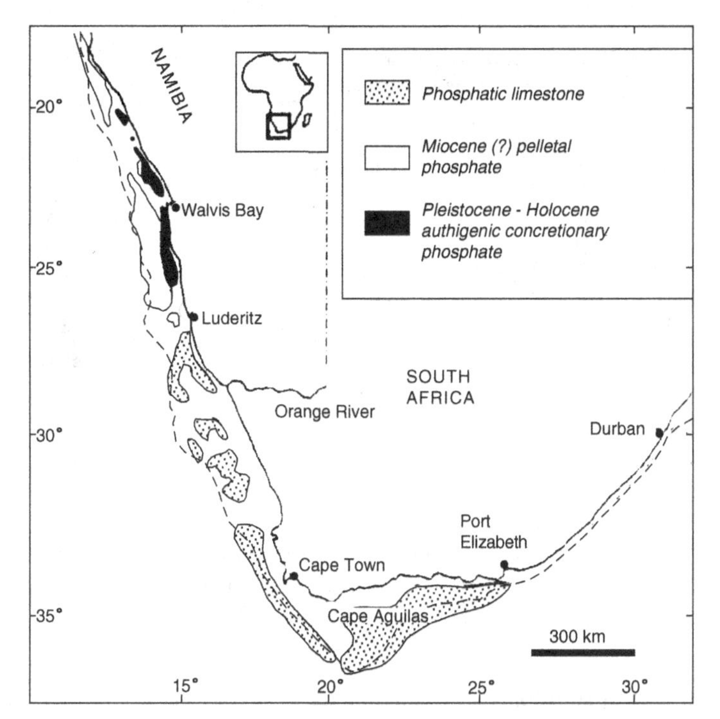

Figure 5.14 Distribution of phosphate accumulations on the continental shelf of southwest Africa. Phosphate-rich sediment is presently accumulating along the inner shelf off the Namibian coast adjacent to the Namib Desert. Older accumulation is more widespread and to greater water depths. After Thomson *et al.* (1984).

about the origin of phosphate deposits thus comes from studies of these present-day environments and their sediments.

'Phosphogenesis' is the process of precipitation of phosphate minerals in sedimentary environments, in particular apatite, and has been and is a widespread process in shallow sediments on the ocean floor. Most marine sediments have a fairly constant concentration of P_2O_5 of about 0.1%. Phosphate is released from organisms on decay, and becomes dissolved in pore waters in the sediments. Apatite precipitation in shallow sediments is widespread and occurs over time spans of thousands to millions of years and may be both inorganic and modulated by bacteria, most particularly by sulfide-oxidising bacteria. Apatite accumulates where rates of release of phosphorus from decaying organic matter exceed the combined rates of organic recycling of phosphorus and release upwards into ocean bottom waters.

The accumulation of apatite-rich sediments (phosphorites) is occurring in a number of areas in the oceans, particularly along the Peru–Chile continental margin and the Namibian margin (Figure 5.14). The presently and recently active sites of accumulation of phosphorite contain most conspicuously nodules in unconsolidated sediments on the ocean floor in shallow shelfal waters (< 400 m depth).

The Peru–Chile and Namibian coasts are the sites along which deep ocean water upwells as a result of ocean current and atmospheric circulation patterns, in particular

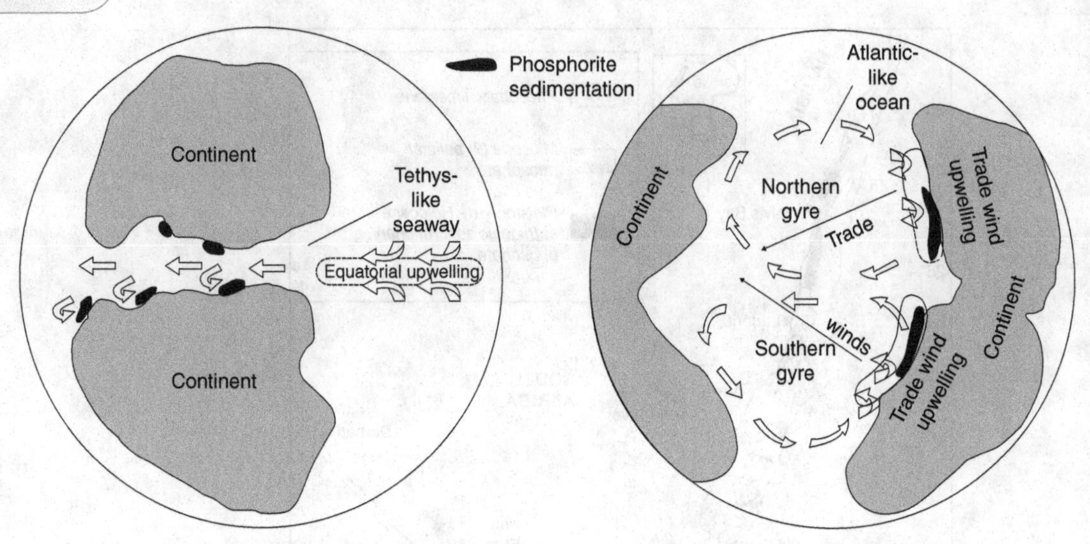

Figure 5.15 Theoretical patterns of ocean currents and continent distributions that would give rise to upwelling of nutrient-rich bottom waters and consequent enhanced precipitation of apatite and accumulation of phosphorite (after Sheldon, 1980).

of 'divergent' flow at sub-tropical latitudes along west coasts of continents (Figure 5.15). Because of the rain-shadow effect of mountain chains and continental highlands, these areas of deep ocean water upwelling are also areas of very limited river input and hence clastic sediment input onto a shelf. Shallow east–west channelways in tropical latitudes would also be sites of upwelling, and may be the site of formation of early Phanerozoic phosphorite units in the Georgina Basin of interior Queensland, Australia. Other areas of less strong upwelling in oceans are the result of 'dynamic' flow associated with irregular sea-floor topography and strong sea-bottom currents on east coasts of continents, for instance over seamounts and along eastern Australia.

Upwelling ocean bottom waters are oxygen-depleted but are rich in a number of nutrient elements, most importantly P and N (Figure 5.2). These nutrients have accumulated in deeper ocean water from the descent of organic matter derived from decaying organisms through the ocean column. Upwelling therefore brings nutrients to near-surface waters, promoting in some cases extreme organic productivity, for instance dinoflaggelate blooms that give rise to 'red tides' and mass fish kills. The critical factors for the formation of phosphorite thus appear to be (Figure 5.16): (i) high organic productivity of surface waters and high rates of sedimentation of organic matter, (ii) limited dilution by terrestrial clastic sediments and (iii) reworking of sediment by bottom currents. The reworking of the sediments by ocean-floor currents leads to the commonly observed oolitic, pisolitic and hard-ground facies of phosphorite (Figure 5.16).

Reconstruction of the palaeogeography of the Permian Phosphoria Formation of western USA shows that it fits many aspects of the models developed to explain present-day ocean-floor phosphate precipitation. The formation was deposited in desert latitudes on the west coast of continental margin of Laurasia (Figure 5.17). Unlike present sites of phosphorite accumulation the formation was deposited in a relatively shallow marine semi-restricted basin sited behind volcanic island arcs rather than on an open shelf.

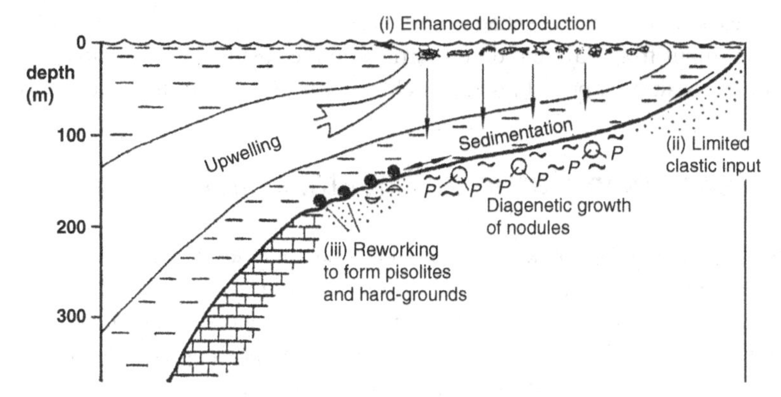

Figure 5.16 Model for the accumulation of phosphates in shelfal sediments in areas of deep-ocean upwelling (Baturin, 1989). P = phosphate nodules. The text explains further the critical factors in the accumulation of phosphorite.

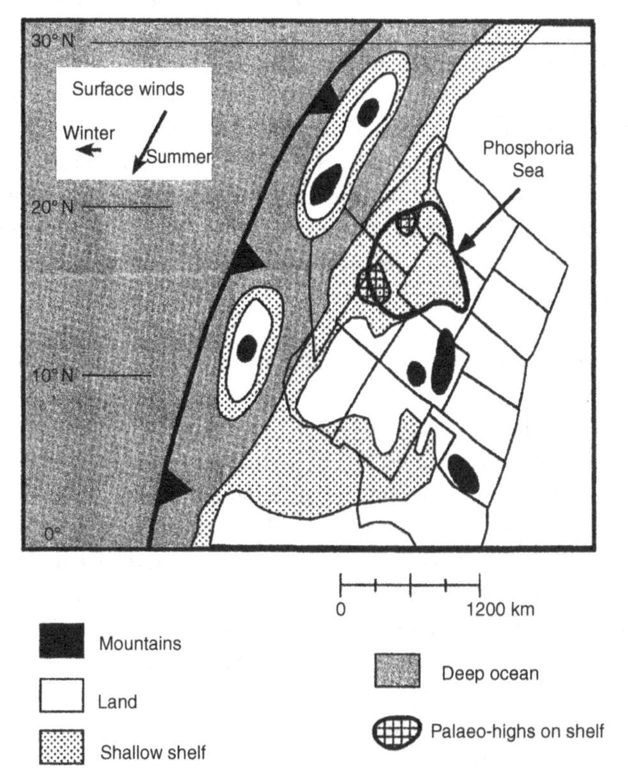

Figure 5.17 Palaeogeographic reconstruction of the depositional setting of the Permian Phosphoria Formation (Hiatt and Budd, 2003). Thin lines show the positions of the palaeo-environmental features relative to the present state boundaries in the USA.

Processes that may have enhanced primary organic productivity include wind-blown sediment input from adjacent desert terrains to the east and evaporation on the landward side of the basin that pulled nutrient-rich water into the semi-restricted marine basin.

5.2 Ore deposits in clastic sedimentary environments

Deposits of a number of metals and industrial minerals are mineral concentrations in sediments and sedimentary rocks that formed through mechanical sorting and separation within a mass of sediment. We distinguish **placer** deposits that formed by sedimentary sorting in a present-day sedimentary environment, and **palaeoplacer** deposits that formed in a sedimentary unit that has been buried, and in some cases deformed and metamorphosed.

Processes in the formation of placer deposits

(a) Properties of minerals concentrated in clastic sedimentary environments

In order to be sorted by mechanical processes in a clastic sedimentary environment, a mineral must have one or more distinct physical properties. Most typically, the distinct property is density relative to the dominant sediment component, which in almost all sand-sized and coarser sediment is quartz and feldspars and has a density of about 2.7 g cm^{-3}. Minerals mined from placer deposits range in density from diamond (about 3.5 g cm^{-3}) to gold and platinum-group minerals (PGMs) with densities up to about 20 g cm^{-3} (see Table 5.1).

In order to survive transport and repeated reworking in a clastic sedimentary environment a mineral must be mechanically stable. Stability may be a result either of physical hardness, such that it will not shatter on impact with quartz and other minerals during sediment transport, or of high malleability, in which case it will deform on impact. In addition, the mineral must be either chemically stable or chemically unreactive in the sedimentary environment.

(b) Physical processes involved in mechanical concentration in sedimentary environments

Placer concentrations of minerals develop over the range of scales from individual laminae in fore-sets of dunes to sedimentary units tens of kilometres in extent. The most important medium for sorting at all scales is flowing water, although aeolian, colluvial and glacial processes are recognised as contributory to the development of some placer deposits.

Although we can derive equations for the transport behaviour of individual grains of idealised shape in water or in a sediment mass, and make experiments in flumes of the behaviour of sediment masses containing mixtures of grains of differing sizes and densities under controlled flow conditions, sediment behaviour in natural environments is dependent on a large number of, in part, interdependent processes, and is hence difficult to predict. Transport of grains will be dependent on unique characteristics of the sediment mass, such as the relative abundances of clasts of different size and density, and on the unique characteristics of the transport and depositional environments. In addition, clastic sedimentary environments are characterised by variable energy of flow in time and space such that there may be clast entrainment into the flow, different modes of clast transport (bed load, saltation, suspension), and deposition of clasts multiple times during the history of a sedimentary system. Despite these complexities, specific processes in the cycle of entrainment,

Table 5.1 Minerals mined from placer and palaeoplacer deposits, their physical properties, common placer settings of ores, and the types of rocks that provide primary sources of the minerals.

Mineral	Chemical formula	Density (g cm^{-3})	Hardness (Moh's scale)	Setting of deposits	Major hard-rock source of mineral
Rutile	TiO_2	< 5.5	6–6.5	Strandline	Widespread
Ilmenite	$FeTiO_3$	4.75	5–6	Strandline	Mafic igneous rocks
Zircon	$ZrSiO_4$	4.6	7.5	Strandline	Felsic igneous rocks
Monazite	$(REE)PO_4$	5.1	5	Strandline	Felsic igneous rocks
Garnet	$X_2Y_3Si_3O_{12}$*	3.35–4.3	6–7.5	Strandline and fluvial	Garnet-rich gneisses and skarns
Chromite	$FeCr_2O_4$	5.1	7.5–8	Fluvial	Ores
Gold	Au	19.3	2.5–3	Fluvial	Ores
Cassiterite	SnO_2	7.0	6–7	Fluvial	Ores
Uraninite	UO_2	8.7–10.9	5–6	Fluvial	Granites
PGMs	(various)	6.7–22.9	3–7	Fluvial	Ores
Diamond	C	3.5	10	Fluvial and strandline	Ores

*X and Y in the chemical formula for garnet refer to crystal sites that hold dominantly 2+ and 3+ cations respectively.

transportation and deposition appear to contribute to sorting of particles by density and size. In generalised terms the important processes of sorting are likely to be the following:

(i) Differences in particle hydraulics: the rate of settling, and hence of deposition, of a particle from either laminar or turbulent flow is related to its density, its size and also to its shape. Denser and larger particles will be deposited closer to a source, or earlier during waning of flow. We can define settling or hydraulic equivalence of particles that will settle at the same velocity through a water column based on the ratio of size to density. Settling hydraulics leads to sorting during sedimentation, hence during aggradation.

(ii) Differential particle entrainment from a bed: the lower boundary of flowing water is influenced by flow resistance or boundary shear stresses imposed from the bed. Where the bed is relatively planar, a viscous or laminar sublayer of flow may be developed below turbulent water. The ease of entrainment of a particle from the bed into suspension in the flow is related to a balance of shear forces imposed by the flow in this sublayer to inertial forces of the particle. In general a small, dense particle will remain in the bed or in the transported bed load because of its lower profile and higher inertial stresses, and also because it may remain 'hidden' by larger, less-dense particles. Sorting thus occurs through preferential entrainment and transport downstream of lighter components. Differential entrainment may build up a concentration of heavy minerals in a lag during degradation, or in a bed load.

Additional processes that may affect sorting include: settling of fine particles through a coarse-grained bed; the destruction of the viscous sublayer where there are irregularities of flow set up, for instance, by effectively permanent irregularities on a bed, such as potholes or large immobile clasts; sorting in sheer flow, either in water with dense concentrations of suspended load or in moving bed load; and the dynamics of turbulent flow.

As is discussed further below, large concentrations of minerals with particular density occur where there are repeated cycles of clast entrainment into a current, transport and deposition, hence through alternation of sorting by aggradational and lag processes. The whole geomorphic history of a basin may contribute to the development of a resource of mineable size.

(c) Sites of placer formation

The most important sites of placer deposit formation are:

1. Shorelines or strandlines:

   (i) On sand-dominated shorelines, *heavy-mineral sand* deposits occur along dune-sand shorelines (Section 5.2.1). These occur along present-day shorelines and along palaeoshorelines at sea-level high-stands of the Neogene and that are broadly parallel to the present shoreline but at altitudes up to about 150 m above current sea level. These deposits are concentrations of minerals that are significantly denser than quartz or feldspar and that are widely dispersed as trace minerals in many rock types. Similar heavy-mineral concentrations are known in older sandstones and metamorphosed equivalents, but none are economic, in part because of the additional costs of mining indurated sedimentary rocks.

   (ii) On bedrock shorelines, accumulations specifically of diamonds can form in the surf zone and in fossil surf zones. The diamonds are most strongly concentrated in depressions on eroded bedrock beneath present-day beaches and flooded Pleistocene beaches. This style of placer is almost uniquely developed along the coast of southern Namibia and northern Cape Province, South Africa, north of the mouth of the Orange River, which drains the area of diamondiferous kimberlites and orangeites of interior South Africa (Cronan, 2000).

2. Rivers:

Accumulations of ore minerals in alluvial sediments of present-day rivers and in sediments and sedimentary rocks of past rivers from late-Archaean to Pleistocene age (Section 5.2.2). In most cases the accumulations are in coarse-grained, bed-load sediments of high-energy rivers that drain catchments with exposed primary ('hard-rock') deposits or sub-economic concentrations of ore minerals. The majority of these fluvial placer deposits are at most 10–20 km downstream of a primary ore source. Ore grades in a placer can be similar to those of the eroded hard-rock resources. However, the primary hard-rock ores may be sub-economic, whereas the ease of mining of placers and collection of ore minerals from multiple hard-rock ore bodies that are each too small to be mined economically contribute to economic viability of a placer deposit.

5.2.1 Heavy-mineral sand deposits on shorelines and palaeoshorelines

Sandy sediments of shoreline and palaeoshoreline placers are the major world resource of titanium minerals (rutile, ilmenite, leucoxene), zircon (as source of zirconium and hafnium) and REE-bearing minerals (e.g. monazite, xenotime), and are locally exploited for

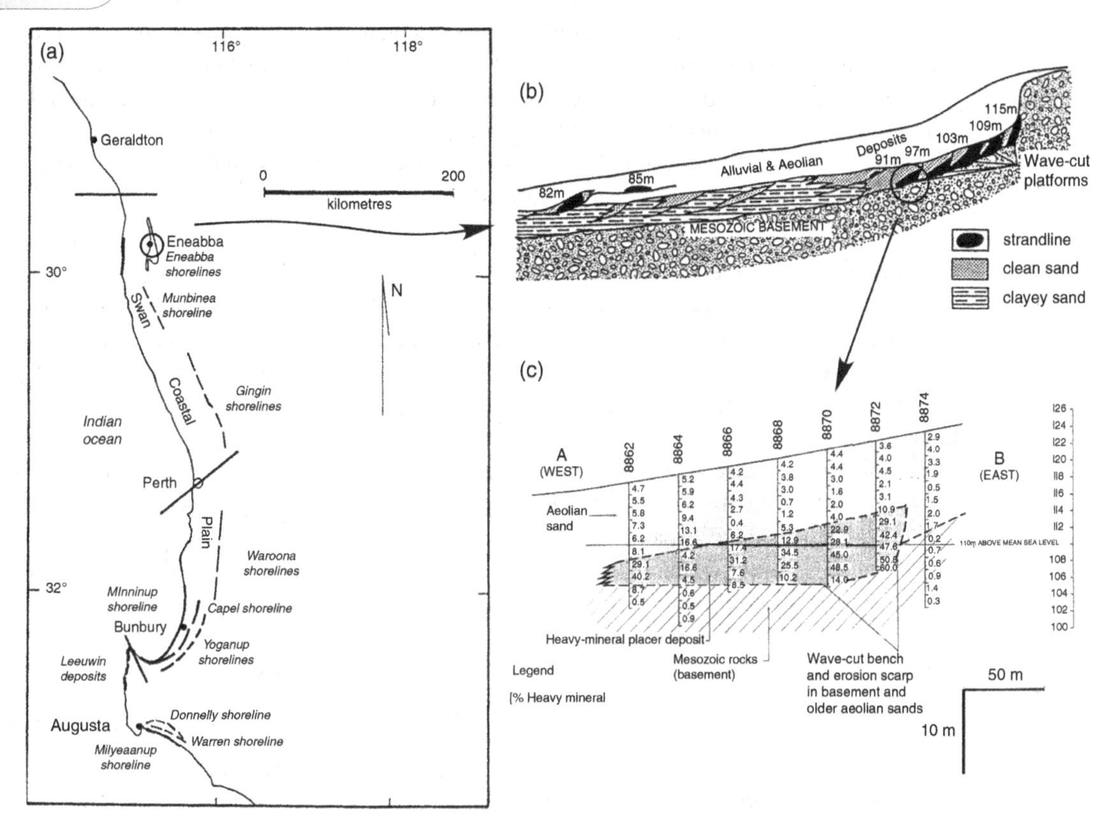

Figure 5.18 Setting of Pliocene heavy-mineral sands at Eneabba, Western Australia (Shepherd, 1990). The sands are along multiple palaeoshorelines at between 60 and 110 m above present sea level. Each deposit forms a ribbon with lenticular cross section sited along a wave-cut platform on the underlying Mesozoic sedimentary rocks. Reprinted with the permission of the Australasian Institute of Mining and Metallurgy.

garnet. Important deposits include those along the west, south and east coasts of Australia, in southern India, along the southeast coast of South Africa, and in southeastern USA.

Characteristics of heavy-mineral sands

Typical heavy-mineral concentrations in a beach sand are less than 1%. Heavy-mineral-sand ores are those parts of sand strandline dunes or fossil strandline dunes which have heavy-mineral contents of 5–50%. The deposits are ribbon-shaped ore bodies up to a few kilometres long and a few hundred metres wide that are parallel to the shoreline or palaeoshoreline (Figures 5.18 and 5.19). In cross section, the concentrations are lenticular and can be up to a few metres thick. The ore bodies are sited towards the base of the sand body (Figure 5.18c), although the shapes of a body can be relatively complex where the ore has formed through multiple processes and cycles of transport and deposition. On present-day strandlines, heavy minerals are typically most strongly concentrated at the berm, that is, the level to which storm waves reach (Figure 5.19b).

The deposits are most characteristically well-sorted fine to medium sands: a high concentration of finer particles, e.g. clay minerals, between sand-sized clasts, reduces the value of the ore because of the additional processing that is required to separate the ore minerals. In general, multiple economic minerals are present in any heavy-mineral sand deposit and each is separated and extracted. A typical make-up of the heavy-mineral fraction is 35–55% ilmenite, 15–30% rutile, 10–25% zircon and about 3% monazite. Different ores have different relative abundances of these minerals, and some ore bodies are zoned with respect to the concentrations of the ore minerals.

Settings of heavy-mineral sand deposits

Strandline heavy-mineral concentrations are best developed along microtidal- to mesotidal-wave-dominated coasts, that is, along coasts with less than about a 2-m tidal range that are exposed to open oceans and on which wave energy rather than the energy of tidal currents is the dominant source of energy for sediment transport. A further requirement is that there is a relatively low rate of input of sediment from rivers. Deposits form along shorelines that fringe land masses with relatively subdued topography (Western Australia, southeast India) or with closely spaced, short-length rivers (northern New South Wales, Australia; Natal, South Africa).

There can be multiple parallel fossil strandlines along any one stretch of coast, each corresponding to a different stand of sea level (Figure 5.18b). Shorelines formed during sea-level high-stands before, or early during, the present cycle of build-up of glacial ice are particularly important. The deposits of southeastern USA (Florida, Georgia, North and South Carolina), for instance, are fossil strandlines formed during Pliocene sea-level high-stands.

Ore bodies are not continuous along a strandline. Irregularities of the coastline and prevailing wind and current systems are important controls on deposit setting. Many deposits are developed adjacent to promontories (Figure 5.19a), especially on the up-drift side relative to littoral drift. Other deposits are down-drift of river months.

Processes of ore formation

Heavy-mineral sand deposits are the result of sorting of the approximately 1% content of heavy minerals from the lighter minerals ('lights') in a mass of sediment, with the latter being transported and concentrated elsewhere in the sedimentary system. Heavy minerals are concentrated by each wave at the limit of wave reach. The major sorting process at a small scale on a beach system is thus the difference in sediment transport and entrainment into incoming waves and into backwash (Figure 5.20), and hence the difference in particle transport in high-energy turbulent flow relative to a low-energy, low-flow regime, effectively laminar water flow.

Wave action by itself does not sort sediment laterally along a strandline, nor does it sort minerals by density over distances larger than a few metres. Processes that contribute to larger-scale sorting include:

(a) Littoral (long-shore) drift. At a promontory, for instance, the 'lights' may be carried by backwash down the beach to a water depth at which they can then be carried by littoral drift around the promontory (e.g. Figure 5.19a) whereas a portion of the 'heavies' are trapped at the berm on the up-current side. Over time, there will thus be an increase in the concentrations of heavy minerals on the up-current side of the promontory.

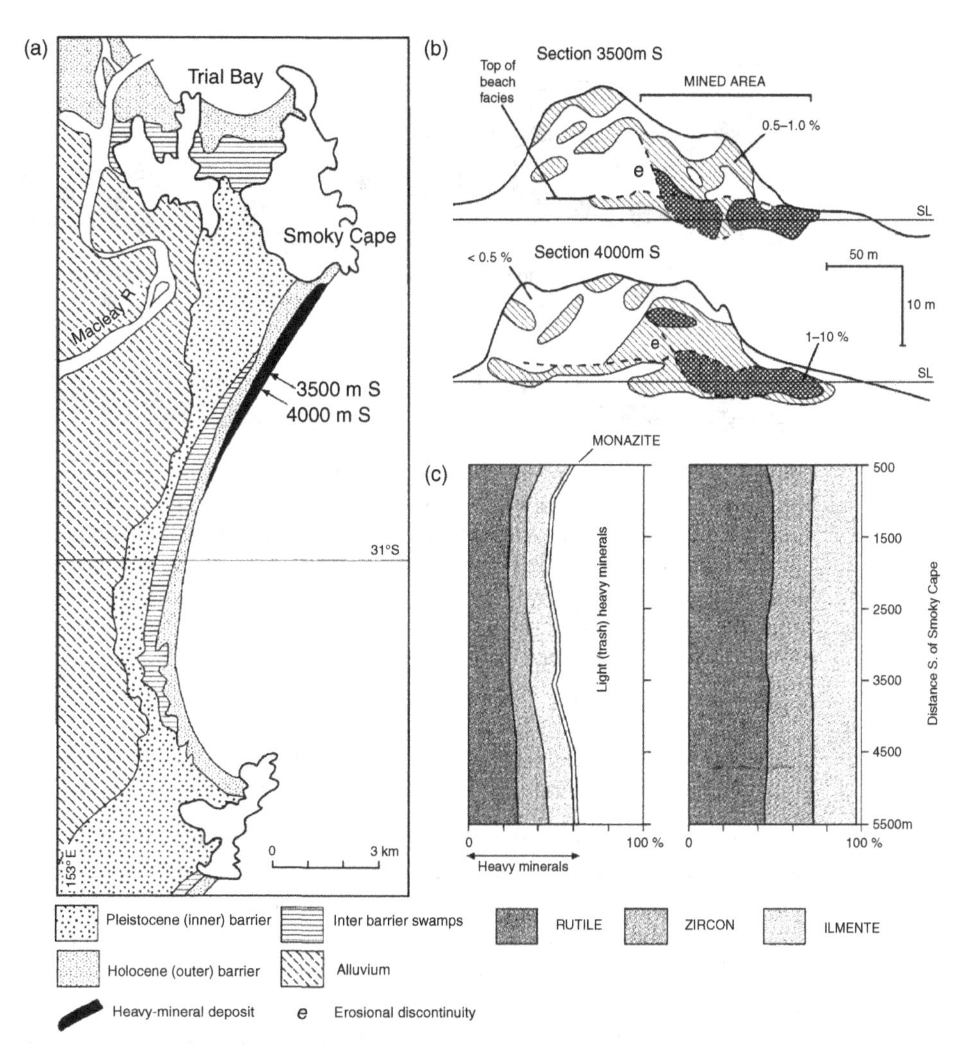

Figure 5.19 Position (a) and cross-sectional shape (b) of the heavy-mineral deposit on a Holocene beach at Smoky Cape, New South Wales, Australia. The heavy minerals are most strongly concentrated at the berm at and just above the level that waves most commonly reach at high tide. The erosional unconformity marks the position of truncation of older Pleistocene dunes. In plan view, the deposit is on the up-current side of the rocky promontory of Smoky Cape relative to the south-to-north littoral drift (Roy, 1999). The graphs (c) show the components of the heavy-mineral fraction of the sand, and show the uniformity of composition along the length of the deposit. Light (trash) heavy minerals are minerals of little economic value and in this deposit include leucoxene, magnetite, tourmaline, hornblende and garnet.

(b) Transgression–regression cycles on a stable shelf. During transgression, heavy minerals are progressively swept up-slope as the level of wave reach moves upwards across the beach with time. The heavy minerals at a sea-level high-stand will therefore be concentrated from a wide strip of the shelf. Similar, but less-efficient, mechanisms operate during regression, with winnowing of light minerals from a wide

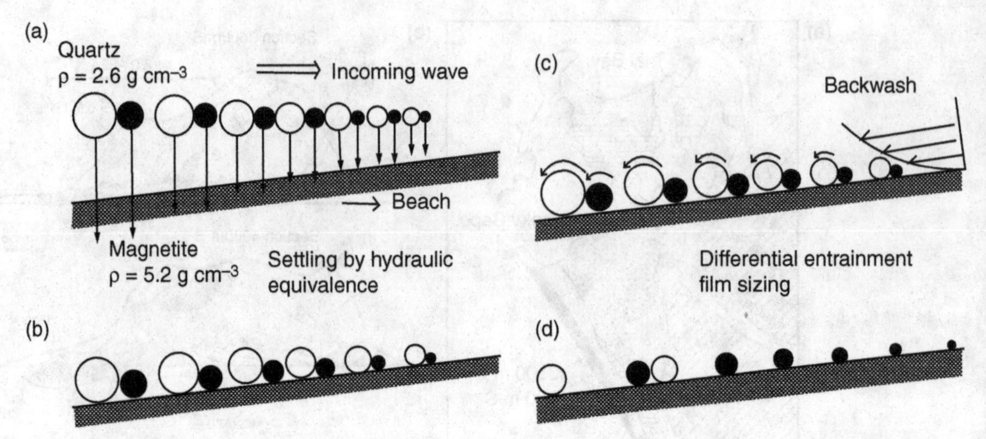

Figure 5.20 Processes in the development of heavy-mineral concentrations at the wave front or berm along wave-dominated beaches. Turbulent flow of incoming waves entrains available sand-sized sediment. (a, b) As incoming waves progressively lose energy, particles are sorted as a function of their hydraulic equivalence and hence by the product of density and size. (c, d) Re-entrainment in the backwash is controlled by shear stress in approximately laminar flow. Large, light grains are thus preferentially entrained leaving behind small, dense grains at the wave front (after Siebold and Berger, 1996).

strip of the shelf. Repeated episodes of concentration through geological history may be a contributing factor in the development of major heavy-mineral-sand resources.

The heavy minerals are derived as part of the mixed sediment input from rivers or from local older sedimentary basins that are being eroded along the shoreline. Ultimately, the minerals are derived from crystalline rocks of, for instance, orogenic belts or Precambrian shield areas, and the mineralogical make-up of a deposit is controlled by the make-up of source regions. The input into the shoreline system may be several hundred kilometres up-drift of a deposit. A low rate of sediment supply may be important to allow the most efficient sorting during both processes: the input sediment must be processed fast enough to remove the non-heavy-mineral component from the site of heavy-mineral concentration.

5.2.2 Fluvial placer and palaeoplacer deposits

Fluvial placers in coarse-grained sandy to pebbly sediment (gravels) of modern river channels and palaeoplacers of older but unconsolidated alluvial gravels have been at many times in history the most important source of gold, and are continuing to be mined for gold on a small scale in a number of parts of the world. Gold placers and unconsolidated palaeoplacers were sought ores, particularly early in the exploitation of a new ore field, because of the ease of mining of loose, near-surface sediments and the ease of extraction of the gold with minimal capital investment. Alluvial palaeoplacer deposits in indurated sedimentary rocks are and have been an important source of gold in gold fields such as in the Archaean Witwatersrand basin of South Africa and the Palaeoproterozoic Tarkwaian rocks of Ghana.

Similar placer and palaeoplacer deposits have been or are less commonly mined for diamonds (Orange River and its tributaries, South Africa), for uraninite as a source uranium (Blind River–Elliot Lake, Ontario, Canada) and also for PGMs, for instance in Siberia and Alaska. Alluvial and related placer ores have also been a major source of cassiterite as an ore mineral of tin, such as from the Kinta Valley in Malaysia and from adjacent Indonesia and southern Thailand.

In almost all cases, the ore minerals of fluvial placers and palaeoplacers are derived from eroding hard-rock ores. In some of these cases, the primary hard-rock ores are uneconomic, either because of low grade, or, especially in the case of gold, because they are high-grade veins too thin to be mined efficiently. The majority of fluvial placers of all commodities are proximal to the eroded ore, typically at most 10–20 km downstream.

There are a number of exceptions to the generality that fluvial placers are proximal to source. Diamonds are transported more than 1500 km in the Orange River of South Africa. Some alluvial gold fields such as those on the east flanks of the Southern Alps in the South Island of New Zealand are up to 200 km from likely sources. Long transport distances affect the mineralogical characteristics of these deposits. In addition to generally finer grain size of ore minerals, the proportion of gem to industrial diamonds increases with transport distance, because of preferential abrasion of flawed stones, and distal gold placers are characterised by strongly flattened and rolled gold grains with smoothed impacted surfaces and by gold-rich–silver-poor rims of the grains.

Major Cenozoic fluvial gold placer and palaeoplacer deposits

In a number of important gold fields in the world, the production from the alluvial sections of the fields is comparable to and in some cases a few times greater than the production from hard-rock mines, including in the Sierra Nevada region of California, USA, the Otago gold field of the South Island of New Zealand, and the Ballarat gold field of Victoria, Australia. In all of these areas, the hard-rock ores are orogenic deposits, although placer production has also been a major component of gold production around some fields of low-sulfidation epithermal deposits.

The primary ores of the large gold placer fields are sited in areas of relatively high relief. Gravel bed-load streams are in all cases a major source of placer ore. There are two generalised settings of the placer deposits in the rivers or alluvial sediments. Many occur in relatively narrow valleys with steep gradients. In these settings much of the gold may be reworked from relicts of gravel terraces at higher elevations than the current stream bed. Other high-tonnage placer deposits occur in broader gravel terraces that formed where there was a sharp decrease in stream gradient, most commonly at a mountain front.

In the Ballarat gold field, for instance, ore production from placers and palaeoplacers is about four times that of hard-rock mining from 'orogenic' or 'quartz-vein' deposits, even though the field has been intensely explored and mining extended to about 700 m depth. Placers are traceable about 10–15 km downstream from the primary source. Ore grades in the placer deposits were generally in the range of 3–40 ppm, hence similar to those of the hard-rock ores. The ore field shows many typical characteristics. The terrain has a relief of a few hundred metres. Multiple episodes of gold accumulation are recognised by multiple geomorphological settings of ore (Figure 5.21). The oldest placers are in broad alluvial gravel terraces of probable early-Cenozoic age that formed over a weathered landscape and are locally preserved at about 200 m higher elevation than the current stream

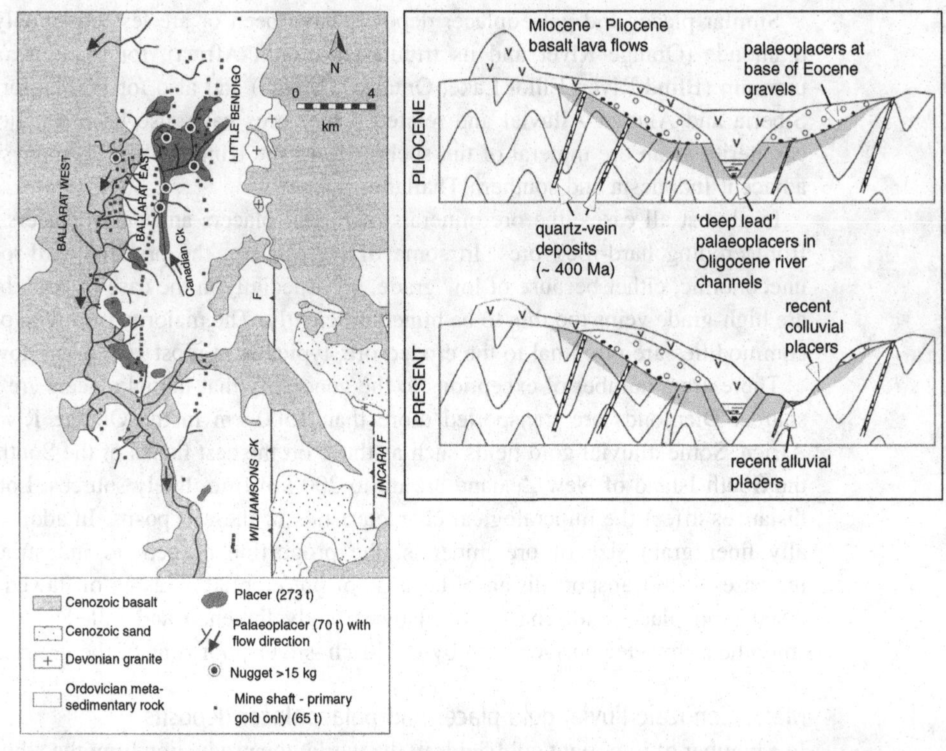

Figure 5.21 Distribution and interpretation of gold placer and palaeoplacer deposits at Ballarat, Victoria, Australia (after Taylor and Gentle, 2002 and Hughes *et al.*, 2004). The 'deep lead' palaeoplacers are stream channels of probably Oligocene to Pliocene age buried beneath Pliocene to Miocene basalt lavas. The cross sections illustrate the importance of reworking of gold in landscape through the Cenozoic. Note that (i) almost all the placer and palaeoplacer gold was mined from within a few kilometres of the primary vein deposits, even where it has been reworked, the positions of which are indicated by the mine shafts, and (ii) that the amount of gold mined from placers and palaeoplacers has been greater than that mined from hard rock, even though mining was down to about 700 m at the deepest shafts.

channels. These ores were partly reworked in the late Oligocene or early Miocene into gravels at the base of sediment fill of incised narrow valleys. Pliocene basalts covered and preserved the lower gravels to form the important 'deep leads' in the field. More recent erosion of outcropping vein deposits has produced gold accumulations in slope colluvium. Erosion of gold from the ancient gravel terraces and re-accumulation into the presently active river channels appears to have also been important.

The presence of large placer deposits such as at Ballarat demonstrate that the processes of mechanical concentration of gold from eroding veins can be very efficient. There are a number of possible explanations for the dominance of placer over hard-rock gold. There may have been higher gold abundances in the eroded rock than at and beneath the present surface. Much of the gold may have been present in small, uneconomic hard-rock deposits. The gold may have been concentrated from a slice of eroded crust much thicker than the depth of mining. Whatever is the relative importance of each of these three factors, we can infer the following: from the presence of multiple placer accumulations in gravels as old as early Cenozoic in many of the large fields that placers may have accumulated gold from a thick slice of now-eroded crust; and, from the dominance of placer over hard-rock production and the short distances of transport down-slope, that over the time scale of denudation, and on the length scale of the river system, gold is concentrated in a heavy-mineral lag and remains close to the source whereas the dominant light minerals are transported downstream.

The Witwatersrand palaeoplacer Au–U deposits, South Africa

Greater than about 50 000 tonnes of gold, or about 35% of the world's total production through history, has come from deposits in low-metamorphic-grade sedimentary rocks of the late-Archaean Witwatersrand Basin. Uranium from uraninite has been a by-product from many of the mines. With respect to size, extent and grade these deposits are unique. Smaller and lower-grade deposits that have very similar stratigraphic and host-rock settings do, however, occur in sedimentary basins of slightly younger, early-Palaeoproterozoic age (Tarkwa, Ghana; Blind River–Elliot Lake, Ontario, Canada; Jacobina, Brazil), of which the Blind River–Elliot Lake deposits were a major source of uranium from uraninite.

The Witwatersrand deposits are interpreted as palaeoplacer deposits, although hydrothermal genesis of the deposits has alternatively been proposed. Some of the arguments for and against this alternative proposal for ore genesis are discussed in Box 5.2.

(a) Geology of the Witwatersrand Basin

The Witwatersrand Basin is the best-preserved large, late-Archaean, dominantly terrestrial clastic sedimentary basin. It developed over about three hundred million years ($\approx$ 3.0–2.7 Ga) and contains approximately 7 kilometres of overall coarsening-upwards quartz arenites, lesser shales, minor, thin conglomeratic beds, and some volcanic rocks that were deposited over an older granite–gneiss basement terrain (Figure 5.22). Basin sedimentation was terminated on the eruption of the extensive thick sequence of the Ventersdorp basalt and andesite lava flows. Present-day outcrop is in a basin-like form with the stratigraphically highest units generally outcropping towards the centre of the preserved basin.

The major depositional environments during the period of and in the area of the Au deposits are interpreted to be alluvial braid-plain, terrestrial braid-delta and lacustrine. The rivers flowed from highlands to the north and west of the presently outcropping extent of the basin with shoreline and marine sediments recognised in the centre and along the opposing, southeast side of the preserved basin. Internal unconformities show that there was tectonic activity during sedimentation and repeated or progressive uplift along the western and northern margins of the basin across which sediment was supplied. This uplift also gave rise to multiple episodes of reworking of the sediment within the basin during its development, and reworking included both the formation of near-planar unconformity surfaces and local channelised incision into older units. The geometries of all of the unconformity surfaces are such as to mark relative uplift and erosion on the up-slope side of the basin margin and progressively basin-ward movement of positions of the break in slope at the basin edge. The upper boundary of the basin sequence is also marked by an unconformity and erosion before cover by the regionally extensive Ventersdorp lavas.

Geochemical and isotope compositions and the ages of detrital zircons of the sediment in the basin show that it was largely derived from eroded granite–greenstone and gneissic terrains that evolved with abundant felsic igneous activity from up to a few hundred million years earlier to just before the onset of sedimentation in the basin (3.5 to 2.95 Ga). These sources may have been either immediately adjacent to the basin or in more distal orogenic belts. The adjacent older terrain to the north and west is largely covered by younger rocks and is poorly exposed. Its characteristics, however, show that it could have provided sediment to the basin. It does, for instance, host a few known orogenic (quartz-vein) gold deposits.

Major fault systems cut the rocks, but none displace them to the extent that basin geometry is masked. The Palaeoproterozoic terrestrial impact structure of the Vredefort Dome affected the rocks throughout the basin. The sequence also underwent regional metamorphism at up to lower greenschist-facies assemblages in the Palaeoproterozoic such that ores and host-rocks are thus at least in part recrystallised.

(b) Setting of Au ores in the basin

The ores are hosted in the Central Rand Group, which forms the upper 2–3 km of basin fill. There are about 10 major stratiform Au-bearing horizons (called **reefs**) developed to different extents in the seven or eight gold fields that lie along the north and west sides of the limit of outcrop of the Central Rand Group (Figure 5.22). Each of these gold fields is considered to mark the site of a major river and hence of fluvial sediment input in the basin from adjacent highlands. Many of the reefs can, however, be correlated between the gold fields and over distances of over 100 km (Figure 5.23). These reefs are not continuously ore along their strike length, but within six major gold fields one or more of the stratiform reefs is mined extensively over blocks of up to 20 km along strike parallel to the basin margin and down dip to the east or south over about 10 km in plan view.

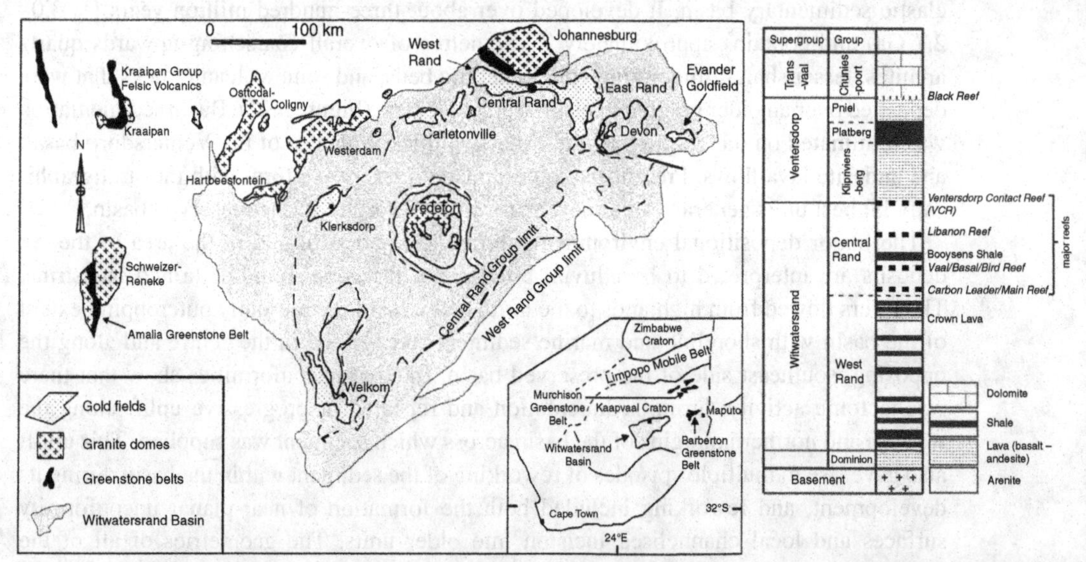

Figure 5.22 Map of the ≈ 3.0–2.7 Ga Witwatersrand Basin, South Africa showing the distribution of the major gold fields around the northern and western margins of the outcrop of the Central Rand Group of the basin. Greenstone belts in the surrounding terrains are highlighted as they are interpreted to be the most likely source terrains of sedimentary gold from orogenic gold deposits. The simplified stratigraphic column of the ≈ 7-km-thick sequence of the basin shows the position of the major reefs in the upper half of the sequence (Barnicoat *et al.*, 1997; Poujol *et al.*, 1999).

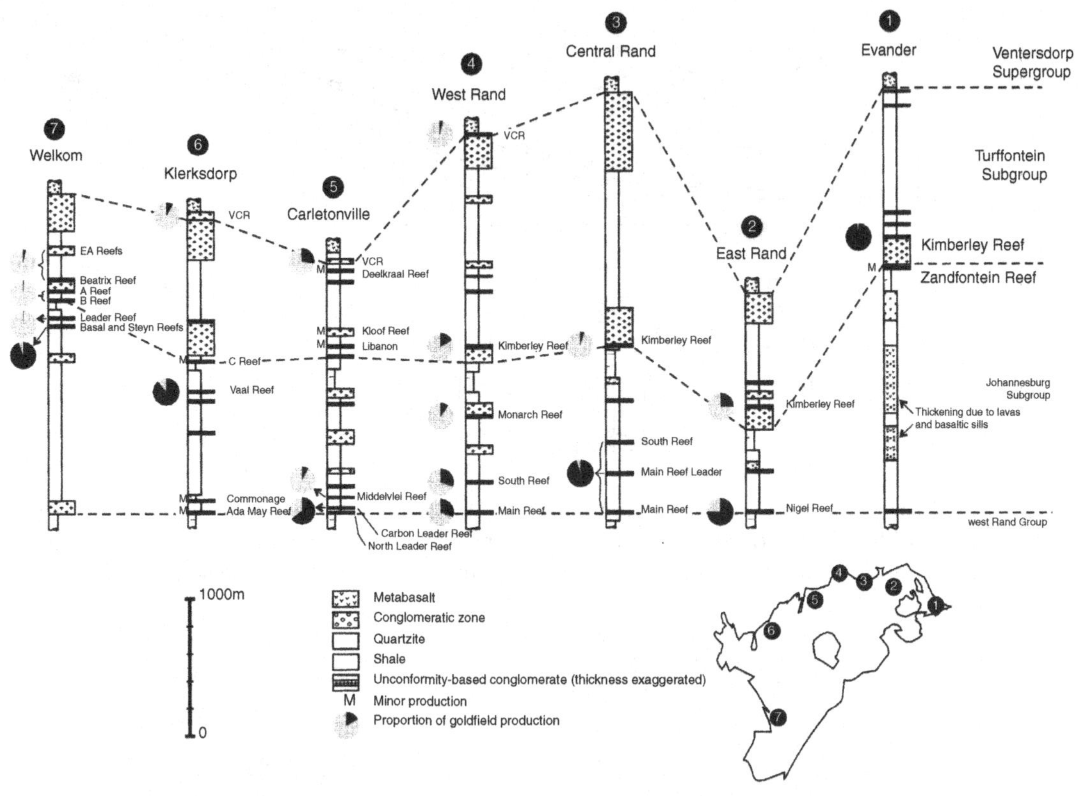

Figure 5.23 Stratigraphic columns of the Central Rand Group of the Witwatersrand gold fields (see Figure 5.22) showing the position, correlations between, and relative Au production from each of the major reefs. Note that the stratigraphically lowest reef has produced most of the gold in each gold field (Tankard *et al.*, 1982, Phillips and Law, 2000).

Although increasing depth of the reefs towards the centre of the basin hinders economic mining, Au content does also generally decrease down dip. Gold deposition was thus over a restricted section of the palaeoslope, and the equivalent horizons to the reefs lack ore where they outcrop on the southern side of the basin.

The reefs are most typically thin (5 cm–2 m) beds of conglomerate or pebbly sandstone in the predominantly quartz–arenite sequence (Figure 5.24). Not all conglomerates in the succession are mineralised, and thicker beds are generally less productive than thinner. The reefs are laterally continuous but have marked lateral facies variations such as channel and overbank facies. One distinct reef facies is the 'carbon-leader' reef. These reefs have high concentrations of fine gold grains together with U minerals within organic matter, which occur as millimetre-thick bedding-parallel 'seams', and have provided about 40% of the total produced gold. Carbon-leader reefs may mark 'overbank' settings, but in some cases occur at the base of conglomerate units. Overall the different reef facies define braided patterns with anastomosing sub-parallel channels directed towards the basin centre (Figure 5.25).

Many reefs, especially those with the highest gold grades, are immediately above the low-angle erosional unconformities in the sequence. Others reefs appear to be conformable, but may mark the time of formation of unconformities up-slope of the reef position.

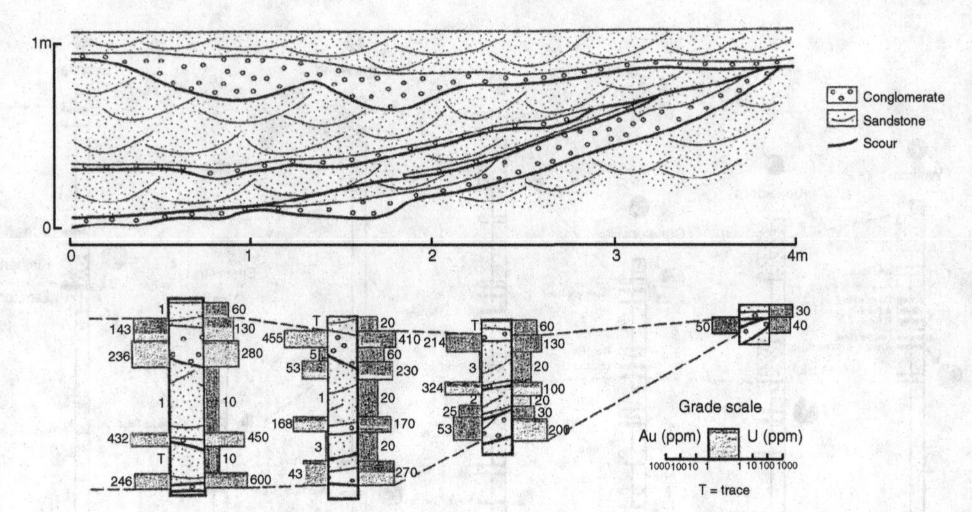

Figure 5.24 Sketch of part of the Basal Reef, Welkom gold field and distribution of Au and U within the reef (after Minter, 1978). Note the rapid variations in sedimentary facies, typical of fluvial sediments, and the close association of Au and U with the conglomerate facies.

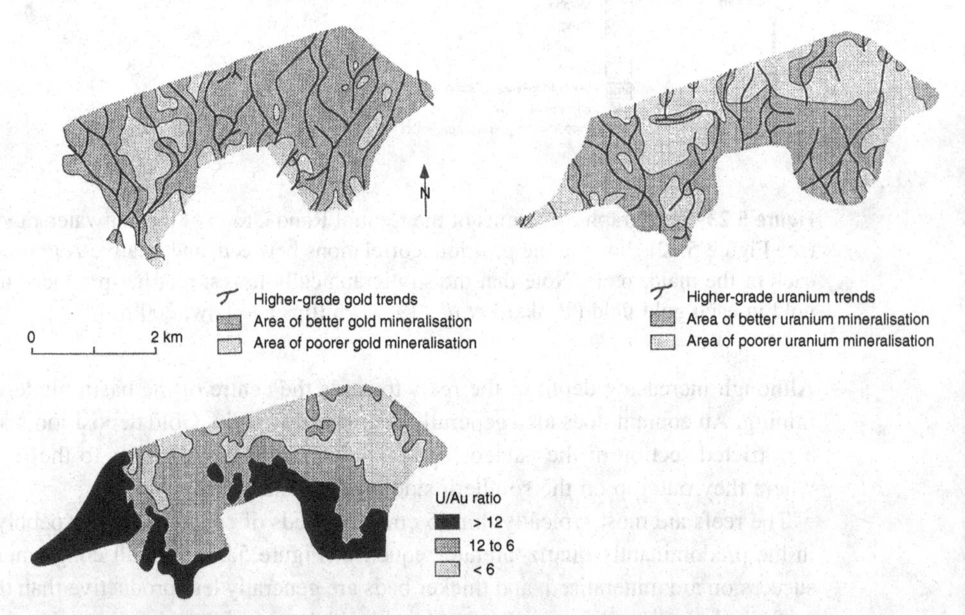

Figure 5.25 Zonation of Au and U in the Carbon Leader Reef, Witwatersrand. Palaeoslope is interpreted to be from north to south. Higher Au and U concentrations occur in sinuous trends which are interpreted as channels in a braided river system. High U/Au ratios occur down the interpreted palaeoslope and are interpreted as a result of hydraulic sorting (Buck and Minter, 1985).

(c) Characteristics of the ores

Well-rounded quartz-vein and chert pebbles dominate the conglomerate clast population. Common heavy minerals in the conglomerates include uraninite, pyrite, zircon, leucoxene and chromite, all of which occur as rounded clasts. Pyrite forms typically about 3% of the ore and is present variably as rounded grains and as euhedral grains. The sizes of the different heavy-mineral grains indicate hydraulic equivalence with the dominant quartz-vein clasts of the reefs. Systematic spatial variations of the size of clasts and of U/Au ratios, for instance, down the palaeoslope along reefs, indicate that sorting was active during deposition (Figure 5.25).

Gold is present as fine grains up to a couple of millimetres diameter in the conglomerate matrix. It has a wide range of morphologies including flakes smeared on pebble surfaces and fine crystalline growths. Some gold grains show a flattened toroidal cross-sectional morphology, which is characteristic of abrasion in an aeolian environment. Individual gold grains are chemically homogeneous, but grain composition, especially with respect to fineness (a measure of Au/Ag ratio of gold grains), varies between grains in any sample.

Average mined gold grades on the reefs have been between 5 and 20 ppm, but these grades understate the true concentration of gold because mining generally includes some barren sandstone from above and below the reef horizon. Gold grades within the conglomerate reefs can be as high as hundreds of ppm. In detail the gold is very irregularly distributed and much higher grades may be present over thin intervals of the reef. In many exposures, gold is very strongly concentrated within relatively thin conglomerate horizons, especially those that immediately overlie channel scours (Figure 5.24), and hence erosional surfaces. In plan view, the reefs contain 'pay shoots' (= ribbon-shaped bodies of higher-grade ore), which are typically elongate approximately perpendicular to the basin margin and form anastomosing patterns which correspond to the anastomosing pattern of channel facies on the reefs (Figure 5.25), although high-grade shoots are not in all cases hosted in the channel facies.

(d) Interpretation of Witwatersrand ores as palaeoplacer deposits

Sediment transport into the intracratonic foreland basin was along six or more major rivers draining from the northwest (Figure 5.26). The source terrains for the material in the reefs was the same as that for the basin as a whole and probably included the Archaean granite–gneiss–greenstone terrain immediately to the north and northwest of the original extent of the basin and more distal terrains from a larger hinterland area, including the Limpopo Mobile Belt, a major orogenic belt that was actively developing a few hundreds of kilometres to the north.

The chemistry of the gold grains and the characteristics of some of the quartz-vein pebbles are consistent with derivation from orogenic Au deposits (Section 3.3.1). By analogy with other Archaean terrains, the source terrains could have hosted a relatively high density of orogenic Au deposits. Uraninite was in contrast derived from eroded granites. Much of the pyrite was also detrital, and was derived from a range of sources, including vein deposits and diagenetic grains in sediments. The presence both of detrital uraninite (in which uranium is present in the reduced U(IV) oxidation state) and detrital pyrite in fluvial deposits has been an important piece of evidence used to argue for an oxygen-free atmosphere in the Archaean (see Box 5.1).

The stratigraphic position of the reefs approximately 3–4 km above the base of the basin fill is unusual for fluvial gold placer deposits as the placers are neither above nor immediately downstream of a primary source. The depositional environment of the host sequence is interpreted as a braid-plain, on which there were generally relatively low lateral and down-slope gradients over which there was dispersed flow along braided rivers

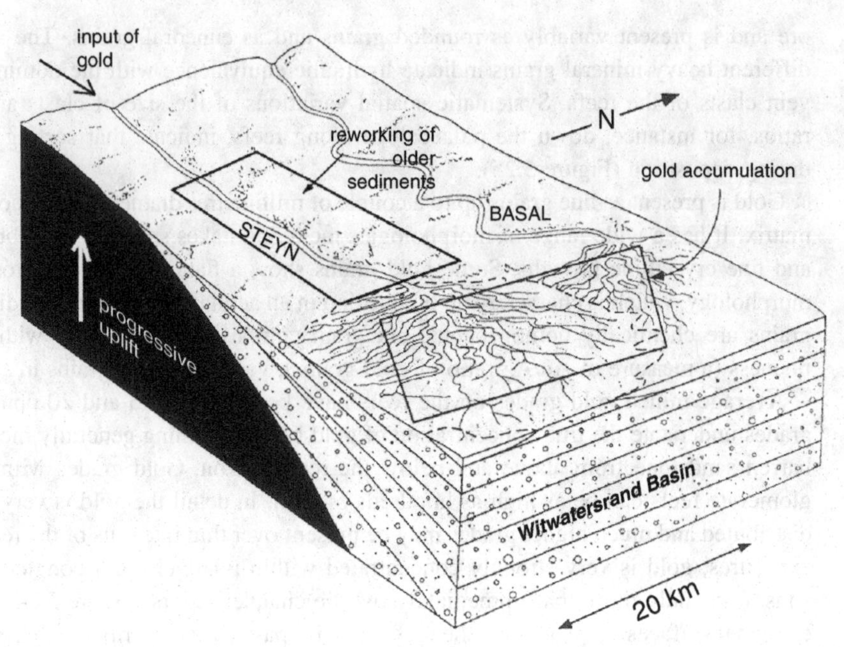

Figure 5.26 Braid-plain/braid-delta setting at the foot of a progressively rising highland area with braided rivers downstream of bed-rock rivers as the environment of placer deposition of the Witwatersrand deposits, specifically the coalesced Basal and Steyn Reefs of the Welkom Goldfield (after Roscoe and Minter, 1993). The unconformities within the Witwatersrand Supergroup are the result of uplift along the basin margin over the period of basin fill, with resultant reworking of the clastic sediment. The reworking may be either onto near-planar unconformities, as shown, or in some cases with significant channelised incision (e.g. Els, 1991; 2000).

that were not strongly incised (Figure 5.26). The setting of the ores may have been at the edge of a rising mountain chain below the line of a sharp drop in river gradient. The lack of terrestrial vegetation in the Archaean would have favoured the development of braided rivers, as is the case in parts of the world with high rainfall but which are poorly vegetated, such as in Iceland. The lack of terrestrial vegetation would also have allowed wind action on sediment whatever the climatic regime. The anastomosing pay shoots broadly follow the coarsest sediments and mark anastomosing channels of the rivers (Figure 5.25).

The relatively high grade of the reefs that are above unconformities in the basin show that sediment reworking within the basin was at least a process of local Au concentration. The eroded older sedimentary rocks would have been dominated by quartz–arenite, and the sand-sized light-mineral component was presumably transported downstream, leaving a deflation conglomerate lag. These unconformities mark times of tilting of the basin strata, presumably as a result of renewed uplift in the adjacent terrain. Repeated episodes of sediment reworking in the basin may have been an important factor in achieving high Au concentrations in the conglomerates. The high concentrations of Au in the organic seams of the carbon-leader reefs are explained as the result of trapping of Au particles from suspension onto microbial mats.

Post-depositional deformation and metamorphism has influenced the textures and mineralogy of ore and host-rocks. The presence of both rounded grains of pyrite and euhedral epigenetic grains and of both flattened flakes and crystalline gold indicates variable degrees of mineral dissolution and recrystallisation, induced most probably during the deformation and low-grade metamorphism.

Both the setting of the Witwatersrand ores and the geological history of the host basin resemble those of large Cenozoic gold placer ores that are hosted in broadly developed alluvial gravels at a mountain front of an actively rising mountain chain, such as in the Otago region of the South Island of New Zealand. In all of these terrains there was repeated down-cutting such that contained gold was reworked, possibly repeatedly, into younger sediments over time periods of at least a few tens of million years. The amounts of contained gold in the Witwatersrand are however about two orders of magnitude higher than that in analogous Cenozoic gold fields, grades of ores are significantly higher and the lateral extents of the mineralised reefs are much greater. Factors that may have contributed to the unique resource of the Witwatersrand include:

(i) The filling history and geometry of the basin. Gold eroded from the hinterland may have been very efficiently collected below a mountain front. The repeated internal reworking of sediment as a result of progressive uplift of the basin margin, erosion and redeposition of heavy minerals at progressively basin-ward new positions of the break in slope may have promoted high gold grades.

(ii) The nature of the source terrain. Archaean granite–greenstone terrains are **prospective** worldwide for orogenic Au deposits. The processes leading to the formation of these deposits may have acted more strongly in Mesoproterozoic terrains such as the terrain that is inferred to be the source of sediments in the Witwatersrand Basin. This terrain may thus have had a high density of Au deposits.

(iii) Limited chemical reworking of gold in reducing surficial environments of the late Archaean. Although Au has a low solubility in surface waters, it is more soluble in oxidising than in reducing waters. Anoxic conditions may have allowed gold grains to survive on average longer in surficial environments. Gold grains in the Witwatersrand ores lack the high-fineness rims that are characteristic of Cenozoic placer ores presumably because the lower oxidation state also inhibited dissolution of silver.

Box 5.2 Debate over the genesis of the Witwatersrand gold deposits

There has been debate over the genetic origins of the Witwatersrand gold ores since early in the history of their exploitation and study in the late 1800s and early 1900s. The phrase 'modified placer deposit' has been often used to express the fact that ore minerals have been largely recrystallised after deposition. Ore genesis through placer processes in a fluvial environment was, however, more thoroughly questioned in the late 1980s in view of observations of the distribution of grade and ore on particular conglomerate reefs (Phillips *et al.*, 1987). It was noted that ore in at least one field may be better described as a rectilinear distribution of high-grade ore shoots rather than anastomosing shoots following braid-plain channels, and that ore grade on a minor reef decreases over a distance of a few tens of metres below the truncation of the reef by an unconformably overlying major reef. Both of these features would be easily explained if gold was deposited by hydrothermal fluids that migrated along both faults and major reef horizons.

Phillips *et al.* (1987) and in subsequent works (see summary by Phillips and Law, 2000) further described mineralogical, petrological and geological features of the deposits that had been overlooked and that could be explained as the result of hydrothermal activity. This hydrothermal activity had apparently taken place at temperatures similar to those that are typical of orogenic gold deposits (see Section 3.3.1). The widespread presence of epigenetic, euhedral pyrite and gold was re-emphasised and it was newly shown that:

- Previous work had overlooked the presence of a metamorphic event in the rocks of the basin, and that mineral assemblages in pelitic horizons include muscovite, chlorite, pyrophyllite and chloritoid, and hence indicate that the deposits are hosted in rocks that were regionally metamorphosed to lower greenschist-facies conditions.
- Unusually aluminous mineral assemblages are widespread in pelites in the Central Rand Group, and also in the matrices of reef conglomerates, and indicate that the rocks have been metasomatically altered.
- Quartz veins and structures such as pressure solution seams are widespread in the reefs and indicate material movement in solution in aqueous fluids in the rocks.

It was further argued that:

- The rounded pyrite grains that are characteristic of the deposits are not transported clastic grains, but are pseudomorphs of iron hydroxide pisolites such as form in lateritic weathering environments.
- The Au resource in the Witwatersrand Basin can be more easily explained as being introduced by hydrothermal fluids than as a result of erosion and transport from a nearby source. The amount of Au in ores in the basin is several times greater than could be eroded from deposits in comparable Archaean terrains of the size of those that surround the basin. Greater than 50 000 t of Au has been extracted from the Witwatersrand deposits, compared to production to date of 4000 t from the Yilgarn Craton of Western Australia and a probable total resource of about 10 000 t.

None of the new observations or new arguments nullifies a placer origin of the ores if placer minerals had recrystallised during tectonism and low-grade metamorphism. However, a fundamental re-evaluation of an ore genetic model such as was proposed by Phillips *et al*., (1987) can be of value to groups exploring for undiscovered deposits and for additional ore in known deposits. We can never be absolutely certain that our interpretations of geological events and past processes are correct and complete, and exploration strategies should take the uncertainty of our geological interpretations into account. From a financial investment perspective, it can be worth taking the risk to 'test' a radical but well-founded and argued proposal for the genesis of an ore deposit type by investigating and drilling 'targets' that are identified by this proposal. Targets may be identified in parts of the Witwatersrand Basin that had not previously been explored, or in basins elsewhere in the world that had similar geological histories. In order to aid such exploration mining companies sponsored follow-up studies to re-evaluate the large-scale geometry and history of deformation in the basin (e.g. Coward *et al.*, 1995) and the petrological setting of gold (e.g. Barnicoat *et al.*, 1997). The study of the deformation history identified multiple generations of large faults in the basin and its surroundings and showed that these faults were active both during sedimentation and during events of later extensional and compressional tectonism. These faults had been previously overlooked, even along the mineralised and well-explored north-west margin of the basin. The latter study documented a dense network of microscopic faults and bedding-parallel fractures in the reefs, and that many small fractures had fillings of organic matter together with uraninite and gold grains. The authors further showed that hydrocarbons had migrated in the reef horizons, probably at the same time as growth of epigenetic gold grains, and they thus re-interpreted the carbon-leader seams as solidified migrated oil rather than as *in situ* relicts of microbial organisms.

The controversy over the genesis of such economically important resources also prompted much other research on the ores and their genesis from a large number of research groups and gave the topic priority as a site for application of new analytical technology in geochemistry and petrology. Almost all new data have however provided support for the placer origin of these ores, for instance:

(a) Metasomatism of the rocks to give aluminous compositions can equally be interpreted as a result of deep weathering, presumably soon after deposition (e.g. Sutton *et al.*, 1990). There is no increase in the degree of alteration near faults such as would indicate hydrothermal fluid flow along faults.
(b) Electron microscopy imaging and chemical analysis of rounded pyrite grains in the reefs shows that the rounded edges truncate internal structures, and the rounded shapes are thus more likely the result of abrasion in a sedimentary environment than epigenetic growth in place in the rock (England *et al.*, 2002b; Fleet, 1998). Elemental and isotope compositions of pyrite grains within any conglomerate sample vary, which is as expected if the pyrites were transported in from a number of sources.
(c) Improved techniques of mineral separation have demonstrated the widespread presence of gold grains with the morphologies and textures that are essentially identical to those in modern fluvial placer deposits (e.g. Minter, 1999).
(d) Additional field observations of relations between faults and ore have shown that unmineralised feeder dykes that fed the immediately overlying Ventersdorp lavas cross-cut ore horizons and are not cut by faults along the ore horizons. These relations demonstrate both that Au deposition predated the eruption of the immediately overlying lavas, and that some segments of the mineralised reefs are not fault planes (Meier *et al.*, 2009). Syn-sedimentary minor faults cut the carbon seams in carbon-leader reefs and thus indicate that these seams are indigenous and not residues of migrated hydrocarbons (Mossman *et al.*, 2008).
(e) Organic geochemical and petrographic analysis has shown that although there is some migrated oil in the rocks, the majority of organic matter in the carbon-leader reefs is indigenous kerogen, for instance granular kerogen, derived from

sedimented organic matter (England *et al.*, 2002a; Mossman *et al.*, 2008). Hydrogen isotope ratios of organic matter indicate that, despite its high Au content, the organic matter on the carbon-leader reef was not strongly altered by interaction with a large volume of hydrothermal water derived from outside of the basin, for instance (Grove and Harris, 2010).

(f) Radiometric dating of gold grains in reefs using Re–Os systematics indicates that gold is older than the host Central Rand Group sedimentary rocks by at least 100 million years (Kirk *et al.*, 2001) and hence was presumably transported in from an older eroded ore. This evidence is not however definitive. Rhenium and Os are only present in analysable concentrations in a minority of gold grains. The analyses may not have been of elements in the lattice of the gold, but may have been submicroscopic grains of a PGM that was, for instance, included in gold as it grew.

The source terrain, or terrains, of the gold has yet to be positively identified, and may have been sufficiently strongly reworked by later tectonic events to be unidentifiable. The elemental compositions of gold grains are consistent with derivation largely from eroded greenstone-belt-hosted orogenic Au deposits. The isotope composition of detrital zircons is consistent with crystallisation of these grains in granites in a relatively juvenile segment of crust, hence consistent with granite–greenstone terrains in the source region. Further, the argument that the gold endowment of the basin is too large to be derived by erosion is questionable in view of the recognition that placer processes can produce a stronger concentration of gold than hydrothermal processes, and that many large placer gold fields contain more gold than is apparently available in local hard-rock resources (see Section 5.2.2). Loen (1992) argued that the gold content of a placer gold field should be compared against the mass of gold that would be derived from the total mass of the eroded terrain, assuming that this eroded rock had an average crustal abundance of gold (~ 4 ppb). By this measure, the mass of crustal rock that was eroded to provide the sediment in the basin would have been easily sufficient to provide the gold contained in these unusual ores.

Questions and exercises

Review questions and problems

5.1 Mn deposits and one type of Pb–Zn deposit (Chapter 4) can form in a sedimentary basin on a continental shelf where there is a chemically layered ocean water column with an oxidised layer over a reduced layer. For each of these deposit types provide labelled sketches showing:

(i) the likely shape of the basin in cross section;

(ii) at approximately what depths the deposits will form at, relative to the redox interface;

(iii) arrows showing the movement of ocean waters or hydrothermal waters required to form the deposits.

The labelling should be sufficient to explain the major chemical processes involved in ore formation.

5.2 Chemical precipitation of minerals in shallow marine environments is an essential factor in the formation of both the world's major Fe deposits and the world's major Mn deposits.

(i) For each of Fe and Mn, describe the Fe- and Mn-rich rocks formed by chemical precipitation.

(ii) Explain the chemical processes that give rise to precipitation of the metals.

(iii) Explain why Fe deposits formed in the Earth's history only before about 1850 Ma whereas Mn deposits have continued forming essentially up until the present. (The youngest large Mn deposits are Cretaceous.)

Discussion questions

5.3 What may the nature of early-Proterozoic sedimentary Mn ores tell us about the oxidation state of the atmosphere and the upper layers of the oceans? The important characteristics of the deposits are that:

- There are oscillations in the host sequence between iron and manganese formations.
- Fe/Mn ratios in the iron formations below the Hozatel formation are similar to those in the iron formations interbedded with Mn ore.

Explain your reasoning and the assumptions you have made in your analysis.

5.4 An exploration geologist tells you that strandline heavy-mineral sands are structurally controlled (Box 3.7) because he observes a relationship between the positions of the deposits and lineaments on geophysical images such as maps of gravity and magnetic intensity. These lineaments likely mark the positions of boundaries in the rocks below the sequence that host the deposits. Are the relations he observes chance relations, or is there a rational geological explanation for them?

5.5 The Au of large placer deposit fields has been derived from orogenic Au deposits or, less commonly, low-sulfidation epithermal deposits. Despite containing approximately equal amounts of Au, there are no major placer fields with Au derived from porphyry, high-sulfidation epithermal deposits or Carlin-type Au deposits. The enormous endowment of Au in Carlin-type deposits of northern Nevada, USA was not recognised by generations of prospectors crossing the region, for instance. What are the contributory reasons for this fact?

Further readings

The following articles give background to some of the processes in the formation of ore deposits in sedimentary environments, and also debates on the enrichment of iron formations to iron ore.

(1) Iron deposits in BIFs – debate over hypogene versus supergene formation of high-grade haematitic ores:

Morey, G. B. (1999). High-grade iron ore deposits of the Mesabi Range, Minnesota – Product of a continental scale Proterozoic ground-water flow system. *Economic Geology* **94**, 133–141.

Morris, R. C. (1993). Genetic modelling for banded iron-formation of the Hamersley Group, Pilbara Craton, Western Australia. *Precambrian Research* **60**, 243–286.

Morris, R. C. and Kneeshaw, M. (2011). Genesis modelling for the Hamersley BIF-hosted iron ores of Western Australia: a critical review. *Australian Journal of Earth Sciences* **58**, 417–451.

Powell, C. M., Oliver, N. H. S., Li, Z. X., Martin, D. M. and Ronaszeki, J. (1999). Synorogenic hydrothermal origin for the giant Hamersley iron oxide ores. *Geology* **27**, 175–178.

Taylor, D., Dalstra, H. J., Harding, A. E., Broadbent, C. G. and Barley, M. E. (2001). Genesis of high-grade hematite orebodies of the Hamersley province, Western Australia. *Economic Geology* **96**, 837–873.

(2) Phosphate deposits – a modern coverage of phosphate precipitation in ocean-floor sediments:

Filipelli, G. M. (2011). Phosphate rock formation and marine phosphorus geochemistry: The deep time perspective, *Chemosphere* **84**, 759–766.

(3) Placer deposits – theoretical background to the behaviour of sediment grains in fluvial systems:

Saxton, J., Fralick, P., Panu, U. and Wallace, K. (2008). Density segregation of minerals during high-velocity transport over a rough bed: implication for the formation of placers. *Economic Geology* **103**, 1657–1664.

Slingerland, R. and Smith, N. D. (1986). Occurrence and formation of water-laid placers, *Annual Review of Earth and Planetary Science* **14**, 113–147.

(4) Placer deposits – settings of the formation of large Cenozoic gold placer deposits:

Craw, D. (2010). Delayed accumulation of placers during exhumation of orogenic gold in southern New Zealand, *Ore Geology Reviews* **37**, 224–235.

Craw, D., Youngson, J. H. and Koons, P. O. (1999). Gold dispersal and placer formation in an active oblique collisional mountain belt, Southern Alps, New Zealand. *Economic Geology* **94**, 605–614.

Hughes, M. J., Phillips, G. N. and Carey, S. P. (2004). Giant placers of the Victorian Gold Province. *Society of Economic Geologists Newsletter* **56**, 1–18.

6 Supergene ores and supergene overprinting of ores

Supergene implies genesis at or near the Earth's surface. The word is used to contrast with **hypogene**, or genesis at depth. In most cases, supergene ores are formed as a result of action of meteoric waters on rocks through the chemical processes and mineral reactions of weathering, in contrast to hypogene ores formed at depth by ascending waters.

Types of supergene ores

Supergene ores can be (a) formed *in situ* from rocks that are being actively weathered in the **regolith**, or (b) through mineral precipitation from meteoric groundwaters circulating above or at the level of the permanent water table. The term 'supergene ore' is also applied to (c) those parts of originally hypogene ore bodies that have been modified by chemical and mineralogical changes of weathering so as to significantly affect both their nature and their economic value.

(a) Ores formed predominantly *in situ* in the regolith

These ores are 'residual' concentrations of elements. The ore minerals are secondary minerals that form during weathering and that are relatively insoluble and hence become concentrated where more soluble components are removed. Major residual ore types are:

- Bauxite (Section 6.1.1). This is a regolith material that is dominantly composed of a mixture of aluminium hydroxide minerals. This is the only important ore of Al. Bauxite ores form either in lateritic weathering profiles over Al-rich rock types such as granites, mafic rocks and arkosic sandstones; or as insoluble residue in irregular pockets overlying karst-weathered carbonates (Jamaica, and examples in the rock record, e.g. Baux, France).
- Laterite Ni–Co ores (Section 6.1.2). These are concentrations of Ni- and Co-bearing clay minerals, serpentine and Ni-bearing iron hydroxide minerals in lateritic regolith profiles above ultramafic rocks. Some examples are fossil laterites formed as palaeosol horizons.
- Iron and Mn ores. These are either ores formed by residual enrichments of Fe in iron formations and Mn in Mn-rich sediments, as described in Sections 5.1.1 and 5.1.2; or Fe-rich haematite or limonite residue in residual karst pockets on limestone. The latter were historically important sources of iron in Europe, for instance the Bohnerz of Germany and Switzerland, and ores in Cumbria, England.
- REE red clays or ion adsorption ores. These ores are relatively weak concentrations of rare-earth elements (REEs), in particular of the heavy rare-earth elements (HREEs) and

Y, in specific horizons in thick, *in situ* regolith over weathered granite or felsic volcanic rocks (Chi, 2008). They provide the majority of the world's supply of HREEs. The ores occur where one or more kandite group mineral (kaolinite, halloysite) is dominant on hills in humid sub-tropical terrains. To date they have only been identified in southern China (e.g. Hunan and Jiangxi provinces). Summed concentrations of the REEs in the ores are up to about 0.05–0.3%, compared to concentrations in the underlying unweathered granite of about 0.03%. The ores are economic because the REEs are adsorbed as ionic species onto surfaces of clay minerals and the majority can be extracted at a low cost through desorption into solution in suitable solvents. The solvent is either added to mined ore that is piled on dumps or is added to the ore *in situ* and collected down-slope and hence without the need of mining.

(b) Ores formed by precipitation from shallow groundwaters

There are also some ores that are the result of flow of water at lower temperatures than typical hydrothermal systems (< 50 °C) and groundwater flow in weathered near-surface rock.

- Uranium deposits. These include deposits in peatlands and **calcrete** or palaeochannel deposits in currently active or recently buried alluvium-filled channels of ephemeral drainages in desert and semi-desert terrains, e.g. Honeymoon Well, South Australia; Yeelirrie, Western Australia (Cameron, 1990).
- Magnesite deposits. These are concentrations of magnesium carbonate (magnesite) in concretions and nodules in soils and alluvium formed through dissolution of Mg from weathering ultramafic rocks, transport in solution in shallow groundwaters and reprecipitation. These are the so-called 'nodular/cryptocrystalline' ores for magnesite that are mined as a feedstock industrial mineral and as a source of magnesium metal. The Kunwarara deposit in Queensland, Australia is hosted as an extensive tabular body at a few metres depth in river gravels and sands down-slope of an extensive exposure of actively weathering serpentinite (Milburn and Wilcock, 1998).

(c) Ore bodies formed by supergene upgrading of primary ores

The changes to mineralogy and the redistribution of ore elements in an ore body as a result of weathering can cause the ore grade of a 'hard-rock' ore body to be increased or the economic value of ore body to be improved, particularly for ores of Cu or Au. Some ore bodies are only economical to mine because the ore has been enriched or modified by supergene processes.

All sulfide-bearing ore bodies are to some degree affected by weathering processes. Which physical and chemical processes bring about the most significant changes and how they affect ore depends both on ore deposit types and on climatic and weathering environments. Two examples of the development of specific deposit types in weathering profiles in specific climates are discussed in detail in Sections 6.2.1 and 6.2.2.

6.1 *In situ* supergene ores

General characteristics of chemical weathering

The driving agent for chemical weathering can be considered to be the downward percolation of meteoric waters into near-surface rock and hence the downward infiltration of both water and oxygen from the Earth's surface. Rainfall or snowmelt has low concentrations of solutes and is undersaturated with respect to essentially all minerals. All minerals will thus dissolve to an extent into fresh meteoric water, in amounts dependent on dissolution rates and on solubility. The mineral dissolution reactions may be congruent, such that all components are dissolved, or incongruent, such that some of the chemical components of the primary mineral precipitate as secondary, generally hydrous minerals. Example dissolution reactions for common rock-forming minerals are the congruent dissolution of, for instance, calcite to calcium and bicarbonate ions:

$$CaCO_{3(Cal)} + H^{+}{}_{(aq)} \rightarrow Ca^{2+}{}_{(aq)} + HCO_{3}{}^{-}{}_{(aq)}; \tag{6.1}$$

and incongruent dissolution through hydrolysis reactions of albite to kaolinite and sodium ions and silicic acid:

$$2NaAlSi_{3}O_{8(Al)} + 9H_{2}O + 2H^{+}{}_{(aq)} \rightarrow Al_{2}Si_{2}O_{5}(OH)_{4(Kao)} + 2Na^{+}{}_{(aq)} + 4H_{4}SiO_{4(aq)}. \tag{6.2}$$

In most environments, a proportion of the dissolved material is carried away from the site of weathering in solution, ultimately in river waters to the oceans. In general, therefore, weathering involves leaching of the more soluble chemical components from rock (see Figure 3.3 for guidelines as to which elements are likely to be soluble), and hence reduction of the land surface and formation of porosity. Solubilised components may be reprecipitated in the regolith, for instance at sites of groundwater evaporation or where there are changes in groundwater chemistry.

In addition to partial dissolution of a rock, weathering involves oxidation. Although the solubility of molecular oxygen in water is relatively low (a few ppm), there is a contrast between the high oxidation state of surface-derived waters, which are in approximate equilibrium with the pressure of oxygen in the atmosphere, and waters in rocks at depth, which are generally significantly more reduced. Many chemical reactions between rock and downward-percolating surface waters therefore involve oxidation of redox-sensitive elements, that is, those elements that have multiple possible stable oxidation states over the range of possible chemical conditions. The most important redox reactions in the regolith involve Fe(II) to Fe(III), Mn(II) to Mn(IV), sulfide (oxidation state of minus 2) to sulfate (oxidation state of plus 6) and reduced organic carbon to oxidised carbonate. In addition, many ore elements that are present in trace concentrations are susceptible to changing oxidation state and undergo redox transformations over the range of natural environments, including Ag, Au, Cu and As. A redox transformation will change the solution and bonding behaviour of the element. In general, weathered rocks at the surface and above the water table are modified so as to be in redox equilibrium with the atmosphere and redox-sensitive elements in the rock are thus oxidised, whereas those in rocks below the water table generally remain reduced or are at an intermediate oxidation state.

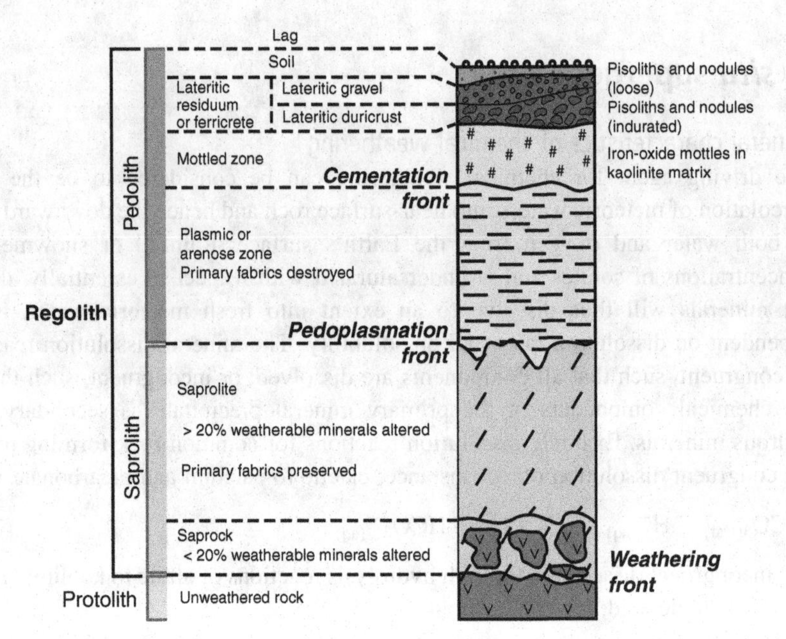

Figure 6.1 Laterite terminology and the layers of a typical lateritic weathering profile (Anand and Paine, 2002). See text for further discussion.

Processes and products of lateritic weathering

The formation of large, high-grade ore bodies through *in situ* concentration of ore elements is most important in terrains where intense chemical weathering takes place to relatively great depths. Most *in situ* supergene ores are thus formed in warm, humid climates of tropical, humid sub-tropical and warm temperate climatic regimes, or in areas where such conditions prevailed in the past. These are the climatic regimes in which lateritic weathering occurs and **lateritic weathering profiles** are developed.

The term 'laterite' refers to red-coloured iron oxide- and alumina-rich subsoil that is leached of bases and is characteristic of weathering of essentially all rock types in climatic zones of intense chemical weathering. The word refers to its use in brick-making in many sub-tropical areas. 'Lateritic profile' and 'lateritic weathering' refer to the style of weathering that is developed below laterite. In tectonically stable areas of low relief and low erosion rates, lateritic weathering can extend to 50 to 100 m depth or greater.

Lateritic profiles extend from the surface to fresh rock and are typically layered with different mineral components and textures reflecting increasing intensity of oxidation, hydration and leaching of soluble components towards the surface (Figure 6.1). The units in a typical profile are, from top to base:

(i) Laterite (sensu stricto) – loose Fe-rich subsoil formed as a result of residual enrichment and precipitation of iron hydroxide and oxide minerals from Fe derived deeper in the profile. This horizon can be composed largely of **pisolites** – gravel-sized sub-spherical accretionary concentrations, similar in structure to oolites in limestones but between 2 and 10 mm in diameter.

(ii) Ferruginous zone – formed of concentrations of iron(III) oxides and hydroxides near the water table where these precipitate as a result of evaporation and drying-out during seasonal lowering of the water table. This zone can be cemented by precipitation of silica and iron hydroxide minerals to form a **duricrust** that is resistant to later erosion.

(iii) Plasmic (clay) zone – in which textures of the original rock are overprinted by plasmic masses of clays, typically dominated by kandite group minerals. The upper part of this zone is often mottled red–white as a result of localised precipitation or dissolution of iron(III) oxide minerals.

(iv) **Saprolite** (= rotten rock) – in which original rock textures are largely preserved, but most weatherable minerals such as mafic minerals and feldspars are replaced by mixtures of clay minerals, including smectites, illites, kaolinite and other kandite group minerals. Quartz and other less weatherable minerals are largely preserved.

(v) **Saprock** – in which less than 20% of weatherable minerals are weathered. Carbonate minerals and sulfides are characteristically dissolved or replaced at the deepest horizons of a weathering profile.

The exact make-up of a profile and the strength of development of the different horizons is dependent on the parent rock type. It is also dependent on climatic regime, especially on seasonality of rainfall. Iron-rich gravel and duricrust are better developed where there is marked seasonality of rainfall and consequential seasonal drops and rises of the water table through the profile, often over several metres. In contrast, Fe is not as strongly enriched near the surface in regimes with year-round rainfall and perpetually high water table.

In a tectonically stable terrain, the depth of weathering, as marked by the weathering front, increases with time, and each of the layers in the lateritic weathering profile thickens (Figure 6.2). The land surface becomes deflated or lowered over time, largely as a result of the loss of mass with the leaching and transport away of soluble components.

6.1.1 Bauxite in lateritic weathering profiles

Bauxites are regolith materials composed of mixtures of fine-grained aluminium hydroxide minerals, most commonly the hydrated mineral gibbsite ($Al_2O_3{\cdot}3H_2O$) and semi-hydrated boehmite ($AlO{\cdot}OH$), or more rarely diaspore ($AlO{\cdot}OH$). Most bauxites include both gibbsite and boehmite, although one of the two is generally dominant. Bauxite ores in lateritic regolith profiles contribute almost all of the world's Al resources.

The settings and genesis of most bauxites are related to the present land surface and the formation of the ore was in the Cenozoic or Quaternary. Almost all bauxite ores are within about 35 degrees latitude of the equator in warm humid climate zones (e.g. Ghana; Venezuela; Weipa, Queensland, Australia; Darling Range, Western Australia). Bauxite may be actively forming in many of the ores. Bauxites also occur, however, along fossil regolith horizons (e.g. Baux, France), and many of these examples are preserved from past climatic regimes and formed when the host terrain was at tropical or sub-tropical latitudes.

Characteristics of bauxite ores

Economic bauxites are dominated by aluminium hydroxide minerals such that concentrations of Al_2O_3 are greater than about 30% and locally up to 60% (Figure 6.3). Fe_2O_3 is generally the second most abundant chemical component and may be dominant above and below the ore zone. Waste or gangue minerals in ore include iron oxide and hydroxide minerals (haematite, goethite), together with kaolinite, silica minerals, and grains of

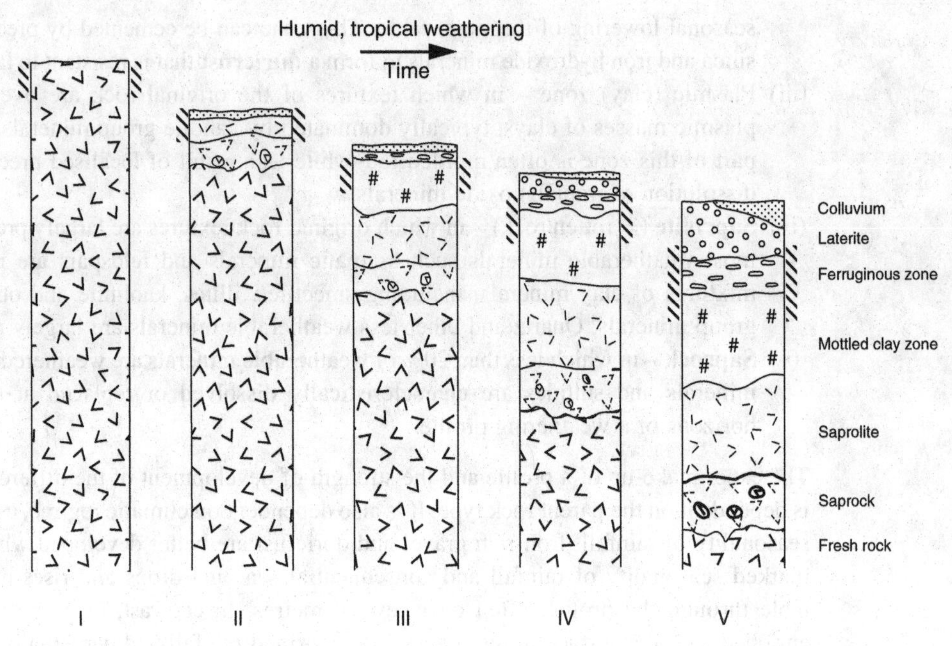

Figure 6.2 Schematic development through time of a lateritic weathering profile. The water table is indicated by hatching. Weathering involves substantial leaching of soluble components, hence there is mass and volume loss and reduction in thickness of the rock column with time (McFarlane, 1976).

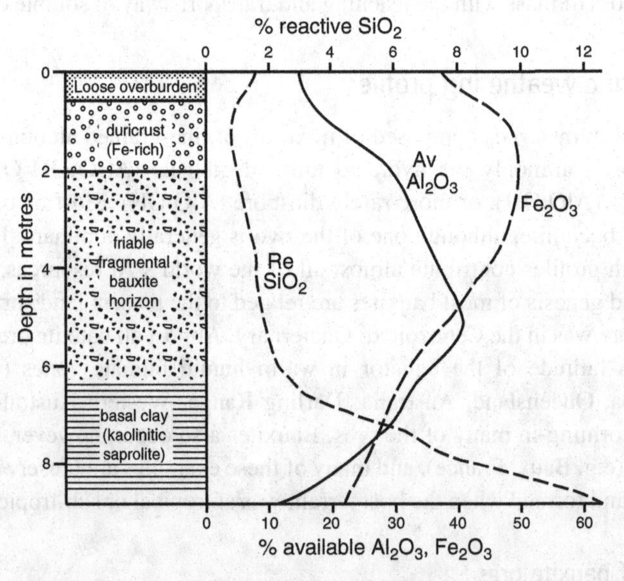

Figure 6.3 Schematic section through a typical bauxite deposit showing variations in oxide concentrations with depth (Smurthwaite, 1990). Av Al_2O_3 = alumina that is available for extraction through the Bayer process, i.e. excludes Al in clay and mica minerals. Re SiO_2 = reactive silica = the silica of the clay minerals that reacts during alumina refining. This measure excludes residual quartz grains, which can be mechanically separated before chemical processing.

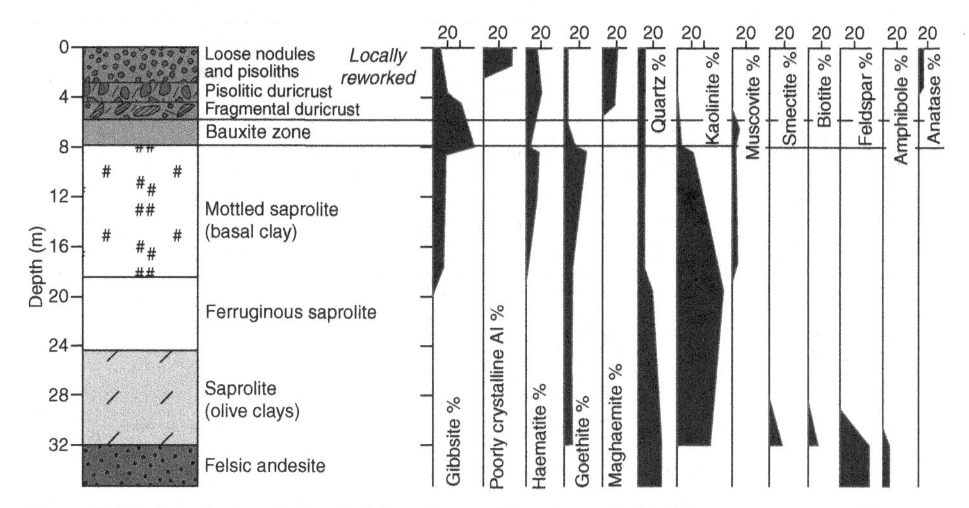

Figure 6.4 Mineralogy of a typical bauxite profile over andesite in a warm, seasonally humid climate, example of Boddington in the Darling Ranges of of southwestern Western Australia (after Anand, 1994).

incompletely dissolved residual quartz and other poorly soluble minerals. Some poorly soluble trace minerals such as the titanium oxide anatase (TiO_2) and zircon are present at concentrations up to a few times greater than in parent rock (Figure 6.4).

Bauxite ore can be a strong red colour or white-coloured depending on the concentration of dispersed iron oxide and hydroxide minerals. The structures and textures of ore are variable within any ore body and there is typically no consistent distribution of the textural types. The major textural types are porous ores, earthy and friable massive ores, and pisolitic, nodular or, less commonly, oolitic ores. Oolites and pisolites are distinguished by size (less than or greater than 2 mm). They are typically sub-spherical with nuclei such as fragments of earlier-formed bauxite or relict quartz grains and with multiple concentric fine growth layers or cortices composed of fine-grained precipitates of iron oxide and bauxite minerals. Nodules are more irregular concretions greater than 10 mm diameter. The oolitic and pisolitic textures result from physical and chemical mobility of material in the upper regolith through processes such as collapse into dissolution cavities and into tree root casts.

The ore horizons are in all cases close to the top of the lateritic regolith profile. They occur most typically as layers up to a few metres thick (4–15 m) that extend continuously or discontinuously over tens of kilometres. The tops of the ore horizons are typically 2–5 m below the current land surface. They may be overlain by subsoil or an iron oxide-dominated horizon and are typically underlain by the saprolite or plasmic zone in which kaolin is dominant. However, the exact position and relations of the ore horizon varies between deposits. In the Darling Range of southern Australia (Figure 6.4) it is beneath an Fe-rich duricrust with higher concentrations of iron oxide and hydroxide minerals, whereas at Weipa, in northern Queensland, Australia, ore is immediately below the soil horizon and overlies an Fe-rich duricrust. There is generally a transition downwards over several metres of increasing kaolinite to bauxite minerals into upper saprolite, which is known as the basal clay.

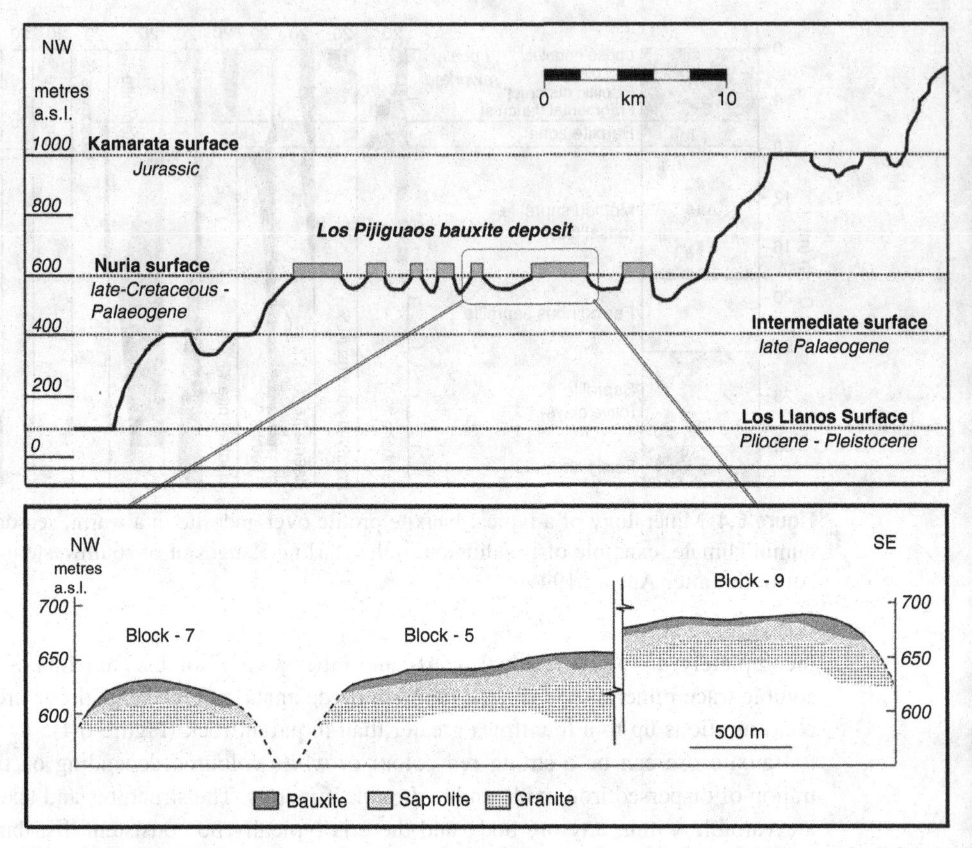

Figure 6.5 Schematic cross section through the Pijiguaos bauxite deposit, Venezuela (Meyer *et al.*, 2002, after Menendez and Sarmentero, 1984). The ore is on mesa-like hill tops on a dissected Palaeogene plateau.

Settings of bauxite ores

Lateritic bauxite ores occur in regolith above a variety of primary rock types including granites and granitic gneisses (see Figures 6.3 and 6.4), mafic to intermediate rocks and arkosic sandstones. Bauxite can form over any parent rock type that has a relatively high Al_2O_3 content of greater than about 12%, although this concentration may be lower where bauxite forms over calcareous bedrock. The different parent rocks are reflected in part by differing iron hydroxide content such that ores over intermediate rocks tend to have higher concentrations of Fe.

Where bauxites are developed they constitute one layer near the top of a much thicker lateritic weathering profile. They are not developed uniformly over a lateritically weathered terrain but are most characteristically best developed with thickest bauxite layers and highest grades under the tops of relatively flat but dissected plateaus (e.g. Pijiguaos, Venezuela, Figure 6.5) or under slopes in gently undulating terrains (Darling Range, Western Australia). Ore is fairly uniformly developed over suitable bedrock and in suitable topographic environments, although the variability of parent rock, for instance mafic dykes cutting granite, can affect ore quality.

Genesis of bauxites

Suitable bedrock for the development of bauxite is any silicate rock with greater than about 12% Al_2O_3, or more rarely a carbonate rock with lower concentrations of Al-bearing silicate minerals. The concentration of Al_2O_3 to greater than 30% is a residual concentration that results from leaching of greater than about 65% of the original rock mass.

The zonation of the regolith profile from base to top can be considered as equivalent to the time history of a rock as it is progressively weathered. The most strongly *in situ* weathered horizon of a typical lateritic profile such as shown in Figure 6.1 is the upper saprolite or plasmic zone dominated by kandite group clay minerals and from which all bases (e.g. Na_2O, CaO, MgO, etc.) have been leached. Bauxite is the result of weathering of this already strongly weathered material. The chemical difference between typical upper saprolite and bauxite is the SiO_2 to Al_2O_3 ratio. The formation of bauxite requires the replacement of kandite group clays by aluminium hydroxide minerals and removal of silica in solution in groundwater, for example:

$$Al_2Si_2O_5(OH)_{4(kaolin)} + 5H_2O \rightarrow Al_2O_3{\cdot}3H_2O_{(gibbsite)} + 2H_4SiO_{4(aq)}. \quad (6.3)$$

This replacement reaction is apparent from petrographic observations of many bauxite ores. Although silica is relatively weakly soluble in solution at millimolar levels at surface temperatures, with saturation with respect to quartz at a few ppm, it is about five orders of magnitude more soluble than any of the aluminium hydroxide minerals at near-neutral pH and the cumulative effect of small amounts of leaching over time will therefore be to increase the Al_2O_3 to SiO_2 ratio of saprolite.

Sufficient leaching of silica from saprolite to form bauxite requires a warm humid climate to enhance rates of mineral dissolution reactions, and long periods of landscape and tectonic stability to prevent continual erosion of the bauxite after it forms. The climatic requirements do not restrict bauxite formation to one climatic zone so long as past or present rainfall is sufficient (greater than about 1200 mm per year), and ores have formed in humid tropical (West Africa), seasonally humid tropical (Venezuela, northern Australia) and warm humid Mediterranean (southwestern Western Australia) climates. The dominance of gibbsite or boehmite in ores is related to climate, with the less-hydrated boehmite being dominant for ores in climate zones with stronger seasonality.

The setting of some bauxite ores on dissected erosion surfaces (Figure 6.5) implies that ore formation processes may have been operating since the onset of dissection. Dissection commenced in many cases in the Cenozoic, and ore formation may thus have been operative over time spans of many million years. Continued formation at the present day is implied by, for instance, higher concentrations of silica in waters seeping from bauxitic regolith than in local rainfall.

The topographic settings of the highest-grade and thickest bauxite ores on plateaus (Figure 6.5) or on gentle slopes (for instance, the Darling Range deposits) can be understood as a result of the requirement for effective removal of silica in solution out of the saprolite. This requires efficient drainage of infiltrated meteoric water downwards through the regolith, and is thus enhanced in terrains with a few tens of metres of relief.

6.1.2 Ni–Co laterite deposits

These are Ni–Co ores that form through lateritic weathering of silicate ultramafic rocks and provide about 30–40% of the world's supply of Ni. Major deposits occur for instance in New Caledonia, the Dominican Republic and Australia. Ore grades are similar to those of Ni sulfide ores (of order 1%). The ease of mining a near-surface tabular ore body of weathered rocks makes these ores economically prospective. In contrast, however, the metallurgical extraction of Ni metal from silicate and oxide minerals that are stable at the Earth's surface requires significantly greater energy than extraction from Ni-bearing sulfide minerals.

Chemical and mineralogical behaviour of Ni and Co in the regolith

Nickel and, to a lesser degree, Co are strongly compatible elements in magmatic systems. The highest Ni concentrations of common rock types are thus in ultramafic rocks. Primary ultramafic rocks of various origins, especially dunites and serpentinites derived from dunites or olivine-rich peridotites, including those of the mantle section of ophiolites and of komatiite lavas, have whole-rock Ni contents of greater than about 2000 ppm and Co contents of about 100 ppm, both dominantly substituting for Mg in solid solution in forsterite or serpentine.

Excluding carbonate rocks, ultramafic rocks have the highest percentage of oxides that are soluble in environments of intense weathering (SiO_2, MgO, CaO, etc.) – approximately 90% of the mass of an ultramafic rock can be readily or fairly readily dissolved in the most intensely weathered upper horizons of a lateritic profile, although SiO_2 may be reprecipitated in the profile as cryptocrystalline silica. Laterite profiles over ultramafic rocks have a sharp Mg front above which the majority of Mg is leached. The high degree of leaching allows for extreme residual enrichments of the insoluble components, which will include Fe_2O_3, TiO_2 and Cr_2O_3.

Nickel forms a divalent cation under natural conditions (see Figure 3.3), but is relatively poorly soluble in the geochemical environment of weathering, especially under the weakly alkaline conditions likely to develop on ultramafic rocks. It can be stabilised in the weathering profile by incorporation into Ni-bearing Mg–Fe-bearing clays, e.g. Ni-nontronite ($(NiFe)_2Si_4O_{10}(OH)_4$), in nepouite (Ni-serpentine = $Ni_3Si_2O_5(OH)_4$), garnierite (a poorly crystalline, mixed-layer silicate with relations to talc and serpentine) and in Ni-bearing iron hydroxides (goethite). As it is a largely insoluble element it can be concentrated ten-fold from primary concentrations (e.g. 2000 ppm) to about 2% in laterites which have been deflated by 90%. Cobalt is also a relatively insoluble trace element, especially in oxidising environments in which it is present as Co(III) and Co(IV), and hence undergoes a similar degree of enrichment as Ni from parent-rock concentrations of about 100 ppm to greater than 0.1%.

Nature and genesis of Ni–Co laterite deposits

A Ni–Cu laterite deposit, like bauxite, forms as a laterally extensive sub-horizontal layer in a laterite regolith profile. These deposits form over magnesian and silica-poor ultramafic rock types of any origin, including dunites of ophiolites and komatiites. The range of climates in which they form is similar to those of bauxites, but ores occur in slightly drier climates in the interior of Australia and in slightly cooler climates such as

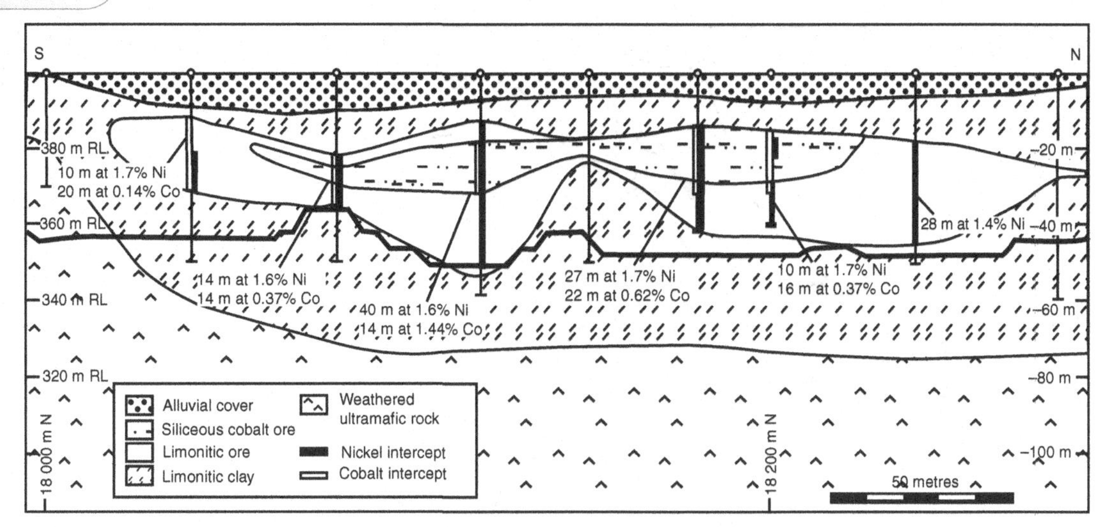

Figure 6.6 Sectional view of a typical laterite Ni–Co oxide deposit over serpentinised peridotite showing the distribution of ore and its relationship to regolith horizons (Bunyip Dam, WA, Australia) (Hellston *et al.*, 1998). Reprinted with the permission of the Australasian Institute of Mining and Metallurgy.

in Oregon, USA. Efficient drainage through the regolith does not appear to be a necessary prerequisite for the formation of these ores, and many nickel laterites are beneath topographic lows that develop because deflation of the host ultramafic rocks was greater than surrounding rock. Landscape stability also appears to be less critical, and nickel laterite ores are developed in areas of relatively rapid erosion in, for instance, New Caledonia.

Three different types of Ni–Co laterite ores are distinguished based on mineralogy and the dominant mineralogical setting of Ni.

(i) Hydrous silicate ores – in which Ni is present dominantly in Ni-rich serpentine (nepouite) and in the poorly crystalline, sheet-silicate mineral garnierite. Garnierite may contain up to 40% NiO. In these deposits, the Ni ore is relatively deep in the regolith profile in the lower saprolite. Garnierite is most abundant in ores formed over ultramafic rocks which have not been pervasively serpentinised.

(ii) Clay silicate ores – in which Ni is present dominantly in secondary Ni-bearing clay minerals including nontronite and smectite. The highest Ni grades are in the mid to upper horizons of the saprolite. These appear to form more readily from clinopyroxene-bearing ultramafic rocks, which have higher concentrations of Ca, Na and Al.

(iii) Oxide ores – in which Ni is present dominantly with iron hydroxide minerals, most importantly goethite (Figure 6.6). Manganese oxide minerals may also be present and have high concentrations of Ni and Co. Ore is relatively high in the regolith profile. These ores form most readily over dunite rather than peridotite. In some cases, oxide ores overlie hydrous silicate ores.

In the hydrous silicate and clay silicate ore types, the distributions of Ni and Co with depth show that enrichment of Ni and Co is not only a residual enrichment that results

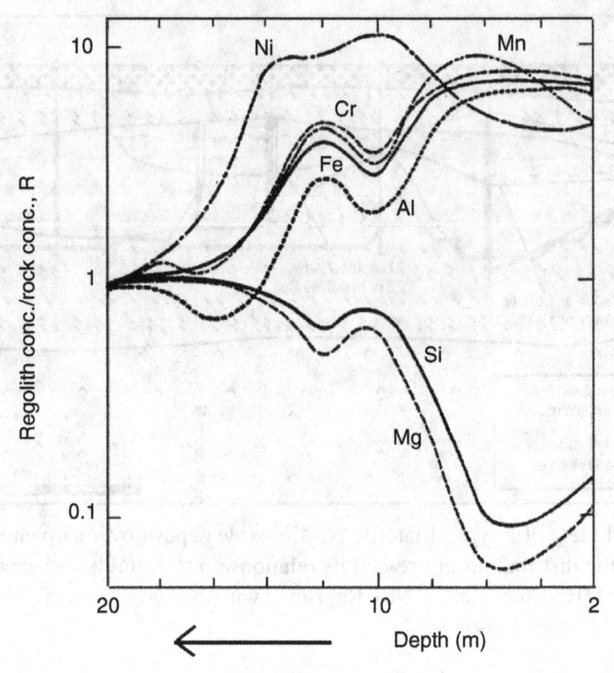

Figure 6.7 Concentrations of elements with depth in a lateritic weathering profile over ultramafic rocks. The concentrations are recorded as 'R', the ratio relative to the concentration in unweathered rocks. Note the mirror-image patterns of soluble and insoluble elements (indicating *in situ* enrichment), but that Ni is enriched over and above the factor of residual enrichment in the lower (clay) zone, and hence has been transported downwards through the profile (Golightly, 1981).

from leaching of soluble components. Rather both elements have migrated in solution vertically, most typically downwards through the profile during weathering (Figure 6.7). The highest concentrations occur where minerals that can potentially host Ni are stable in the profile, and as these are Mg–Si bearing clays or serpentine, the ore horizons are thus below the Mg and clay front, and hence in the saprolite zone.

6.2 Supergene ores formed by overprinting of hypogene ores

Many ore minerals, including all sulfide minerals, are chemically unstable in high-oxidation-state environments at or near the Earth's surface and are thus dissolved completely or at least partially from rocks during weathering. The ore elements that are released into solution may be leached out of the ore body or may be precipitated in secondary minerals, generally oxides, carbonates, sulfates or members of more unusual mineral groups such as phosphates and native metals. The secondary minerals form either at the site of dissolution of primary ore minerals or elsewhere in or around the ore body. Native Ag, for instance, is characteristic of Ag-bearing ores only where they have been partially oxidised above the water table, whereas Ag is present as a component of

Ag-bearing sulfide minerals in unweathered deeper ore. The presence of many secondary minerals as **efflorescences** at the surface has been a guide in prospecting.

In a sulfide-bearing ore body that crosses the weathering front, we often distinguish **oxide ores** in which the sulfide minerals have been largely dissolved and replaced by secondary minerals, from **sulfide ores** in which the dominant ore minerals are unweathered sulfides. The metallurgical behaviour and ore grade of the two ore bodies may be different, and in consequence the economics and requirements of mining of each ore body may need to be assessed independently.

6.2.1 Supergene gold ores in lateritic weathering profiles

Gold is present in unweathered ore in the various common types of primary deposits in one or multiple mineralogical settings: as grains of native Au in the rock or vein matrix ('free' gold), as grains of native Au included in sulfide minerals and in solid solution in sulfide minerals. In the last case, the gold is most commonly in either or both of pyrite and arsenopyrite, or more rarely in telluride minerals. Where Au was in inclusions in sulfide minerals, the breakdown of the sulfides in weathering 'releases' the Au and this alone can simplify metallurgical processing required for its extraction. The Au can be similarly 'released' from where it was in solution in sulfide minerals. Gold is not precipitated to any degree as part of any secondary compound mineral, and gold released from solution in sulfide minerals dominantly crystallises as fine-grained, newly grown euhedral grains of metallic gold within weathered rock, in some cases in grains of nanometre scale or as nanoparticles.

The common distribution of ore in lateritically weathered veins of lode-style Au ore, for instance an orogenic or a low-sulfidation epithermal deposit, is shown schematically in Figure 6.8. Sulfide minerals are dissolved and replaced above the weathering front and Au extraction is thus often easier from ores mined above this depth. A blanket of ore spreading out from the primary ore body at or just below the surface is characteristic. These blankets can have similar or slightly lower grades than the primary ore, but are more cost-effective to mine.

Gold is weakly soluble in most oxidised surficial waters as various aqueous complexes, including some organic complexes. It is much less soluble in reduced waters. In consequence:

- Where groundwaters are oxidising (above the groundwater table), some transport in solution can occur. Newly crystallised grains of gold are most abundant in the ferruginous zone, presumably because this is the water table and the level at which groundwater becomes more reducing with depth.
- Where groundwaters are moderately reducing, native Au will generally act as an insoluble residual mineral in lateritic weathering, but minor amounts of solution and reprecipitation give rise to some lateral spreading of the ore zone (Figure 6.8).

6.2.2 Supergene copper ores in arid and semi-arid climates

The ore mineralogy and grade and distribution of ore in Cu deposits show variable degrees of secondary modification as a result of weathering. Oxide ores are present in

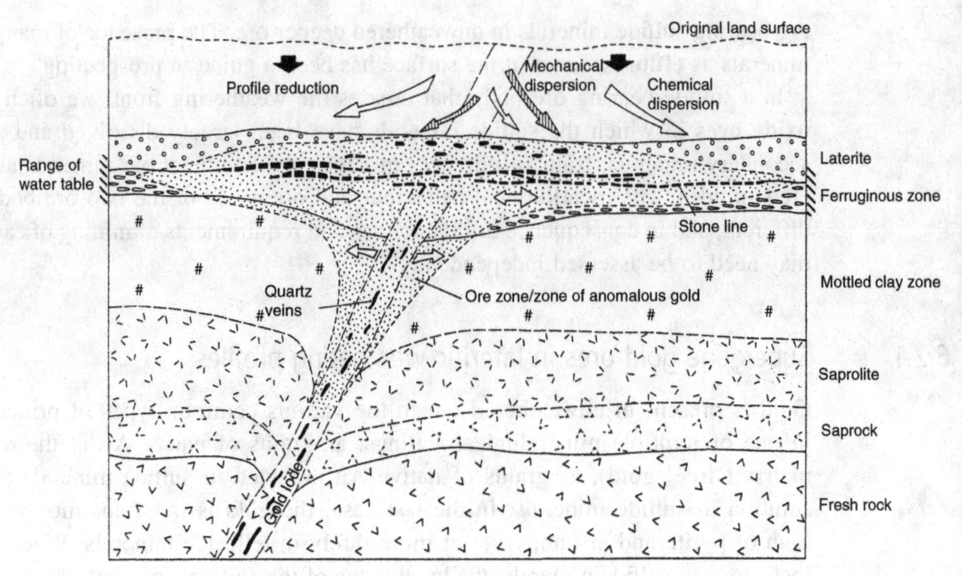

Figure 6.8 Distribution of gold about a primary ore zone in lateritic weathering and schematic representation of behaviour. Gold is typically dispersed over a few tens of metres laterally from a weathered ore zone. The Au released on weathering of sulfide minerals is either physically transported or is chemically transported and is precipitated near the water table in the ferruginous (iron oxide-rich) upper layer of the laterite profile where it may be adsorbed onto newly precipitated oxides (after Lawrance, 1990). This results in a readily mineable Au ore zone at shallow depths.

the weathered zones of some ore bodies. Many weathered copper ores are also marked by a horizon of relatively high-grade sulfide ore known as the enrichment blanket, which is also formed as a result of supergene processes. The most significant modifications occur to ores in semi-arid climatic environments in which the water table is at relatively great depth, in some cases at more than 100 m. This is the environment of the important porphyry copper belts of both southwestern USA and of the central Andes of Peru and northern Chile, and oxide ores are an important component of deposits in both of these areas. Over much of these regions the climate has remained arid or semi-arid since at least the Eocene. The present distribution of ores is thus a result of evolution during a long period of exposure and weathering. The degree of leaching and consequently also the strength of development of an enrichment blanket is generally much weaker in other climatic regimes.

Mineralogical and chemical behaviour of Cu in weathering

The major primary sulfide minerals of porphyry Cu deposits are chalcopyrite ($CuFeS_2$), bornite (Cu_5FeS_4) and pyrite (FeS_2). There are a large number of Cu-bearing phases of different mineral classes that are stable in contact with the atmosphere, including relatively soluble copper carbonates and sulfates, e.g. malachite ($Cu_2(OH)_2(CO_3)$) and azurite ($Cu_3(OH)_2(CO_3)_2$). The acidity of the environment is a major control on which of these minerals form. There are also a large number of Cu-rich phases, including some sulfide minerals, which can form at intermediate oxidation states in the presence of sulfur. These include native copper (Cu), chalcocite (Cu_2S), digenite and djurleite ($Cu_{2-x}S$), covellite (CuS) and cuprite (Cu_2O).

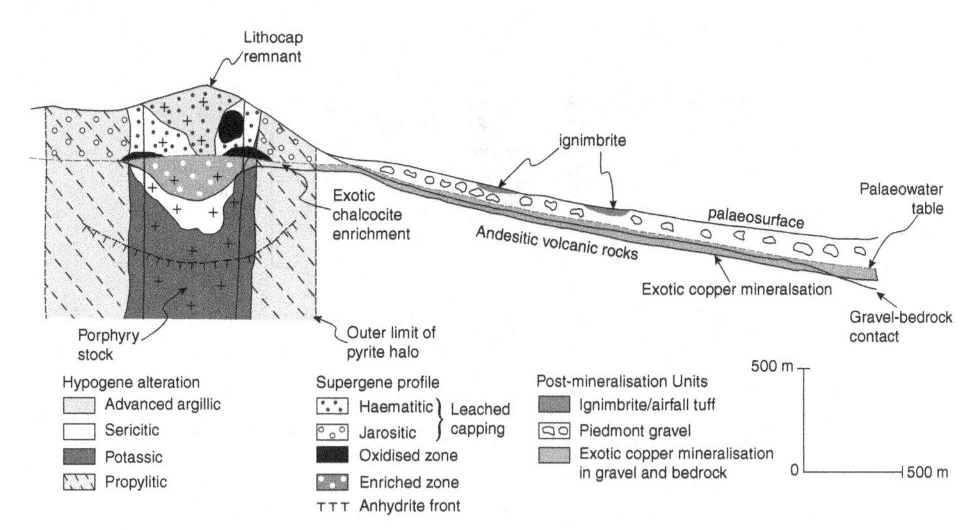

Figure 6.9 Settings of secondary Cu ores around a porphyry deposit in a desert based on Palaeogene examples in the Chilean Andes (after Sillitoe, 2008).

Copper is generally leached from ore above the water table by downward percolating oxygenated groundwater and surface outcrops and upper levels of ore are thus depleted in Cu in a so-called 'leached capping'. As Cu solubility is dependent on pH, the degree of leaching is controlled by acidity. In the case of porphyry Cu deposits it is thus controlled by the abundance of pyrite, the oxidative weathering of which produces sulfuric acid, and on the acid-neutralisation capacity of the host-rock.

Three settings and types of supergene modified ore are distinguished around porphyry Cu deposits (Figure 6.9) – enrichment blanket ores or supergene blankets, perched supergene ores (oxidised ores) and exotic ores. These different settings reflect different degrees of Cu dissolution by percolating meteoric waters, and different sites of Cu reprecipitation around the Cu porphyry ores above or at the water table.

In a similar fashion as for Ni–Co laterite deposits, ores at shallow depth in weathered rock are economically interesting ores because of the ease of mining. Ores in which the Cu is held in many of the secondary oxide minerals are further preferred because they are amenable to acid solvent-extraction metallurgical processing techniques such as SX/EW (solvent extraction/electrowinning) where the Cu is first dissolved from crushed ore on a leach dump. In contrast to the case of Ni laterite ores, the secondary Cu minerals are relatively soluble and the leaching is an energy-efficient technique for extraction of Cu.

Enrichment blanket ores

Many ores are marked by a layer or blanket up to a few tens of metres thick at and below the water table in which Cu is present mainly in copper sulfide minerals and is at higher grades than either above or below (Figures 6.10 and 6.11). This blanket marks the interface between oxide mineralogy above and hypogene sulfide mineralogy below, and in arid terrains can be at a depth of hundreds of metres below the surface. The Cu sulfide

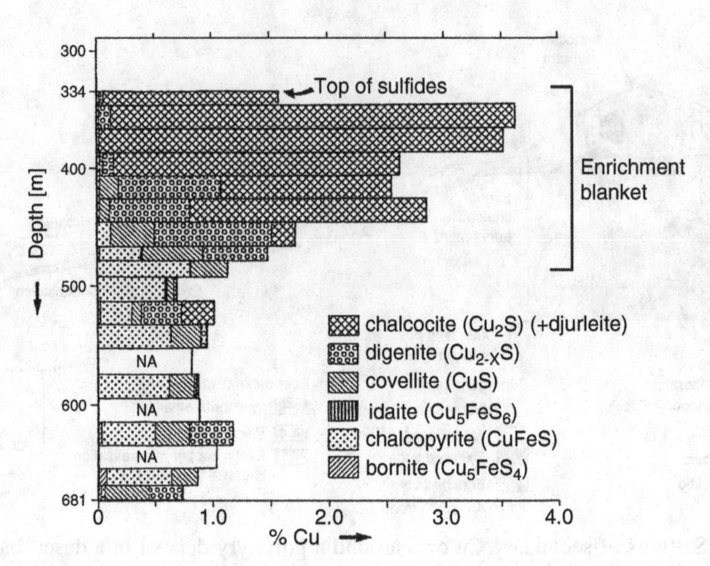

Figure 6.10 Copper minerals in a profile through an enrichment blanket of the La Escondida porphyry Cu deposit, Chile. The primary ore minerals are predominantly chalcopyrite with lesser bornite. The enrichment blanket is dominated by sulfide minerals that are stable to higher oxidation levels than most sulfides, e.g. digenite and chalcocite (Alpers and Brimhall, 1989).

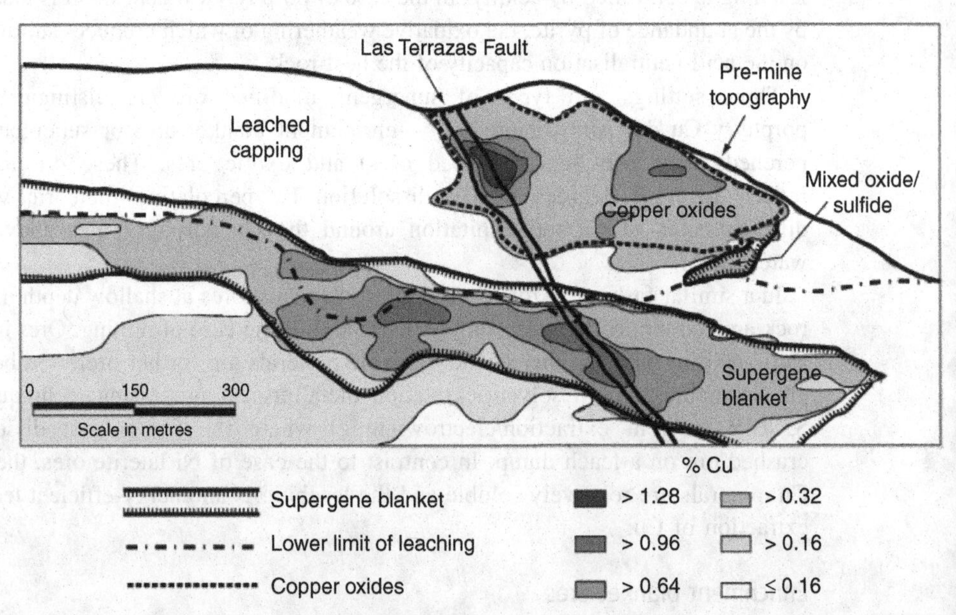

Figure 6.11 Cross section showing ore distribution and ore type in part of the deeply weathered Morenci Cu porphyry deposit of Arizona, USA (after Melchiorre and Enders, 2003). The supergene blanket lies at and below the depth of leaching. The perched copper oxide ore body has been largely preserved from leaching above the water table, probably in part because it is a remnant of an earlier supergene blanket that was leached of pyrite but not Cu sulfide minerals.

minerals in the enriched ore are assemblages of secondary Cu-rich sulfides such as chalcocite, digenite, djurleite and covellite. The enriched ore may also include horizons of native copper, particularly at its upper boundary.

The setting of enrichment blankets at and below the water table and underlain by unaltered ore indicates that they form as a result of reprecipitation of Cu leached from overlying ore into downward percolating oxidised water. Reprecipitation is as secondary sulfide minerals at the more reducing conditions at or beneath the water table. Sulfide for the precipitation of secondary sulfides comes from dissolution and replacement of pyrite and the primary Cu sulfide minerals such as chalcopyrite and bornite (Figure 6.10). The replacement reactions are oxidation reactions that are favoured at intermediate oxidation states in which the Cu-rich sulfide minerals are stable. The enrichment blanket has a higher grade because the Cu/Fe ratio in the sulfide component of the rock is higher, and in most environments there is sufficient S to allow precipitation of the Cu derived by leaching of overlying ore.

Supergene or oxide ores

Supergene ore bodies are bodies with ore-grade concentrations of supergene (oxide) ore minerals within the overall leached upper levels of the deposits above the regional water table (Figure 6.11). These ores may occur in different positions relative to the water table, including as perched ore bodies separated from the enrichment blanket and as oxide ores just above the upper limits of the enrichment blanket. A large number of different ore minerals can occur in these Cu ore bodies, including copper oxides (tenorite and cuprite), Cu-bearing limonite (iron hydroxide minerals), chrysocolla (hydrated copper silicate), azurite and malachite (hydrated copper carbonates) and copper sulfate minerals. Most oxide ores are amenable to SX/EW Cu extraction.

These bodies form where a portion of the Cu has not been leached from above the water table on oxidation. Relatively mild acidity is probably a critical factor in the formation of these ores. Some are relict enrichment blankets preserved from an earlier period of weathering when the weathering front and water table were higher (e.g. Figure 6.11). Such ores are inherently pyrite-poor before oxidation, and hence are more likely to avoid being leached if the water table drops.

Exotic Cu ores

These are ore bodies outside, but within a few kilometres, of the primary porphyry ore body. Most typically they are oxide ores with chrysocolla, copper sulfate and carbonate minerals as precipitates at the surface, or at a few metres depth in the pore space of coarse gravels that have deposited in ephemeral drainage channels, or within pediment sheets near the primary ore bodies (Figure 6.12). These ores are generally amenable to SX/EW Cu extraction.

These ore bodies form where not all of the Cu that is leached from the capping is reprecipitated at and below the water table (flux $\neq 0$, Figure 6.13), but where groundwater flow paths carry dissolved Cu away from the primary ore. The volcanic edifices beneath which many of the Cenozoic deposits in the Andes formed remain topographic highs and groundwater flow is thus generally down-slope from weathered ore (Figure 6.12). The ephemeral drainage channels are suitable aquifers for groundwater flow. Copper minerals are precipitated either on neutralisation of acidity as the groundwater interacts with gravel clasts, or where the water ponds and evaporates.

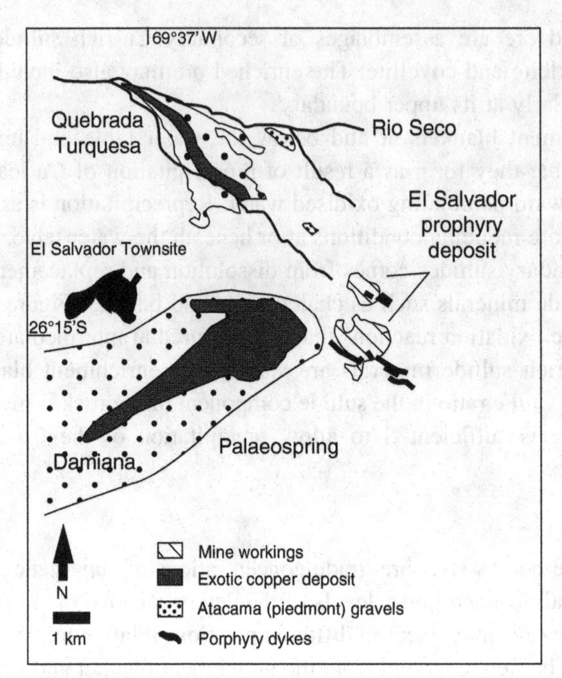

Figure 6.12 Map of the area of the El Salvador Cu porphyry deposit, Chile, showing the distribution of exotic deposits at a few metres depth in piedmont gravels in drainages around the eroded volcanic centre at El Salvador (after Mote *et al.*, 2001a). The area is arid with intermittent surface drainage.

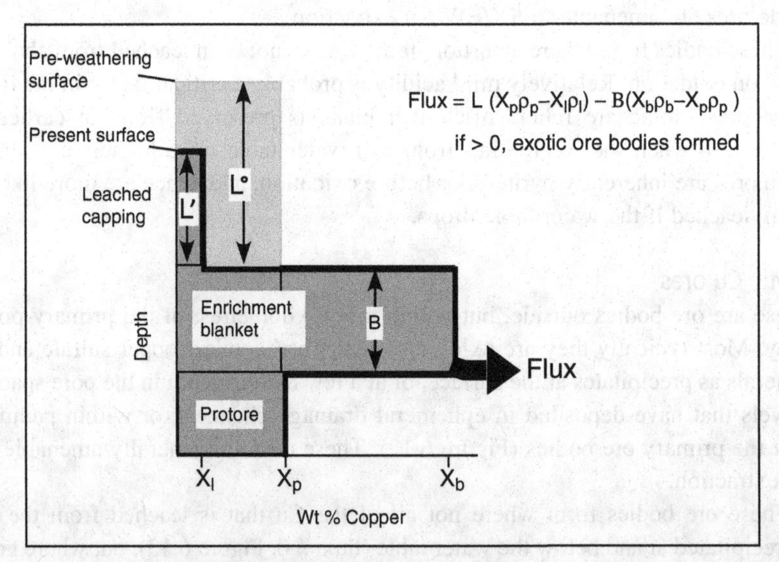

Figure 6.13 Mass balance of weathering of Cu in a dry climate with a deep water table (after Mote *et al.*, 2001b). Copper leached from above the water table is reprecipitated in part or fully in the enrichment blanket. Two parameters vary between deposits: (i) the proportion of primary Cu leached in the oxidised zone, such that the oxidised zones may still contain enough Cu to be ore; (ii) the flux of Cu out of the ore body. If the latter is large, exotic secondary ore bodies are expected to form in the regions of groundwater flow around the primary deposit.

Questions and exercises

Review questions and problems

6.1 Based on Figure 3.3, what economic commodities other than Al, Ni and Co are likely to be residually enriched in residual regolith of weathered profiles?

6.2 Mass-balance calculation of the formation of a supergene enrichment blanket over a porphyry Cu deposit.

Assuming that the primary sulfide ore contains 3% pyrite and 1.2% chalcopyrite by mode, what is the maximum likely percentage of Cu in the supergene enrichment zone if no sulfur is leached from the rocks and all is incorporated into chalcocite (Cu_2S)?

What is the minimum thickness of leached ore needed to supply a 50-m thickness of enriched ore?

6.3 Mass-balance calculation for Ni laterite ores.

Based on Figure 6.7:

(i) Divide the profile into 2-m-depth slices and estimate the percentage of rock that was leached in each slice in the profile based on the residual enrichment of Al, Fe and Ti. The best result will come from calculations based on each of the three elements and comparison of valves.

(ii) Using the results from (i) estimate how much Ni has been leached or added to each slice of the profile and determine whether Ni has been conserved in the profile or whether a proportion has been removed.

Discussion questions

6.4 Supergene enrichment of porphyry Cu deposits has been described and analysed for deposits in the terrains of semi-arid to arid climate and high topographic relief in which many of these deposits in the American cordilleras occur. What would we expect in other climates and environments? Consider, for instance, tropical climates in an area of low relief, and humid cool temperate climates. Summarise the general factors that you need to consider if you are exploring for and evaluating deposits in a specific part of the world.

Further readings

(1) Bauxites – descriptions of example bauxite deposits and fields:

Eggleton, R. A. and Taylor, G. (eds.) (2008). Weipa Bauxite, northern Australia. *Australian Journal of Earth Sciences Supplement* **55**, S1–S103.

Soler, J. M. and Lasaga, A. C. (2000). The Los Pijiguaos bauxite deposit (Venezuela): a compilation of field data and implications for the bauxitization process. *Journal of South American Earth Sciences* **13**, 47–65.

(2) Ni–Co laterite deposits – what may control the formation of the three different mineral types of these deposits?

Brand, N. W., Butt, C. R. M. and Elias, M. (1998). Nickel laterites: classification and features. *AGSO Journal of Australian Geology and Geophysics* **17**(4), 81–88.

(3) Supergene modification of ore deposits – the geochemistry of regolith processes:

Anand, R. R. and Butt, C. R. M. (2010). A guide for mineral exploration through the regolith in the Yilgarn Craton, Western Australia. *Australian Journal of Earth Sciences* **57**, 1015–1114.

Brimhall, G. H., Alpers, C. N. and Cunningham, A. B. (1985). Analysis of supergene ore-forming processes and ground-water solute transport using mass balance principles. *Economic Geology* **80**, 1227–1256.

Mote, T. I., Brimhall, G. H., Tidy-Finch, E., Muller, G. and Carrasco, P. (2001). Application of mass-balance modeling of sources, pathways and sinks of supergene enrichment to exploration and discovery of the Quebrada Turquesa exotic copper orebody, El Salvador District, Chile. *Economic Geology* **96**, 367–386.

GLOSSARY

The glossary includes terms used in the book, and terms that will likely be encountered in professional literature on ore deposits and ores deposit geology. Terms are classified as follows:

A: Common terms of ore deposit classification, description and interpretation of the shapes, nature, forms and components of ore bodies.
B: Terms originating from other sciences or other parts of the geosciences that have specific uses in ore deposit geology.
C: Terms used to describe the size of ore bodies and economic processing of ore deposits.

The following references were used in compiling this glossary and are useful as sources of definitions in ore deposit geology:

IUPAC Compendium of Chemical Terminology (Gold Book); see http://goldbook.iupac.org/.
Neuendorf, K. K. E., Mehl, J. P. Jr, and Jackson, J. A. (2005) *Glossary of Geology, Fifth Edition*. Alexandrai, VA, AGI.
Thrush, P. W. (comp.) (1968). *A Dictionary of Mining, Mineral, and Related Terms*. US Bureau of Mines Special Publication. (Also available as XML version at http://xmlwords.infomine.com/xmlwords.htm.)

	A	B	C	
Alteration	*			A change in the mineralogical and/or chemical composition of a rock. Most commonly used to express a change that results from hydrothermal activity but may alternatively result from weathering. In many instances the word is synonymous with **metasomatism**. The word is loosely used to describe an altered volume of rock.
Alteration facies	*			Used in a similar sense as metamorphic or mineral facies: a set of mineral assemblages in altered rock defined by the presence of specific index minerals or mineral assemblages.
Anomaly/anomalous	*			Loosely used to indicate a rock or area in which the concentration of an element is significantly

(*cont.*)

	A	B	C	
				different, especially higher, than normal for the rock type or **regolith** material.
Aqueous		*		Of water; in particular of a solution in which water is the solvent.
Authigenic		*		For a mineral – one that crystallised in place, especially in a sediment or **regolith**.
Base metal		*		As opposed to **precious metal** – of low value. Most commonly used for Pb, Zn, Ni and Cu.
Beneficiation			*	Raising the grade or value of an ore before sale, for instance by **milling**, **flotation**, blending, forming a **concentrate**, etc.
Blanket	*			A flat ore body which is relatively thin compared to its length and breadth.
Bonanza			*	A body of unusually high grade ore.
Brownfields exploration			*	Exploration in the surrounds of known ore deposits, especially exploration in known **ore fields** and ore **camps**.
Bulk mining/bulk tonnage			*	Any process of mining in which no effort is made to separate small volumes of high-grade ore or ore from intervening sub-economic ore or waste.
By-product			*	An additional economic product from an ore, the extraction and sale of which does not significantly influence the economics of the mining operation.
Calcrete		*		Interstitial cement of carbonate in the regolith, often as nodules. This is a common component of regolith in semi-arid warm climates.
Calc-silicate		*		A rock consisting largely of silicate in which calcium is an essential component: calcium silicate minerals. Used in ore deposit geology most commonly to describe altered rock that contains a high concentration of calcium silicate minerals
Camp (mining camp)			*	Originally 'a mining town'. Now used in the sense of 'a cluster of ore deposits or mines that could have been served from a single town'.
Canga	*			Hard blocks of iron formation cemented with iron hydroxide minerals.
Chimney	*			Effectively synonymous with **pipe** for an ore body of approximately cylindrical shape and steep plunge.

(cont.)

	A	B	C	
Clarke		*		The average abundance of an element in the crust.
Clarke of concentration		*		Factor of concentration of an element in a rock or ore relative to its Clarke.
Compatible		*		An element that readily substitutes into one or more minerals that is stable in the geological environment of interest.
Complex (aqueous)		*		A chemical coordination entity: an assembly in aqueous solution consisting of a central atom (usually metallic) to which is attached a surrounding array of other groups of atoms (ligands). The bonding between the components is normally weaker than in a covalent bond.
Component		*		Constituent of a chemical mixture.
Concentrate			*	Enriched ore after removal of waste and at least a portion of **gangue** minerals.
Co-product			*	In contrast to by-product – an additional economic product from an ore, the extraction and sale of which significantly influences the economics of the mining operation.
Country rock	*			The rock into which an ore-bearing intrusion has intruded, or a hydrothermal ore has precipitated.
Cupola		*		Approximately dome-shaped upward protrusion of the top of a large igneous intrusion.
Cut-off grade			*	The **grade** of ore in an ore body below which it is not economical to mine and extract. Ore below cut-off grade is in most cases left in the ground.
Devolatilisation		*		Of a mineral reaction – one that releases volatile chemical components such as water or carbon-dioxide.
Dilatant/dilatancy		*		Increasing volume, in particular of a rock during deformation.
Disseminated	*			Fine-grained ore minerals uniformly distributed in an ore.
Duricrust		*		Hard crust at the surface or as a layer within the upper **regolith**.
Efflorescence		*		Fine-grained mineral that has precipitated on a rock surface, particularly at the Earth's surface.
Endo-	*			Prefix meaning 'within': added to various terms to contrast with '**exo**'.

(*cont.*)

	A	B	C	
Enrichment	*			A process in which the grade of an ore-body is increased after its original formation: part of an ore body in which the grade has been increased by some process.
Epigenetic	*			Formed after lithification of the host-rock.
Epithermal	*			A hydrothermal ore formed at low temperatures and shallow depths (original definition, 50–200° C at ≈ 0–1.5 km).
Euxinic		*		Describes a surficial or near-surface environment in which hydrogen sulfide is the stable form of sulfur in solution in waters.
Exhalative	*			Adjective describing seepages of hydrothermal solutions from sediment or rock into a water body, normally on the sea floor.
Exo-	*			Prefix meaning 'outside': added to various terms in contrast to '**endo**'.
Exsolution/exsolved		*		The process by which a solution becomes two or more phases. Commonly used for separation of one **fluid** phase (e.g. an aqueous fluid) from a second fluid phase (e.g. silicate melt).
Ferricrete		*		Fragmented surficial material cemented by iron oxide or hydroxide minerals.
Field (ore field)	*			A geographical or geological region that is characterised by a specific type of ore deposit or multiple related types of ore deposit.
First boiling		*		**Exsolution** of an aqueous phase or other volatile phase from magma during rise of the magma through the crust.
Flotation			*	Separation of finely crushed minerals in a froth created in water with reagents.
Fluid		*		Matter that can flow. Used in ore deposit geology when there is not a clear distinction between liquid, vapour and **supercritical fluid**.
Foot-wall	*			The underlying side of an ore body, especially the stratigraphically underlying side.
Gangue	*			A rock that is valueless in and around ore. Used as shorthand for gangue minerals – the minerals within an ore which are of no economic value.
Geothermal		*		Related to unusually hot rock at depths of at most a few kilometres in the crust.

(cont.)

	A	B	C	
Geothermal system		*		All geological and hydrogeological components of a volume of rock in which there is elevated temperature within a few kilometres of the Earth's surface.
Giant ore deposit			*	Ore body containing mass of metal equivalent to that in 10^{11} t of average crust.
Grade/ore grade	*		*	The concentration of a metal or ore mineral within ore or mineralised rock.
Greenfields exploration			*	As opposed to **brownfields exploration** – exploration in an area in which there is no known ore deposit.
Greenstone belt		*		Typically elongate or belt-like area of a deformed and metamorphosed sequence of rocks in which mafic and ultramafic volcanic and sub-volcanic rocks are a major component. Characteristic of Precambrian shields.
Halo/alteration halo	*			Altered rock or low-**grade** ore surrounding and enveloping an ore body.
Hanging-wall	*			The overlying side of an ore body, especially the stratigraphically overlying side.
Host rock	*			Rock type or unit that an ore body or occurrence is in.
Hydrofracture/ hydrofracturing		*		A fracture in rock formed or enlarged as a result of fluid pressure/the process by which a fluid forms or enlarges a fracture in rock as a result of its pressure.
Hydrostatic pressure		*		Fluid pressure at depth equal to the weight of a (hypothetical) column of fluid extending up to the water table or surface.
Hydrothermal	*			Adjective meaning: related to hot aqueous solutions; formed by the action of hot aqueous solutions.
Hydrothermal system	*			All components and geological processes in the formation, migration and dispersion of a hydrothermal **fluid**.
Hypogene	*			In contrast to **supergene** – formed below the surface of the Earth.
Hypothermal	*			Of high temperatures and depths. Originally used for hydrothermal deposits with minerals such as amphibole, pyroxenes, garnets, biotite etc. in ores and altered rock, and to contrast with **mesothermal** and **epithermal**.

(*cont.*)

	A	B	C	
Immiscible		*		Coexist stably as two phases. Generally used for two similar phases, for instance, two liquids.
Incompatible		*		An element that does not readily substitute into one or more minerals that are stable in the geological environment being studied.
Industrial mineral			*	A rock or mineral extracted as a raw material for purposes other than a source of energy, a gemstone or a source of a metal.
Komatiite		*		Ultramafic lava: extrusive or sub-volcanic equivalent of peridotite
Laterite		*		Originally defined as – Fe-rich regolith material that can be cut into bricks. More generally – reddish, iron-rich **saprolite** in a deep weathering profile.
Laterite weathering profile		*		Deep weathering profile of warm to hot humid climates characterised by a laterite layer in the upper levels of the profile.
Ligand		*		In an inorganic coordination entity or **complex**, the atoms or groups bonded to the central atom.
Lithocap	*			**Stratabound** body of advanced-argillic and residual silica alteration.
Lithostatic pressure		*		Fluid pressure at depth equal to the weight of the column of rock extending up to the surface.
Lode	*			Originally a mining term for a planar ore body of finite thickness, e.g. a vein or a shear zone.
Magmatic-hydrothermal	*			Hot aqueous solutions that are derived by **exsolution** from magma.
Manto	*			Approximately horizontal pipe or ribbon of ore, often stratabound.
Massive	*			An adjective describing ore comprised of close to 100% ore minerals, minimum value normally taken as 50% by volume.
Mesothermal	*			Of intermediate temperatures and depths (original definition, 200–300° C at 1.2–4.5 km).
Metallogenic province	*			A large area, normally of distinct geological evolution, characterised by one or more types of ore deposit. The deposits may have formed in a number of periods of geological history.
Metallurgy/extractive metallurgy			*	The science of separating metals from ore.

(*cont.*)

	A	B	C	
Metamorphogenic	*			Formed by processes related to metamorphism.
Metasomatism		*		Deposition of new mineral grains by replacement of old in a rock so as to change the rock chemical composition. The process takes place by dissolution into and precipitation from interstitial fluid. Generally synonymous with **alteration**.
Milling			*	Crushing of ore to a fine fragment size as a precursor to separating the ore minerals or extracting the metals.
Mineralise/ mineralised rock	*			General term for precipitation and growth of ore minerals/a rock unit or volume of rock that contains ore minerals.
New metal			*	Most commonly used for a metal that is used in newly developed technology and for which demand has increased because of the development of new technology.
Occurrence	*			A concentration of ore minerals that is not large enough or of high enough grade or is not known to be large enough to be considered economic ore.
Ophiolite		*		Oceanic crust and underlying upper mantle that has been tectonically emplaced onto or into continental crust. An ophiolite may be tectonically dismembered such that not all layers of the ocean crust are present in one area of outcrop.
Ore body	*			A fairly continuous mass of ore.
Ore district	*			A number of ore deposits in close proximity – normally used where all deposits are within a few kilometres of each other.
Ore dressing			*	Removing some or all valueless components of ore, normally by mechanical means.
Ore mineral	*			A mineral that carries the valued constituent of an ore deposit.
Ore shoot	*			Area on a lode or vein of higher grade or economic ore.
Orthomagmatic	*			Adjective describing an ore formed of ore minerals that crystallised from a melt.
Overpressure		*		Normally used to indicate fluid pressure higher than hydrostatic pressure in rock.

(*cont.*)

	A	B	C	
Oxide ore	*			Term used in contrast to **sulfide ore** in partially weathered ore bodies. Ore in which the ore elements are dominantly in oxide, hydroxide and similar mineral classes.
Palaeoplacer	*			An ore deposit that formed as a **placer**, that is as mechanically concentrated ore minerals in loose surficial material, but since buried and consolidated.
Paragenesis	*			Associations and co-occurrences of minerals in an ore deposit. Interpretation of which minerals grew at the same time or in equilibrium. See Box 3.5.
Paragenetic sequence	*			Time sequence of mineral growth.
Partition coefficient		*		The ratio of concentrations of an element in two coexisting solutions.
Parts per million (ppm)			*	Concentration measurement of an element, especially a trace element, as proportion by mass.
Petroleum system		*		All components of rock and geological process involved in the maturation, migration and trapping of a hydrocarbon resource.
Pelletoidal		*		Composed of pellets – rounded to subrounded aggregates up to a few centimetres in diameter that have accumulated through accretion onto a nuclei.
Phase		*		Matter in a specific form – melt, **fluid**, mineral **species** etc.
Pipe	*			Sub-cylindrical or funnel-shaped form of a rock unit (e.g. a breccia) or an ore body.
Pisolite/pisolith		*		A concretionary growth of minerals in sediments or **regolith** larger than oolith (> 2 mm diameter).
Placer	*			Any ore deposit in loose material on or near the surface. An ore deposit formed of mechanically concentrated ore minerals in loose surficial material.
Precious metal			*	Metal of high value, specifically Au, Ag and PGEs.
Pre-mining resources			*	The resource of an ore body before commencement of mining. It includes the current resource and ore that has been mined.
Primary	*			Adjective describing an ore or alteration mineral – formed at the same time as initial accumulation of the ore.

(*cont.*)

	A	B	C	
Prospect			*	An **occurrence** which may possibly be converted to economic ore through additional exploration and evaluation.
Prospectivity/ prospective			*	Concept of the likelihood and amount of ore in an area or a terrain/an adjective indicating that an area or terrain has a high likelihood of containing ore.
Rare element		*		An element which has a concentration of less than about 100 **ppm** in the crust.
Rare metal			*	Less common and generally more expensive metal.
Recovery			*	The proportion of a metal or mineral in ore that is extracted by the mining and metallurgical processing used.
Reef	*			Similar to **lode**, but most commonly used where the dip of the blanket-like ore body is relatively low. Originally used for outcropping quartz veins.
Regolith		*		The layer of fragmented and loose material at the Earth's surface. Can be either residual or transported.
Replacement	*			An ore body or zone of hydrothermal alteration in which ore and gangue minerals have replaced earlier minerals.
Reserve			*	The tonnage of an ore deposit that is economical to mine with present technology and at present prices. Reserves are split into proven, probable and potential, dependent on the amount of exploration.
Resource			*	The total tonnage of mineralised rock defined from geological criteria, but not necessarily all economical to extract.
Restite		*		Unmelted minerals of a partially melted rock. May be left behind if the melt is efficiently extracted or may be carried in the magma.
Run-of-mine			*	Ore of average or typical grade produced from a mine.
Saddle reef	*			A vein or **reef** that has the form of an upright anticline.
Saprock		*		Incipiently weathered rock. Typically a proportion of minerals have been dissolved or partially replaced by clay minerals in saprock.
Saprolite		*		Intensely weathered rock. Typically clay-rich and with original structures preserved.

(*cont.*)

	A	B	C	
Secondary	*			Adjective describing an ore or alteration mineral that has crystallised after the initial formation of the deposit, e.g. during metamorphism or weathering.
Second boiling	*			**Exsolution** of an aqueous phase or other volatile phase from magma as a result of crystallisation of the magma.
Selective mining			*	Opposite of **bulk mining**: any process of mining in which high-grade ore or ore is mined and intervening low-grade ore or waste left unmined.
Shear-hosted	*			An ore body that is tabular and essentially confined to a shear- or fault zone.
Sheeted veins/ sheeting	*			A set of sub-parallel veins, generally closely spaced.
Species (chemical species)		*		Specific form of an element defined as to electronic or oxidation state, and/or complex or molecular structure.
Stockwork	*			A set of veins that form a three-dimensional network. A stockwork ore body may be the whole body of rock in which the stockwork is present.
Stratabound	*			Adjective describing an ore body that is enclosed within a specific part of the stratigraphic column. The body may be discordant or concordant.
Stratiform	*			Adjective describing an ore body which in three dimensions has the form of a sediment bed.
Stringer	*			A discontinuous approximately planar **veinlet** of ore minerals.
Stripping ratio			*	The ratio of the amount of waste that must be removed in order to mine an amount of ore.
Structural control/ structurally controlled	*			In the broadest sense – that there is a spatial or shape relationship between an ore body and any tectonic structure. The term may be used to specifically imply that the genesis of the ore body was related to the presence or the formation of the structure.
Sulfidation/sulfidation state		*		Analogous to oxidation and oxidation state. The degree of bonding of cations with sulfur. Measured by the relative fugacity of sulfur gas (S_2).

(*cont.*)

	A	B	C	
Sulfide ore	*			Term used in contrast to **oxide ore** in partially weathered ore bodies. Ore in which the ore elements are dominantly in sulfide minerals.
Sulfosalt		*		A sulfide mineral in which both a metal and a semi-metal (e.g. As, Sb) are essential components.
Supercritical fluid		*		A **fluid** at temperatures and pressures higher than its critical point. Such a fluid will neither boil nor condense if either of temperature or pressure is changed.
Supergene	*			An adjective used to describe ores or minerals formed by processes at the Earth's surface or within the **regolith**. The term is commonly used for weathering of earlier ore deposits, especially where this leads to increased value of the ore.
Super-giant			*	Of an ore body – one containing mass of metal equivalent to that in 10^{12} t of average crust.
Syngenetic	*			Formed at the same time as the host-rock.
Tailings			*	Milled ore from which the ore minerals have been extracted.
Tenor			*	Used in two senses: (i) As a synonym for **grade**, and (ii) For the concentration of an ore element in the sulfide component of an ore that is a mixture of sulfide and silicate minerals.
Vein	*			As a historical mining term, a vein was an approximately planar body of mineralised rock with distinct boundaries against unmineralised rock. The term is most commonly used for a planar body of minerals that were precipitated from hydrothermal solution.
Veinlet	*			A small vein, most typically less than a few centimetres wide.
Vein type	*			An ore body in which ore is composed of veins and in some cases also immediately adjacent country rock.
Volatile (element or chemical compound)		*		An element or compound that partitions into a gas phase from magma, or escapes as gas into the atmosphere from lava.
Vugh (or vug)	*			Open space in a rock or a **vein**, often lined with crystals.

(*cont.*)

	A	B	C	
Wall-rock	*			The **host-rock** specifically to a vein or other hydrothermal precipitate.
World-class			*	An ore deposit that is within the top ten per cent of deposits by contained metal in its type.
Zonation/zone	*			In general, any spatial distribution pattern of elements or minerals. The term is specifically used to describe a volume of rock of one **alteration facies** within or around an **ore body**.
Zoning/zonation	*			Any concentric set of zones, for instance of elements or minerals.

REFERENCES

Adamides, N. G. (2010). Mafic-dominated volcanogenic sulphide deposits in the Troodos ophiolite, Cyprus Part 2 – a review of genetic models and guides for exploration. *Applied Earth Science (Transactions of the Institution of Mining and Metallurgy)* **B 119**, 193–204.

Ahrens, L. H. (1954). The lognormal distribution of the elements (A fundamental law of geochemistry and its subsidiary). *Geochimica et Cosmochimica Acta* **5**, 49–73.

Albarede, F. (2009). *Geochemistry, An Introduction, 2nd edition.* Cambridge, Cambridge University Press.

Alpers, C. N. and Brimhall, G. H. (1989). Paleohydrologic evolution and geochemical dynamics of cumulative supergene metal enrichment at La Escondida, Atacama Desert, northern Chile. *Economic Geology* **84**, 299–255.

Anand, R. R. (1994). *Regolith-Landform Evolution and Geochemical Dispersion from the Boddington Gold Deposit, Western Australia.* CRC LEME Open File Report **3**, Perth, CRC LEME,

Anand, R. R. and Paine, M. (2002). Regolith geology of the Yilgarn Craton, Western Australia: implications for exploration. *Australian Journal of Earth Sciences* **49**, 3–162.

Annen, C., Blundy, J. D. and Sparks, R. S. J. (2006). The genesis of intermediate and silicic magmas in deep crustal hot zones. *Journal of Petrology* **47**, 505–539.

Appold, M. S. and Garven, G. (1999). The hydrology of ore formation in the southeast Missouri district: numerical models of topography-driven fluid flow during the Ouachita orogeny. *Economic Geology* **94**, 913–936.

Arai, S. (1997). Origin of podiform chromitites. *Journal of Asian Earth Sciences* **15**, 303–310.

Arribas, A. J., Hedenquist, J. W., Itaya, T. *et al.* (1995). Contemporaneous formation of adjacent porphyry and epithermal Cu–Au deposits over 300 ka in northern Luzon, Philippines. *Geology* **23**, 337–340.

Atkinson, W. W. Jr. and Einaudi, M. T. (1978). Skarn formation and mineralization in the contact aureole at Carr Fork, Bingham, Utah. *Economic Geology* **73**, 1326–1365.

Audétat, A. and Pettke, T. (2003). The magmatic-hydrothermal evolution of two barren granites: a melt and fluid inclusion study of the Rito del Medio and Cañada Pinabete plutons in northern New Mexico (USA). *Geochimica et Cosmochimica Acta* **67**, 97–121.

Audétat, A., Günther, D. and Heinrich, C. A. (2000). Causes for large-scale zonation around mineralized plutons: fluid inclusion LA-ICP-MS evidence from the Mole Granite, Australia. *Economic Geology* **95**, 1563–1582.

Audétat, A., Pettke, T., Heinrich, C. A. and Bodnar, R. J. (2008). The compositions of magmatic-hydrothermal fluids in barren and mineralized intrusions. *Economic Geology* **103**, 877–908.

Audétat, A., Dolejs, D. and Lowenstern, J. B. (2011). Molybdenite saturation in silicic magmas: occurrence and petrological implications. *Journal of Petrology* **52**, 891–904.

Ayres, D. E., Wray, E. M., Farstad, J. and Ibrahim, H. (1983). Geology of the Midwest uranium deposit. *Geological Survey of Canada Paper* **82–11**, 33–40.

Ballhaus, C. G. and Stumpfl, E. F. (1986). Sulfide and platinum mineralization in the Merensky Reef: evidence from hydrous silicates and fluid inclusions. *Contributions to Mineralogy and Petrology* **94**, 193–204.

Barnicoat, A. C., Henderson, I. H. C., Knipe, R. J. *et al.* (1997). Hydrothermal gold mineralization in the Witwatersrand Basin. *Nature* **386**, 820–824.

Barton, C. A., Zoback, M. D. and Moos, D. (1995). Fluid-flow along potentially active faults in crystalline rock. *Geology* **23**, 683–686.

Barton, P. B. (1991). Ore textures: problems and opportunities. *Mineralogical Magazine* **55**, 303–315.

Baturin, G. N. (1989). Origin of marine phosphorite. *International Geology Review* **31**, 327–342.

Bekker, A., Slack, J. F., Planavsky, N. *et al.* (2010). Iron formations: the sedimentary product of a complex interplay among mantle, tectonic, oceanic, and biospheric process. *Economic Geology* **105**, 467–508.

Bentley, H. W., Phillips, F. M., Davis, S. N. *et al.* (1986). Chlorine 36 dating of very old groundwater. The Great Artesian Basin, Australia. *Water Resources Research* **22**, 1991–2001.

Berning, J., Cooke, R., Hiemstra, S. A. and Hoffman, U. (1976). The Rössing uranium deposit, South West Africa. *Economic Geology* **71**, 351–368.

Berry, A. J., Harris, A. C., Kamenetsky, V. S., Newville, M. and Sutton, S. R. (2009). The speciation of copper in natural fluid inclusions at temperatures up to 700 °C. *Chemical Geology* **259**, 2–7.

Bethke, P. M. (1988). The Creede, Colorado ore-forming system: a summary model. *US Geological Survey Open-File Report* **88–403**.

Bethke, P. M., Rye, R. O., Stoffregen, R. E. and Vikre, P. G. (2005). Evolution of the magmato-hydrothermal acid-sulfate system at Summitville, Colorado: integration of geological, stable isotope, and fluid-inclusion evidence. *Chemical Geology* **215**, 281–315.

Beukes, N. J. and Gutzmer, J. (1996). A volcanic-exhalative origin for the world's largest (Kalahari) manganese field. A discussion of the paper by D. H. Cornell and S. S. Schütte. *Mineralium Deposita* **31**, 242–245.

Binns, R. A., Barriga, F. J. A. S. and Miller, D. J. (2007). Leg 193 synthesis: anatomy of an active felsic-hosted hydrothermal system, Eastern Manus Basin, Papua New Guinea. In *Proceedings ODP Scientific Results 193*, Barriga, F. J. A. S., Binns, R. A., Miller D. J. and Herzig, P. M. (eds.), College Station, TX, Ocean Drilling Program, pp. 1–71.

Birch, G. J. and Buchanan, D. L. (1989). Controls on the distribution of nickel sulphide mineralization associated with the Madziwa mafic intrusion, Zimbabwe. In *Magmatic Sulphides – The Zimbabwe Volume*, Prendergast, M. D. and Jones, M. J. (eds.), London, Institution of Mining and Metallurgy, pp. 21–42.

Blevin, P. L. (2004). Redox and compositional parameters for interpreting the granitoid metallogeny of eastern Australia: implications for gold-rich ore systems. *Resource Geology* **54**, 241–252.

Blevin, P. L. (2010). Eastern Australian granites: origins and metallogenesis. Conference presentation, SMEDG, Wines and Mines, Mudgee, New South Wales, 24–25 September 2010. See: http://smedg.org.au/M&W%202010/Blevin%20Eastern%20Australia%20Granites.pdf.

Böhlke, J. K. (1999). Mother Lode gold. *Geological Society of America Special Paper* **338**, 55–67.

Bolton, B. R., Barents, H. W. and Frakes, L. A. (1990). Groote-Eylandt manganese deposit. In *Geology and Mineral Deposits of Australia and Papua New Guinea*, Hughes, F. E. (ed.), Melbourne, Australasian Institute of Mining and Metallurgy, Monograph **14**, pp.1575–1579.

Boni, M. (1985). Les gisements de type Mississippi Valley du sud-ouest de le Sardaigne (Italie), une synthèse. *Chronique de la Recherche Minière* **479**, 7–34.

Boudreau, A. (2008). Modeling of the Merensky Reef, Bushveld Complex, Republic of South Africa. *Contributions to Mineralogy and Petrology* **156**, 431–437.

Boudreau, A. E., Mathez, E. A. and McCallum, I. S. (1986). Halogen geochemistry of the Stillwater and Bushveld complexes: evidence for transport of the platinum group elements by Cl-rich fluids. *Journal of Petrology* **27**, 967–986.

Brauhart, C., Groves, D. I. and Morant, P. (1998). Regional alteration systems associated with volcanogenic massive sulfide mineralization at Panorama, Pilbara, Western Australia. *Economic Geology* **93**, 292–302.

Brauhart, C. W., Huston, D. L., Groves, D. I., Mikucki, E. J. and Gardoll, S. J. (2001). Geochemical mass-transfer patterns as indicators of the architecture of a complete volcanic-hosted massive sulfide hydrothermal system, Panorama district, Pilbara, Western Australia. *Economic Geology* **96**, 1263–1278.

Broadbent, G. C., Myers, R. E. and Wright, J. V. (1998). Geology and origin of shale-hosted Zn–Pb–Ag mineralization at the Century deposit, Northwest Queensland, Australia. *Economic Geology* **93**, 1264–1294.

Brown, A. C. (1971). Zoning in the White Pine copper deposit, Ontonagon County, Michigan. *Economic Geology* **66**, 543–573.

Brown, A. C. (2006). Genesis of native copper lodes in the Keweenaw district, northern Michigan: a hybrid evolved meteoric and metamorphogenic model. *Economic Geology* **101**, 1437–1444.

Brown, K. L. (1986). Gold deposition from geothermal discharges in New Zealand. *Economic Geology* **81**, 979–983.

Buck, S. G. and Minter, W. E. L. (1985). Placer formation by fluvial degradation of an alluvial fan sequence: the Proterozoic carbon leader placer, Witwatersrand Supergroup, South Africa. *Geological Society of London Journal* **142**, 757–764.

Burrows, D. R., Spooner, E. T. C., Wood, P. C. and Jemielita, R. A. (1993). Structural controls on formation of the Hollinger-McIntyre Au quartz vein system in the Hollinger shear zone, Timmins, southern Abitibi greenstone-belt, Ontario. *Economic Geology* **88**, 1643–1663.

Cagnioncle, A-M., Parmentier, E. M. and Elkins-Tanton, L. T. (2007). Effect of solid flow above a subducting slab on water distribution and melting at convergent plate boundaries. *Journal of Geophysical Research* **112**, article B090402.

Calabrese, S., Aiuppa, A., Allard, P. *et al.* (2011). Atmospheric sources and sinks of volcanogenic elements in a basaltic volcano (Etna, Italy). *Geochimica et Cosmochimica Acta* **75**, 7401–7425.

Cameron, E. (1990). Yeelirrie uranium deposit. In *Geology and Mineral Deposits of Australia and Papua New Guinea*, Hughes, F. E. (ed.), Melbourne, Australasian Institute of Mining and Metallurgy, Monograph **14**, pp. 1625–1629.

Campbell, I. H., Naldrett, A. J. and Barnes, S. J. (1983). A model for the origin of platinum-rich sulfide horizons in the Bushveld and Stillwater complexes. *Journal of Petrology* **24**, 133–165.

Carman, G. D. (2003). Geology, mineralization, and hydrothermal evolution of the Ladolam Gold Deposit, Lihir Island, Papua New Guinea. *Society of Economic Geologists Special Publication* **10**, 247–284.

Carr, H. W., Groves, D. I., and Cawthorn, R. G. (1994). Controls on the distribution of Merensky Reef potholes at the Western Platinum Mine, Bushveld Complex, South Africa:

implications for disruptions of layering and pothole formation in the complex. *South African Journal of Geology* **97**, 431–441.

Castor, S. B. (2008). The Mountain Pass rare-earth carbonatite and associated ultrapotassic rocks, California. *Canadian Mineralogist* **46**, 779–806.

Catchpole, H., Kouzmanov, K., Fontbote, L., Guillong, M. and Heinrich, C. A. (2011). Fluid evolution in zoned Cordilleran polymetallic veins – insights from microthermometry and LA-ICP-MS of fluid inclusions. *Chemical Geology* **281**, 293–304.

Cathles, L. M. (1977). An analysis of the cooling of intrusives by ground water convection which includes boiling. *Economic Geology* **72**, 804–826.

Cathles, L. M. (1991). The importance of vein salvaging in controlling the intensity and character of subsurface alteration in hydrothermal systems. *Economic Geology* **86**, 466–471.

Cathles, L. M. (1993). Mass balance evaluation of the late diagenetic hypothesis for Kupferschiefer Cu mineralization in the Lubin Basin of southwestern Poland. *Economic Geology* **88**, 948–956.

Cathles, L. M. and Adams, J. J. (2005). Fluid-flow and petroleum and mineral resources in the upper (< 20-km) continental crust. In *Economic Geology, 100th Anniversary Volume*, Hedenquist, J. W., Thompson, J. F. H., Goldfarb, R. J. and Richards, J. P. (eds.), Colorado, Society of Economic Geologists, pp. 77–110.

Cathles, L. M. and Shannon, R. (2007). How potassium silicate alteration suggests the formation of porphyry ore deposits begins with the nearly explosive but barren expulsion of large volumes of magmatic water. *Earth and Planetary Science Letters* **262**, 92–108.

Cawthorn, R. G. (2011). Geological interpretations from the PGE distribution in the Bushveld Merensky and UG2 chromitite reefs. *Journal of the Southern African Institute of Mining and Metallurgy* **111**, 67–79.

Cerny, P., Ercit, T. S. and Vanstone, P. T. (1996). Petrology and mineralization of the Tanco rare-element pegmatite, southeastern Manitoba. In *Field Trip Guidebook A-4*, Winnipeg, Geological Association of Canada–Mineralogical Association of Canada, 1996.

Chai, G. and Naldrett, A. J. (1992). Characteristics of Ni–Cu–PGE mineralization and genesis of the Jinchuan Deposit, Northwest China. *Economic Geology* **87**, 1475–1495.

Chang, Z., Hedenquist, J. W., White, N. C. *et al.* (2011). Exploration tools for linked porphyry and epithermal deposits: example from the Mankayan intrusion-centered Cu–Au district, Luzon, Philippines. *Economic Geology* **106**, 1365–1398.

Charlier, B., Duchesne, J.-C. and Vander Auwera, J. (2006). Magma chamber processes in the Tellnes ilmenite deposit (Rogaland Anorthosite Province, SW Norway) and the formation of Fe–Ti ores in massif-type anorthosites. *Chemical Geology* **234**, 264–290.

Chen, J., Halls, C. and Stanley, C. J. (1992). Tin-bearing skarns of south China – geological setting and mineralogy. *Ore Geology Reviews* **7**, 225–248.

Chi, R. (2008). *Weathered Crust Elution Deposited Rare Earth Ores*. New York, Nova Science Publications.

Chouinard, A., William-Jones, A. E., Leonardson, R. W. *et al.* (2005). Geology and genesis of the multistage high-sulfidation epithermal Pascua Au–Ag–Cu deposit, Chile and Argentina. *Economic Geology* **100**, 463–490.

Cline, J. S. and Bodnar, R. J. (1991). Can economic porphyry copper mineralization be generated by a typical calc-alkaline melt? *Journal of Geophysical Research* **B 96**, 8113–8126.

Cloud, P. (1973). Paleoecological significance of banded iron-formation. *Economic Geology* **68**, 1135–1143.

Clout, J. M. F. and Simonson, B. M. (2005). Precambrian iron formations and iron formation-hosted iron ore deposits. In *Economic Geology, 100th Anniversary Volume*, Hedenquist, J. W., Thompson, J. F. H., Goldfarb, R. J. and Richards, J. P. (eds.), Colorado, Society of Economic Geologists, pp. 643–679.

Clout, J. M. F., Cleghorn, J. H. and Eaton, P. C. (1990). Geology of the Kalgoorlie gold field. In *Geology and Mineral Deposits of Australia and Papua New Guinea*, Hughes, F. E. (ed.), Melbourne, Australasian Institute of Mining and Metallurgy, Monograph **14**, pp. 411–431.

Conmou, D., Driesner, T. and Heinrich, C. A. (2008). The structure and dynamics of mid-ocean ridge hydrothermal systems. *Science* **321**, 1825–1828.

Constantinou, G. and Govett, G. J. S. (1978). Geology, geochemistry, and genesis of Cyprus sulfide deposits. *Economic Geology* **68**, 843–858.

Cooke, D. R. and Simmons, S. F. (2000). Characteristics and genesis of epithermal gold deposits. *Reviews in Economic Geology* **13**, 221–244.

Coward, M. P., Spencer, R. M. and Spencer, C. E. (1995). Development of the Witwatersrand Basin, South Africa. *Geological Society Special Publication* **95**, 243–269.

Cowden, A. and Roberts, D. E. (1990). Komatiite-hosted nickel sulphide deposits, Kambalda. In *Geology of the Mineral Deposits of Australia and Papua New Guinea*, Hughes, F. E. (ed.), Melbourne, Australasian Institute of Mining and Metallurgy, Monograph **14**, pp. 567–581.

Cox, S. F. (2005). Coupling between deformation, fluid pressures, and fluid flow in ore-producing hydrothermal systems at depth in the crust. In *Economic Geology 100th Anniversary Volume*, Hedenquist, J. W., Thompson, J. F. H., Goldfarb, R. J. and Richards, J. P. (eds.), Colorado, Society of Economic Geologists, pp. 39–76.

Cox, S. F. and Ruming, K. (2004). The St Ives mesothermal gold system, Western Australia – a case of golden aftershocks? *Journal of Structural Geology* **26**, 1109–1125.

Craig, J. R. and Vaughan D. J. (1994). *Ore Microscopy and Ore Mineralogy, 2nd Edition*, New York, John Wiley.

Crerar, D., Wood, S., Brantley, S. and Bocarsly, A. (1985). Chemical controls on solubility of ore forming minerals in hydrothermal solutions. *Canadian Mineralogist* **23**, 333–352.

Cronan, D. S. (2000). *Handbook of Marine Mineral Deposits.* Baton Rouge, CRC Press.

Dalton, J. A. and Presnall, D. C. (1998). Carbonatitic melts along the solidus of model lherzolite in the system $CaO–MgO–Al_2O_3–SiO_2–CO_2$ from 3 to 7 GPa. *Contributions to Mineralogy and Petrology* **131**, 123–135.

Dasgupta, R., Hirschmann, M. M. and Smith, N. D. (2007). Partial melting experiments of peridotite + CO_2 at 3 GPa and genesis of alkalic ocean island basalts. *Journal of Petrology* **48**, 2093–2124.

Dasgupta, R., Hirschmann, M. M., McDonough, W. F., Spiegelman, M. and Withers, A. C. (2009). Trace element partitioning between garnet lherzolite and carbonatite at 6.6 and 8.6 GPa with applications to the geochemistry of the mantle and of mantle-derived melts. *Chemical Geology* **262**, 57–77.

Davidson, G. J., Paterson, H., Meffre, S. and Berry, R. F. (2007). Characteristics and origin of the Oak Dam East breccia-hosted, iron oxide Cu–U–(Au) deposit: Olympic Dam region, Gawler Craton, South Australia. *Economic Geology* **102**, 1471–1498.

Deb, M. and Goodfellow, W. D. (eds.) (2004). *Sediment-Hosted Lead–Zinc Sulphide Deposits: Attributes and Models of Some Major Deposits in India, Australia and Canada*, New Delhi, Narosa Publishing House, p. 367.

DeMatties, T. A. (1994). Early Proterozoic volcanogenic massive sulfide deposits in Wisconsin: an overview. *Economic Geology* **89**, 1122–1151.

Deming, D. and Nunn, J. A. (1991). Numerical simulation of brine migration by topographically driven recharge. *Journal of Geophysical Research* **96**, 2485–2499.

de Ronde, C. E. J., Spooner, E. T. C., de Wit, M. J. and Bray, C. J. (1992). Shear zone-related, Au quartz vein deposits in the Barberton greenstone belt, South Africa: field and petrographic characteristics, fluid properties, and light stable isotope geochemistry. *Economic Geology* **87**, 366–402.

de Ronde, C. E. J., Faure, K., Bray, C. J. and Whitford, D. J. (2000). Round Hill shear zone-hosted gold deposit, Macraes Flat, Otago, New Zealand: evidence of a magmatic fluid. *Economic Geology* **95**, 1025–1048.

Dickinson, W. R. (2004). Evolution of the North American Cordillera. *Annual Review of Earth and Planetary Sciences* **32**, 13–45.

Dietz, R. S. (1964). Sudbury structure as an astrobleme. *Journal of Geology* **72**, 412–434.

Dilles, J. H. and Einaudi, M. T. (1992). Wall-rock alteration and hydrothermal flow paths about the Ann-Mason porphyry copper deposit, Nevada – a 6-km vertical reconstruction. *Economic Geology* **87**, 1963–2001.

Distler, V. V., Kryachko, V. V. and Yudovskaya, M. A. (2008). Ore petrology of chromite-PGE mineralization in the Kempirsai ophiolite complex. *Mineralogy and Petrology* **92**, 31–58.

Dong, G., Morrison, G. and Jaireth, S, (1995). Quartz textures in epithermal veins, Queensland – classification, origin and implication. *Economic Geology* **90**, 1841–1856.

Duke, J. M. (1983). Ore deposit models; 7, Magmatic segregation deposits of chromite. *Geoscience Canada* **10**, 15–24.

Eastoe, C. J. and Gustin, M. M. (1996). Volcanogenic massive sulfide deposits and anoxia in Phanerozoic oceans. *Ore Geology Reviews* **13**, 179–197.

Eckstrand, O. R. and Hulbert, L. (2007). Magmatic nickel–copper–platinum group element deposits. *Geological Association of Canada Mineral Deposits Division Special Publication* **5**, 205–222.

Eilu, P. and Mikucki, E. J. (1998). Alteration and primary geochemical dispersion associated with the Bulletin lode-gold deposit, Wiluna, Western Australia. *Journal of Geochemical Exploration* **63**, 73–103.

Einaudi, M. T. (1982). Descriptions of skarns associated with porphyry copper plutons. In *Advances in Geology of the Porphyry Copper Deposits Southwestern North America*, Titley, S. R. (ed.), Tucson, University of Arizona Press, pp. 139–183.

Einaudi, M. T., Hedenquist, J. W. and Esra Inan E. (2003). Sulfidation state of fluids in active and extinct hydrothermal systems: transitions from porphyry to epithermal environments. *Society of Economic Geologists Special Publication* **10**, 285–313.

Els, B. G. (1991). Placer formation during progradational fluvial degradation: the late Archean Middlevlei gold placer, Witwatersrand, South Africa. *Economic Geology* **86**, 261–277.

Els, B. G. (2000). Unconformities of the auriferous, Neoarchaean Central Rand Group of South Africa: application to stratigraphy. *Journal of African Earth Sciences* **30**, 47–62.

Emsbo, P., Hofstra, A. H., Lauha, E. A., Griffin, G. L. and Hutchinson, R. W. (2003). Origin of high-grade gold ore, source of ore fluid components, and genesis of the Meikle and neighboring Carlin-type deposits, northern Carlin trend, Nevada. *Economic Geology* **98**, 1069–1106.

England, G. L., Rasmussen, B. Krapez, B., and Groves, D. I. (2002a). Archaean oil migration in the Witwatersrand Basin of South Africa. *Journal of the Geological Society London* **159**, 189–201.

England, G. L., Rasmussen, B., Krapez, B. and Groves, D. I. (2002b). Palaeoenvironmental significance of rounded pyrite in siliciclastic sequences of the late Archaean Witwatersrand Basin: oxygen-deficient atmosphere or hydrothermal alteration? *Sedimentology* **49**, 1133–1156.

Ericksen, G. E. (1981). Geology and origin of the Chilean nitrate deposits. *US Geological Survey Professional Paper* **1188**.

Farquhar, J. and Wing, B. A. (2003). Multiple sulfur isotopes and the evolution of the atmosphere. *Earth and Planetary Science Letters* **213**, 1–13.

Farquhar, J., Bau, H. M. and Thiemens, M. (2000). Atmospheric influence of Earth's earliest sulfur cycle. *Science* **289**, 756–758.

Farquhar, J., Wu, N., Canfield, D. E. and Oduro, H. (2010). Connections between sulfur cycle evolution, sulfur isotopes, sediments, and base metal sulfide deposits. *Economic Geology* **100**, 509–533.

Farrow, C. E. G. and Lightfoot, P. C. (2002). Sudbury PGE revisited: toward an integrated model. *Canadian Institute of Mining and Metallurgy, Special Volume* **54**, 273–297.

Field, M., Stiefenhofer, J., Robey, J. and Kurszlaukis, S. (2008). Kimberlite-hosted diamond deposits of southern Africa: a review. *Ore Geology Reviews* **34**, 33–75.

Fleet, M. E. (1998). Detrital pyrite in Witwatersrand gold reefs: X-ray diffraction evidence and implications for atmospheric evolution. *Terra Nova* **10**, 302–306.

Forster, C. and Smith, L. (1990). Fluid flow in tectonic regimes. In *Crustal Fluids*, Nesbitt, B. E. (ed.), MAC Short Course Handbook **18**, Québec, Mineralogical Association of Canada, pp. 1–47.

Fournier, R. O. (1985). The behaviour of silica in hydrothermal solution. *Reviews in Economic Geology* **2**, 45–62.

Francheteau, J., Needham, H. D., Choukroune, P. *et al.* (1979). Massive deep-sea sulphide ore deposits discovered on the East Pacific Rise. *Nature* **277**, 523–528.

Garven, G. (1995). Continental scale groundwater flow and geologic processes. *Annual Review of Earth and Planetary Sciences* **23**, 89–117.

Garven, G., Ge, S., Person, M. A. and Sverjensky, D. A. (1993). Genesis of stratabound ore deposits in the midcontinent basin of North America I. The role of regional groundwater flow. *American Journal of Science* **293**, 497–568.

Gauthier-Lafaye, F. and Weber, F. (1989). The Francevillian (Lower Proterozoic) uranium ore deposits of Gabon. *Economic Geology* **84**, 2267–2285.

Giggenbach, W. F. (1992). Magma degassing and mineral deposition in hydrothermal systems along convergent plate boundaries. *Economic Geology* **87**, 1927–1944.

Gill, J. (1981). *Orogenic Andesites and Plate Tectonics*, New York, Springer.

Godel, B., Seat, Z., Maier, W. D. and Barnes, S-J. (2011). The Nebo-Babel Ni–Cu–PGE sulfide deposit (West Musgrave Block, Australia): Part 2. Constraints on parental magma and processes, with implications for exploration. *Economic Geology* **106**, 557–584.

Goff, F., Stimac, J. A., Larocque, A. C. L. *et al.* (1994). Gold degassing and deposition at Galeras Volcano, Colombia. *GSA Today* **4**, 243–247.

Golightly, J. P. (1981). Nickeliferous laterite deposits. *Economic Geology, 75th Anniversary Volume*, Skinner, B. (ed.), Lancaster, PA, Economic Geology Publishing, pp. 710–735.

Grant, J. N., Halls, C., Avila Salinas, B. W. and Avila, G. (1977). Igneous geology and the evolution of hydrothermal systems in some sub-volcanic tin deposits of Bolivia. In *Volcanic Processes in Ore Genesis*, London, Institution of Mining and Metallurgy, pp. 117–126.

Grauch, V. J. S., Rodriguez, B. D. and Wooden, J. L. (2003). Geophysical and isotopic constraints on crustal structure related to mineral trends in north-central Nevada and implications for tectonic history. *Economic Geology* **98**, 26–286.

Gray, J. E. and Coolbaugh, M. F. (1994). Geology and geochemistry of Summitville, Colorado: an epithermal acid-sulfate deposit in a volcanic dome. *Economic Geology* **89**, 1906–1923.

Gresham, J. J. and Loftus-Hills, G. D. (1981). The geology of the Kambalda nickel field, Western Australia. *Economic Geology* **76**, 1373–1416.

Grove, D., and Harris, C. (2010). O- and H-isotope study of the carbon leader reef at the Tau Tona and Savuka mines (western deep levels), South Africa: implications of the origin and evolution of Witwatersrand basin fluids. *South African Journal of Geology* **113**, 75–88.

Groves, D. I. and Vielreicher, N. M. (2001). The Phalabowra (Palabora) carbonatite-hosted magnetite–copper sulfide deposit, South Africa: an end-member of the iron oxide–copper–gold–rare earth element deposit group? *Mineralium Deposita* **36**, 189–194.

Groves, D. I., Bierlein, F. P., Meinert, L. D. and Hitzman, M. W. (2010). Iron oxide copper–gold (IOCG) deposits through Earth history: implications for origin, lithospheric setting, and distribution from other epigenetic iron oxide deposits. *Economic Geology* **105**, 641–654.

Gruen, G., Heinrich, C. A. and Schroeder, K. (2010). The Bingham Canyon porphyry Cu–Au–Mo deposit. II. Vein geometry and ore shell formation by pressure driven rock extension. *Economic Geology* **105**, 69–90.

Grunder, A. L., Klemetti, E. W., Feeley, T. C. and McKee, C. M. (2008). Eleven million years of arc volcanism at the Aucanquilcha volcanic cluster, northern Chilean Andes: implications for the life span and emplacement of plutons. *Transactions of the Royal Society of Edinburgh: Earth Sciences* **97**, 415–436.

Gustafson, L. B. and Hunt, J. P. (1975). The porphyry copper deposit at El Salvador, Chile. *Economic Geology* **70**, 857–912.

Gutzmer, J. and Beukes, N. J. (1996). Mineral paragenesis of the Kalahari manganese field, South Africa. *Ore Geology Reviews* **11**, 405–428.

Hagemann, S. G., Groves, D. I., Ridley, J. R. and Vearncombe, J. R. (1992). The Archean lode gold deposits at Wiluna, Western Australia: high-level brittle-style mineralization in a strike-slip regime. *Economic Geology* **87**, 1022–1053.

Hamilton, P. J. (1979). Sr isotope and trace element studies of the Great Dyke and Bushveld mafic phases and their relation to Proterozoic magma genesis in southern Africa. *Journal of Petrology* **18**, 24–52.

Hannington, M. T. and Barrie, C. T. (eds.) (1999). The Giant Kidd Creek Volcanogenic Massive Sulfide Deposit, Western Abitibi Subprovince, Canada. *Economic Geology Monograph* **10**.

Hannington, M., Jamieson, J., Monecke, T., Petersen, S. and Beaulieu, S. (2011). The abundance of seafloor massive sulfide deposits. *Geology* **39**, 1155–1158.

Hanor, J. S. (1994). Origin of saline fluids in sedimentary basins. *Special Publications of the Geological Society* **78**, 151–174.

Harris, A. C., Golding, S. D. and White, N. C. (2005). Bajo de la Alumbrera copper–gold deposit: stable isotope evidence for a porphyry-related hydrothermal system dominated by magmatic aqueous fluids. *Economic Geology* **100**, 863–886.

Harshman, E. N. (1972). Geology and uranium deposits, Shirley Basin area, Wyoming. *US Geological Survey Professional Paper* **745**.

Haxel, G. B. (2005). Ultrapotassic mafic dikes and rare earth element- and barium-rich carbonatite at Mountain Pass, Mojave Desert, Southern California: summary and field trip localities. *US Geological Survey Open-File Report* **2005–1219**.

Haydon, R. C. and McConachy, G. W. (1987). The stratigraphic setting of Pb–Zn–Ag mineralization at Broken Hill. *Economic Geology* **82**, 826–856.

Hedenquist, J. W. and Lowenstern, J. B. (1994). The role of magmas in the formation of hydrothermal ore deposits. *Nature* **370**, 519–527.

Hedenquist, J. W., Simmons, S. F., Giggenbach, W. F. and Eldridge, C. S. (1993). White Island, New Zealand, volcanic-hydrothermal system represents the geochemical environment of high-sulfidation Cu and Au ore deposition. *Geology* **21**, 731–734.

Hedenquist, J. W., Matsuhisa, Y., Izawa, E. *et al.* (1994). Geology, geochemistry, and origin of high-sulfidation Cu–Au mineralization in the Nansatsu district, Japan. *Economic Geology* **89**, 1–30.

Hedenquist, J. W., Arribas, A. Jr. and Reynolds, T. J. (1998). Evolution of an intrusion-centered hydrothermal system; far Southeast–Lepanto porphyry and epithermal Cu–Au deposits, Philippines. *Economic Geology* **93**, 373–404.

Hedenquist, J. W., Arribas, A. R. and Gonzalez-Urien, E. (2000). Exploration for epithermal gold deposit. *Reviews in Economic Geology* **13**, 245–277.

Heidrick, T. L. and Titley, S. R. (1982). Fracture and dike patterns in Laramide plutons and their structural and tectonic implications. In *Advances in Geology of the Porphyry Copper Deposits Southwestern North America*, Titley, S. R. (ed.), Tucson, University of Arizona Press, pp. 73–91.

Hein, J. R., Yen, H.-W., Gunn, S. H., Gibbs, A. E. and Wang, C.-H. (1994). Composition and origin of hydrothermal ironstones from central Pacific seamounts. *Geochimica et Cosmochimica Acta* **58**, 179–189.

Hellston, K., Lewis, C. R. and Denn, S. (1998). Cawse nickel–cobalt deposit. In *Geology of Australian and Papua New Guinean Mineral Deposits*, Berkman D. A. and Mackenzie, D. H. (eds.), Australasian Institute of Mining and Metallurgy, Monograph **22**, pp. 335–338.

Henry, C. D., Elson, H. B., McIntosh, W. C., Heizler, M. T. and Castor, S. B. (1997). Brief duration of hydrothermal activity at Round Mountain, Nevada, determined from $^{40}Ar/^{39}Ar$ geochronology. *Economic Geology* **92**, 807–826.

Herrington, R. J. and Wilkinson, J. J. (1993). Colloidal gold and silica in mesothermal vein systems. *Geology* **21**, 539–542.

Hewitt, W. P. (1968). Geology and mineralization of the main mineral zone of the Santa Eulalia district, Chihuahua, Mexico. *American Institute of Mining Engineers Transactions* **241**, 228–260.

Hiatt, E and Budd, D. A. (2003). Extreme paleoceanographic conditions in a Paleozoic oceanic upwelling system; productivity and widespread phosphogenesis in the Permian Phosphoria sea. *Geological Society of America Special Paper* **370**, 245–264.

Hitzman, M. W. and Beaty, D. W. (1996). The Irish Zn–Pb–(Ba) orefield. *Society of Economic Geologists Special Publication* **4**, 112–143.

Hitzman, M. W., Oreskes, N. and Einaudi, M. T. (1992). Geological characteristics and tectonic setting of Proterozoic iron oxide (Cu–U–Au–REE) deposits. *Precambrian Research* **58**, 241–287.

Hitzman, M., Kirkham, R., Broughton, D., Thorson, J. and Selley, D. (2005). The sediment-hosted stratiform copper-ore systems. *Economic Geology, 100th Anniversary Volume*, Hedenquist, J. W., Thompson, J. F. H., Goldfarb, R. J. and Richards, J. P. (eds.), Colorado, Society of Economic Geologists, pp. 609–642.

Hodgson, C. J. (1989). Uses (and abuses) of ore deposit models in mineral exploration. *Ontario Geological Survey Special Volume* **3**, 31–45.

Hofstra, A. H. and Cline, J. S. (2000). Characteristics and models for Carlin-type gold deposits. *Reviews in Economic Geology* **13**, 163–220.

Hofstra, A. H., John, D. A. and Theodore, T. G. (2003). Preface: a special issue devoted to gold deposits in northern Nevada: part 2. Carlin-type deposits. *Economic Geology* **98**, 1063–1068.

Holland, H. D. (1984). *The Chemical Evolution of the Atmosphere and Oceans*, Princeton, Princeton University Press.

Holliday, J. R., Wilson, A. J., Blevin, P. L. *et al.* (2002). Porphyry gold–copper mineralisation in the Cadia district, eastern Lachlan Fold Belt, New South Wales, and its relationship to shoshonitic magmatism. *Mineralium Deposita* **37**, 100–116.

Horan, M., Morgan, J. W., Walker, R. J. and Cooper, R. W. (2001). Re–Os isotopic constraints on magma mixing in the Peridotite Zone of the Stillwater Complex, Montana, USA. *Contributions to Mineralogy and Petrology* **141**, 446–457.

Hronsky, J. M. A, Groves, D. I., Loucks, R. R. and Begg, G. C. (2012). A unified model for gold mineralisation in accretionary orogens and implications for regional-scale exploration targeting methods. *Mineralium Deposita* **47**, 339–358.

Hsieh, P. A. and Bredehoeft, J. D. (1981). A reservoir analysis of the Denver earthquakes – a case of induced seismicity. *Journal of Geophysical Research* **B 86**, 903–920.

Hughes, M. J., Phillips, G. N. and Carey, S. P. (2004). Giant placers of the Victorian Gold Province. *Society of Economic Geologists Newsletter* **56**, 1–18.

Ibaraki, K. and Suzuki, R. (1993). Gold–silver quartz-adularia veins of the Main, Yamada and Sanjin deposits, Hishikari gold mine; a comparative study of their geology and ore deposits. *Resource Geology Special Issue* **14**, 1–11.

Ildefonse, B. and Mancktelow, N. S. (1993). Deformation around rigid particles – the influence of slip at the particle matrix interface. *Tectonophysics* **221**, 345–359.

Ingebritsen, S., Sanford, W. and Neuzil, C. (2006). *Groundwater in Geologic Processes, 2nd Edition*, Cambridge, Cambridge University Press.

Ingebritsen, S. E., Geiger, S., Hurwitz, S. and Driesner, T. (2010). Numerical simulation of magmatic hydrothermal systems. *Reviews of Geophysics* **48**, RG1002.

Ionov, D. A., Dupuy, C. and O'Reilly, S. Y. (1993). Carbonated peridotite xenoliths from Spitsbergen – implications for trace-element signature of mantle carbonate metasomatism. *Earth and Planetary Science Letters* **119**, 283–297.

Irvine, T. N. (1977). Origin of chromitite layers in the Muskox intrusion and other stratiform intrusions: a new interpretation. *Geology* **5**, 273–277.

Janse, A. J. A. (1994). Is Clifford's Rule still valid? Affirmative examples from around the World. In *Proceedings of the Fifth International Kimberlite Conference, Araxa, Brazil, 1991, Volume 2*, Meyer, H. O. A. and Leonardos, O. H. (eds.), CPRM Special Publication 1B, Oxford, Blackwell, pp. 215–235.

Jefferson, C. W., Thomas, D. J., Gandhi, S. S. *et al.* (2007). Unconformity-associated uranium deposits of the Athabasca Basin, Saskatchewan and Alberta. *Geological Association of Canada Mineral Deposits Division Special Publication* **5**, 273–306.

Jelsma, H., Barnett, W., Richards, S. and Lister, G. (2009). Tectonic setting of kimberlites. *Lithos* **112S**, 155–165.

Jowett, E. C. (1992). Role of organics and methane in sulfide ore formation, exemplified by Kupferschiefer Cu–Ag deposits, Poland. *Chemical Geology* **99**, 51–63.

Jupp, T. and Schultz, A. (2000). A thermodynamic explanation for black smoker temperatures. *Nature* **403**, 880–883.

Kelley, K. D. and Jennings, S. (2004). A special issue devoted to barite and Zn–Pb–Ag deposits in the Red Dog district, Western Brooks Range, Northern Alaska, Preface. *Economic Geology* **99**, 1267–1280.

Kesler, S. E. and Wilkinson, B. H. (2008). Earth's copper resources estimated from tectonic diffusion of porphyry copper deposits. *Geology* **36**, 255–258.

Kesler, S. E. and Wilkinson, B. H. (2009). Resources of gold in Phanerozoic epithermal deposits. *Economic Geology* **104**, 623–633.

Khashgerel, B-E., Rye, R. O., Hedenquist, J. W. and Kavalierts, I. (2006). Geology and reconnaissance stable isotope study of the Oyu Tolgoi porphyry Cu–Au system, South Gobi, Mongolia. *Economic Geology* **101**, 503–522.

King, C.-Y., Zhang, W. and Zhang, Z. (2006). Earthquake-induced groundwater and gas changes. *Pure and Applied Geophysics* **163**, 633–645.

Kinnaird, J., Kruger, F. J., Nex, P. A. M. and Cawthorn, R. G. (2002). Chromitite formation – a key to understanding processes of platinum enrichment. *Transactions of the Institution of Mining and Metallurgy* **B 111**, 23–35.

Kinnaird, J. A., Hutchinson, D., Schurmann, L., Nex, P. A. M. and de Lange, R. (2005). Petrology and mineralization of the southern Platreef: northern limb of the Bushveld Complex, South Africa. *Mineralium Deposita* **40**, 576–597.

Kirk, J., Ruiz, J., Chesley, J., Titley, S. and Walshe, J. (2001). A detrital model for the origin of gold and sulfides in the Witwatersrand Basin based on Re–Os isotopes. *Geochimica et Cosmochimica Acta* **65**, 2149–2159.

Kitto, P. A., Evans, D. A. and Mrozcek, C. R. (1997). Renison – new advances in the geological understanding of a world-class ore deposit. *Australasian Institute of Mining and Metallurgy Publication Series* **97**, 31–39.

Kogel, J. E., Trivedi, N. C., Barker, J. M. and Krukowski, S. T. (eds.) (2006). *Industrial Minerals and Rocks, Commodities, Markets, and Uses, 7th Edition*, Littleton, Society for Mining, Metallurgy and Exploration Inc.

Kogiso, T. and Hirschmann, M. H. (2006). Partial melting experiments of bimineralic eclogite and the role of recycled mafic oceanic crust in the genesis of ocean island basalts. *Earth and Planetary Science Letters* **249**, 185–199.

Kruger, F. J. (1994). The Sr-isotopic stratigraphy of the Western Bushveld Complex. *South African Journal of Geology* **97**, 393–398.

Kruger, F. J. (2005). Filling of the Bushveld Complex magma chamber: lateral expansion, roof and floor interaction, magmatic unconformities, and the formation of giant chromitite, PGE and Ti-V-magnetitite deposits. *Mineralium Deposita* **40**, 451–472.

Kruger, F. J. and Marsh, J. S. (1982). The significance of $^{87}Sr/^{86}Sr$ ratios in the Merensky cyclic unit of the Bushveld Complex. *Nature* **298**, 53–55.

Kuleshov, V. N. (2011). Manganese deposits: communication 1. Genetic models of manganese ore formation. *Lithology and Mineral Resources* **46**, 473–493.

Kwak, T. A. P. and Tan, T. H. (1981). The geochemistry of zoning in skarn minerals at the King Island (Dolphin) mine. *Economic Geology* **76**, 468–497.

Lalou, C., Munch, U., Halbach, P. and Reyss, J. L. (1998). Radiochronological investigation of hydrothermal deposits from the MESO zone, Central Indian Ridge. *Marine Geology* **149**, 243–254.

Landtwing, M. R., Pettke, T., Halter, W. E. *et al.* (2005). Copper deposition during quartz dissolution by cooling magmatic-hydrothermal fluids: the Bingham porphyry. *Earth and Planetary Science Letters* **235**, 229–243.

Landtwing, M. R., Furrer, C., Redmond, P. B. *et al.* (2010). The Bingham Canyon porphyry Cu–Au–Mo deposit. III. Zoned copper–gold ore deposition by magmatic vapor expansion. *Economic Geology* **105**, 91–118.

Lange, I. M., Nokleberg, W. J., Plahuta, J. T., Krouse, H. R. and Doe, B. R. (1985). Geological setting, petrology, and geochemistry of stratiform sphalerite–galena–barite deposits: Red Dog Creek and Drenchwater Creek areas, Northwestern Brooks Range, Alaska. *Economic Geology* **80**, 1896–1926.

Langmuir, D. (1997). *Aqueous Environmental Geochemistry*, Upper Saddle River, NJ, Prentice Hall.

Large, D. and Walcher, E. (1999). The Rammelsberg massive sulphide Cu–Zn–Pb–Ba-deposit, Germany: an example of sediment-hosted, massive sulphide mineralisation. *Mineralium Deposita* **34**, 522–538.

Large, R. R. (1992). Australian volcanic-hosted massive sulfide deposits: features, styles, and genetic models. *Economic Geology* **87**, 471–510.

Large, R. R., Bull, S. W., Cooke, D. R. and McGoldrick, P. J. (1998). A genetic model for the HYC deposit, Australia: based on regional sedimentology, geochemistry and sulfide-sediment relationships. *Economic Geology* **93**, 1345–1368.

Large, R., McGoldrick, P., Bull, S. and Cooke, D. (2004). Proterozoic stratiform sediment-hosted zinc–lead–silver deposits of Northern Australia. In *Sediment-hosted Lead-Zinc Sulphide Deposits*, Deb, M. and Goodfellow, W. D. (eds.), New Delhi, Narosa Publishing House, pp. 1–23.

Lasky, S. G. (1950). How tonnages and grade relations help predict ore reserves. *Engineering Mining Journal* **151**, 81–85.

Lawrance, L. M. (1990). Supergene gold mineralization. In *Gold Deposits of the Archaean Yilgarn Block, Western Australia: Nature, Genesis and Exploration Guides*, Ho, S. E., Groves, D. I. and Bennett, J. M. (eds.), Crawley, University of Western Australia, Publication **20**, pp. 299–314.

Leach, D. L., Viets, J. G., Kozlowski, A. and Kibitlewski, S. (1996). Geology, geochemistry, and genesis of the Silesia–Cracow zinc–lead district, southern Poland. *Society of Economic Geologists Special Publication* **4**, 144–170.

LeBlanc, M. and Nicolas, A. (1992). Ophiolitic chromites. *International Geology Review* **34**, 653–686.

Lesher, C. M. (1989). Komatiite-associated nickel sulfide deposits, ore deposits associated with magmas. *Reviews in Economic Geology* **4**, 45–102.

Lesher, C. M. (2007). Ni–Cu–(PGE) deposits in the Raglan area, Cape Smith belt, New Quebec. *Geological Association of Canada Mineral Deposits Division Special Publication* **5**, 351–386.

Lewis, B. L. and Landing, W. M. (1992). The investigation of dissolved and suspended particulate trace-metal fractionation in the Black Sea. *Marine Chemistry* **40**, 105–141.

Li, C., Maier, W. D. and de Waal, S. A. (2001). The role of magma mixing in the genesis of PGE mineralization in the Bushveld Complex: thermodynamic calculation and new interpretations. *Economic Geology* **96**, 653–662.

Loen, J. S. (1992). Mass balance constraints on gold placers: possible solutions to "source area problems". *Economic Geology* **87**, 1624–1634.

Logan, R. G., Murray, W. J. and Williams, N. (1990). HYC silver–lead–zinc deposit, McArthur River. In *Geology and Mineral Deposits of Australia and Papua New Guinea*, Hughes, F. E. (ed.), Melbourne, Australasian Institute of Mining and Metallurgy, Monograph **14**, pp. 907–911.

Long, K. R., Van Gosen, B. S., Foley, N. F. and Cordier, D. (2010). The principal rare earth elements deposits of the United States – a summary of domestic deposits and a global perspective. *US Geological Survey Scientific Investigations Report* **2010–5220**.

Lott, D. A., Coveney, R. M., Murowchick, J. B. and Grauch, R. I. (1999). Sedimentary exhalative nickel–molybdenum ores in south China. *Economic Geology* **94**, 1051–1066.

Lottermoser, B. G. (1995). Ore minerals of the Mt Weld rare-earth element deposit, Western Australia. *Transactions of the Institute of Mining and Metallurgy* **B 104**, 203–209.

Lowell, J. D. and Guilbert, J. M. (1970). Lateral and vertical alteration-mineralization zoning in porphyry ore deposits. *Economic Geology* **65**, 373–408.

Lowenstein, T. K., Timofeeff, M. N., Kovalevych, V. M. and Horita, J. (2005). The major ion composition of Permian seawater. *Geochimica et Cosmochimica Acta* **69**, 1701–1719.

Lydon, J. W. (1988). Ore deposits models; volcanogenic massive sulphide deposits. *Geosciences Canada* **15**, 43–65.

McCracken, S. R., Etminan, H., Connor, A. G. and Williams, V. A. (1996). Geology of the Admiral Bay carbonate-hosted zinc–lead deposit, Canning Basin, Western Australia. *Society of Economic Geologists Special Publication* **4**, 330–349.

McFarlane, M. J. (1976). *Laterite and Landscape*. London, Academic Press.

Macumber, P. G. (1992). Hydrological processes in the Tyrrell Basin, southeastern Australia. *Chemical Geology* **96**, 1–18.

Maier, W. D., Arndt, N. T. and Curl, E. A. (2000). Progressive crustal contamination of the Bushveld Complex: evidence from Nd isotopic analyses of the cumulate rocks. *Contributions to Mineralogy and Petrology* **140**, 316–327.

Maier, W. D., Barnes, S.-J., Chinyepi, G. *et al.* (2008). The composition of magmatic Ni–Cu–(PGE) sulfide deposits in the Tati and Selebi-Phikwe belts of eastern Botswana. *Mineralium Deposita* **43**, 37–60.

Mandal, N., Misra, S. and Samanta, S. K. (2004). Role of weak flaws in nucleation of shear zones: an experimental and theoretical study. *Journal of Structural Geology* **26**, 1391–1400.

Marschik, R. and Fontboté, L. (2001). The Candelaria–Punta del Cobre iron oxide Cu–Au (–Zn–Ag) deposits, Chile. *Economic Geology* **96**, 1799–1826.

Martin, H. J. (1963). The Bikita Tinfield. *Southern Rhodesia Geological Survey Bulletin* **58**.

Mathieson, G. A. and Clark, A. H. (1984). The Cantung E-zone scheelite skarn orebody, tungsten, Northwest Territories – a revised genetic model. *Economic Geology* **79**, 883–901.

Matthai, S. K., Heinrich, C. A. and Driesner, T. (2004). Is the Mount Isa copper deposit the product of forced brine convection in the footwall of a major reverse fault? *Geology* **32**, 357–360.

Mavrogenes, J. A. and O'Neill, H. S. (1999). The relative effects of pressure, temperature and oxygen fugacity on the solubility of sulfide in mafic magmas. *Geochimica et Cosmochimica Acta* **63**, 1173–1180.

Maynard, J. B. (2010). The chemistry of manganese ores through time: a signal of increasing diversity of Earth-surface environments. *Economic Geology* **100**, 535–552.

Megaw, P. K. M., Ruiz, J. R. and Titley, S. R. (1988). High-temperature, carbonate-hosted Ag–Pb–Zn(Cu) deposits of northern Mexico. *Economic Geology* **83**, 1856–1885.

Meier, D. L., Heinrich, C. A. and Watts, M. A. (2009). Mafic dikes displacing Witwatersrand gold reefs: evidence against metamorphic-hydrothermal ore formation. *Geology* **37**, 607–610.

Meinert, L. D., Dipple, G. M. and Nicoescu, S. (2005). World skarn deposits. *Economic Geology 100th Anniversary Volume*, Hedenquist, J. W., Thompson, J. F. H., Goldfarb, R. J. and Richards, J. P. (eds.), Colorado, Society of Economic Geologists, pp. 299–336.

Melcher, F., Grum, W., Thalhammer, T. V. and Thalhammer, O. A. R. (1999). The giant chromite deposit at Kempirsai, Urals: constraints from trace element (PGE, REE) and isotope data. *Mineralium Deposita* **34**, 250–272.

Melchiorre, E. B. and Enders, M. S. (2003). Stable isotope geochemistry of copper carbonates at the Northwest Extension Deposit, Morenci District, Arizona: implications for conditions of supergene oxidation and related mineralization. *Economic Geology* **98**, 607–622.

Menendez, A. and Sarmentero, A. (1984). Geology of the Los Pijiguaos bauxite deposit, Venezuela. In *Proceedings of the 1984 Bauxite Symposium, Los Angeles*, Jacobs, L. Jr. (ed.), New York, American Institute of Mining, Metallurgy and Petroleum Engineering, pp. 387–407.

Mertig, H. J., Rubin, J. N. and Kyle, J. R. (1994). Skarn Cu–Au orebodies of the Gunung Bijih (Ertsberg) district, Irian Jaya, Indonesia. *Journal of Geochemical Exploration* **50**, 179–202.

Meyer, C., Shea, E., Goddard, C. *et al.* (1968). Ore deposits at Butte, Montana. In *Ore Deposits of the United States 1933–1967 (Graton-Sales Volume)*, Ridge, J. D., (ed.), New York, American Institute of Mining, Metallurgy and Petroleum Engineers, pp. 1363–1416.

Meyer, F. M., Happel, U., Hausberg, J. and Wiechowski, A. (2002). The geometry and anatomy of the Los Pijiguaos bauxite deposit, Venezuela, *Ore Geology Reviews* **20**, 27–54.

Milburn, D. and Wilcock, S. (1998). Kunwarara magnesite deposit. In *Geology of Australian and Papua New Guinean Mineral Deposits*, Berkman D. A. and Mackenzie, D. H. (eds.), Australasian Institute of Mining and Metallurgy, Monograph **22**, pp. 815–819.

Miller, A. R., Densmore, C. D., Degens, E. T. *et al.* (1966). Hot brines and recent iron deposits in the deeps of the Red Sea. *Geochimica et Cosmochimica Acta* **30**, 341–359.

Millero, F. J. and Söhn, M. L. (1992). *Chemical Oceanography,* Caldwell, NJ, Telford Press.

Minter, W. E. L. (1978). A sedimentological synthesis of placer gold, uranium and pyrite concentrations in Proterozoic Witwatersrand sediments. *Canadian Society of Petroleum Geology Memoir* **5**, 801–829.

Minter, W. E. L. (1999). Irrefutable detrital origin of Witwatersrand gold and evidence of eolian signatures. *Economic Geology* **94**, 665–670.

Mitchell, R. H. (1986). *Kimberlites: Mineralogy, Geochemistry and Geology*. New York, Plenum Press.

Moore, D. W., Young, L. E., Modene, J. S. and Plahuta, J. T. (1986). Geologic setting and genesis of the Red Dog zinc–lead–silver deposit, Western Brooks, Range, Alaska. *Economic Geology* **81**, 1696–1727.

Morris, R. C. (1993). Genetic modelling for banded iron-formation of the Hamersley Group, Pilbara Craton, Western Australia. *Precambrian Research* **60**, 243–286.

Morris, R. C. and Ramanaidou, E. R. (2007). Genesis of the channel iron deposits (CID) of the Pilbara region, Western Australia. *Australian Journal of Earth Sciences* **54**, 733–756.

Morris, R. C. and Kneeshaw, M. (2011). Genesis modelling for the Hamersley BIF-hosted iron ores of Western Australia: a critical review. *Australian Journal of Earth Sciences* **58**, 417–451.

Mossman, D. J., Minter, W. E. L., Dutkiewicz, A. *et al.* (2008). The indigenous origin of Witwatersrand "carbon". *Precambrian Research* **164**, 173–186.

Mote, T. I., Becker, T. A., Renne, P. and Brimhall, G. H. (2001a). Chronology of exotic mineralization at El Salvador, Chile, by $^{40}Ar/^{39}Ar$ dating of copper wad and supergene alunite. *Economic Geology* **96**, 351–366.

Mote, T. I., Brimhall, G. H., Tidy-Finch, E., Muller, G. and Carrasco, P. (2001b). Application of mass-balance modeling of sources, pathways and sinks of supergene enrichment to exploration and discovery of the Quebrada Turquesa exotic copper orebody, El Salvador District, Chile. *Economic Geology* **96**, 367–386.

Mudd, G. M. (2007). *The Sustainability of Mining in Australia: Key Production Trends And Their Environmental Impacts for the Future*. Research Report no. RR5, Department of Civil Engineering, Monash University and Mineral Policy Institute.

Mukasa, S. B., Wilson, A. H. and Carlson, R. W. (1998). A multielement geochronological study of the Great Dyke, Zimbabawe: significance of the robust and reset ages. *Earth and Planetary Science Letters* **164**, 353–369.

Naldrett, A. J. (1973). Nickel sulphide deposits – their classification and genesis with special emphasis on deposits of volcanic association. *Canadian Institute of Mining and Metallurgy, Bulletin* **66**, 45–63.

Naldrett, A. J. (2004). *Magmatic Sulfide Deposits: Geology, Geochemistry and Exploration.* Berlin, Springer.

Naldrett, A. J. (2010). From the mantle to the bank: the life of a Ni–Cu–(PGE) sulfide deposit. *South African Journal of Geology* **113**, 1–32.

Naldrett, A. J. and Li, C. (2007). The Voisey's Bay deposit, Labrador, Canada. *Geological Association of Canada Mineral Deposits Division Special Publication* **5**, 387–408.

Naldrett, A. J., Lightfoot, P. C., Fedorenko, V., Doherty, W. and Gorbachev, N. S. (1992). Geology and geochemistry of intrusions and flood basalts of the Noril'sk region, USSR, with implications for the origin of the Ni-Cu ores. *Economic Geology* **87**, 975–1004.

Naldrett, A. J., Wilson, A., Kinnaird, J. and Chunnett, G. (2009). PGE tenor and metal ratios within and below the Merensky Reef, Bushveld Complex: implications for its genesis. *Journal of Petrology* **50**, 625–659.

Naranjo, J. A., Henriquez, F. and Nystrom, J. O. (2010). Subvolcanic contact metasomatism at El Laco volcanic complex, Central Andes. *Andean Geology* **37**, 110–120.

Nash, J. T., Granger, H. C. and Adams, S. S. (1981). Geology and concepts of genesis of important types of uranium deposits. *Economic Geology, 75th Anniversary Volume*, Skinner, B. (ed.), Lancaster, PA, Economic Geology Publishing, pp. 63–116.

Nelson, J. (1997). The quiet counter-revolution: structural control of syngenetic deposits. *Geoscience Canada* **24**, 91–98.

Northrop, H. R. and Goldhaber, M. B. (eds.) (1990). Genesis of the tabular-type vanadium-uranium deposits of the Henry Basin, Utah. *Economic Geology* **85**, 215–269.

Nyström, J. O. and Henriquez, F. (1994). Magmatic features of iron ores of the Kiruna-type in Chile and Sweden: ore textures and magnetite geochemistry. *Economic Geology* **89**, 820–839.

Oberthür, T., Weiser, T., Amanor, J. A. and Chryssoulis, S. L. (1997). Mineralogical siting and distribution of gold in quartz veins and sulfide ores of the Ashanti mine and other deposits of the Ashanti belt of Ghana. *Mineralium Deposita* **32**, 2–15.

Ohle, E. L. (1996). Significant events in the geological understanding of the southeast Missouri lead district. *Society of Economic Geologists Special Publication* **4**, 1–7.

Okamoto, A., Saishu, H., Hirano, N. and Tsuchiya, N. (2010). Mineralogical and textural variation of silica minerals in hydrothermal flow-through experiments: implications for quartz vein formation. *Geochimica et Cosmochimica Acta* **74**, 3693–3706.

O'Neill, C. J., Moresi, L. and Jaques, A. L. (2005). Geodynamic controls on diamond deposits: implications for Australian occurrences. *Tectonophysics* **404**, 217–236.

Oszczepalski, S. (1999). Origin of the Kupferschiefer polymetallic mineralization in Poland. *Mineralium Deposita* **34**, 599–613.

Palabora Mining Company Limited Mine Geological and Mineralogical Staff (1976). The geology and economic deposits of copper, iron, and vermiculite in the Palabora Igneous Complex: a brief review. *Economic Geology* **71**, 177–192.

Partington, G. A. (1990). Environment and structural controls on the intrusion of the giant rare metal Greenbushes pegmatite, Western Australia. *Economic Geology* **85**, 437–456.

Pavlov, A. A. and Kasting, J. F. (2002). Mass-independent fractionation of sulfur isotopes in Archean sediments: strong evidence for an anoxic Archean atmosphere. *Astrobiology* **2**, 27–41.

Percival, T. J. and Radtke, A. S. (1994). Sedimentary-rock hosted disseminated gold mineralization in the Alsar district, Macedonia. *Canadian Mineralogist* **32**, 649–665.

Peters, S. G., Jianzhan, H., Zhiping, L. and Chenggui, J. (2007). Sedimentary rock-hosted Au deposits of the Dian-Qian-Gui area, Guizhou, and Yunnan Provinces, and Guangxi District, China. *Ore Geology Reviews* **31**, 170–204.

Petersen, S., Herzig, P. M. and Hannington, M. D. (2000). Third dimension of a presently forming VMS deposit: TAG hydrothermal mound, Mid-Atlantic Ridge, 26° N. *Mineralium Deposita* **35**, 233–259.

Petrov, S. V. (2004). Economic deposits associated with the alkaline and ultrabasic complexes of the Kola Peninsula. In *Phoscorites and Carbonatites from Mantle to Mine: the Key Example of the Kola Alkaline Province*, Wall, V. and Zaitsev, A. N. (eds.), London, Mineralogical Society, pp. 469–490.

Pettke, T., Diamond, L. W. and Kramers, J. D. (2000). Mesothermal gold lodes in the northwestern Alps: a review of genetic constraints from radiogenic isotopes. *European Journal of Mineralogy* **12**, 213–230.

Pettke, T., Oberli, F. and Heinrich, C. A. (2010). The magma and metal source of giant porphyry-type ore deposits based on lead isotope microanalysis of individual fluid inclusions. *Earth and Planetary Science Letters* **296**, 267–277.

Pettke, T., Oberli, F., Audetat, A. *et al.* (2012). Recent developments in element concentration and isotope ratio analysis of individual fluid inclusion by laser ablation single and multiple collector ICP-MS. *Ore Geology Reviews* **44**, 10–38.

Phillips, G. N. and Hughes, M. J. (1996). The geology and gold deposits of the Victorian gold province. *Ore Geology Reviews* **11**, 255–302.

Phillips, G. N. and Law, J. D. M. (2000). Witwatersrand gold fields: geology, genesis, and exploration. *Reviews in Economic Geology* **13**, 439–500.

Phillips, G. N., Meyers, F. M. and Palmer, J. A. (1987). Problems with the placer model for Witwatersrand gold. *Geology* **15**, 1027–1030.

Pieczonka, J., Piestrzynski, A., Mucha, J. *et al.* (2008). The red-bed-type precious metal deposit in the Sieroszowice–Polkowice copper mining district, SW Poland. *Annales Societas Geologorum Poloniae* **78**, 151–280.

Pollard, P. J. and Taylor, R. G. (2002). Paragenesis of the Grasberg Cu–Au deposit, Irian Jaya, Indonesia: results from logging section 13. *Mineralium Deposita* **37**, 117–136.

Porter, T. M. (ed.) (2000). *Hydrothermal Iron Oxide Copper–Gold and Related Deposits: A Global Perspective*, PGC Publishing, Adelaide.

Poujol, M., Robb, L. J. and Respaut, J. P. (1999). U–Pb and Pb–Pb isotopic studies relating to the origin of gold mineralization in the Evander Goldfield, Witwatersrand Basin, South Africa. *Precambrian Research* **95**, 167–185.

Prendergast, M. D. (2003). The nickeliferous late Archean komatiitic event in the Zimbabwe Craton – magmatic architecture, physical volcanology, and ore genesis. *Economic Geology* **98**, 865–891.

Prendergast, M. D. and Wilson, A. H. (1989). The Great Dyke of Zimbabwe – II: mineralization and mineral deposits. In *Magmatic Sulphides – The Zimbabwe Volume*, Prendergast, M. D. and Jones, M. J. (eds.), London, Institution of Mining and Metallurgy, pp. 21–42.

Prescott, J. R. and Habermehl, M. A. (2008). Luminescence dating of spring mound deposits in the southwestern Great Artesian Basin, northern South Australia. *Australian Journal of Earth Sciences* **55**, 167–181.

Profett, J. M. (2003). Geology of the Bajo de la Alumbrera porphyry copper-gold deposit, Argentina. *Economic Geology* **98**, 1535–1574.

Rackley, R. I. (1976). Origin of Western-States type uranium mineralization. In *Handbook of Strata Bound and Stratiform Ore Deposits. Part II. Regional Studies and Specific Deposits*, Wolf, K. H. (ed.), Amsetrdam, Elsevier, pp. 89–156.

Radtke, A. S. (1985). Geology of the Carlin gold deposit, Nevada, *US Geological Survey Professional Paper* **1267**.

Raffensperger, J. P. and Garven, G. (1995). The formation of unconformity-type uranium ore deposits. I: coupled groundwater flow and heat transport modeling. *American Journal of Science* **295**, 581–636.

Ramdohr, P. (1980). *The Ore Minerals and Their Intergrowths.* Oxford, Pergamon Press.

Reeve, J. S., Cross, K. C., Smith, R. N. and Oreskes, N. (1990). Olympic Dam copper–uranium–gold–silver deposit, In *Geology and Mineral Deposits of Australia and Papua New Guinea*, Hughes, F. E. (ed.), Melbourne, Australasian Institute of Mining and Metallurgy, Monograph **14**, pp. 1009–1035.

Rentzsch, J. (1974). The Kupferschiefer in comparison with the deposits of the Zambian copperbelt. In *Gisements Stratiforme et Provinces, Cuprifères*, Bartholomé, P. (ed.), Liège, Société Géologique de Belgique, pp. 395–418.

Ressel, M. W. and Henry, C. D. (2006). Igneous geology of the Carlin trend, Nevada: development of the Eocene plutonic complex and significance for Carlin-type gold deposit. *Economic Geology* **101**, 347–383.

Reynolds, L. J. (2001). Geology of the Olympic Dam Cu–U–Au–Ag–REE deposit. *MESA Journal* **23**, 4–11.

Rice, C. M., Harmon, R. S. and Shepherd, T. J. (1985). Central City, Colorado: the upper part of an alkaline porphyry molybdenum system. *Economic Geology* **80**, 1769–1796.

Richards, J. P. (2009). Post subduction porphyry Cu–Au and epithermal Au deposits: products of remelting of subduction modified lithosphere. *Geology* **37**, 247–250.

Richards, J. P. (2011). High Sr/Y arc magmas and porphyry Cu±Mo±Au deposits: just add water. *Economic Geology* **100**, 1075–1082.

Richards, J. P. and Tosdal, R. M. (eds.) (2001). Structural controls on ore genesis. *Reviews in Economic Geology* **14**.

Ridley, J. R. (1993). The relations between mean rock stress and fluid flow in the crust with reference to vein- and lode-style deposits. *Ore Geology Reviews* **8**, 23–37.

Ridley, J. and Mengler, F. (2000). Lithological and structural control on the form and setting of vein stockwork orebodies at the Mount Charlotte gold deposit, Kalgoorlie. *Economic Geology* **95**, 85–98.

Risacher, F. and Fritz, B. (2009). Origins of salts and brine evolution of Bolivian and Chilean salars. *Aquatic Geochemistry* **15**, 123–157.

Robert, F. (1990). Structural setting and control on gold–quartz veins of the Val d'Or area, southeastern Abitibi Subprovince. In *Gold and Base Metal Mineralization in the Abitibi Subprovince, Canada, with Emphasis on the Quebec Segment*, Ho, S. E., Robert, F. and Groves, D. I. (eds.), University of Western Australia, Geology Key Centre and University Extension Publication **24**, pp. 164–209.

Robert, F. and Brown, A. C. (1986). Archean gold-bearing quartz veins at the Sigma mine, Abitibi greenstone belt, Quebec. *Economic Geology* **81**, 578–616.

Roedder, E. (1984). Fluid inclusions. *Reviews in Mineralogy* **12**.

Rojstaczer, S. S. and Wolf, S. (1994). Hydrologic changes associated with the earthquake in the San Lorenzo and Pescadero drainage basins. *US Geological Survey Professional Paper* **1551-E**, 51–64.

Rollinson, H. (1993). *Using Geochemical Data*. Oxford, Blackwell.

Roscoe, S. M. and Minter, W. E. L. (1993). Pyritic paleoplacer gold and uranium deposits. *Special Paper, Geological Association of Canada* **40**, 103–124.

Rousell, D. H., Fedorowich, J. S. and Driessler, B. O. (2003). Sudbury Breccia (Canada): a product of the 1850 Ma Sudbury Event and hosts to footwall Cu–Ni–PGE deposits. *Earth-Science Reviews* **60**, 147–174.

Rowe, M. C., Kent, A. J. R. and Nielsen, R. L. (2009). Subduction influence on oxygen fugacity and trace and volatile contents in basalts across the Cascade Volcanic arc. *Journal of Petrology* **50**, 61–91.

Rowland, J. V. and Simmons, S. F. (2012). Hydrologic, magmatic, and tectonic controls on hydrothermal fluid flow, Taupo Volcanic Zone, New Zealand: implications for the formation of epithermal vein deposits. *Economic Geology* **102**, 427–459.

Roy, P. S. (1999). Heavy mineral beach placers in southeastern Australia: their nature and genesis. *Economic Geology* **94**, 567–588.

Rubin, J. N. and Kyle, J. R. (1997). Precious metal mineralogy in porphyry-, skarn-, and replacement-type ore deposits of the Ertsberg (Gunung Bijih) district, Irian Jaya, Indonesia. *Economic Geology* **92**, 535–550.

Rubright, R. D. and Hart, O. J. (1968). Non-porphyry ores of the Bingham district, Utah. In *Ore Deposits of the United States, 1933–67 (Graton-Sales Volume)*, Ridge, J. D. (ed.), New York, AIME, pp. 886–907.

Rusk, B. G., Reed, M. H. and Dilles, J. H. (2008). Fluid inclusion evidence for magmatic-hydrothermal fluid evolution in the porphyry copper–molybdenum deposit at Butte, Montana. *Economic Geology* **103**, 307–334.

Sanderson, D. J. and Zhang, X. (1999). Critical stress localization of flow associated with deformation of well-fractured rock masses, with implications for mineral deposits. *Geological Society of London, Special Publication* **155**, 69–81.

Sanford, R. F. (1994). A quantitative model for ground-water flow during formation of tabular uranium sandstone uranium deposits. *Economic Geology* **89**, 341–360.

Sangster, A. L. and Smith, P. K. (2007) Metallogenic summary of the Meguma gold deposits, Nova Scotia. *Geological Association of Canada Mineral Deposits Division Special Publication* **5**, 723–732.

Sangster, D. F. (2002). The role of dense brines in the formation of vent-distal sedimentary-exhalative (SEDEX) lead–zinc deposits: field and laboratory evidence. *Mineralium Deposita* **37**, 149–157.

Sato, T. (1977). Kuroko deposits: their geology, geochemistry and origin. In *Volcanic Processes in Ore Genesis*, Anon, (ed.), London, Institution of Mining and Metallurgy, pp. 153–161.

Saunders, J. A. (1990). Colloidal transport of gold and silica in epithermal precious-metal systems: evidence from the Sleeper deposit, Nevada. *Geology* **18**, 757–760.

Saunders, J. A. and Brueseke, M. E. (2012). Volatility of Se and Te during subduction-related distillation and the geochemistry of epithermal ores of the western United States. *Economic Geology* **107**, 165–171.

Scherba, G. N. (1970). Greisens. *International Geology Review* **12**, 114–151, 239–254.

Schmitt, A. K., Stockli, D. F., Lindsay, J. M. *et al.* (2010). Episodic growth and homogenization of plutonic roots in arc volcanoes from combined U–Th and (U–Th)/He zircon dating. *Earth and Planetary Science Letters* **295**, 91–103.

Schoenberg, R., Nagler, T. F., Gnos, E., Kramers, J. D. and Kamber, B. S. (2003). The source of the Great Dyke, Zimbabwe, and its tectonic significance: evidence from Re–Os isotopes. *Journal of Geology* **111**, 565–578.

Schouwstra, R. P., Kinloch, E. D. and Lee, C. A. (2000). A short geological review of the Bushveld Complex. *Platinum Metals Review* **44**, 33–39.

Scott, S. D. (1997). Submarine hydrothermal systems and deposits. In *Geochemistry of Hydrothermal Ore Deposits*, Barnes, H. L. (ed.), New York, John Wiley, pp. 797–875.

Seedorff, E. and Einaudi, M. T. (2004). Henderson porphyry molybdenum system, Colorado: I. Sequence and abundance of hydrothermal mineral assemblages, flow paths of evolving fluid, and evolutionary style. *Economic Geology* **99**, 3–37.

Seeger, C. M., Nuelle, L. M., Day, W. C. *et al.* (2001). Geologic maps and cross sections of mine levels at the Pea Ridge iron mine, Washington County, Missouri. *US Geological Survey Miscellaneous Field Studies Map* **MF-2353**.

Selley, D., Broughton, D., Scott, R. *et al.* (2005). A new look at the geology of the Zambian Copperbelt. *Economic Geology 100th Anniversary Volume*, Hedenquist, J. W., Thompson, J. F. H., Goldfarb, R. J. and Richards, J. P. (eds.), Colorado, Society of Economic Geologists, pp. 965–1000.

Shannon, J. R., Nelson, E. P. and Golden, R. J. (2004). Surface and underground geology of the world-class Henderson molybdenum porphyry mine, Colorado. *Geological Society of America Field Guide* **5**, 207–218.

Sharp, Z. (2007). *Principles of Stable Isotope Geology*, New York, Pearson/Prentice Hall.

Sheldon, R. P. (1980). Episodicity of phosphate deposition and deep ocean circulation: a hypothesis. *Society of Economic Paleontologists and Mineralogists Special Publication* **29**, 239–247.

Shepherd, M. S. (1990). Eneabba heavy mineral sand placers. In *Geology and Mineral Deposits of Australia and Papua New Guinea*, Hughes, F. E. (ed.), Melbourne, Australasian Institute of Mining and Metallurgy, Monograph **14**, pp. 1591–1594.

Shepherd, T. J., Rankin, A. H. and Alderton, D. H. M. (1985). *A Practical Guide to Fluid Inclusion Studies*. Glasgow, Blackie.

Sibson, R. H. (1987). Earthquake rupturing as a mineralizing agent in hydrothermal systems. *Geology* **15**, 710–714.

Sibson, R. H. (1996). Structural permeability of fluid-driven fault fracture meshes. *Journal of Structural Geology* **18**, 1031–1042.

Sibson, R. H., Robert, F. and Poulsen, K. H. (1988). High-angle reverse faults, fluid-pressure cycling, and mesothermal gold deposits. *Geology* **16**, 551–555.

Sidder, G. B., Day, W. C., Nuelle, L. M. *et al.* (1993). Mineralogic and fluid-inclusion studies of the Pea Ridge iron–rare-earth-element deposit, southeast Missouri. *US Geological Survey Bulletin* **2093**, 205–216.

Siebold, E. and Berger, W. H. (1996). *An Introduction to Marine Geology*. Heidelberg, Springer.

Sillitoe, R. H. (2003). Iron oxide–copper–gold deposits: an Andean view. *Mineralium Deposita* **38**, 787–812.

Sillitoe, R. H. (2008). Major gold deposits and belts of the North and South American Cordillera: distribution, tectonomagmatic settings and metallogenic considerations. *Economic Geology* **103**, 663–687.

Sillitoe, R. H. (2010). Porphyry copper systems. *Economic Geology* **105**, 3–41.

Sillitoe, R. H. and Burrows, D. R. (2002). New field evidence bearing on the origin of the El Laco magnetite deposit, northern Chile. *Economic Geology* **97**, 1101–1109.

Simmons, S. F. and Brown, K. L. (2006). Gold in magmatic hydrothermal solutions and the rapid formation of a giant ore deposit. *Science* **314**, 288–291.

Simmons, S. F. and Brown, K. L. (2007). The flux of gold and related metals through a volcanic arc, Taupo Volcanic Zone, New Zealand. *Geology* **35**, 1099–1102.

Simmons, S. F. and Browne, P. R. L. (2000). Hydrothermal minerals and precious metals in the Broadlands–Ohaaki geothermal system: implications for understanding low-sulfidation epithermal environments. *Economic Geology* **95**, 971–999.

Simmons, S. F., White, N. C. and John, D. A. (2005). Geological characteristics of epithermal precious and base metal deposit. *Economic Geology, 100th Anniversary Volume*, Hedenquist, J. W., Thompson, J. F. H., Goldfarb, R. J. and Richards, J. P. (eds.), Colorado, Society of Economic Geologists, pp. 485–522.

Simon, A. C., Pettke, T., Candela, P. A., Piccoli, P. M. and Heinrich, C. A. (2006). Copper partitioning in sulphur bearing magmatic systems. *Geochimica et Cosmochimica Acta* **70**, 5583–5600.

Simpson, M. P. and Mauk, J. L. (2007). The Favona epithermal gold–silver deposit, Waihi, New Zealand. *Economic Geology* **102**, 817–840.

Simpson, M. P., Mauk, J. L. and Simmons, S. F. (2001). Hydrothermal alteration and hydrologic evolution of the Golden Cross epithermal Au–Ag deposit, New Zealand. *Economic Geology* **96**, 773–796.

Skinner, B. J. (1976). A second iron age ahead? *American Scientist* **64**, 258–269.

Smurthwaite, A. J. (1990). Alumina, geology and mineral resources of Western Australia. *Geological Survey of Western Australia Memoir* **61**, 5–624.

Solomon, M. (1990). Subduction, arc reversal, and the origin of porphyry copper–gold deposits in island arcs. *Geology* **18**, 630–633.

Solomon, M. and Groves, D. I. (2000a). Volcanic-hosted massive sulphide deposits of the Tasman Fold Belt System. In *The Geology and Origin of Australia's Mineral Deposits*, Hobart, Centre for Ore Deposit Research, pp. 580–723.

Solomon, M. and Groves, D. I. (2000b). Proterozoic sediment-hosted, stratiform (sedex), lead-zinc deposits. In *The Geology and Origin of Australia's Mineral Deposits*, Hobart, Centre for Ore Deposit Research, pp. 168–239.

Solomon, M. and Groves, D. I. (2000c). Proterozoic uranium–platinum group element–gold deposits of the Pine Creek Inlier and the Murphy Inlier. In *The Geology and Origin of Australia's Mineral Deposits*, Hobart, Centre for Ore Deposit Research, pp. 287–343.

Spandler, C., Mavrogenes, J. and Arculus, R. (2005). Origin of chromitites in layered intrusions: evidence from chromite-hosted melt inclusions from the Stillwater Complex. *Geology* **33**, 893–896.

Spiridonov, E. M. (2010). Ore-magmatic systems of the Noril'sk ore field. *Russian Geology and Geophysics* **51**, 1059–1077.

Stimac, J. and Hickmott, D. (1996). Ore metal partitioning in intermediate-to-silicic magmas: PIXE results on natural mineral assemblages. In *Giant Ore Deposits – II: Controls on the Scale of Orogenic Magmatic-Hydrothermal Mineralization*, Clark, A. H. (ed.), Ontario, Kingston, pp. 197–235.

Strom, S., Shane, P., Schmitt, A. K. and Lindsay, J. M. (2012). Decoupled crystallization and eruption histories of rhyolite magmatic systems at Tarawera volcano revealed by zircon ages and growth rates. *Contributions to Mineralogy and Petrology* **163**, 505–519.

Sutton, S. J., Ritger, S. D. and Maynard, J. B. (1990). Stratigraphic control of chemistry and mineralogy in metamorphosed Witwatersrand quartzite. *Journal of Geology* **98**, 329–341.

Sverjensky, D. A. (1986). Genesis of Mississippi Valley-type lead–zinc deposits. *Annual Reviews of Earth and Planetary Sciences* **14**, 177–199.

Sverjensky, D. A. (1989). Chemical evolution of basinal brines that formed sediment-hosted Cu–Pb–Zn deposits. *Geological Association of Canada Special Paper* **36**, 127–134.

Tankard, A. J., Jackson, M. P. A., Eriksson, K. A. *et al.* (1982). *Crustal Evolution of southern Africa*. New York, Springer-Verlag.

Taylor, D., Dalstra, H. J., Harding, A. E. *et al.* (2001). Genesis of high-grade hematite orebodies of the Hamersley province, Western Australia. *Economic Geology* **96**, 837–873.

Taylor, D. H. and Gentle, L. V. (2002). Evolution of deep-lead palaeodrainages and gold exploration at Ballarat, Australia. *Australian Journal of Earth Sciences* **49**, 869–878.

Teal, L. and Benavides, A. (2010). History and geologic overview of the Yanacocha Mining District, Cajamarca, Peru. *Economic Geology* **105**, 1173–1190.

Thompson, T. B. and Arehart, G. B. (1990). Geology and origin of ore deposits in the Leadville district, Colorado: part I. Geological studies of orebodies and wall rocks. *Economic Geology Monograph* **7**, 130–155.

Thompson, T. B., Trippel, A. D. and Dwelley, P. C. (1985). Mineralized veins and breccias in the Cripple Creek district, Colorado. *Economic Geology* **80**, 1669–1688.

Thomson, J., Calvert, S. E., Mukherjee, S., Burnett, W. C. and Bremner, J. M. (1984). Further studies of the nature, composition and ages of contemporary phosphorite from the Namibian shelf. *Earth and Planetary Science Letters* **69**, 341–353.

Thorne, W. S., Hagemann, S. G. and Barley, M. (2004). Petrographic and geochemical evidence for hydrothermal evolution of the North Deposit, Mt Tom Price, Western Australia. *Mineralium Deposita* **39**, 766–783.

Titley, S. R. (ed.) (1982). *Advances in Geology of the Porphyry Copper Deposits Southwestern North America*, Tucson, University of Arizona Press.

Titley, S. R. (1993). Characteristics of porphyry copper occurrence in the American southwest. *Geological Association of Canada Special Paper* **40**, 433–463.

Titley, S. R. (2001). Crustal affinities of metallogenesis in the American southwest. *Economic Geology* **96**, 1323–1342.

Tompkins, L. A., Rayner, M. J. and Groves, D. I. (1994). Evaporites: *in situ* sulfur source for rhythmically banded ore in the Cadjebut Mississippi-Valley type Zn–Pb deposit, Western Australia. *Economic Geology* **89**, 467–492.

Tompkins, L. A., Eisenlohr, B., Groves, D. I. and Raetz, M. (1997). Temporal changes in mineralization style at the Cadjebut Mississippi Valley-type deposit, Lennard Shelf, W. A. *Economic Geology* **92**, 843–862.

Tornos, F. (2006). Environment of formation and styles of volcanogenic massive sulphides: the Iberian Pyrite Belt. *Ore Geology Reviews* **28**, 259–306.

Tosdal, R. M. and Richards, J. P. (2001). Magmatic and structural controls on the development of porphyry Cu±Mo±Au deposits. *Reviews in Economic Geology* **14**, 157–181.

Turner-Peterson, C. E. (1985). Lacustrine–humate model for primary uranium ore deposits, Grants Uranium Region, New Mexico. *American Association of Petroleum Geologists Bulletin* **69**, 1999–2020.

Ulmer, P. (2001). Partial melting in the mantle wedge – the role of H_2O in the genesis of mantle-derived "arc-related" magmas. *Physics of Earth and Planetary Interiors* **127**, 215–232.

Ulrich, T. and Heinrich, C. A. (2002). Geology and alteration geochemistry of the porphyry Cu–Au deposit at Bajo de la Alumbrera, Argentina. *Economic Geology* **97**, 1865–1888.

Ulrich, T., Günther, D. and Heinrich, C. A. (1999). Gold concentrations in magmatic brines and the metal budget of porphyry copper deposits. *Nature* **399**, 676–679.

Urabe, T. and Sato, T. (1978). Kuroko deposits of the Kosaka Mine, Northeast Honshu, Japan – products of submarine hot springs on Miocene ocean floor. *Economic Geology* **73**, 161–179.

Van Gosen, B. S. (2009). The Iron Hill (Powderhorn) carbonatite complex, Gunnison County, Colorado; a potential source of several uncommon mineral resources. *US Geological Survey Open-File Report* **2009–1005**.

Van Houten, F. B. and Bhattacharyya, D. P. (1982). Phanerozoic oolitic ironstones – geological record and facies model. *Annual Review of Earth and Planetary Sciences* **10**, 441–457.

Vearncombe, S., Barley, M. E., Groves, D. I. *et al.* (1995). 3.26 Ga black smoker-type mineralization in the Strelley Belt, Pilbara Craton, Western Australia. *Journal of the Geological Society, London* **152**, 587–590.

Volkov, A. V., Serafimovski, T., Kochneva, N. T., Tomson, I. N. and Tasev, G. (2006). The Alshar epithermal Au–As–Sb–Tl deposit, southern Macedonia. *Geology of Ore Deposits* **48**, 175–192.

von Quadt, A., Emiu, M., Martinek, K. *et al.* (2011). Zircon crystallization and the lifetimes of ore-forming magmatic hydrothermal systems. *Geology* **39**, 731–734.

Wagner, P. A. (1929). *The Platinum Deposits and Mines of South Africa*. Capetown, C. Struik (Pty) Ltd.

Waite, K. A., Keith, J. D., Christiansen, E. H. *et al.* (1998). Petrogenesis of the volcanic and intrusive rocks associated with the Bingham Canyon porphyry Cu–Au–Mo deposit, Utah. *Society of Economic Geologists Guidebook Series* **29**, 69–90.

Warren, J. K. (2000). Evaporites, brines and base metals: low-temperature ore emplacement controlled by evaporite diagenesis. *Australian Journal of Earth Sciences* **30**, 179–208.

Webb, S. J., Cawthorn, R. G., Nguuri, T., and James, D. (2004). Gravity modeling of Bushveld Complex connectivity supported by Southern African Seismic Experiment results. *South African Journal of Geology* **107**, 207–218.

White, W. S. (1968). The native copper deposits of northern Michigan. In *Ore Deposits of the United States 1933–1967*, Ridge, J. D. (ed.), New York, American Institute of Mining, Metallurgy and Petroleum Engineers, vol. 1, pp. 303–325.

Whitmeyer, S. J. and Karlstrom, K. E. (2007). Tectonic model for the Proterozoic growth of North America. *Geosphere* **3**, 220–259.

Wilkinson, J. J. (2001). Fluid inclusions in hydrothermal ore deposits. *Lithos* **55**, 229–272.

Williams, P. J., Barton, M. D., Johnson, D. A. *et al.* (2005). Iron oxide copper–gold deposits: geology, space-time distribution, and possible modes of origin. *Economic Geology, 100th Anniversary Volume*, Hedenquist, J. W., Thompson, J. F. H., Goldfarb, R. J. and Richards, J. P. (eds.), Colorado, Society of Economic Geologists, pp. 71–405.

Wilson, A. J., Cooke, D. R., Stein, H. J. *et al.* (2007). U–Pb and Re–Os geochronological evidence for two alkalic porphyry ore-forming events in the Cadia district, New South Wales, Australia. *Economic Geology* **102**, 3–26.

Wodzicki, A. and Piestrzynski, A. (1994). An ore genetic model for the Lubin–Sieroszowice mining district, Poland. *Mineralium Deposita* **29**, 30–43.

Yang, K-F., Fan, H-R., Santosh, M., Hu, F-F. and Wang, K-Y. (2011). Mesoproterozoic carbonatitic magmatism in the Bayan Obo deposit, Inner Mongolia, North China: constraints for the mechanism of super accumulation of rare earth elements. *Ore Geology Reviews* **40**, 122–131.

Zajacz, Z., Hanley, J. J., Heinrich, C. A., Halter, W. E. and Guillong, M. (2009). Diffusive reequilibration of quartz-hosted silicate melt and fluid inclusions: are all metal concentrations unmodified? *Geochimica et Cosmochimica Acta* **73**, 3013–3027.

Zaw, K., Peters, S. G., Cromie, P., Burrett, C. and Hou, Z. (2007). Nature, diversity of deposit types and metallogenic relations of South China. *Ore Geology Reviews* **31**, 3–47.

Zientek, M. L., Cooper, R. W., Corson, S. R. and Geraghty, E. P. (2002). Platinum-group element mineralization in the Stillwater Complex, Montana. *Canadian Institute of Mining, Metallurgy and Petroleum Special Volume* **54**, 459–481.

Zierenberg, R. A., Fouquet, Y., Miller, D. J. *et al.* (1998). The deep structure of a sea-floor hydrothermal deposit. *Nature* **392**, 485–488.

INDEX

Bold entries indicate a section dedicated to the topic, italic entries indicate a reference to a figure.

Printed in the United States
by Baker & Taylor Publisher Services